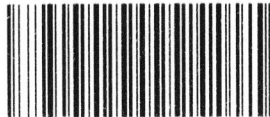

CW01430032

Water and Food

History of Water Series
Series Editor: Terje Tvedt.

As concerns over global water resources continue to grow, the pioneering *History of Water* series brings a much needed historical perspective to the relationship between water and society. Covering all aspects of water and society – social, cultural, political, religious and technological – the volumes reveal how water issues can only be fully understood when all aspects are properly integrated. Unprecedented in its geographical coverage and unrivalled in its multidisciplinary span, the *History of Water* series makes a unique and original contribution to a key contemporary issue.

Series I
Volume 1: Water Control and River Biographies
T. Tvedt and E. Jakobsson (eds), 2006
ISBN 978 1 85043 445 0

Volume 2: The Political Economy of Water
R. Coopey and T. Tvedt (eds), 2006
ISBN 978 1 85043 446 7

Volume 3: The World of Water
T. Tvedt and T. Oestigaard (eds), 2006
ISBN 978 1 85043 447 4

Series II
Volume 1: Ideas of Water from Ancient Societies to the Modern World
Terje Tvedt and Terje Oestigaard (eds), 2010
ISBN 978 1 84511 980 5

Volume 2: Rivers and Society
Terje Tvedt and Richard Coopey (eds), 2010
ISBN 978 1 84885 350 8

Volume 3: Water, Geopolitics and the New World Order
Terje Tvedt, Graham Chapman and Roar Hagen (eds), 2011
ISBN 978 1 84885 351 5

Series III
Volume 1: Water and Urbanization
Terje Tvedt and Terje Oestigaard (eds), 2014
ISBN 978 1 78076 447 4

Volume 2: Sovereignty and International Water Law
Terje Tvedt, Owen McIntyre and Tadesse Kassa Woldetsadik (eds), 2015
ISBN 978 1 78076 448 1

Volume 3: Water and Food: From Hunter-Gatherers to Global Production in Africa
Terje Tvedt and Terje Oestigaard (eds), 2016
ISBN 978 1 78076 871 7

A History of Water

Series III

Volume 3: Water and Food
From Hunter-Gatherers to Global Production in Africa

Edited by

Terje Tvedt *and* **Terje Oestigaard**

I.B. TAURIS

LONDON · NEW YORK

We would like to thank the Norwegian Research Council, University of Bergen, Norway, the Nordic Africa Institute, Uppsala, and the 'Increased research collaboration and outreach at the regional Africa level' programme financed by SIDA (Swedish International Development Cooperation), Sweden.

First published in 2016 by
I.B.Tauris & Co Ltd
London • New York
www.ibtauris.com

Copyright Editorial Selection and Introduction © 2016 Terje Tvedt and Terje Oestigaard
Copyright Individual Chapters © 2016 Matthew V. Bender, Tor A. Benjaminsen, Louise Bertini, Atakilte Beyene, Katherine Blouin, Jean Charles Clanet, Brock Cutler, Gessesse Dessie, Maurits W. Ertsen, Awa-Niang Fall, Elena A.A. Garcea, Randi Haaland, Tobias Haller, Dean Kampanje-Phiri, Jessica Kampanje-Phiri, Jeppe Kolding, Jacques Lemoalle, Alan Mikhail, Pierre Morand, Ketlhatlogile Mosepele, Terje Oestigaard, Andrew Ogilvie, Andrew Reid, Fridtjov Ruden, Marcel Rutten, Emil Sandström, Georges Serpantié, Famory Sinaba, Johamm W.N. Tempelhoff, Raphael M. Tshimanga, Paul A.M. van Zwieten

The right of Terje Tvedt and Terje Oestigaard to be identified as the editors of this work has been asserted by them in accordance with the Copyright, Designs and Patents Act 1988.

All rights reserved. Except for brief quotations in a review, this book, or any part thereof, may not be reproduced, stored in or introduced into a retrieval system, or transmitted, in any form or by any means, electronic, mechanical, photocopying, recording or otherwise, without the prior written permission of the publisher.

Every attempt has been made to gain permission for the use of the images in this book. Any omissions will be rectified in future editions.

References to websites were correct at the time of writing.

ISBN: 978 1 78076 871 7
eISBN: 978 1 78672 138 9
ePDF: 978 1 78673 138 8

A full CIP record for this book is available from the British Library
A full CIP record is available from the Library of Congress

Library of Congress Catalog Card Number: available

Typeset in Garamond Three by OKS Prepress Services, Chennai, India
Printed and bound in Great Britain by CPI Group (UK) Ltd, Croydon, CR0 4YY

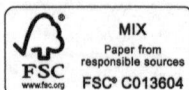

MIX
Paper from
responsible sources
FSC
www.fsc.org   FSC® C013604

Contents

Part IV: Contemporary Water and Food Regimes

Part V: The Hidden Waters of Africa

List of Figures

Chapter 1

Chapter 2

Chapter 3

Chapter 4

List of Tables

Chapter 14

Chapter 17

Chapter 20

Chapter 21

Chapter 22

List of Boxes

Approaches to African Food Production from a Water System Perspective

Terje Tvedt and Terje Oestigaard

FOOD AND AGRO-WATER VARIABILITY AND RELATIONS

This volume will show that food production in general, and Africa's history in particular, cannot be understood properly without locating them within particular water systems. From the early evolutionary history of mankind to the future global challenges of feeding 9 billion people, the relationship between water and food production is fundamental.[1] The developments of complex societies and civilizations were to a large extent based on the wealth generated by surplus agricultural production in natural or artificially irrigated land, and the revolutionary population growth during the last century was due to more efficient food production, which again – and this has been tended to be overlooked in many analyses of the Green revolution – was premised on radically more and different uses of water. Future pressure on water resources and water management in order to increase food production will thus most likely increase. Very few now share the widespread optimism of the early 1970s, when the world's population turned 4 billion and the then US Secretary of State, Henry Kissinger, proclaimed that 'no child will go to bed hungry within ten years'.[2] As development continues and the number of people increases, meeting the world's demand for food, and thus for water, will be one of the most important challenges for the world community in the twenty-first century.

The intimate connection between poverty and food and water is, of course, not a recent phenomenon. In the Old Testament, Genesis 41:30–32 tells the infamous story from Pharaonic Egypt:

> Seven years of great plenty will come throughout all the land of Egypt; but after them seven years of famine will arise, and all the plenty will be forgotten in the land of Egypt; and the famine will deplete the land. So the plenty will not be known in the land because of the famine following, for it *will be* very severe.

The seven successive rich years followed by seven years of famine should partly be read metaphorically, highlighting that even the richest and most fertile areas for agricultural production, generating immense wealth in good years, may suffer from food insecurity and famine in years when the life-giving rains or the annual floods fail or are insufficient, or if there is too much water at the wrong time for cultivation. This story in Genesis should also be seen as a story expressing the knowledge people at that time had of water in general and the hydrological cycles of the Nile in particular. They knew the connections between water and food and life and death, and they knew that the amount of water carried by the Nile was never the same, and periodically changed radically with dire consequences, but at that time they had no means to control this water. In a long-term perspective this Biblical prophecy can be seen as a commentary to the anxieties of what can be termed 'The Age of Water Insecurity'.[3] This is an era where uncertainty about the future waterscape – whether there will be more droughts or more floods or whether the sea will rise because the ice will melt – dominates, with immense consequences for food production.

The world population is predicted to grow from 6.9 billion in 2010 to 8.3 billion in 2030 and 9.1 billion in 2050, and food demand is predicted to increase by 60 per cent by 2050.[4] Currently, it is estimated that some 860 million people go hungry in the world despite the fact that globally there is enough food to feed everybody. Calculations suggest that in mere volume there will be enough food for 9 billion people in 2050. In Africa, the population is expected to increase from around 1 billion today to 2 billion in 2050. Economic growth and individual wealth result in shifting diets from predominantly starch-based to meat and dairy, which require more water. The production of 1 kg of beef may require five times more water than the production of 1 kg of grain. Hence, not only population growth but the rapid growth in the global middle class puts more pressure on water for food production. Still, it has been asserted that there is no food crisis, only a crisis of just and efficient distribution. We will argue that it is an issue of distribution, of market efficiency, but that regional and local deficiencies in food production cannot be reduced to social variables alone, due to extreme (and in the future unknown) differences in agro-water variability and food–water systems across the globe.

There is hardly any topic that has been discussed so widely, and for so long a time, as food production regimes and their consequences for societal development. Archaeologically, agriculture is the origin of civilization and the basis for economic development. Without surplus food production, stratified societies could only emerge to a limited degree. The whole chronological development of prehistory worldwide is based on agriculture. The Danish scholar, Christian Jürgensen Thomsen, was proposing the Three-Age system in 1836, classifying the past into Stone Age, Bronze Age and Iron Age. This was based on the raw material of the tools used and did not relate directly to the different modes of subsistence:

hunting and gathering or farming. In 1865, John Lubbock divided the Stone Age into two categories: an older Palaeolithic (Old Stone) period and a later Neolithic (New Stone) period.[5] The Neolithic period was characterized by agriculture and the origins of agriculture have been an ever-reoccurring theme in archaeology. Crucial in the debates on the origins of agriculture is the question of domestication of plants and animals and the relation to different degrees of sedentary practices using non-domesticated species.[6] The focus here will not be the raw materials of tools, or overall modes of subsistence practices, but different methods of food production and their shifting relationships to agro-water variabilities. We suggest that major periodizations of agricultural history can be based on shifts and differences in such agro-water relations.

Food production represents a form of adaptation to the local and regional characteristics of the physical and engineered waterscapes. Wheat and rice, goats and camels, fish and fowl – they all thrive in different waterscapes according to tolerance to drought and waterlogged soil, degrees of water salinity and stream velocity, and so on. Precipitation will in many areas be the ultimate source of water for food production, and therefore the seasonal variations of how water runs in the landscape and through agricultural lands – annual rains or floods, the absence or presence of which types of water at what time of the year, discharge curves of rain-fed rivers – are physical premises for food production. A nomad society in semi-arid regions or in deserts is structured differently from farming and fishing communities living in wetland areas. This difference is very easy to observe, but it is at the same time a clear illustration of a more complex and universal phenomenon: how can confluences between water and society structuring food and agricultural production be framed?

While agro-water relations tend to structure food-producing regimes, different food systems are also intrinsically interwoven into the social matrix of any society; political, economic, cultural and religious premises influence what is grown and how it is harvested, and also what is culturally accepted as food. Societal development in the past was not merely a matter of subsistence or surplus production for exchange or trade enabling elites: the very modes of production and types of food produced enabled distinctive development trajectories. Specific waterscapes that were often continuously modified over time, encouraged the creation of certain agricultural activities, cropping patterns, plot sizes, and crops' complementarity. These in turn influenced certain types of societies and food-production systems. All long-term agricultural history can therefore be fruitfully analysed in an agro-water variability perspective, since these relationships and interconnections have created and re-created the context for the farmers' choices and agency in fundamental ways.

Still, the more complex relation between water and food has not yet been explored in depth within a historical and comparative perspective. This volume aims to provide a modest input to this process. In order to deepen our understanding of water and food systems, it focuses on forms of food production as both a livelihood strategy and an economic process, from an individual level to state organization and distributions within a country and beyond, and how they relate to and impact on each other.

WHY FOOD AND WATER IN AFRICA?

There are several reasons why Africa as a continent has been chosen as the focus of this volume on water and food. Not only did modern human evolutionary history start in Africa, but many agricultural innovations developed on the African continent in relation to the wider world.[7] Many of the most important general practices and processes regarding water and food systems can be studied in Africa: hunters and foragers in the past and the present, fisheries from small-scale to industrial scale, rain-fed agriculture and irrigation, nomadism and pastoralists, devastating droughts and floods, climate change adaptation and mitigation, local tenure systems and colonial governance, independence and adaptation to marked liberalism, and interactions and dependencies at various levels from individual households to states and the wider world.

While the focus is on Africa, the case studies will also have relevance to food and water relations in other parts of the world. Africa has a special contemporary relevance, due to the food-shortage problems faced by the continent and many of its countries. In the early 1960s, Africa was more or less food self-sufficient but since then food security has decreased, albeit with regional differences.[8] Hence, being concerned with reconstructing the history of food production in relation to water variability and availability, and its management, this book rests on two premises. On the one hand, historical analyses of interconnections between water and food regimes have an important value in themselves. On the other hand, Africa's current and future food challenges cannot be understood sufficiently without taking a historical perspective and making a comparative analysis of local and regional waterscapes and different water–society relations.

This volume will therefore employ a long historical perspective on the history of African food production, from hunter-gatherers 10,000 years ago to today's large-scale foreign investments in agriculture; from the rainforests to the deserts in some of the most extreme and different climates, and the varying environments and water availabilities in between these extremities, including the great lakes in the Interlacustrine region; from South Africa at the tip of the continent to the delta in Egypt and the Mediterranean Sea, and from Ethiopia in the East to Morocco in the West.

The understanding of Africa is caught up in a paradox: Africa is a rich continent when it comes to natural resources, and yet it is at the same time the continent that still symbolizes and expresses poverty and food insecurity. In particular, areas of sub-Saharan Africa continue to face huge challenges with persistent hunger, and hunger is most widespread in rural areas. Farmers and smallholders account for about 65 per cent of Africa's labour force, but in the last decades there have also been significant changes in rural areas. Rural households have increasingly been involved in non-agricultural income earning businesses, like trade, handicrafts, fishing, forestry and other small-scale activities.[9] These shifts are important since they alter the food production regimes.

In theory, it is argued, there is more than enough water and food for Africa's own food production and it is even claimed that Africa has the potential to feed other parts of the world.[10] Africa's history of water and food thus points in two different directions: on the one hand, it has been plagued by chronic food insecurity and climate variability for millennia, but on the other hand, it has also given rise to major agricultural civilizations, including the ancient Egyptian and Nubian civilizations, and export of food beyond the continent. As this volume illustrates, the food insecurity in recent decades is not a new phenomenon and throughout history from early hunter-gatherers through major civilizations to colonialism to the present day, there have been varying degrees of food security while at the same time Africa's water and land have produced large surpluses and wealth, and in different periods and regimes enabled major exports of food out of the continent. Understanding water and food in Africa in an historic and comparative context may thus enable new insights into the evolution of food systems from the early humans to today's challenges in the global world.

FOOD IN A WATER SYSTEM PERSPECTIVE

This volume presents a number of case studies from African history together with methodological and theoretical arguments for analysing food production regimes using a water system approach. The term should for several reasons be contrasted with the 'ecological system' concept and the theories behind it. (A) The focus on water implies that the term 'ecological system' is considered too broadly, because it is empirically close to impossible to study the relationship between societies and ecological systems due to the fact that ecological systems are almost everything. (B) Since all ecosystems rely on water and all agricultural societies need to manage water, water is a universal factor in all food regimes and at the same time easier to 'follow' empirically and analytically, both when it comes to flora and fauna and in society. (C) The water system approach does not treat the whole social or physical world and their relations as a unity in varying degrees of (distorted) equilibrium. On the contrary, it is more

concerned with conflicts and paradoxes between elements in the
waterscape and between the waterscape and society. (D) More specifically,
a water system approach probes into the heart of the nature–culture
dichotomy, since water does not change from being 'natural' or 'social'
because it can be both at the same time, but by specifying and demarcating
fields of enquiries and presenting analytical perspectives enables
approaches transcending determinism and constructivism.[11] Being an
approach encouraging historical studies of hydrology, hydro-history and
hydro-archaeology and the relevance of the physical world and the physical
aspects of the water system and its relations to society and action, it also
differs from all types of system theories that are only concerned with the
social or with social variables.

A water system approach may open up new avenues to understanding
these mutual relations. It consists of three layers, which should be analysed
in connection with each other:

1. The physical and natural waterscape;
2. The modification of, and adaptation to, different water systems; and
3. The cultural, instrumental and religious ideas about water at a given place.

The use of the terms 'first', 'second', and 'third' layer of the water system
does not signify any ordered hierarchies of importance or categories; they
are separate, distinctive and related layers. The different levels are meant to
designate analytical ways of separating and discussing the complex nature–
culture relationships without falling into the trap of ecological determinism
or social constructivism, by acknowledging that water is both culture and
nature at the same time.[12]

Importantly, all the chapters in this volume include all three levels,
although with different emphases in the analytical and empirical
discussions. Some aspects are highlighted with regards to the respective
layers in the water system but, as will be indicated, all chapters could have
been used to exemplify all layers. Hence, although the layers are
analytically, empirically and theoretically separate and distinctive, they are
integrated in historic and social contexts, which this volume will attempt
to illustrate.

THE FIRST LAYER: THE PHYSICAL AND NATURAL WATERSCAPE AND FOOD

This approach rests on the premise that the physical world exists and has
existed independently of humans, and will continue to do so, even though
humans may also significantly alter the natural world (see Level 2). Today,
the climate change discourse has shaped parts of the premise seeing
nature as very much influenced by humans and consequently not existing

independently of human activity. At the outset, one may point to three aspects. First, when climate scientists are taking deep-core samples from the ice in Antarctica, going back several thousand years, these provide a description of nature and of waterscapes not impacted by societies. Second, although interpreting and understanding this is, of course, a (social) construction, it does not mean that the various bodies of water have never existed prior to our understanding. On the contrary: it is as 'non-impacted' that it becomes relevant data both to understand climate and society. Third – and perhaps the most important from a water system perspective, due to its implications for analysing the consequences of human induced climate change – as far as we know today, the amount of water circulating on earth as part of the hydrological cycle is basically constant. *This* is water as external nature independent of humans; climate change may influence where and when there will be more or less water as rain or floods or the melting of glaciers, but the overall amount of water circulating in the atmosphere and existing on the planet remains the same. Or to use another example: the mere fact that humans may pollute and poison a pristine lake does not 'culturalize' or 'de-naturalize' the water – it is still external nature and water, but in a deteriorated state from a human perspective. Ultimately this water will evade human control and influence, primarily because the sun evaporates water, which then comes back as clean water. The possibility of humans to have an impact (often radical) on nature and waterscapes, and thus also on the relations between societies and water, does not make water, as nature, something of the past. The hydrological cycle is still an external physical fact, while the hydro-social cycle – the way water travels through societies – influences it and is influenced by it.[13] Analyses on this layer can, for example, be concerned with reconstructions of how different food systems in varying and changing waterscapes, from deserts to tropical rainforests, have enabled certain adaptations and technological innovations at different points in history. Concepts like water zones give attention to such relations between systematic differences in waterscapes and flora and fauna and agricultural food regimes.

Fridtjov Ruden highlights these aspects of water as external nature, and at the same time the human ability to severely pollute pristine waters, by analysing the role and importance of African mega-aquifers.[14] Water exists externally to humans and this may represent a huge potential for Africa's future. It is estimated that beneath the Sahara, some 5 per cent of the world's freshwater is found and stored in deep mega-aquifers. Some of these deep waters are probably more than 1 million years old, predating humanity, and have existed independently of humans until their recent discovery. All over Africa there are large aquifers of hidden waters and there are probably many that remain to be discovered. If developed well and exploited carefully, these mega-aquifers may contribute significantly to securing the future water needs of Africa in the long-term. However, in many of these geological contexts there are also large oil reservoirs above the mega-aquifers, and pollution from oil extraction may pollute the pristine waters for eternity, causing ecological disasters of unparalleled proportions.

A long-term perspective on food production, or a hydro-historical and hydro-archaeological approach to food production, makes it clear that what matters is that for all people at all time the presence or absence of water is a reality and not only a social construction. When droughts hit the African continent and the life-giving rains never come, this is an external nature people have to adapt and relate to, whether it is part of the natural annual precipitation variability or accelerated by human induced climate change. Whether caused by the absence of rains for years, or by irregular and devastating floods, throughout history too much or too little water for food production at the right time for cultivation has been a matter of plenty or famine, and life or death. The physical water world is a premise for human existence and adaptation. Importantly, these water worlds have always been changing as part of natural variations or accelerated by climate change and drought, and have regularly haunted the continent throughout history.

Marcel Rutten analyses the Maasai pastoralists and their adaptive strategies to drought, in particular the drought in Kenya in 2008–9.[15] For a Maasai to survive on a diet of milk and meat from cattle, it is estimated that it requires about ten cattle per person; in 2008 the ratio was 1:5, which was partly due to population increase. The number of cattle is dependent upon the rainfall, but social changes and modernization also affect the traditional livelihood. While hydrological and meteor-ological droughts are scientifically defined in amounts of precipitation and water availability, the Maasai have culturally defined perceptions of drought, linking failing rains to hunger, since the cattle become thin producing little milk, with the lack of food resulting in crisis for herds and humans, and subsequent disaster for livestock, ultimately causing death. During the 2008–9 drought as much as 90 per cent of the herds were lost, although many of the cows did not die from the drought itself but rather from disease. Thus drought is not mere absence of water, but includes a broad spectrum of multiplying factors, which lead to extreme adaptive strategies involving great mobility searching for water and fodder, with severe implications for social life and well-being.

From the neighbouring country, Terje Oestigaard discusses the 2011 drought in East Africa, affecting the Sukuma agriculturalists along the southern shores of Lake Victoria in Tanzania.[16] The drought affected as many as 13 million people and an estimated 500,000 were affected in Tanzania. Being dependent upon the erratic and unpredictable rains is a gamble where the farmers cannot afford to make the wrong decisions with regards to what to grow and when and on which agricultural plots, but still they do. However, all strategies are in vain when the rains fail. While pastoralists have an option, no matter how meagre or small it may be, to take with them the cattle and search for water and food elsewhere, agriculturalists do not. As one farmer put it, one cannot move the land – farmers have to stay where their land is and wait for the rain. However, farmers and pastoralists alike need money for food, and during periods of drought it alters the social structure. If they cannot move their land, they can move themselves. Migrant work is an option when money matters, and although the market economy puts other limitations on rain-fed agriculturalists, despite the suffering and hardships the current drought was seen as easier to survive than those in the early-1980s, when it was more difficult to buy food even if they had money.

To what extent these two droughts were worsened by climate change is a matter of debate, but the current climate change challenges are generally expected to affect Africa harder than other continents. This has to be understood from a water perspective, and there are hydrological reasons for this increased future vulnerability. Although there are always hydrological fluctuations, this inter-annual variability and availability has been greater and more unpredictable in Africa, which historically has resulted in extremely robust resilience and risk-management strategies. This can be illustrated through varying rainfall precipitation, though not only in mere measures of millimetres of rain, since the very rainy season has a specific character on different continents. In parts of Asia, for instance, the monsoon winds lead to a more well-watered agriculture than compared with large parts of Africa, directly impacting food production. The particular condition that has framed large parts of Africa is the great agro-ecological variability within even small territories: it may rain sufficiently in one village but not in the neighbouring one, etc.[17] Given that ecology is a very broad concept including numerous understandings and many aspects of 'nature', we would rather propose a more specific term: *agro-water variability*. This directs attention to the role of water in different food production regimes, and by using it as a common denominator, as a sort of methodological entry-point, the greater agro-ecological variability might be easier to reconstruct and analyse.

Starting with mere precipitation, from the tropical rainforests receiving up to 5,000 mm rain each year to the deserts in the Sahara virtually without a single drop of water for years, the agro-water variability of Africa is extreme. But the agro-water variability of Africa is not a gradient, with fading amounts of precipitation from the rainforests to the deserts – from 5,000 mm to 0 mm; if this was the case the predictability and adaptability would have been rather straightforward. If it is one thing that characterizes statistical averages concerning precipitation, it is that the actual rainfall in a given year is hardly ever the average: more often than not it is extreme one way or another. In other words, there is hardly ever a 'normal' year; the erratic and unpredictable rainfall patterns become 'the norm'. Moreover, the total inter-annual amount of precipitation is not always the most relevant measure. If there are two rainy seasons – the long and the short – what matters fundamentally for agriculturalists is that the right amount of water comes at the right time. In particular, rain-fed agriculture is highly vulnerable to erratic precipitation patterns. If the rains fail, or are abundant at the wrong time for the cultivation season, it may have devastating consequences. Moreover, too intensive rains may destroy crops, which also need sun; and the perfect balance between heavy and light rain for some days, then sun and more good rain, and so on, are parameters that nobody can predict but which farmers are dependent upon for a successful harvest. Importantly, as pointed out above, the actual rainfall may also be very specifically localized, even within small areas. As a consequence, not only are what and when to grow fundamentally affected, but also *where* within a village, creating systems of multi-plot approaches aimed at predicting and reducing risk. Understanding the social consequences of this great agro-water variability in time and space is crucial when comparing most African food regimes with, for instance, development and the Green revolution in Asia.

The dependency on which type of water also relates to technological innovation in a given time period, creating varying opportunities and varying social development; and closeness to oases and rivers in dry environments offers other possibilities than dependency on the life-giving rains. Prehistoric subsistence and habitation patterns show a strong correlation with adaptation and the specific utilization of different water bodies.

Elena A.A. Garcea analyses semi-permanent foragers in North and West Africa from an archaeological perspective, going back 10,000 years.[18] This was a time when there were huge ecological changes. Lake Mega-Chad had a maximum surface area of about 340,400 km^2 in two periods from 7,400 to 8,500 years ago. Still, the semi-permanent hunter-fisher-gathers were living in aridity-prone areas and the desert foragers adapted aquatic livelihood subsistence patterns along the main bodies of water. Comparing two sites only 20 km apart along the Nile in today's Sudan with a site in the Ténéré desert in Niger in the Sahara, the settlement patterns show clear differences in adaptation where the various bodies of water in dry environments were like 'islands in the sea'. Although hunters and gatherers are commonly believed to be on the move, and they were, livelihood strategies in arid environments were also more complex. The access and closeness to water resources enabled greater investment in territoriality and semi-permanent settlements, with thick layers of continuous habitation for a long time, which also included heavier tools and equipment, like non-portable pottery and grinding stones.

Louise Bertini addresses dependency and adaptation to the fluctuating Nile from another perspective, namely swine husbandry practices in ancient Egypt.[19] Domesticated animals accounted for more than 50 per cent of the faunal assemblages by the start of the Early Dynastic period (c.3000 BC). In ancient Egypt, cattle were highly valued, but remains of pigs show from settlement sites that pigs were one of the most common domesticated animal. Pigs were for subsistence, and, as a vital source of protein, were exempt from taxation. Based on meticulous analyses of dental defects of the pigs' teeth – since tooth growth is an indication of different degrees of stress (environmental stress included) – there were two different swine management practices: free-ranging and perennial. In particular, with regards to free-ranging pigs, different tooth growths indicate different stress factors related to ecological changes caused by fluctuating Nile levels (high or low floods) impacting on the pigs' environment and foraging range. Not only may the degree of agriculture determine the extent of swine husbandry practices, but, perhaps more importantly, the actual annual flood – and hence pigs and their environmental stresses – may be an indicator of the fragility of the ancient Egyptian agricultural adaptation despite the fact that the flood came each and every year, although in fluctuating levels.

A long historical perspective will allow us to explain why it is crucial to go beyond the social reductionism inherent in much of the social sciences when studying societies' development trajectories. Johann W.N. Tempelhoff analyses the history of water, migration and settlement in southern Africa from a long-term perspective (200–1850 CE).[20] In the latter part of the nineteenth century, settlers became aware of water shortages and the perception of a sub-continent that was in the process of dying and 'drying out' was put forward. Through a meticulous historic analysis Tempelhoff shows how varying water availability was the key value to land, and he shows how water created both opportunities and limitations to societal development. What becomes clear is the non-reductionist ways people have adapted to the changing water-world in various societies, including mobility patterns among hunters and gatherers to the civilization of Great Zimbabwe. The history of the Iron Age in southern Africa is not a narrative of the rise and fall of societies in relation to water. On the contrary, it is a story of how water availability has been a fundamental part, where human ingenuity has created resilient and resistant processes, and the collapse of some structures and societies has led to the emergence of new forms of developments in dynamic ways.

From another perspective, Andrew Ogilvie, Jean Charles Clanet, Georges Serpantié and Jacques Lemoalle analyse water and agriculture in the Niger Basin through the twentieth century.[21] Throughout the century drought has been a recurring problem, to varying degrees, whilst at the same time in the southern areas, with their abundance of water, the excessively moist climate has created yet other problems like the tsetse fly attacking livestock. The irregularity of weather fluctuations and the dependency of the annual rains have structured livelihood and livestock practices. While pastoralist herders in the Sahelian semi-desert grasslands have to balance the herds' need for water and food on the one hand, with the physical effort to find where water and fodder are available on the other, agriculturalists aim to reduce vulnerability by diversifying crops. But, when the rain fails over a number of years the consequence has been severe famine, like the Sahelian famines of 1972–3 and 1983–4. Yet failing rains may also, paradoxically, create more runoff in rivers, at least initially. Changes in land use and land cover, including forest degradation, have reduced the water-holding capacity of the soil from the 1970s and, as a consequence, the runoff in some of the Sahelian catchments increased despite of the reduction in rainfall.

As a last example, Raphael Tshimanga analyses contemporary hunters and gatherers in the Congo rainforest.[22] In tropical climates a rainforest is generally defined as an area receiving more than 1,600 mm of precipitation annually. Although it is commonly believed that there is plenty of water the year around in rainforests, Tshimanga shows that even in rainforests there is great seasonality, impacting human adaptation. These seasons where the rain is unevenly distributed – and there are even dry seasons, relatively speaking – are structuring the life and subsistence patterns of hunters and gatherers. The appropriate and preferred season for hunting certain types of game depends on the rains. In other seasons, gathering is the dominant subsistence practice, and local variations in the ecology, even in rainy areas like rainforests, create specific opportunities and limitations.

This volume has pointed out some of the extreme variations in precipitation patterns across the African continent from year to year, affecting inter-annual fluctuations in river inundations, but also within a year at different seasons at a given place. It has also shown some of the extreme variability and unpredictable changes in the actual natural and physical water world and how it has impacted rural life. Different food systems in varying and changing waterscapes, from deserts to tropical rainforests, have enabled certain adaptations and technological innovations at different points in history. We suggest that terms such as 'water zones' will be useful in analyzing patterns of food regimes,[23] and specific agro-water variability within water zones, all the time acknowledging that these adaptive innovations and developments have also been integral to specific societal organization. Since variation and fluctuating patterns of water zones are products both of nature and human ingenuity, this term also directs our attention to the second layer of analyzing food within a water perspective: human modifications of the waterscape.

THE SECOND LAYER: HUMAN MODIFICATION OF THE WATERSCAPE AND FOOD PRODUCTION

The human modification of particular waterscapes is important with regards to food production from the smallest to the largest interventions, but in Africa and elsewhere it is very important to analyse how these interventions are framed and influenced by the structural properties of the physical waterscape. In sub-Saharan Africa today, more than 90 per cent of the agriculture is rain-fed, and even though the rains fall from above, the waterscape has to be modified intensively to secure a bountiful harvest. This sole dependency on the seasonal rains necessitates the utmost preparation of fields for a successful harvest – from tilling the soil to the careful supervision of the fields throughout the season. Dependency on one single, and at times highly variable, source of water may require more intensive and thorough modification of the fields and the waterscape, since the margin between a successful and failed harvest, depending as it does on the fluctuating rains, is an act of balancing on the edge.

The most common form of modifying the waterscape for extensive agriculture is irrigation. Matthew V. Bender analyses the traditional irrigation system among the Chagga farmers on Kilimanjaro in Tanzania from a historical perspective.[24] Once heralded by the early colonialists as indigenous ingenuity, from the 1920s onwards the irrigation system was seen as wasteful and indeed as increasing water scarcity on the plains. By the turn of the twentieth century, about 80,000 lived on the mountain, and although the mountain is renowned for its icecap at the top, water scarcity and erratic rain fluctuations were prevalent, and the furrow-based irrigation system was a sophisticated adaptation to the unpredictable nature of water. The change in colonial perceptions of the irrigation system was due to several factors: there were a number of prolonged droughts, the American Dust Bowl (1930–6) generated fear that the mountain would be drier, population increase created greater pressure on resources, and there were conflicting interests with regards to water use. British settlers endeavoured to transform the water regime on Kilimanjaro, a policy that post-independence Tanzania continued. But the resilience of the agricultural practice persisted throughout much of the twentieth century, and today's decline is mainly due to land shortage, socio-economic factors and job opportunities away from the mountain, and not due to centralized policies.

While only a limited amount of the total feasible irrigation potential is fully developed in Africa, the possible expansion of irrigation is generally seen as limited. However, one should treat the different statistics with caution because there are grading differences between extensive rain-harvesting techniques, traditional irrigation and full-scale industrial projects. Globally, about 20 per cent of cultivated land is irrigated. This land produces 40 per cent of the world's food. As opposed to rain-fed agriculture, irrigated agriculture is largely dependent upon dams. In the period from 1945 to 1990 more than a 1,000 large dams – i.e., with a height at least 15 m or with a reservoir capacity of 3 million m³ or more – were built in Africa for irrigation and electricity purposes.[25] Many of the dams were built solely for hydropower, whereas in those cases where they are multipurpose dams, there are often tensions and internal domestic discussions regarding the amounts of water that should be used for electricity (industry) and agriculture (irrigation) and at which time of the year, directly affecting the overall agricultural production.

Pierre Morand, Famory Sinaba and Awa-Niang Fall study the traditional fishers, herders and rice-farming communities of the inner-Niger Delta and the consequences of lower floods, mainly due to recent dam building.[26] In the delta, there is what is called a 'social ecological system', where different ethnic groups have specialized in different food products: milk, fish and rice. Although the system among the herders, fishers and farmers is not egalitarian, it has enabled cooperation and exchange between the groups utilizing different resources in the same ecosystem. All these production systems are dependent upon the annual flood, which is weakened by several dam constructions upstream. It is estimated that a 1 cm reduction in the water height during the flood peak implies a loss of about 65 km² being inundated. Currently, the existing dams result in losses on inundated areas of between 1,200 km² and 2,850 km², and this is expected to increase to between 3,500 km² and 6,500 km² with new dams. Consequently, the fishers, herders and farmers are negatively affected in different ways by the reduction of the flood, and although the dams enable hydropower and irrigation, it stresses the fact that large-scale water structures have huge social and economic consequences for the better or worse.

Maurits W. Ertsen discusses the French colonial irrigation plans and schemes in West Africa and demonstrates a global context and influence of these schemes from the early ideas and plans to today's successors.[27] From the early nineteenth century, the French colonial administrators developed schemes in Senegal inspired by their experiences in the Mekong valley. Based on the results in Senegal, the French were convinced that even larger projects could be developed in the central delta of the Niger River in Mali. French engineers had visited and learnt from British India, and later from the Gezira scheme in Sudan, aiming to turn the Niger River into a 'French Nile'. While the irrigation schemes were not as successful as their British counterparts, the purpose of the schemes was not only colonial exploitation by France for export of cash crops, but also to serve the food needs of the colonies and to be a grain basket for French West Africa. After independence, including the period from the 1980s with development donors and market liberalism, the overall structures of the colonial schemes continued, although with significant changes. From 2005 onwards, these former colonial irrigation schemes have been allocated to national and international investors, being part of what has been described as the current wave of 'land-grabbing' or 'land-acquisitions'.

Turning to another French colonial area, Brock Cutler studies urban–rural relations in the modern Maghreb.[28] Common throughout many areas and regions in Africa, absence or presence of water is not only a matter of hydrological parameters and rainfall patterns, but also literally man-made constructions on the ground, part of explicit policies whether colonial or not. Moreover, there is often a transfer of water, food and resources from rural areas to urban areas. By using historic examples from Morocco, Algeria and Tunis, Cutler shows that the most relevant question is not always the total amount of water, but who *controls* and *transfers* the water through large-scale infrastructural projects. From the French colonial period onwards, water has been directed to urban areas, and even there only privileging certain sectors of towns, while the rural areas and the agricultural needs for water have been marginalized. In the twentieth-century this has been implemented further within the frame of modernization, where the challenge has been that, although agriculture has been the largest water consumer, it has nourished fewer and fewer people while at a same time there has been an explosive urban growth demanding more water.

This points to the fact that water has always been transported from one location to another, either in the form of water or as food. Although jars, pottery and today's plastic bottles and jerry cans are often seen as simple tools, the importance of these in water and food analyses cannot be underestimated. Smallholder agriculture is predominantly a female domain and carrying water is usually solely the work of women. Moreover, food preparation and cooking in domestic households is traditionally the task of women. Not only are gender relations involved at all levels, but it also points to the outcome of agriculture and food production: food and consumption of food, but also redistribution of food and transport over long distances whether these transactions are embedded in social relationships or merely export for cash.

In a wider perspective the different means and techniques for transport of food have to be included in this second level of a water system approach. Food may be transported on the head while walking, or via long-distance transport on the backs of camels, on boats along the main watery arteries or on modern lorries and even in planes, connecting distant localities in the rural hinterland to the global world.[29]

Although food is usually what one thinks about when it comes to agricultural production, throughout history water and land have provided resources for other types of harvest – non-food items – which sometimes are in competition with food production if the water and land resources are limited. This raises the fundamental questions of what water and land are used for, and how food systems are part of other economic systems which are dependent upon the same resources. The colonial projects were, to a large extent, preoccupied with cotton production – agricultural practices that have continued to today. Cultivation of flowers is another practice which is highly water intensive. Gessesse Dessie focuses on a type of agricultural production that is not much discussed, namely using water and land for drugs.[30] Ethiopia is one of the world's main producers of khat. Among smallholders, khat production amounts to about 20 per cent of the average land holding, and it is the most lucrative cash crop by far and among the top five most important foreign currency earners. Khat production is highly water intensive and the replacement of food with khat production represent a huge loss of calorie production on the one hand but also enables a greater income which may secure food purchases on the other hand. Thus, replacing food with drugs may increase food security at an individual household, at least in the short run.

Whereas khat production for the international market in Ethiopia is an individual household strategy among Ethiopian smallholder farmers, there is another recent process taking place on a grand scale on the African continent, which has been labelled 'land-grabbing' or, more neutrally, termed 'land acquisition'. These investments are to a large extent foreign, although there is increasingly domestic investment in parallel. Atakilte Beyene and Emil Sandström present an overview over recent land and water acquisitions in Africa and its relation to food production.[31] Whereas there are varying estimates, in the media and the academic literature, a thorough investigation reveals that the actual numbers are probably significantly less than previously reported, and that in the period 2007–14 the land deals in Africa accounted for about 20 million hectares. Moreover, there is a huge gap between intended and operational deals. In their discussion, they focus on the water question. Most of the crops cultivated as part of the land deals are water intensive, but recently there has been a gradual shift to a greater focus on food production than, for instance, bio-fuel cultivation. Large-scale investment often requires irrigation to safeguard the profits of investors, and although most of these deals are not transparent, the water question is one of the main drivers for foreign (and national) land acquisition, often facilitated by changing land laws.

Economic profit motives are major drivers for much of the recent agricultural development and investment. National and international actors invest in African water and land for export of food and cash crops both within and beyond the continent. Although food has always been integrated into spheres of redistribution, exchange and sale, agricultural products are increasingly becoming seen and exchanged as a commodity on the global finance market, with subsequent implications for investment in African water and land. To what extent these processes are increasing or jeopardizing Africa's food security is an ongoing debate, but more and more of the most fertile agricultural land is used for cash crops for foreign export instead of supplying national food needs. In practice, this also means that much of the water used in agriculture flows out of Africa.

Thus, the second level analysing the relationship between food production regimes and water consists of the human modifications of the waterscape and adaptation to changing water worlds. This includes the preparation and tilling of fields in areas dependent upon rain-fed agriculture up to mega-dams providing water for large-scale irrigation schemes. It also includes the small pottery jars and fishing hooks up to large industrial fishing boats and mechanized farming and fertilized schemes, as well as modes of transport both within and outside the continent. In one way or another, all these practices and equipment relate to the actual physical waterscape and the ways in which people at all times have modified and adapted to their actual water worlds, but this happens in social and political contexts structured and governed by ideas and laws.

THE THIRD LAYER: IDEAS AND MANAGERIAL CONCEPTS OF WATER AND FOOD PRODUCTION

All human activities and ideas are obviously part of social, cultural, political and religious spheres in varying degrees. The third analytical layer of the water system approach addresses the social and human context in which practices related to water take place. Tilling the soil with a hoe is a practice shared by most smallholder farmers, being an adaptation to and modification of how the water runs through and across the fields, but prayers to the gods to let it rain when rain is needed are also among the wide arrays of practices that can be studied within this analytical perspective. These social practices may be embedded in centuries-old traditions and belief systems or modern conceptual frames. Life-giving rains have been intimately connected with ancestors and rainmaking rituals, and have been seen as precious gifts from the Christian God or Allah, whilst the absence of rain has been viewed as a penalty by the gods (or ancestors) for sinful conduct in the community, or the result of human induced climate change. The choices of what to grow, and thus what kind of water management is necessary – food for the family or cash crops for sale – also depend on a wide range of factors: individual taste and preferences, access to markets, and economic stress (for instance, how many children attend school), bridewealth and social commitments in the wider family, etc.

Traditional farming is also regulated by laws and land and water stewardship. Land tenure is the customary practice of managing and using the land and the way it might be transferred through generations or shared within and among families, villages and beyond. Land tenure systems may not be formalized from a state perspective, and in countries like Tanzania and Ethiopia the state is the ultimate owner of all land. Yet customary practices define user rights, and these are often embedded in other social relations in the extended family. Access to land and natural resources is crucial, but in many customary systems the introduction of free liberalism and the market economy is not necessarily applicable, and centuries-old traditions regulate who has access to which type of water and land.

In all cases, behind the basic human need for food, the decisions regarding what to grow where, when and by whom, and not the least who should be the consumers in which regions and at what time, are deeply embedded in cultural, social, political, juristically and religious domains. These are not merely recent conditions or contexts in which food production is situated, for throughout history all food-producing processes have been deeply rooted and structured by socio-cultural and juridical structures regulating social and political organizations. The Nile Delta in Egypt may illustrate such historic structures and processes from different perspectives.

Katherine Blouin discusses the generally held perception that the Nile Delta was a breadbasket in antiquity and the Roman period.[32] The size of the delta is approximately 26,000 km^2 compared to the 9,900 km^2 of the Nile Valley. Although the written papyri texts are scarce due to the humid conditions in the delta compared with the drier conditions in Upper Egypt, there are some historic sources and these indicate highly developed local water management practices in combination with a state-organized system supervising and integrating taxation, storage and transport facilities. The agricultural land was divided into three types of land categories: grain land; vineyards and gardens; and pastures – all of them relating to the local ecology and the ways in which the Nile annually inundated the areas. Grain cultivation was the main agricultural crop, constituting more than 90 per cent of the arable land, but areas not well suited to this cultivation were intensively utilized for other crops and husbandry. This extensive intensification in combination with diversification testifies to the important role the delta had in antiquity, thus supporting the notion of being a 'breadbasket' not only for the region but also for the Roman Empire.

Alan Mikhail analyses the role of the Nile and food export in the early modern Ottoman period.[33] After the Ottomans conquered Egypt in 1517, Egypt became the main source of food supply and the breadbasket for the whole Ottoman Empire. The Ottomans had a particular interest in, and profit from, controlling and maintaining the irrigation systems of Egypt. Egypt literally produced the energy empowering the muscles of the political powers in Istanbul, controlling the provinces from Morocco, Syria and Yemen. After Istanbul, the first destination of Egyptian grain, came Mecca and Medina. The annual pilgrimage and festival held in Mecca – the ḥajj – was secured by food where Egypt was the main supplier.

The practical importance for the Empire in facilitating this utmost religious pilgrimage is evident by the fact that Istanbul sent one of the most trusted functionaries to oversee the work and transport carried out in Egypt. Thus, the Nile waters and the fertile fields of Egypt increased the food security of the Ottomans, enabling the Empire to control vast areas, physically as well as spiritually.

Politics and power relations play fundamental roles in food production. Tobias Haller discusses the role of institutions and power relations for water governance and food resilience in the African floodplains.[34] Although floodplains and wetlands have often been a common pool resource, this does not mean that they have not been managed by local communities in pre-colonial times. Local institutions regulated and controlled who could use which type of resource and for what purpose at a given time at a certain place – for instance, drinking, fishing, pastures, wildlife, forestry and irrigation. Today, this would have been sustainable water use, but the aim was to secure livelihoods and reduce risk, although the outcome proved to be sustainable. Thus, contrary to Hardin's theory of the 'tragedy of the commons', common pool resources were not depleted. However, with colonial and later independent states' centralized control over these recourses, the belonging and ownership to these resources were alienated from the local people, who were no longer able to continue their customary practices protecting and preserving their common resources for the betterment of them all. This is turn has led to a situation which resembles Hardin's argument where people started to behave as if the water and the resources were open and freely accessible to all, and given the mistrust in the state and the weakening of local traditions and institutions, people started profiting as much as possible, and preferably before everybody else.

From another perspective, Tor A. Benjaminsen analyses land-use dynamics along the Niger River in Mali in the region where the river 'bends'.[35] This has historically been a scarcely populated area dominated by pastoralism. While it has generally been held that there has been an increasing desertification in the Sahel due to overuse of local resources, it seems that the fluctuating changes are closely related to the annual rainfall rather than to livestock numbers. Contrary to another postulate, namely that the Sahel is 'overpopulated', Benjaminsen argues that it is underpopulated and therein are some of the challenges to agricultural production causing other conflicts. This challenges Malthus' thesis, since there are not enough people for intensive agriculture, hence supporting Boserup's thesis that population increase is a prerequisite for agricultural development and improvement. In this region, pastoralists reluctantly adopted farming as a livelihood because it was seen as too risky and generating too little income. However, when the land was open and vacant for longer periods, other sedentary groups moved in and started cultivating the land, causing conflict when the pastoralists returned.

Politics and law are also involved in other ways with ecological implications as well with regards to food security. Jeppe Kolding, Paul van Zwieten and Ketlhatlogile Mosepele start with the obvious – but still often neglected or under-communicated fact – that fish is food, and in particular inland fish plays a fundamental role in diets.[36] In Africa, as elsewhere, fisheries have been regulated, with the aim of protecting the small fishes from over-fishing, based on the assumption that if the small fishes are caught, it may jeopardize the whole ecosystem. This assumption is challenged from new theoretical and empirical approaches. While most fishermen will argue that their catches are reducing, this is most likely because there are more fishermen and not fewer fish, while at the same time the total catch is the same or increasing. Given that there are seemingly fewer big catches, fishermen reduce the net sizes and are 'fishing down', going for the smaller catches. This is generally interpreted as a sign of over-fishing and crisis, but it may be the other way around – a sign of a healthy ecosystem. In the nutritional pyramid, big fish are far above predators in the animal world, while the greatest biomass production takes place at the bottom. Hence, a focus on small fish may not only enable more sustainable ecosystems, contrary to current policies enforced by law, but also offer an untapped resource of more food.

The Great Lakes like Lake Victoria have obviously been important for fishing throughout history, but subsistence practices are yet structured by other practices and belief systems. Andrew Reid analyses the importance of bananas in the Buganda kingdom in today's Uganda.[37] Situated by the shores of Lake Victoria, the second largest lake in the world, water played a prominent role not only in the subsistence, but also in the cosmology of the Bugandas. The lake was called *Nalubaale*, which means 'the place of mother *lubaale*'; *lubaale* probably meaning spirits or even ancestors. Reid shows the complex dynamics and relations between rain-fed farming on land and lake adaptation. The year-round rainfall pattern created optimal conditions for banana cultivation. While bananas are highly nutritious, they lack proteins, and fish was fundamental in the diet. According to tradition, bananas were associated with the spirit Kintu, or the creator of Buganda, or with the Ssese Islands. Throughout the history of the Buganda, royal and religious powers were contested. About 75 per cent of all Ganda shrines were located at the Ssese Islands in Lake Victoria and the most important spirits and ancestors were related to water. Thus food, and in particular bananas, were intricately interwoven in the religious and spiritual realms of different water bodies with emphasis on rain and the lake.

From another, and contemporary, perspective Jessica Kampanje-Phiri and Dean Kampanje-Phiri eloquently point out that food is not just food or substances of caloric value, even in times of crisis and during drought.[38]

Among the Chewa of Malawi, white maize has a special social and symbolic value in the community, although this crop is not indigenous in the traditional sense. White maize symbolizes high status and wealth, whereas dark coloured maize is associated with poverty and low rank. During the 2001–6 hunger crisis donors provided cheap and subsidized American maize as food aid. This maize, however, was not the white variety but the coloured one. Not only was this food aid perceived as contradictory to the cultural and cosmological structure among the Chewa, but some even claimed that it smelled like 'dirt'. While foreign aid workers were distressed that the food aid was not perceived as appropriate food even in times of crises, it also challenges what 'food security' means, since in today's accepted definition it is stated that it includes 'food preferences for an active and healthy life'. Food is intrinsically linked to social and cosmological life; not only is it basic to human life and survival but it is also part of the foundations of society and religion, and in many cases these perceptions are stronger and more pervasive than the need for mere calories.

Food is not only something to eat in order to survive, but a way of living in a broad sense. Intriguingly, this is also possible to trace in archaeological material dating back to the very origins of domestication in Africa. Randi Håland analyses the early aquatic sites along the Nile going back 10,000 years. Domestication and sedentism were closely related to different food systems. While Egyptian cuisine was related to the Near Eastern cuisine based on bread and baking in ovens, Southern Sudanese cuisine was similar to the African savannah cuisine based on porridge/beer and boiling technology. Rather than using grains for bread, in the southern areas aquatic resources and wild grains were cooked into a stew, or made as beer, which not only enabled a growing population as it reduced the vulnerable infant stage when young children passed from breast milk to solid food, but it was also a social process, most likely driven by women, challenging other theories of societal hierarchy. Moreover, in an area where different groups with distinct languages were living, it seems like the pastoral Cushitic-speaking people developed a fish taboo in relation to their fish-eating Nilo-Saharans. Thus, from the earliest times, cultural and religious perceptions of food have been integral to the very evolution and development of these food systems.[39]

PAVING THE WAY FOR A NON-REDUCTIONIST APPROACH

Whether the context of study is hunters and gatherers in the rainforest, rain-fed agriculturalists in semi-dry environments, or mega-dams and irrigation projects based on market liberalism, a water system approach opens up possibilities for analysing water and food relations as both culture and nature as well as how these cultural and natural factors mutually influence and impact on each other, creating possibilities and limitations for food production and different societal developments. We suggest that studies of food production and food-producing regimes will benefit from a water system perspective.[40]

By separating the distinctive processes at work, while at the same time showing how the physical, managerial and political and cultural are related and interconnected at various levels and in different contexts, both environmental determinism and social reductionism can be evaded. All three levels do not have equal explanatory strength and importance in a given study. It is very important to realize that the approach opens up and allows emphasis for only one or two of the levels, as with many of the chapters in this volume.

This volume shows that Africa's history and food production cannot be understood properly without locating the practices within the continent's particular water systems. While it has never been the intention of this volume to come up with practical policies for Africa's current and future food insecurity (and as this volume shows, there can never be one template for all, given the extremely varied ecologies and historically specific adaptations), historical studies do, however, offer insights into processes and practices that have worked in the past and still do. Despite the challenges that Africa as a continent has faced with regards to food security throughout millennia, she has also prospered and managed extreme situations remarkably well, despite all the hardships and suffering. As many of the chapters in this volume illustrate, the knowledge and experience based on centuries of traditions have enabled a highly functioning and well adaptive structure to extreme, seasonal and varied ecological conditions and water availabilities in time and space. Rapid and dramatic interventions and changes in these resilient structures, whether colonial or not, have had deep impact on society and development, in some cases radically increasing local and national food security, in other cases not.

Whether within a past or present context, water has been, and will be, at the centre of societal organization and development. The chapters in this volume thus present a different history of Africa by showing the intimate relation between water and food production in the rise and resilience, but also the decline and challenges, of societies in highly diverse and changing water worlds. What becomes clear is that although the actual agro-water variability factor is fundamental to the development of different food-producing regimes, the physical water world at any given time and at any given place creates possibilities and limitations that through human ingenuity have given birth to a wide variety of rural and food adaptations and societal organizations.

NOTES

1 We would like to thank Professor Kjell Havnevik for his constructive comments.
2 Clausen, T.J. 2012. 'Introduction'. In Jägerskog, A., Clausen, T.J. (eds), *Feeding a Thirsty World – Challenges and Opportunities for a water and Food Secure Future*. Report No. 31. Stockholm: SIWI, pp. 6–12.
3 Tvedt, T. 2012. *A Journey into the Future of Water*. London: I.B.Tauris.
4 According to the Food and Agriculture Organization.
5 Trigger, B. 1994. *A History of Archaeological Thought*. Cambridge University Press. Cambridge, pp. 75–94.
6 For African developments, see Mitchell, P. & Lane, P. (eds.). 2013. *The Oxford Handbook of African Archaeology. Oxford Handbooks in Archaeology*. Oxford University Press. Oxford.
7 See Mitchell, P., Lane, P. (eds). 2013. *The Oxford Handbook of African Archaeology. Oxford Handbooks in Archaeology*. Oxford: Oxford University Press.
8 Luan, Y., Cui, X., Ferrat, M. 2013. 'Historical trends of food self-sufficiency in Africa'. *Food Security* 5: 393–405.
9 Bryceson, D.F. (ed.). 2010. *How Africa Works: Occupational change, identity and morality in Africa*. London: Practical Action Publishing.
10 Juma, C. 2011. *The New Harvest: Agricultural Innovation in Africa*. Oxford: Oxford University Press.
11 See, for example, Tvedt, T. 2010a. 'Water systems, environmental history and the deconstruction of nature'. *Environment and History* 16(2): 143–66; Tvedt, T. 2010b. 'Why England and not China and India? Water systems and the history of the industrial revolution', *Journal of Global History* 5: 29–50.
12 For a more thorough discussion of this approach, see Tvedt, T. 2016. *Water and Society: Geopolitics, Scarcity, Security*. London: I.B.Tauris.
13 See Tvedt 2015 for a discussion of this set of interconnected concepts.
14 See Ruden, Chapter 22 of this volume.
15 See Rutten, Chapter 12 of this volume.
16 See Oestigaard, Chapter 13 of this volume.
17 Kjekshus, H. 1996. *Ecology Control and Economic Development in East African History: The Case of Tanganyika 1850–1950*. Second edition, London: James Curry.
18 See Garcea, Chapter 1 of this volume.
19 See Bertini, Chapter 3 of this volume.
20 See Tempelhoff, Chapter 5 of this volume.
21 See Ogilvie, Clanet, Serpantié and Lemoalle, Chapter 10 of this volume.
22 See Tshimanga, Chapter 14 of this volume.
23 For discussion and definition of this term, see Tvedt 2016.
24 See Bender, Chapter 7 of this volume.
25 Hoag, H.J. 2013. *Developing the Rivers of East and West Africa. An Environmental History*. London: Bloomsbury, p. 177.
26 See Morand, Siniba and Fall, Chapter 17 of this volume.
27 See Ertsen, Chapter 9 of this volume.

28 See Cutler, Chapter 8 of this volume.
29 Transport of water in the form of food has been labelled as 'virtual water' by the like of Tony Allan in a number of works, but this concept with its implicit theoretical and political premises is highly context-dependent and not particularly useful in historical analyses.
30 See Dessie, Chapter 20 of this volume.
31 See Beyene and Sandström, Chapter 21 of this volume.
32 See Blouin, Chapter 4 of this volume.
33 See Mikhail, Chapter 6 of this volume.
34 See Haller, Chapter 15 of this volume.
35 See Benjaminsen, Chapter 16 of this volume.
36 See Kolding, van Zwieten and Mosepele, Chapter 18 of this volume.
37 See Reid, Chapter 11 of this volume.
38 See Kampanje-Phiri and Dean Kampanje-Phiri, Chapter 19 of this volume.
39 See Håland, Chapter 2 of this volume.
40 As an example, when the International Food Policy Research institute wrote their *25 Years of Food Policy Research* in 2000, not a word on water was mentioned. Instead the report focused on global food trends, food subsidies, markets under structural adjustments, agricultural linkages to other sectors, biases against agriculture, household food security, environment, agricultural science and technology policy, and trade and globalization (Pinstrup-Andersen 2000). While the role of water to a large extent has been omitted or given at best a secondary significance in the development and constitution of society, today it is incorporated at full strength particularly in policy-based research, but often with an emphasis on the negative aspects or the consequences of climate change or environmental degradation. In the recent decade there has been an explosion of studies on water and food. From 2007–11 the worldwide publication of articles on water resources and food and water grew to almost 5 to 10 per cent annually, and this trend does not seem to stop. In 2011 more than 6,000 articles on water resources and 4,000 articles on water and food research were published (SIWI & Elsevier 2012: 7). A great number of these studies are conducted within the framework of 'Integrated Water Resource Management' (IWRM), but we will argue that a water systems perspective captures the processes at work in a better way without being political normative.

REFERENCES

Bryceson, D.F. (ed.). 2010. *How Africa Works: Occupational Change, Identity and Morality in Africa.* London: Practical Action Publishing.
Clausen, T.J. 2012. 'Introduction'. In Jägerskog, A., Clausen, T.J. (eds). *Feeding a Thirsty World – Challenges and Opportunities for a Water and Food Secure Future.* Report No. 31. Stockholm: SIWI, pp. 6–12.
Hoag, H.J. 2013. *Developing the Rivers of East and West Africa. An Environmental History.* London: Bloomsbury.
Juma, C. 2011. *The New Harvest: Agricultural Innovation in Africa.* Oxford: Oxford University Press.

Kjekshus, H. 1996. *Ecology control & economic development in East African history: the case of Tanganyika 1850–1950*. Second edition, London: James Curry.

Luan, Y., Cui, X., Ferrat, M. 2013. 'Historical trends of food self-sufficiency in Africa'. *Food Security* 5: 393–405.

Mitchell, P., Lane, P. (eds). 2013.*The Oxford Handbook of African Archaeology. Oxford Handbooks in Archaeology*. Oxford: Oxford University Press.

Pinstrup-Andersen, P. 2000. *25 Years of Food Policy Research*. Washington: IFPRI.

SIWI & Elsevier. 2012. *The Water and Food Nexus. Trends and Developments of the Research Landscape*. Stockholm: SIWI & Elsevier.

Trigger, B. 1994. *A History of Archaeological Thought*. Cambridge: Cambridge University Press.

Tvedt, T. 2010a. 'Water systems, environmental history and the deconstruction of nature', *Environment and History* 16(2): 143–66.

——— 2010b. 'Why England and not China and India? Water systems and the history of the industrial revolution', *Journal of Global History* 5: 29–50.

——— 2012. *A Journey into the Future of Water*. London: I.B.Tauris.

——— 2016. *Water and Society: Geopolitics, Scarcity, Security*. London: I.B.Tauris.

Part I   Water and Early Food Regimes

1 Semi-Permanent Foragers in North and West Africa: An Archaeological Perspective

Elena A.A. Garcea

INTRODUCTION

This chapter takes into consideration the archaeological indicators and vestiges of past human adaptational skills, settlement organizations, technologies and economies in relation to the existing natural environments of the last foragers in North and West Africa. These populations were semi-permanent hunter-fisher-gatherers living in aridity-stricken environments between approximately 10,000 and 7,000 years ago. At that time, social, cultural and economic transformations were also partly determined by the changes of the natural environment, and partly led to environmental changes, as is highlighted by the crucial role that water had.

Instead of offering a blunt description of the manifold sites and cultural complexes associated with the last foragers in North and West Africa, it is more useful to present a selection of case studies based on recent archaeological investigations in the Middle Nile Valley in Upper Nubia, Sudan and in the Ténéré desert, in the southern Sahara, Niger, where semi-sedentism became the successful solution for foragers in both the Nile valley and the Sahara desert, although it was performed with different strategies.

THE NORTH AFRICAN ENVIRONMENT IN THE EARLY HOLOCENE

Most of North Africa is presently occupied by the Sahara desert which extends over 9.4 million km². Arid environments have been, and still are, an ordinary condition, which has had profound impacts on human behaviour. Nowadays, in spite of the supposedly adverse conditions of arid zones for humans, they still support over a fifth of the world's population, and arid and semi-arid lands together sustain over a third of the world's population (Barker 2002).

In the early Holocene, between around 11,000 and 8,000 years ago, several lakes existed in the Sahara. In the Fezzan region, Libya, which is now in the central Sahara, Lake Megafezzan was an exceptional hydrogeographic feature, having a maximum expansion of over 76,000 km^2 and being the only lake entirely fed by rivers draining the Sahara (Armitage et al. 2007). The vegetation in the central Saharan mountain ranges (Tadrart Acacus in Libya, Hoggar in Algeria, and Tibesti and Air in Niger) exhibited a Sahelian enclave within a Saharan environment including relict Mediterranean species (see, for example, Schultz 1987, di Lernia 1999, Mercuri & Garcea 2007).

In the eastern Sahara, where the climate was drier than in the rest of the Sahara, increased precipitation supported a higher groundwater table and the formation of the West Nubian Palaeolake, which had an extension of 5,500 km^2 (Pöllath & Peters 2007). During the Holocene optimum from 9,500 years ago, monsoonal summer precipitations reached latitudes located 800 km farther north than in the present day (Neumann 1989, Kuper & Kröpelin 2006, Bubenzer & Riemer 2007, Darius & Nussbaum 2007). A diversified geographic setting of drylands exhibited a mosaic of habitats, with open woodlands, thorn savanna, bush vegetation, and riverine wooded areas close to water courses (Pöllath & Peters 2007).

In the western and southern Sahara, the tropical rainfall belt was farther north at 20°N–24°N, the 400-mm isohyet moved approximately to 21°N, and the frontier between the Sahara and the Sahelian savanna went up to about 18°N (Gasse & Roberts 2004). Consequently, watertable outcropped in several interdunal depressions and formed small permanent lakes. Furthermore, Lake Mega-Chad formed between 8,500 and 8,200 years ago and underwent a second expansion between 7,900 and 7,400 years ago (Maley 2004). During both periods, it reached its maximum expansion of about 340,400 km^2, a depth of 160 m, and a volume of 13,500 km^3 (Leblanc et al. 2006a, 2006b).

THE IMPACT OF ARID ENVIRONMENTS ON HUMAN BEHAVIOUR

Kuper and Kröpelin (2006) provided detailed evidence on climate-controlled spreads and contractions of past human occupation in the Sahara throughout the Holocene. Populations have constantly faced the challenges of oscillations between humid and dry conditions and have developed the skills to live efficiently also during arid periods.

Arid zones feature a number of challenging natural conditions their inhabitants have to deal with and influence their decision making and problem solutions. First of all, rapid climatic change is a common feature to deserts, implying a fragile environmental balance where aridity, desertification, and drought can be alternately present. The term 'desertification' has been proposed to describe the ecological degradation of arid lands into semi-deserts or deserts, brought into areas where such conditions did not previously exist (Aubréville 1949, McGregor & Nieuwolt 1998).

Drought, in turn, is considered as a normal recurrent event in arid and semi-arid lands (Le Houérou 1996).

The second condition of drylands is that they are not homogeneous geographic settings with a broadly uniform environment, but change at various temporal and spatial scales (Spellman 2000). Consequently, food and water resources occur in patchy niches, interspersed within less productive landscapes (Smith et al. 2005). As natural resources depend on unreliable seasonal and altitudinal changes, they provide foragers with a high variability and low predictability of food.

A third challenge for human adaptation, deriving from the other two, is that the boundaries of drylands are neither permanent nor static, as the North African environment in the early Holocene described above has shown. They can move, making drylands encroach or retreat, with strong impacts on human population. Furthermore, their boundaries are not abrupt and transitional zones may extend over large areas. For these reasons, their ecosystems are subject to cycles, which have been defined as 'pulse and reserve' (Noy-Meir 1973) or 'boom and bust' (Smith et al. 2005).

Fourth, as water resources are limited and patchy, Smith et al. (2005) described, in a very suggestive way, foragers living in arid zones as 'water-tethered'. These authors also effectively observed that small water accesses acted like islands in a sea, allowing 'navigation' across drier lands (see below).

RESPONSES OF PAST HUMAN POPULATIONS

After considering the impact that arid environments can have on human behaviour, some of the major responses that past human populations produced should be reviewed. Although foragers developed specific local responses to the various drylands in the world, certain widespread outcomes, including long historical continuity, resilience, and techno-logical innovation are common denominators to foragers' behaviour in drylands.

Yellen (1977) was among the first archaeologists to draw attention to the challenges of biogeographical conditions for foragers adapted to dry environments. Stress management is a primary need of populations living in generally nutrient-poor environments. As environmental conditions frequently fluctuate, the unpredictability of resources increases, while species diversity and biomass diminish. As a response, drylands foragers tend to develop long historical continuity and conservatism, which Yellen explained as the result of reduced competition and few cultural differences among foragers living in arid zones. Moreover, the ability to endure severe environmental conditions and unpredictable resources favoured their resilience skills.

Woodburn (1982, 1988) distinguished between 'immediate-return' and 'delayed-return' hunter-gatherers, the latter practicing intensification of resource exploitation and accumulation of processed and stored food in fixed dwellings in order to defer and schedule the consumption of only seasonally available resources (see also Marshall & Hildebrand 2002, Garcea 2006, 2013b). The reasons for the success of delayed-return foraging are because, in arid and semi-arid environments, resources are patchy, irregular and limited and their availability is often unpredictable. In order to cope with this stress, delayed-return foraging can provide for a more reliable schedule of the availability and predictability of the food supplies. On the other hand, the activities involved in delayed-return systems demand considerable investments of capital, labour, and skills, which not only may be better performed in sedentary than highly mobile settlements, but also require a set of ordered, differentiated, and precise social relationships of the group structure (Woodburn 1988). Furthermore, increased sedentism can have a strong positive impact on the social organization of human groups as it can protect their weaker members, such as elders, ill people and children, and facilitate women during pregnancy, birthing, nursing, and child care (Haaland & Magid 1995).

Furthermore, Torrence (2001) observed that risk management is a key factor in hunter-gathering subsistence economy, and that technological solutions can be more predictable under environmental constraints. In addition, human groups who adapt to arid environments not only live in restricted locations, but are surrounded by drier and ecologically poorer areas. This favours the need to invest in high territoriality, manifested in semi-sedentism, storage, and management of wild animals, plants, and water (Smith et al. 2005).

Foragers' mobility is profoundly affected by the risks of going in search of resources away from the base camp (Casimir 1992). Predictable and widely available resources are able to support high mobility. By contrast, a territory with patchy and restricted resources offers high risks of unsuccessful food procurement. In this case, low mobility is more effective because the costs of moving are not worth the energy investment. In fact, mobility requires an evaluation of the risks involved in transferring the group to a new location, especially when resources are diminishing in the original location (Kelly 1992, 1995).

Low mobility with a 'tethered' pattern of residence has been associated with settings where there are only restricted locations that are able to offer critical resources, such as a body of water (Kelly 1992, 1995, Smith et al. 2005). In fact, accessibility to water can be more crucial than the search for food in arid environments with patchy resources. Once foragers have chosen a (relatively) rich available patch of land and body of water, they are forced into sedentism by subsistence stress. In order to cope with decreasing resources, they develop a subsistence economy based on the consumption of a wider range of foods and specialize in elaborate techniques of harvesting and food processing.

Although it is evident that low mobility and semi-sedentism derive from the binding dependence on the presence of water, the irony of fate of desert foragers suggests that these adaptation patterns are typical of aquatic, rather than terrestrial, landscapes. In fact, desert foragers, like aquatic foragers, show higher degrees of economic complexity and more elaborated technology than common terrestrial foragers (Erlandson 2001). The cultural convergences between the forms of landscape enculturation in northern Europe and in North Africa during the early Holocene Mesolithic support this interpretation, indicating comparable solutions. They precisely include reduced mobility with long-distance exchange of exotic items, production of pottery, bone tools, and polished stone tools, as well as animal management of wild species that were never domesticated (see, for example, Garcea 2003).

Fishing and the exploitation of other aquatic resources, such as molluscs, contribute to broaden the subsistence base and provide high biomass and high-protein food supplies (Garcea 1996, 2006). Although fishing, like hunting, is an extractive, not a productive, subsistence economy, it can be best practised in sedentary dwellings and requires a specific social organization on land and elaborate techniques of resource acquisition. Furthermore, river banks are able to offer a wide variety of resources in limited areas and more stable environmental conditions even under seasonal or cyclic climatic changes. In these areas, some land animals may also be accessible as they may be attracted by drinking water, although plants are a more constantly available resource than animals, particularly if a wider range of them is exploited through more intensive techniques (Mercuri & Garcea 2007). In fact, when humans become aware of the high risks of mobility and opt for a sedentary lifestyle, animals may not make the same choices and therefore humans' potential prey can diminish.

The subsistence economy of delayed-return foragers is usually specialized in elaborate techniques of harvesting and food processing. Generally, the stone tool-kits produced and used at semi-sedentary sites have a reduced variability and consist of expedient or unspecialized tools, which are normally made on locally available raw materials (Binford 1980). Moreover, sedentism enabled foragers to produce and regularly employ fragile and heavy equipment, such as pottery and grinding tools (Garcea 2006).

Pottery could be used for social purposes and for preserving storing wild foods, as well as water. It could also be used for cooking, which implied a significant change in dietary habits. Cooking allowed foragers to eliminate toxins from previously uneatable plants, and to boil, soften, and prepare wet foods, such as soups, stews, porridges, and sauces, which are more digestible, longer-lasting, and more palatable (Casey 2005). Ultimately, cooked foods played a significant role in the economic and social organization of foragers: first of all, they allowed an accumulation of provisions, improving the quantity and the quality of the subsistence base; secondly, boiled foods enabled infants to wean earlier, and consequently, increased the fertility of women and the survival rate of infants (Haaland & Magid 1995).

THE 'AQUALITHIC'

In consideration of the peculiar settlement system, subsistence economy, and artefactual material of the early Holocene archaeology in the Sahara and the Nile Valley, Sutton (1974, 1977) proposed the term of 'Aquatic civilisation' or 'Aqualithic' to define a broad cultural complex rather than a single culture. This definition aimed at representing the foraging groups who lived in a more humid environment than the present Sahara or Sahel thanks to higher rainfall and more permanent lakes and rivers with aquatic fauna. These groups developed new technologies for the exploitation of riverine resources, such as bone harpoons and fish hooks, geometric microlithic artefacts to form composite tools for fishing, groundstone tools, and pottery. Grinding stones were employed for processing wild grains, as well as for cracking nuts, pounding ochre or clay for pottery production, grinding dried meat and fish, and polishing bone tools.

Haaland and Magid (1995) resumed the Aqualithic concept and proposed to define it as 'aqualithic adaptation'. Their perspective intended to emphasize the adaptational skills of the populations settled by water sources in semi-arid environments, who supplemented their subsistence economy with fishing and exploitation of riverine resources.

Nowadays, the term 'Aqualithic' may still seem an attractive and evocative way to describe the adaptational dynamics of the last foragers in North Africa, and has been occasionally resumed (for example, Holl 2005). However, it appears to be too simplistic, broad, and generic to actually refer to the variegated complexity of the last foraging groups in such a vast region like North Africa, as the case studies below demonstrate.

CASE STUDIES FROM ARCHAEOLOGICAL FIELDWORK IN UPPER NUBIA, SUDAN

Sites occupied by early Holocene delayed-return, sedentary or semi-sedentary hunter-fisher-gatherers have been recognized in different parts of Sudan (for example, among others, Nordström 1972, Mohammed-Ali 1982, Caneva 1983, Caneva & Marks 1990, Marks & Mohammed-Ali 1991, Caneva et al. 1993, Haaland & Magid 1995, Garcea 1993, 2004, 2011–2012, Jesse 2003, 2004a, 2004b, Gatto 2006, Honegger 2005, 2006, 2011, 2012, Usai 2004, 2005,

Salvatori & Usai 2008, Garcea & Hildebrand 2009, Salvatori et al. 2011, Salvatori 2012). They are mainly located along the palaeovalleys of the Nile or of its tributaries.

In Upper Nubia, which extends over the middle tract of the Nile Valley, the latest foraging cultures have been called 'Arkinian' (Schild et al. 1968) and 'Khartoum Variant' (Shiner 1968, Nordström 1972, Gatto 2006). The latter was considered to be a derivation of the 'Early Khartoum' of the Upper Nile Valley to the south (Arkell 1949). These cultures were defined following salvage surveys and excavations at several sites in the Wadi Halfa-Second Cataract area conducted prior to the construction of the Aswan High Dam in the early 1960s (Shiner 1968). The Khartoum Variant lithic tool-kits were distinguished from the previous Arkinian ones because they were made of different types of raw materials and comprised several blade and backed tools and rare borers and groovers. In particular, concave sidescrapers and multiple edged sidescrapers, often made with Egyptian flint imported from the oases in the Western desert, were considered as diagnostic tools of the Khartoum Variant. Also pottery was common and often exhibited impressed decorations with dotted zigzags and dotted wavy lines.

Two areas in Upper Nubia have been recently investigated and are here reported as relevant case studies with regard to hunting-fishing-gathering adaptations in the Middle Nile Valley: Sai Island and the Amara West district (Figure 1.1).

Sai Island is located in the Nile River between the Dal and the Third Cataracts. Fieldwork on the island yielded several Khartoum Variant sites (Garcea 2006, 2011–12, Garcea & Hildebrand 2009). Two of them, 8-B-10C and 8-B-76, were thoroughly investigated by extensive excavations. They have been dated between 9515 ± 15 and 6835 ± 45 years ago (cal BP, Garcea et al. 2016).

Site 8-B-10C is located in a formerly vegetated floodplain (Figure 1.2). Excavations over an area of 105 m² brought to light a long-lived Khartoum Variant sedentary settlement consisting of large concentrations of stone and ceramic artefacts scattered on top and around shallow pits, which proved to be hut floors. Level 1 yielded seven hut floors, 100 post holes, three hearths, and three rubbish pits (Figure 1.3). Level 2 revealed another architectural complex with 75 post holes, suggesting an earlier phase of occupation with a similar hut system with a rather permanent use of the site. The closely-spaced hut floors at the site suggested that its occupants concentrated in a restricted area, instead of spreading out over an extended surface.

Another site on Sai Island, 8-B-76, provided a long stratified sequence with a Khartoum Variant deposit below a later, pastoral, 'Abkan' occupation (Figure 1.2). The Khartoum Variant deposit showed a maximum thickness of about 1 m, thinning towards the south-west slope facing an ancient floodplain and the present course of the Nile. In addition, a consistent horizontal stratigraphy could be detected on the surface with discrete

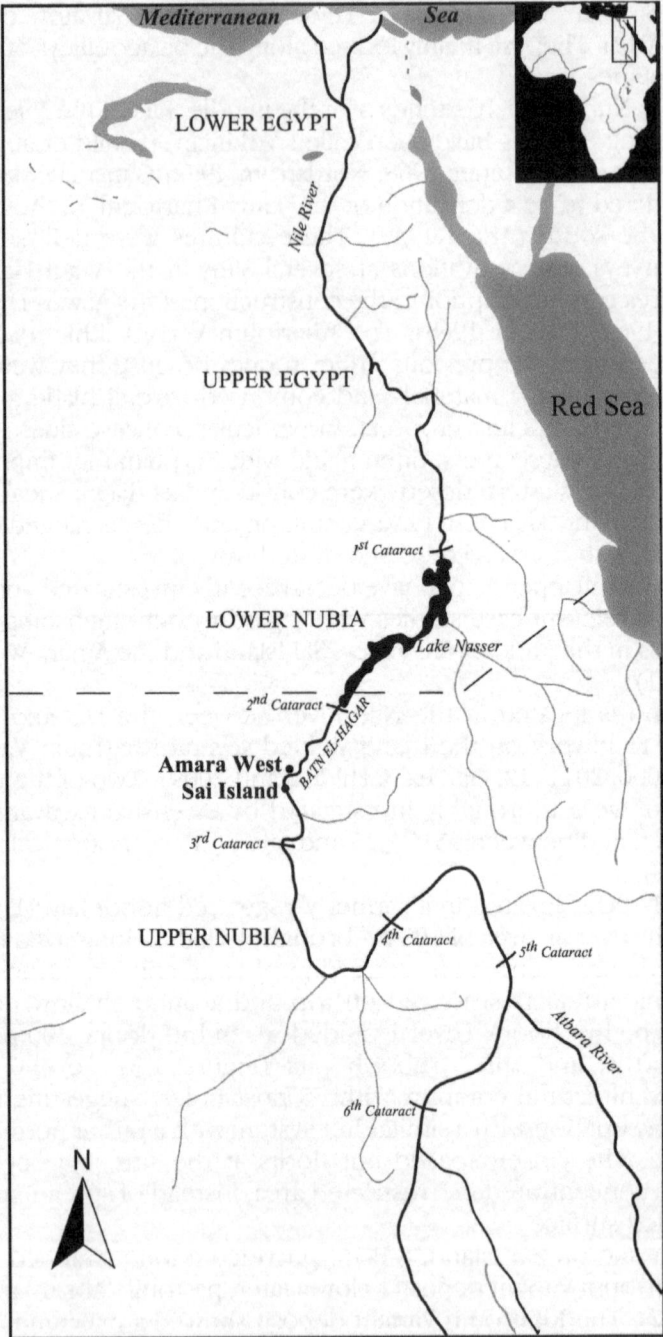

Figure 1.1. Map of Nubia with Sai Island and the Amara West district (© Bruna M. Andreoni).

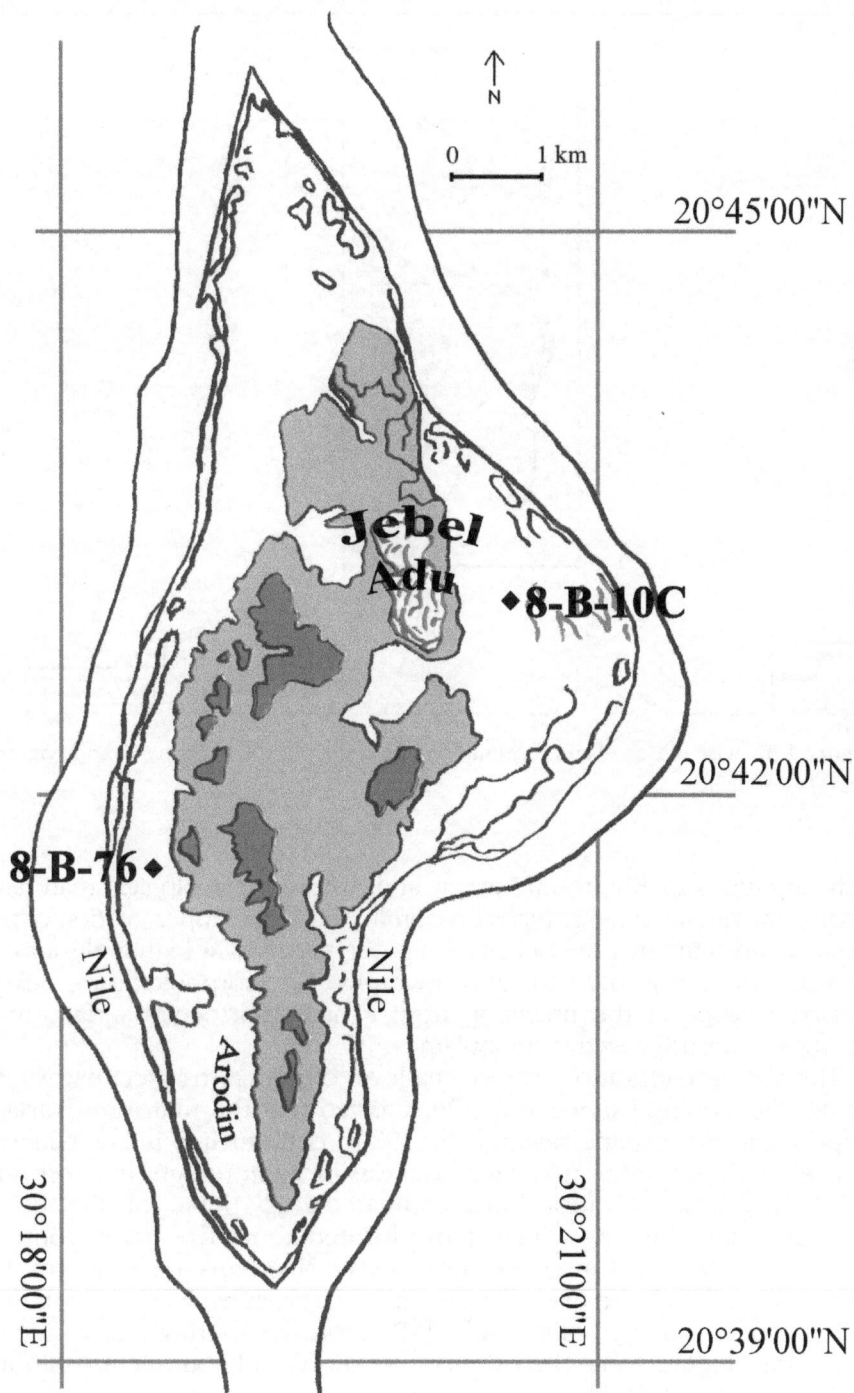

Figure 1.2. Map of sites 8-B-10C and 8-B-76 on Sai Island (© Bruna M. Andreoni).

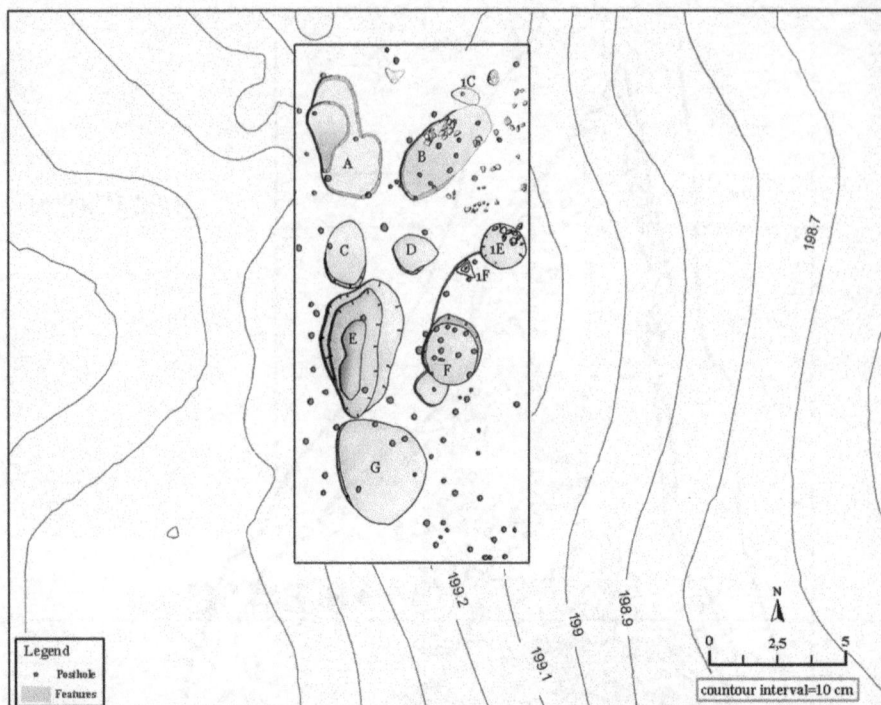

Figure 1.3.   Plan of Level 1 and surrounding area at site 8-B-10C (© Bruna M. Andreoni and Timothy Schilling).

concentrations of Khartoum Variant and Abkan archaeological materials. Khartoum Variant ceramic types predominated in the upper portion of the slope, confirming that the occupation at the time was at higher elevations, as also indicated in the vertical stratigraphy. Geomorphologically, 8-B-76 lies on a slope at the transition from a high Pleistocene surface to a younger, currently inactive floodplain.

The absolute elevations above sea level of a 27 m transect excavated across the sloping surface at 8-B-76, showed that the Khartoum Variant deposit did not expand beyond the 198 m contour line below which a deposit with only later Abkan material was brought to light in a test unit (TU1–2). These elevations are comparable to those of the other excavated site, 8-B-10C, which is also located above the 199 m contour line, that is, about 8 m above the present Nile bank (see Figures 1.3 and 1.4). Although the two sites are rather far from each other – 8-B-10C being located on the eastern side, 8-B-76 on the south-western side of the island (Figure 1.2) – their similar elevations and equivalent positions in inactive floodplains suggest a common topographic patterning of these insular sites.

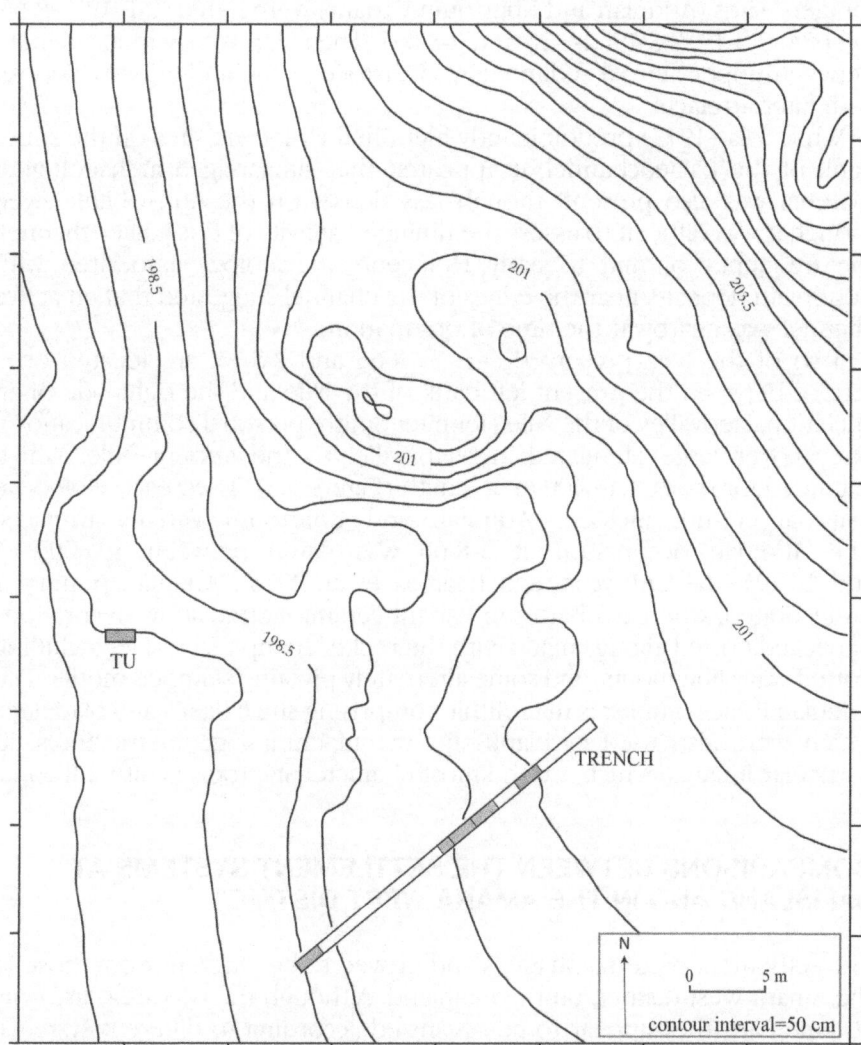

Figure 1.4. Plan of the test unit (TU) and trench excavated at site 8-B-76 (© Bruna M. Andreoni and Timothy Schilling).

The Amara West district is on the mainland, presently on the left bank of the Nile River, 20 km downstream from Sai Island (Figure 1.1). A geoarchaeological survey of the prehistoric sites was recently conducted with the aim of relocating the sites previously recorded by Arkell (1939, 1941) in the 1940s and Vila (1977) in the 1970s, and finding new ones. Different ancient valleys of the Nile could be observed and were surveyed. The early and mid Holocene sites were usually present on terrace treads, several metres above the lowest palaeochannel floor. Among them, several foragers' sites (Arkinian and Khartoum Variant) were found (2-R-12, 2-R-66, 2-R-68A, 2-S-15, 2-S-19, 2-S-20 and 2-S-25). Their deposit lay in a sediment between upper alluvial sediments of Pleistocene age and a lower deposit with later artefacts.

While Vila (1977) predominantly identified Holocene sites on the south bank of the palaeochannel, it appeared that similar-aged archaeological material was also present, though less dense, on the north bank. Even though it was difficult to assess the timing of activity of this palaeochannel, the frequency of mid to early Holocene archaeology associated with floodplain deposits near the edges of the channel suggested that an active channel was nearby at the time of occupation.

Two of the best preserved sites, 2-R-66 and 2-S-15, are located on a terrace between the present left bank of the Nile and the right side of an ancient palaeovalley of the Nile (Jennifer Smith, personal communication). They cover large elongated areas parallel to the ancient Nile, which extended for more than 400 m in length (Figure 1.5). Their early Holocene material culture includes Arkinian and Khartoum Variant artefacts. The Arkinian occupation at 2-R-66 was dated between $10,600 \pm 55$ and $10,495 \pm 65$ CAL years ago (Garcea et al. 2016). Arkinian pottery is undecorated, whereas Khartoum Variant ceramics often show decorations of packed dotted zigzags made with the rocker stamp technique, including dotted wavy line motifs, and some alternately pivoting stamped motifs. The Arkinian lithic complex is microlithic comprising small flakes and bladelets, which were also used as blanks for manufacturing geometric tools, in particular lunates, which are the most characteristic tools at site 2-R-66.

COMPARISONS BETWEEN THE SETTLEMENT SYSTEMS AT SAI ISLAND AND IN THE AMARA WEST DISTRICT

The settlement patterns on Sai Island proved to be different from those in the Amara West district, on the mainland. Although the two areas are only 20 km apart, they appear to be organized according to different systems, both based on long-term occupations, but with different adaptational strategies resulting from dissimilar environmental features, an insular closed zone at Sai, and an open fluvial zone, in the Amara West district. The comparison between the sites at Sai Island and at Amara West shows that the early Holocene delayed-return foragers living there were able to adapt their settlement strategies to the different local ecological conditions.

First, the Khartoum Variant sites at Sai Island, with a system of hut floors, post holes and other architectural features at 8-B-10C, and the thick settlement at 8-B-76 suggest that they were either permanently occupied or systematically reoccupied in the same spot and, for this reason, high investments were made in the intrasite organization of the settlements. On the other hand, the sites at Amara West do not consist of either a

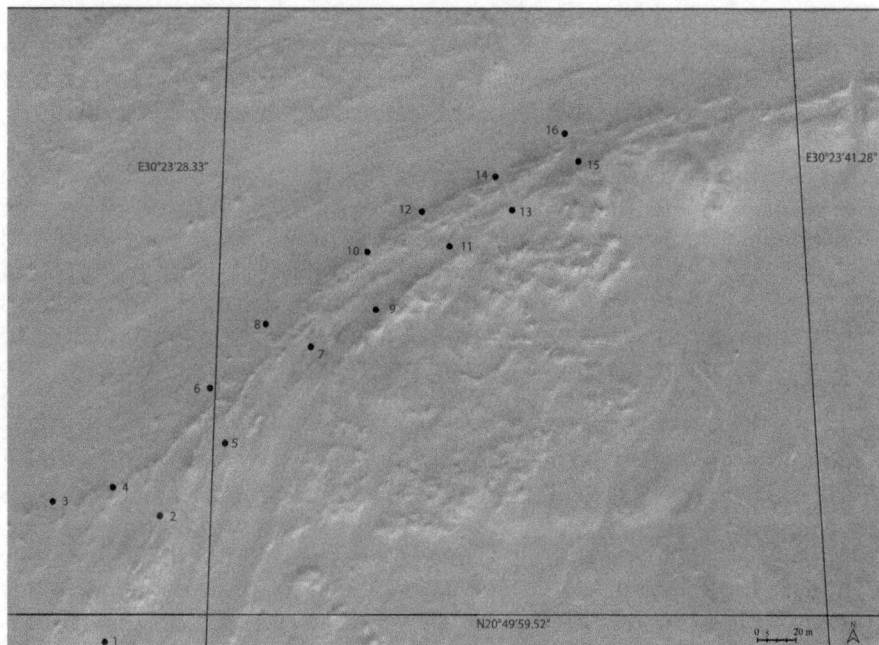

Figure 1.5.   Map of site 2-R-66 (© Bruna. M. Andreoni).

complex internal organization or a thick stratigraphy. As there were no substantial differences along the Nile coast of that region, which was able to offer suitable and undifferentiated places all along the coast, the settlements could be seasonally re-established in the vicinity of the previous ones, resulting in scattered sites extending over several hundreds of metres.

Secondly, at Sai, the sites dating to the different periods (Khartoum Variant and later, Abkan and Pre-Kerma) occupy distinct areas, as indicated by either geographically separated settlements, or discrete surfaces and levels, as in the case of 8-B-76, which was used by both Khartoum Variant and Abkan groups. On the contrary, at Amara West, the same localities were occupied by culturally different groups, as the frequency of mixed artefacts on the surfaces suggests.

Thirdly, the topography of the sites at Sai shows that the two main Khartoum Variant sites lie at the same elevations and in floodplain habitats, whereas the environment at Amara West was not morphologically as stable as on the island. While the river channels of the island were deeply cut and did not vary much through time, different active channels have been observed in the Amara West district, requiring continuous adjustments by its occupants, as well as causing post-depositional reworking of the sediments.

Although both the hunter-gatherers of Sai and those of Amara West can be classified as collectors, and both have a logistical mobility (*sensu* Binford 1980), they differ because only at Sai, but not at Amara West, they are organized according to a 'tethered' pattern of residence (see Kelly 1992, 1995).

Furthermore, the different patterns with concentrated sites on the islands and extended scattered sites on the mainland along the Nile banks, which are archaeologically documented at Sai Island and Amara West, respectively, seem to be comparable with the present topography of modern villages. The tendency to form packed dwellings seems to be a characteristic of islands, where environments suitable to human settlement may be more limited than in the mainland, as also the present concentrations of houses in the northern tip of Sai and on the near-by Ernetta Island show. On the other hand, along the Nile bank on the mainland, also modern settlements seem to extend without any major interruption, at least where the environmental conditions are uniform.

Altogether, the early Holocene foragers in the middle Nile Valley show that, with the same subsistence economy (delayed-return foraging), same material culture, and same habitat (the Nile coast), settlement patterns and occupational strategies can vary even within a small geographic area. If the early Holocene foragers preferred certain places than others on the islands, where resources could be more abundant than elsewhere, on the mainland along the Nile Valley, resources were more homogeneous and their availability was rather undifferentiated and the choice of where to settle was not so relevant.

ARCHAEOLOGICAL EVIDENCE OF FOOD SYSTEMS IN SUDAN

Regarding food systems, animal remains were almost only found on the surface at site 2-R-66 in the Amara West district (Garcea et al. 2016). and therefore only give a partial representation of the available faunal assemblage. Due to the poor preservation condition, several mammal bones were too fragmented to be identified. The identifiable remains include a probable hippopotamus (*Hippopotamus amphibius*), a large Nile perch (*Lates niloticus*) measuring at least 80 cm in length, several freshwater bivalves (*Spathopsis rubens* and *Etheria elliptica*), gastropods (*Pila wernei*), and ostrich eggshells. Ostrich eggshell and freshwater gastropods (*Pila wernei* and *Lanistes carinatus*) were also common at site 8-B-10C on Sai Island. Additionally, a terrestrial gastropod (*Limicolaria* sp.), which presently lives in forests and grasslands south of the 15°N, was found (Garcea et al. 2016).

The diet of Nilotic hunter-fisher-gatherers comprised a major component of riverine resources, with different species of fish and molluscs (Gautier 1983, Peters 1991, Peters & von den Driesch 1993). Fishing of large-size Nile perch was likely to have occurred with the use of boats or rafts during the low water seasons of the Nile as access to the deeper parts of the river was easier, even though fish was an important food component all year round (Gautier 1983, Peters 1991, Caneva et al. 1993). Gathering of molluscs was another relevant activity and source of protein which did not require elaborate equipment as shellfish could be collected by hand. Small and large wild mammals, such as porcupine, African wild ass, gazelles, and antelopes, as well as aquatic reptiles, such as Nile monitor, soft-shell turtle and crocodile, were probably also hunted, as recorded at other early Holocene sites (Gautier 1983, Peters 1991, 1992).

Remains of food plants in the form of desiccated seeds and seed impressions in the pottery have been recovered from some hunting-fishing-gathering sites in Sudan. They suggest that wild berries and fruits were collected for food and eaten raw, including *Celtis integrifolia* and *Ziziphus spina-christi*. Other seed/grain foods, which required processing and/or preparation before consumption, comprised wild *Sorghum*, *Setaria*, and *Panicum* (Magid 1995).

CASE STUDY FROM ARCHAEOLOGICAL FIELDWORK IN THE TÉNÉRÉ DESERT, NIGER

The southern Sahara is another natural laboratory for the study of human/climate interaction, possibly even more than the drier areas of the eastern and central Sahara as the range of climate change has, even nowadays, a wider amplitude extending from extensive floods to severe droughts. Because the border between the Sahara and the Sahelian savanna continued to move latitudinally, people had to cope with periodic, abrupt environmental instability.

The adaptational system of the last hunter-gatherers from the Sahara is generally assigned to the 'Pre-Pastoral' archaeological complex which commonly precedes the successive adoption of animal herding (see, for example, Marshall & Hildebrand 2002, Garcea 2006, 2013a). This complex represents the latest occupations of semi-sedentary fishing populations who settled along riverine and lacustrine environments. The artefactual materials usually comprise impressed pottery with dotted zigzags, microlithic tools, grinding stones, and bone tools, often including harpoons and fish hooks, which are comparable over a wide area extending from the Sahara to the Nile Valley (see, for example, Garcea 2003).

The Gobero archaeological area is located in Niger and represents an intriguing case study from the southern Sahara (Figure 1.6). It comprises eight sites (Figure 1.7), five of which contain both funerary and habitation vestiges, representing both the Pre-Pastoral and the Pastoral periods and

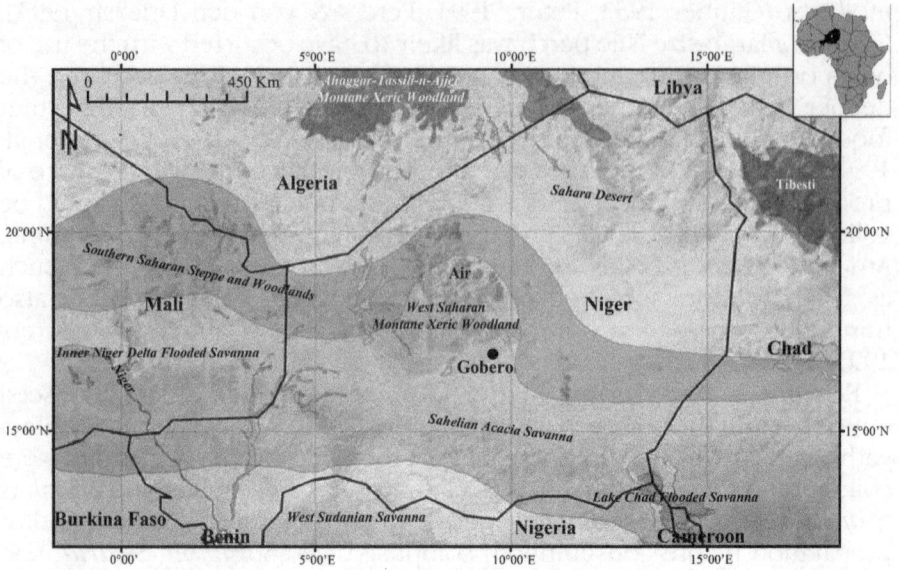

Figure 1.6. Map of Niger with location of Gobero (adapted by E. Cocca).

dating between 9,700 and 4,500 years ago (Sereno et al. 2008, Garcea 2013a). The sites surround the shores of a small palaeolake which underwent alternated phases of increase and decrease of the water level and had a strong impact on the local human population (Garcea et al. 2013, Giraudi 2013). The palaeolake is located in a small endorheic basin and was mainly fed by monsoonal rainfall with consequent groundwater outcrop and surface runoff of ephemeral streams flowing from the north and north-east of the Gobero basin. Thanks to this variety of water supplies, the lake continued to exist for some time after the onset of arid phases. A nearby fault scarp named Mazelet, located to the south, regulated the filling and discharge of the lake (Figure 1.7). When the lake level increased beyond the top of the scarp, water outpoured south of the Gobero basin.

The archaeological area was not always habitable: when the lake reached its maximum extent, of about 30 km^2 and a depth of 9 m, the main sites, G1 and G3, which lay about 4.5 m above the bottom of the lake, were flooded and had to be abandoned by their settlers. Considering these constraints, the sites could only be occupied at the beginning and at the end of the most humid periods.

The Pre-Pastoral period, which is the focus of this paper, ended around 8,100 years ago and was only attested to at site G3 (Figure 1.8), which also revealed Pastoral archaeological vestiges. With regard to the Pre-Pastoral period, the radiocarbon dates indicate that G3 was first employed as a funerary ground, from 9,700 to 9,000 years ago, and

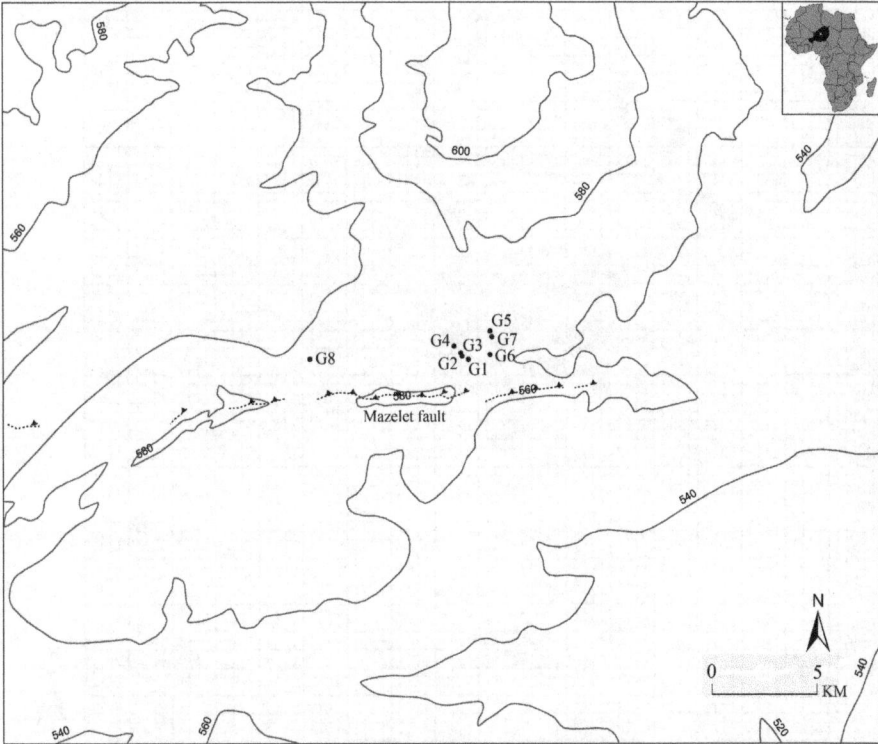

Figure 1.7. Map of the archaeological sites at Gobero (© Davide Mengoli).

subsequently, from 9,000 to 8,200 years ago, as a settlement. A single burial, dated to 8,330–8,160 years ago, marked the end of the Pre-Pastoral period. Then, a severely dry phase occurred around 8,200 years ago not only at Gobero, but in different parts of the Western Sahara and the Sahel (Gasse 2002, Giraudi & Mercuri 2013).

Within the Pre-Pastoral period, two phases have been distinguished and were separated by the highest increase of the lake level that submerged the site, between 9,030–8,790 and 8,720–8,550 years ago, a period that lasted between 70 and 480 years (Giraudi 2013). When the occupants of Gobero had to leave because of the lake inundations, they could have moved either north towards the central or southern Saharan mountain ranges or to the south on the higher reliefs of the Sahelian zone, where more favourable environmental conditions existed.

Although the environment in the early Holocene was wetter than in the middle Holocene, the local landscape included a savanna or grassland vegetation with some tropical, hygrophilous trees and riverine trees near the shores of the lake (Giraudi & Mercuri 2013, Mercuri et al. 2013).

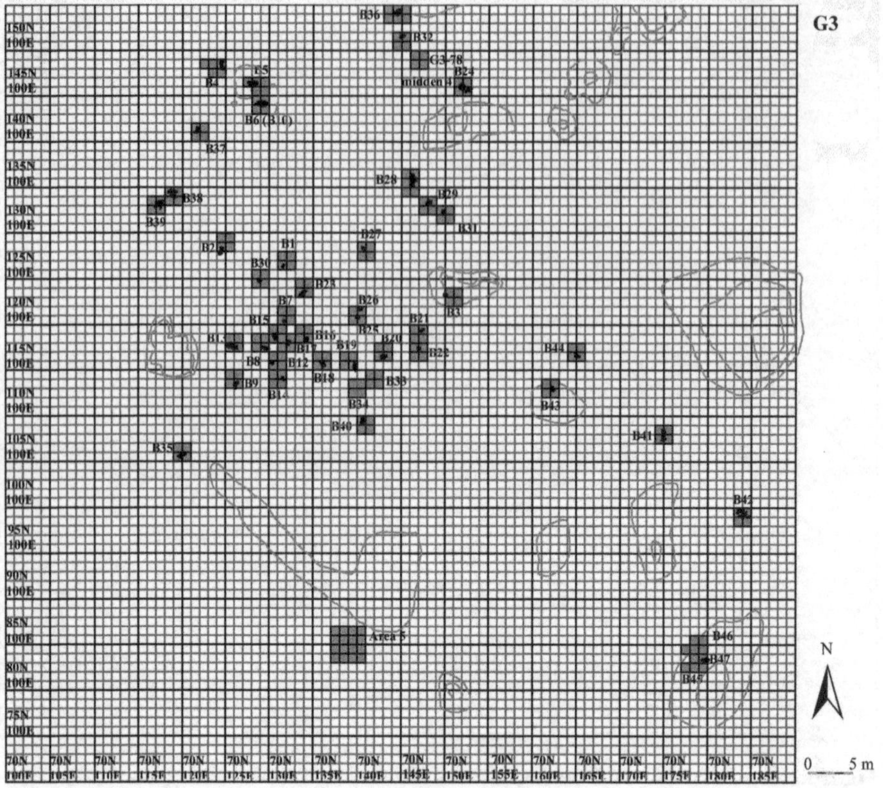

Figure 1.8. Map of site G3 at Gobero (© Davide Mengoli).

Biogeochemical analysis of radiogenic strontium isotopic ratios taken from archaeological human remains, modern and archaeological faunal remains, and soil samples provided information on the mobility patterns and dietary traditions of the Gobero population (Stojanowski & Knudson 2011). In particular, they showed that Pre-Pastoral young and adult individuals had different diets. While the young appeared to have very limited mobility, the adults changed their mobility patterns as they showed to have lived in different environments before becoming sedentary at Gobero.

SETTLEMENT AND FOOD SYSTEMS AT GOBERO

The settlement system of the Gobero foragers is another example, in this case more extreme than Sai Island, of a 'tethered logistical mobility' (Kelly 1992, 1995). No hydro-geological similar features existed in the surroundings of the Gobero palaeolake and, therefore, the lake was able to offer water and other critical resources that could not be easily found elsewhere.

The biogeochemical data confirmed that when the adults found a favourable location in terms of food and water, they permanently settled there and raised their children with no need to move them around, at least as long as the local environment was neither too wet or too dry. The radiogenic strontium isotope data also showed that mobility of the Gobero population was lower than among contemporary foragers living farther north, in the central Sahara (see Tafuri et al. 2006, Stojanowski & Knudson 2011).

Although a low degree of mobility was the most successful settlement strategy, large-scale movements were possible, but remained occasional and often occurred under critical – too dry or too wet – conditions. In fact, in the Sahara, water accesses, including small ones like the Gobero palaeolake, had a major role in sustaining social and economic relations. They appear to have functioned like islands in a sea, allowing 'navigation' across dry lands. Remote sensing and digital elevation models corroborated a system of 'water-tethering' by early Holocene foragers and showed that their settlement patterns appear to be strongly correlated with the palaeohydrological system of the Sahara and the Nile Valley (Bubenzer & Riemer 2007). The wide spread of comparable decorative techniques and impressed motifs on early Holocene ceramic containers indicates that pottery-bearing foragers participated in the trans-Saharan cultural diffusion that – with a few local differentiations – extended from the Atlantic Sahara to the Nile Valley.

In addition to that, as Gobero was at the border between the Sahara and the Sahel, which could shift latitudinally, its inhabitants could benefit from the advantage of living in ecological and social conditions that are typical of edge zones. These areas are able to offer higher biodiversity and social interactions with different groups, providing opportunities to incorporate a wider suite of adaptive responses (Garcea 2013b).

The organic remains from the Pre-Pastoral occupation at Gobero provide a few hints on the food system of the local hunter-fisher-gatherers. Also in this case, fish appear to predominate in the diet, with Nile perch (*Lates niloticus*) and large catfish (Sereno et al. 2008). Other faunal samples, including hippopotamus, reedbuck, small carnivore, soft-shell turtle, and crocodile, were present, but may or may not have been hunted as their skeletal remains were found on the surface near an ancient shoreline of the palaeolake and not in the archaeological deposit (Garcea 2013a).

A few wild cereal impressions were found in the Pastoral ceramics from Gobero, including *Panicum* spp. and *Eleusine* spp. (Fuller 2013). Although these ceramics are chronologically later, gathering of these wild grains could have been started by the earlier foragers, as was observed in the Nile Valley.

CONCLUSION

To sum up, the archaeological case studies from Upper Nubia and the Ténéré desert presented above show several local responses to the impact of dry environments on past human behaviour. All three sites indicate the exploitation of a broad spectrum of food resources, including wild aquatic (fish, molluscs, reptiles) and terrestrial (mammals, birds, other types of molluscs) animal species, and fruits and berries that could be eaten raw, as well as grains that needed preparation before consumption.

The production of new equipment for food processing and storage, including ceramic containers and grinding stones, occurs over a very wide area in North Africa. Furthermore, the spread of comparable products made with similar manufacturing techniques confirms the long historical continuity and conservatism with few cultural differences theorized by Yellen (1977).

All three investigated archaeological areas, namely Sai Island, the Amara West district, and Gobero, demonstrate that a successful economic organization of the late foragers inhabiting those areas involved a delayed-return system (see Woodburn 1992). Given the instability of the environment, sedentism appeared to be a convenient practice. On the other hand, population pressure could be a serious challenge, particularly when food supplies were diminishing and resources became more unreliable.

Arid environments, in particular, require greater investments in territoriality, which implies semi-sedentary settlements, broad-spectrum foraging, high-level management of wild plants and animals, and a high degree of organizational and technological flexibility (Smith et al. 2005). Settlements in restricted locations by the Nile or the Gobero palaeolake with high territoriality were evidenced by the continually occupied sites for long periods of time (Smith et al. 2005). However, some differences could be observed between the insular sites, which indicated a common topographic patterning at both 8-B-10C and 8-B-76 at Sai, and the mainland sites at Amara West, in Upper Nubia. While a 'tethered' pattern of residence (Kelly 1992, 1995, Smith et al. 2005) could be observed at Sai Island, it did not seem to be a requirement along the vast Nile coastline in the Amara West district. On the other hand, the 'tethered' settlement pattern was even more evident at Gobero than on Sai Island, where several Khartoum Variant sites existed. The relatively rich available patch of land and body of water was a fortunate choice for the Gobero foragers, but at the same time forced them into sedentism by subsistence stress.

Various archaeological indicators provided evidence for sedentism: high concentrations of artefacts in restricted areas, thick anthropic deposits, and/or architectural features. Very large quantities of artefacts were found at all investigated sites, whereas a thick stratigraphic deposit was uncovered at 8-B-76, on Sai Island, and complex multiple-phase architectural features, with hut floors, post holes, hearths and rubbish pits was unearthed at 8-B-10C, also on Sai. At Amara West, numerous structured hearths were excavated at 2-R-66, while no formal features could be observed at Gobero, even though the Pre-Pastoral occupation at G3 included both human burials and (slightly later) artefactual materials from the settlement. These differences suggest that distinct investments could be made in the foragers' settlements within their common sedentary or semi-sedentary pattern.

Finally, the use of a heavy, non-portable, and fragile equipment, including grinding stones and pottery, at all sites confirms the frequency of sedentism in North and West Africa during the early Holocene. This settlement organization is further corroborated by the almost exclusive use of local raw materials. A few exceptions exist: flint imported from the Egyptian oases in the Western desert was rarely found at Sai Island, and greenstone, a raw material imported from the Air (see Garcea 2013a), was occasionally employed at Gobero. These imports suggest that some long-distance movements across Saharan water bodies could occur and seemingly contributed to the lowering of cultural differences in such a vast region.

The case studies and considerations on North and West African early Holocene hunter-gatherers presented above may contribute to place Africa in a global context, which should not only take into consideration the global geography of world's interests, but should also take into account the chronological developments of past events which are tightly interlinked to the present.

ACKNOWLEDGEMENTS

I wish to thank the former directors of the Sai Island Archaeological Mission of the University of Lille III, France, F. Geus and D. Devauchelle, for inviting me to be part of the Sai research team, and Neal Spencer of the British Museum, Director of the Amara West Research Project, for encouraging me to study the prehistory in the Amara West district. I co-directed the Gobero Archaeological Project with P. Sereno in 2005 and 2006. I wish to sincerely thank Terje Oestigaard and Terje Tvedt for inviting me to contribute to this volume and to participate at the workshop on 'Water and Food – Africa in a Global Context' held at the Nordic Africa Institute in Uppsala, Sweden. I am grateful to them for their helpful and fruitful comments on a previous draft of this paper.

REFERENCES

Arkell, A.J. 1939. 'Late Sebilian'. *Sudan Antiquities Service Files*, 22–11–39.
———— 1941. 'Levallois type'. *Sudan Antiquities Service Files*, 13–12–41.
———— 1949. *Early Khartoum*. London: Oxford University Press.

Armitage, S.J., Drake, N.A., Stokes, S., El-Hawat, A., Salem, M.J., White, K., Turner, P., McLaren, S.J. 2007. 'Multiple phases of North African humidity recorded in lacustrine sediments from the Fazzan Basin, Libyan Sahara'. *Quaternary Geochronology* 2: 181–6.

Aubréville, A. 1949. *Climats, forêts et désertifications de l'Afrique tropicale*. Paris: Société d'èditions Géographiques et Coloniales.

Barker, G. 2002. 'A tale of two deserts: contrasting desertification histories on Rome's desert frontiers'. *World Archaeology* 33(3): 488–507.

Binford, L.R. 1980. 'Willow smoke and dogs' tails: hunter-gatherer settlement systems and archaeological site formation'. *American Antiquity* 45(1): 4–20.

Bubenzer, O., Riemer, H. 2007. 'Holocene climatic change and human settlement between the central Sahara and the Nile valley: archaeological and geomorphological results'. *Geoarchaeology* 22(6): 607–20.

Caneva, I. 1983. '"Wavy Line" decoration from Saggai I: an essay of classification'. *Origini* 12: 155–90.

Caneva, I., Garcea, E.A.A., Gautier, A., Van Neer, W. 1993. 'Pre-pastoral cultures along the central Sudanese Nile'. *Quaternaria Nova* 3: 177–252.

Caneva, I., Marks, A.E. 1990. 'More on the Shaqadud pottery: Evidence for Saharo-Nilotic connections during the 6th–4th millennium BC'. *Archéologie du Nil Moyen* 4: 11–35.

Casey, J. 2005. 'Holocene occupations of the forest and savanna'. In A.B. Stahl (ed.), *African Archaeology: A Critical Introduction*. Oxford: Blackwell, pp. 225–48.

Casimir, M.J. 1992. 'The Dimensions of Territoriality: An Introduction'. In Casimir, M.J., Rao, A. (eds), *Mobility and Territoriality: Social and Spatial Boundaries among Foragers, Fishers, Pastoralists and Peripatetics*. Oxford: Berg, pp. 1–26.

Darius, F., Nussbaum, S. 2007. 'In search of the bloom – plants as witnesses to the humid past'. In O. Bubenzer, A. Bolten and F. Darius (eds), *Atlas of Cultural and Environmental Change in Arid Africa*. Köln: Heirich-Barth-Institut, pp. 78–81.

di Lernia, S. (ed.), 1999. *The Uan Afuda Cave: Hunter-Gatherer Societies of Central Sahara*. Firenze: All'Insegna del Giglio.

Erlandson, J.M. 2001. 'The archaeology of aquatic adaptations: paradigms for a new millennium'. *Journal of Archaeological Research* 9(4): 287–350.

Fuller, D.Q. 2013. 'Observations on the use of botanical materials in ceramic tempering at Gobero'. In Garcea, E.A.A. (ed.), *Gobero: The No-Return Frontier. Archaeology and Landscape at the Saharo-Sahelian Borderland*. Frankfurt: Africa Magna Verlag, pp. 241–8.

Garcea, E.A.A. 1993. *Cultural Dynamics in the Saharo-Sudanese Prehistory*. Rome: Gruppo Editoriale Internazionale.

——— 1996. 'La culture des pêcheurs d'*Early Khartoum*: un exemple dans la Vallée du Nil'. *Préhistoire Anthropologie Méditerranéennes* 5: 207–14.

——— 2003. 'Cultural convergences of northern Europe and North Africa during the Early Holocene?' In Larsson, L., Kindgren, H., Knutsson, K., Loeffler, D., Åkerlund, A. (eds), *Mesolithic on the Move. Papers presented at the Sixth International Conference on the Mesolithic in Europe, Stockholm 2000*. Oxford: Oxbow Books, pp. 108–14.

———— 2004. 'An alternative way towards food production: the perspective from the Libyan Sahara'. *Journal of World Prehistory*, 18(2): 107–54.

———— 2006. 'Semi-permanent foragers in semi-arid environments of North Africa'. *World Archaeology* 38(2): 197–219.

———— 2011–12. 'Revisiting the Khartoum Variant in its environment'. *Cahiers de Recherches de l'Institut de Papyrologie et d'Egyptologie de Lille* 29: 139–50.

———— (ed.) 2013a. *Gobero: The No-Return Frontier. Archaeology and Landscape at the Saharo-Sahelian Borderland*. Frankfurt: Africa Magna Verlag.

———— 2013b. 'Gobero: the secular and sacred place'. In Garcea, E.A.A. (ed.), *Gobero: The No-Return Frontier. Archaeology and Landscape at the Saharo-Sahelian Borderland*. Frankfurt: Africa Magna Verlag, pp. 271–93.

Garcea, E.A.A., Hildebrand, E.A. 2009. 'Shifting social networks along the Nile: Middle Holocene ceramic assemblages from Sai Island, Sudan'. *Journal of Anthropological Archaeology* 28, 304–322.

Garcea, E.A.A., Mercuri, A.M., Giraudi, C. 2013. 'Archaeological and environmental changes between 9500 BP and 4500 BP: a contribution from the Sahara for the understanding of expanding droughts in the "Great Mediterranean"'. *Annali di Botanica* 3: 115–20.

Garcea, E.A.A., Wang, H., Chaix, L. 2016. High-precision radiocarbon dating application to multi-proxy organic materials from late foraging to early pastoral sites in Upper Nubia, Sudan. *Journal of African Archaeology* 14(1).

Gasse, F. 2002. 'Diatom-inferred salinity and carbonate oxygen isotopes in Holocene waterbodies of the western Sahara and Sahel (Africa)'. *Quaternary Science Reviews* 21: 737–63.

Gasse, F., Roberts, C.N. 2004. 'Late quaternary hydrologic changes in the arid and semiarid belt of northern Africa: implications for past atmospheric circulation'. In Diaz, H.F., Bradley, R.S. (eds), *The Hadley Circulation: Present, Past and Future*. Dordrecht: Kluwer Academic, pp. 313–45.

Gatto, M.C. 2006. 'The Khartoum Variant pottery in context: rethinking the Early and Middle Holocene Nubian sequence'. *Archéologie du Nil Moyen* 10: 57–72.

Gautier, A. 1983. 'Animal life along the prehistoric Nile: the evidence from Saggai I and Geili (Sudan)'. *Origini* 12: 50–1155.

Giraudi, C. 2013. 'Late Upper Pleistocene and Holocene Stratigraphy of the Palaeolakes in the Gobero Basin'. In Garcea, E.A.A. (ed.), *Gobero: The No-Return Frontier. Archaeology and Landscape at the Saharo-Sahelian Borderland*. Frankfurt: Africa Magna Verlag, pp. 67–80.

Giraudi, C., Mercuri, A.M. 2013. 'Early to Middle Holocene environmental variations in the Gobero Basin'. In Garcea, E.A.A. (ed.), *Gobero: The No-Return Frontier. Archaeology and Landscape at the Saharo-Sahelian Borderland*. Frankfurt: Africa Magna Verlag, pp. 114–26.

Haaland, R. and Magid, A.A. 1995. *Aqualithic Sites along the Rivers Nile and Atbara, Sudan*. Bergen: Alma Mater.

Holl, A.F.C. 2005. 'Holocene "Aquatic" adaptations in North Tropical Africa'. In Stahl, A.B. (ed.), *African Archaeology: A Critical Introduction*. Oxford: Blackwell, pp. 174–86.

Honegger, M. 2005. 'El-Barga: un site clé pour la compréhension du Mésolithique et du début du Néolithique en Nubie'. *Revue de Paléobiologie* 10: 95–104.

—————— 2006. 'Habitats préhistoriques en Nubie entre le 8e et le 3e millénaire av. J.-C.: l'exemple de la région de Kerma'. In Caneva, I., Roccati, A. (eds), *Acta Nubica: Proceedings of the X International Conference of Nubian Studies*. Rome: Istituto Poligrafico e Zecca dello Stato, pp. 3–13.

—————— 2011. 'La Nubie et le Soudan: un bilan des vingt dernières années de recherche sur la pré et protohistoire'. *Archéo-Nil* 20: 77–86.

—————— 2012. 'Excavation at Wadi El-Arab and in the eastern cemetery of Kerma'. In Honegger, M. (ed.), *Documents de la mission archéologique suisse au Soudan 4*. Neuchâtel: Universty of Neuchâtel, pp. 3–9.

Jesse, F. 2003. *Rahib 80/87: ein Wavy-Line-Fundplatz im Wadi Howar und die frheste Keramik in Nordafrika*. Köln: Heinrich-Barth Institut.

—————— 2004a. 'The development of pottery design styles in the Wadi Howar region (northern Sudan)'. *Préhistoire Anthropologie Méditerranéennes* 13: 97–107.

—————— 2004b. 'No link between the central Sahara and the Nile Valley? (Dotted) Wavy Line ceramics in the Wadi Howar, Sudan'. In Kendall, T. (ed.), *Nubian Studies: Proceedings of the Ninth Conference of the International Society of Nubian Studies, 1998*. Boston: Museum of Fine Arts and Northeastern University, pp. 296–308.

Kelly, L.R. 1992. 'Mobility/sedentism: concepts, archaeological measures, and effects'. *Annual Review of Anthropology* 21: 43–66.

—————— 1995. *The Foraging Spectrum: Diversity in Hunter-Gatherer Lifestyle*. Washington, DC: Smithsonian Institution Press.

Kuper, R., Kröpelin S. 2006. 'Climate-controlled holocene occupation in the Sahara: motor of Africa's evolution'. *Science* 313: 803–7.

Leblanc, M.J., Favreau, G., Maley, J., Nazoumou, Y., Leduc, C., Stagnitti, F., van Oevelen, P.J., Delclaux, F. Lemoalle, J. 2006a. 'Reconstruction of Megalake Chad using Shuttle Radar Topographic Mission data'. *Palaeogeography, Palaeoclimatology, Palaeoecology* 239: 16–27.

Leblanc, M.J., Leduc, C., Stagnitti, F., van Oevelen, P.J., Jones, C., Mofor, L.A., Razack, M., Favreau, G. 2006b. 'Evidence of Megalake Chad, north-central Africa, during the late Quaternary from satellite data'. *Palaeogeography, Palaeoclimatology, Palaeoecology* 230: 230–42.

Le Houérou, H.N. 1996. 'Climate change, drought and desertification'. *Journal of Arid Environments* 34: 133–85.

Magid, A.A. 1995. 'Plant remains from the sites of Aneibis, Abu Darbein and El Damer and their implications'. In Haaland, R., Magid, A.A. (eds), *Aqualithic Sites along the Rivers Nile and Atbara, Sudan*. Bergen: Alma Mater, pp. 147–77.

Maley, J. 2004. 'Le bassin du Tchad au Quaternaire récent: formations sédimentaires, paléoenvironnements et préhistoire. La question des paléochads'. In Sémah, A.-M., Renault-Miskovsky, J. (eds), *L'évolution de la végétation depuis deux millions d'années*. Paris: Artcom'/Errance, pp. 179–217.

Marks, A.E., Mohammed-Ali, A. (eds), 1991. *The Late Prehistory of the Eastern Sahel*. Dallas: Southern Methodist University.

Marshall, F., Hildebrand, E.A. 2002. 'Cattle before crops: the beginnings of food production in Africa'. *Journal of World Prehistory*, 16(2): 99–143.

McGregor, G.R., Nieuwolt S. 1998. *Tropical Climatology*. Second edition. Chichester: Wiley.

Mercuri, A.M., Garcea, E.A.A. 2007. 'The impact of hunter/gatherers on the vegetation in the Central Sahara during the Early Holocene'. In Cappers, R. (ed.), *Fields of change: Progress in African archaeobotany*. Groningen: Barkhuis & Groningen University Library, pp. 87–104.

Mercuri, A.M., Massamba N'siala, I., Florenzano, A. 2013. 'Environmental and Ethnobotanical Data Inferred from Pollen of Gobero and the Dried Lakebeds in the Surrounding Area'. In Garcea, E.A.A. (ed.), *Gobero: The No-Return Frontier. Archaeology and Landscape at the Saharo-Sahelian Borderland*. Frankfurt: Africa Magna Verlag, pp. 81–104.

Mohammed-Ali, A.S.A. 1982. *The Neolithic Period in the Sudan, 6000–2500 B.C.* British Archaeological Reports S139, Oxford.

Neumann, K. 1989. 'Holocene vegetation of the Eastern Sahara: charcoal from prehistoric sites'. *African Archaeological Review* 7: 97–116.

Nordström, H.-A. 1972. *Neolithic and A-Group Sites*. Uppsala: Scandinavian University Press.

Noy-Meir, I. 1973. 'Desert ecosystems: environment and producers'. *Annual Review of Ecology and Systematics* 4: 25–51.

Peters, J. 1991. 'Mesolithic fishing along the Central Sudanese Nile and Lower Atbara'. *Sahara* 4: 33–9.

—— 1992. 'Late Quaternary mammalian remains from Central and Eastern Sudan and their palaeoecological significance'. *Palaeoecology of Africa* 23: 91–115.

Peters, J., von den Driesch, A. 1993. 'Mesolithic fishing at the confluence of the Nile and the Atbara, Central Sudan'. In Clason, A., Payne, S. and Uerpmann, H.-P (eds) *Skeletons in her Cupboard: Festschrift for Juliet Clutton-Brock*. Oxford: Oxbow, pp. 75–83.

Pöllath, N., Peters, J. 2007. 'Holocene climatic change, human adaptation and landscape degradation in arid Africa as evidenced by the faunal record'. In Bubenzer, O., Bolten, A. and Darius, F. (eds), *Atlas of Cultural and Environmental Change in Arid Africa*. Köln: Heirich-Barth-Institut, pp. 64–7.

Salvatori, S. 2012. 'Disclosing archaeological complexity at the site and regional level'. *African Archaeological Review* 29(4): 399–472.

Salvatori, S., Usai, D. (eds), 2008. *A Neolithic Cemetery in the Northern Dongola Reach: Excavations at Site R12*. London: Sudan Archaeological Research Society.

Salvatori, S., Usai, D., Zerboni, A. 2011. 'Mesolithic site formation and palaeoenvironment along the White Nile (Central Sudan)'. *African Archaeological Review* 28(3): 177–211.

Schild, R., Chmielewska, M., Więckowska, H. 1968. 'The Arkinian and Shamarkian industries'. In Wendorf, F. (ed.), *The Prehistory of Nubia*. Dallas: Fort Burgwin Research Center and Southern Methodist University Press, pp. 651–767.

Schulz, E. 1987. 'Holocene vegetation in the Tadrart Acacus: the pollen record of two early ceramic sites'. In Barich, B.E. (ed.), *Archaeology and Environment in the Libyan Sahara. The Excavations in the Tadrart Acacus, 1978–1983*. Oxford: British Archaeological Reports, pp. 313–26.

Sereno, P.C., Garcea, E.A.A., Jousse, H., Stojanowski, C.M., Saliège, J.-F., Maga, A., Ide, O.A., Knudson, K.J., Mercuri, A.M., Stafford, T.W., Kaye, T.G., Giraudi, C., Massamba N'siala, I., Cocca, E., Moots, H.M., Dutheil, D.B., Stivers, J.P. 2008.

'Lakeside cemeteries in the Sahara: 5000 years of Holocene populations and environmental change'. *PLoS ONE* 3(8): 1–22.

Shiner, J.L. 1968. 'The Khartoum Variant industry'. In Wendorf, F. (ed.), *The Prehistory of Nubia*. Dallas: Fort Burgwin Research Center and Southern Methodist University Press, pp. 768–90.

Smith, M., Veth, P., Hiscock, P., Wallis, L.A. 2005. 'Global deserts in perspective'. In Veth, P., Smith, M., Hillcock. P. (eds), *Desert Peoples: Archaeological Perspectives*. Oxford: Blackwell, pp. 1–13.

Spellman, G. 2000. 'The dynamic climatology of drylands'. In Barker, G., Gilbertson, D. (eds), *The Archaeology of Drylands: Living at the Margin*. London: Routledge, pp. 19–41.

Stojanowski, C.M., Knudson, K.J. 2011. 'Biogeochemical inferences of mobility of early Holocene fisher-foragers from the southern Sahara desert'. *American Journal of Physical Anthropology* 146(1): 49–61.

Sutton, J.E.G. 1974. 'The Aquatic civilization of Middle Africa'. *Journal of African History* 15: 527–46.

——— 1977. 'The African Aqualithic'. *Antiquity*, 51: 25–34.

Tafuri, M.A., Bentley, R.A., Manzi, G., di Lernia, S. 2006. 'Mobility and kinship in the prehistory Sahara: strontium isotope analysis of Holocene human skeletons from the Acacus Mts (southwestern Libya)'. *Journal of Anthropological Archaeology* 25: 390–402.

Torrence, R. 2001. 'Hunter-gatherer technology: macro- and microscale approaches'. In Panter-Brick, C., Layton, R.H., Rowley-Conwy, P. (eds), *Hunter-Gatherers: An Interdisciplinary Perspective*. Cambridge: Cambridge University Press, pp. 73–98.

Usai, D., 2004. 'Early Khartoum and related groups'. In Kendall, T. (ed.), *Nubian Studies: Proceedings of the Ninth Conference of the International Society of Nubian Studies, 1998*. Boston: Museum of Fine Arts and Northeastern University, pp. 419–35.

——— 2005. 'Early Holocene seasonal movements between the desert and the Nile Valley: details from the lithic industry of some Khartoum Variant and some Nabta/Kiseiba sites'. *Journal of African Archaeology* 3(1): 103–15.

Vila, A. 1977. *La prospection archéologique de la vallée du Nil, au sud de la Cataracte de Dal (Nubie soudanaise): Le district d'Amara Ouest*. Volume 7. Paris: Centre National de la Recherche Scientifique.

Woodburn, J.C. 1982. 'Egalitarian societies'. *Man* 17(3): 431–51.

——— 1988. 'African hunter-gatherer social organization: is it best understood as a product of encapsulation?' In Ingold, T., Riches, D., Woodburn, J. (eds), *Hunters and Gatherers. Volume 1: History, Evolution and Social Change*. Oxford: Berg, pp. 31–64.

Yellen, J.E. 1977. 'Long-term hunter-gatherers adaptations to desert environments: a biogeographical perspective'. *World Archaeology* 8(3): 262–74.

2   Origin of Domestication and Aquatic Adaptation: The Nile Valley in Comparative Perspective

Randi Haaland

INTRODUCTION

The conventional wisdom has been that agriculture and pottery developed more or less simultaneously. Artefacts from Africa, however, indicate that pottery was used 2,000 years before the start of agriculture. Ceramics seem closely associated with the heavy reliance on aquatic and plant resources for food boiling. With the advent of cultivation, porridge took over as the main cooked food. In the Near East archaeological evidence shows that domesticated cereals appeared about 2,000 years before ceramics. During the first period of cultivation of cereal no pottery material is found. There is thus a striking contrast between Africa and the Near East with regard to two fundamental cultural features generally assumed as interlinked elements of the Neolithic, namely cultivated cereals and the use of pottery. This clearly identifies the difference between a cooking-pot syndrome in Africa and a baking-oven syndrome in the Near East.

To explore this in further details, I shall ask myself what kind of food was made from cereal. In Africa cereals ground into flour seem to a large extent to have been made into porridge, which was boiled in pots. As regards the Near East, the archaeological material indicates that flour was used for making bread or bread cakes: evidence such as wear pattern on human teeth show that these were severely abraded probably from eating gritty bread (Moore 1995, Akkermans & Schwartz 2003: 74). By making cereal into bread there was not the same need for pots to boil the food.

SYMBOLISM OF FOOD

This chapter will focus on the consumption of food items with reference to their nutritional importance, and also to their symbolic significance.

Food items are not only 'food for the body', but they are also 'food for thought' that determine our relations to 'others' in the world of living people, as well as our relations to cosmological forces. I will then explore how such symbolical uses of food and food-related items (pots, ovens, hearths, etc.) are embedded in material forms. Prepared food has qualities that seem to stimulate symbolic constructions. It is of basic importance for our livelihood. More importantly, the process of transforming natural products into cultural food items can be used as a root metaphor for understanding other transformations. This leads us to the cosmological and ideological importance of food. The importance of food for our bodily survival is obvious and self-evident.

Thus, food items are well suited to serve as metaphors for understanding other important but less well-understood existential aspects of life, solidarity in human relations or dependence on divine forces. The metaphoric potential of food makes it a convenient medium for communicating the importance of values structuring social life. Just as our physical body depends on the continuous intake of physical food, the social body depends on continuous affirmation of the values regulating life among community members. Food is thus not only a biological necessity; it may also be loaded with strong emotions. Good food is a source of pleasure but the act of eating food may be hedged in by taboos that express social distinctions. The incorporation of food into one's body is thus a process that may be understood as fundamental in self-formation. This means that food items may be socially constructed to affect and create social identities such as status, gender, age and ethnicity. Food can therefore be looked upon as a process of lived experience (Curtin 1992: 12). Food ways thus include ideas related to who can eat together and share it (Appadurai 1981).

In the pottery container food was transformed from a natural to a cultural product. The pot is a container like the body is seen as a container – it contains the food that maintains the body. The metaphorical association between pots and bodies is close at hand and it is also manifested in the way we, as archaeologists, classify pots by using bodily traits as diagnostic – we use words from the human body – mouth, neck, shoulder and body. Several ethnographic studies of pottery making show the same bodily terms used in local vocabularies (Barley 1994, Haaland 1997). Through food, social relations between people can be expressed. Not only is eating food a bodily experience that is loaded with meaning that may be further elaborated into standardized cultural symbolism, but the pots in which the food is cooked and served has similar symbolizing potential.

I will in this chapter concentrate on the food ways of Africa, but I will also to some extent contrast it with the food ways of the Near East.

ORIGIN OF POTTERY PRODUCTION IN AFRICA

Close (1995) has made a survey of the earliest sites with remains of ceramics in North Africa and her general conclusion is that pottery was invented somewhere along the southern edge of the Sahara, and that it spread very quickly East–West across a 3,000-km belt of the continent. The distribution of the earliest ceramic-bearing sites in North Africa does not, according to Close, spread to the Mediterranean part of North Africa or east of the river Nile (Close 1995). Close emphasizes that early pottery seem to occur in association with grinders as well as in association with exploitation of aquatic resources. Jesse has more recently (2010) looked at the distribution of early pottery of the dotted wavy-line Khartoum type and she finds the distribution to be similar to the findings of Close from 1995 (Figure 2.1). These early dates of pottery from different areas suggest that pottery was independently invented in Africa, probably as early as 10,000 BP. Most of the finds are from the Sahara, a region that did not open up for this type of resource exploitation until the Early-Holocene humid phase at the beginning of tenth-millennium BP. The savannah vegetational zone appears to have moved further by 500 km to the north (Kuper & Kroplin 2006). The area, which today is a desert, would at that time have consisted of dry savannah type of vegetation in most areas, generally what we today think of as typical for the Sahelian zone of sub-Saharan Africa, further south.

Sutton argued in 1974 that the first ceramic vessels in Africa were part of a parcel of soup, porridge and fish stew revolution. He hypothesized a possible relationship between the emergence of aquatic resource utilization and the invention of pottery technology. I will in this chapter discuss the role of aquatic resources and early plant use in the innovation of pottery.

CONSEQUENCES OF POTTERY MAKING

The adaptive importance of pottery lies in allowing for utilization of a broader and wider range of food resources. Boiling or steaming renders food more digestible and palatable, and pottery expands the range of potential food resources available in the same habitat. Handwerker has argued that pottery and the use of boiled food led to a change in the diet of infants, allowing an early weaning, which would influence and increase the fertility of women and affect the survival rate of infants since the period after weaning is critical (Handwerker 1983: 19). Ethnographic material and archaeological evidence make it reasonable to assume that it was within the female sector of activities that the important innovation of applying fire in order to transform clay jars for storage to clay pots for cooking occurred. This hypothesis has been forcefully argued by Wright, who on the basis of

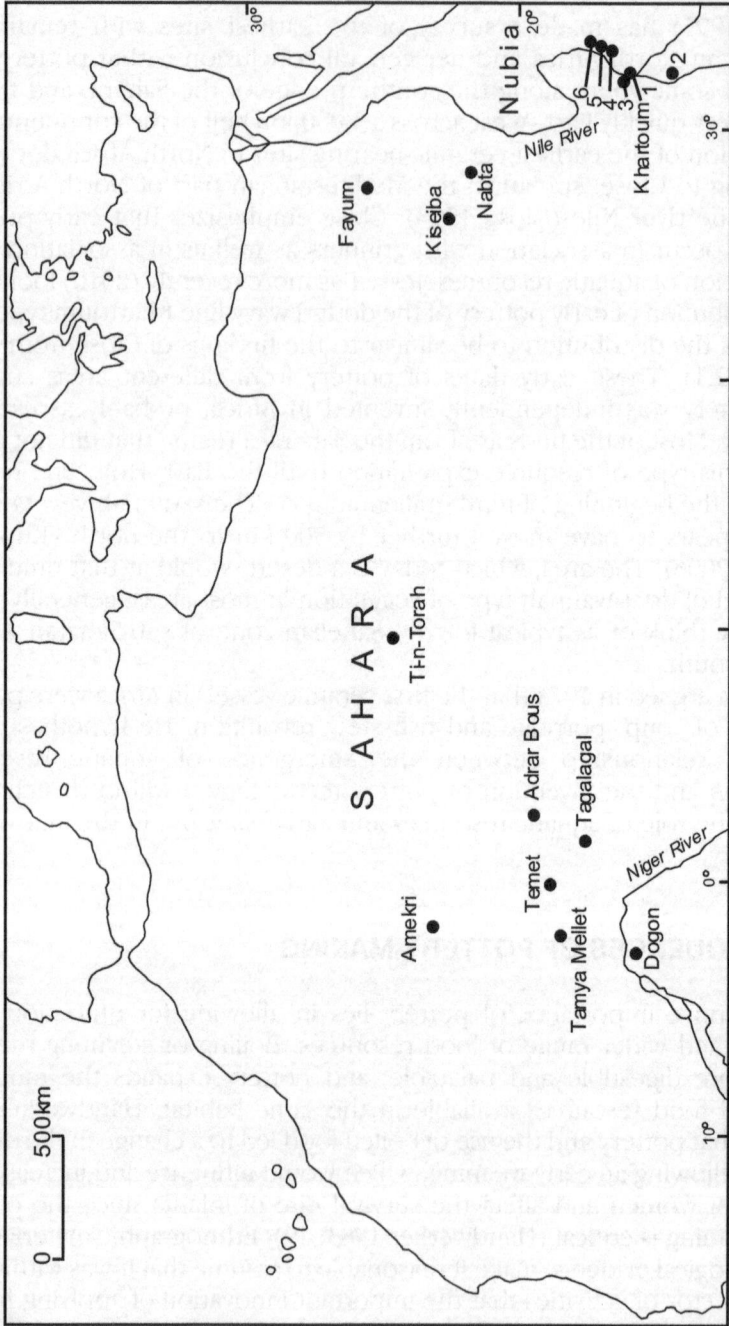

Figure 2.1. Map of Sites in Africa with pottery dated to the tenth and ninth millennia BP (modified from Close, 1995). Sites 1-Kadero, 2-Shabona 3-Sarurab, 4-Abu Darbein, 5-El Damer, 6-Aneibis.

Figure 2.2. A Fur woman is making pottery. Notice her use of grinders and similar tools for grinding grain when preparing food (© Randi Haaland).

the work done by Amiran (1965), pairs the making/cooking of bread/ porridge with pottery production (Wright 1991). Pottery making and food preparation by cooking involve activities, which, in many respects, are similar; grinding, the use of water, kneading and firing (Figure 2.2). Wright argues further 'from its earliest appearance it very likely involved women's labour and its development occurred hand in hand with other economic activities such as the domestication process' (Wright 1991: 214). Women play an important role as a nurturer; through breastfeeding the woman produces milk with her own body and this is a child's first bodily experience of food. After the child has been weaned it is fed solid food from the pot. Pots in most cases were moulded by women and served as recipient to prepare the food by these same women. In Norwegian folk traditions the pot is referred to as the *Gryta hennar mor* ('The mother's pot').

With the advantages of pottery as discussed above, it is even more of a puzzle that pottery was not adopted in the Near East at the beginning of cereal cultivation. However, looking at the early pottery material from these two regions, the evidence does suggest a real gap in time and place. The distribution of the early pottery material in sub-Saharan Africa does not seem to have spread east of the Nile towards the Levant. The earliest pottery in the Near East is found in a belt stretching from Turkey to Syria to

Iraq, Iran and the southern Levant (Moore 1995, Garfinkel 1999). Does this mean that ceramics was invented independently in Africa and the Near East, or did it diffuse from Africa eastward? I shall not engage in this discussion of independent innovation versus diffusion but rather look at pottery in relation to the two different food ways characteristic of Africa and the Levant – one based on cooking and the other on baking.

POTTERY AND AGRICULTURE IN THE SUDANESE NILE VALLEY

To illustrate how the typical African food system based on porridge emerged I will draw on the material that our team excavated in the Central Sudan in the 1980s. Several sites in the eastern and central Sahara as well as in the Sudanese Nile Valley have yielded macro-remains of sorghum, dated to the period between 8000 and 4000 BP (Abdel-Magid 1989, Magid 1995, Wasilikowa et al. 1997). There has been an extensive discussion on sorghum cultivation. However, up to now all macro-remains found from this period is morphologically wild. The oldest remains of domesticated sorghum, dated to 4000 BP, come from sites in the Gash area (Beldados & Constantini 2011).

I will here present material from sites that show the long history of utilization of wild grain: Abu Darbein, El Damer and Aneibis (Haaland & Magid 1995). The pottery found is similar to the Early Khartoum (dotted-wavy line tradition discussed earlier). These sites are dated to the ninth- and eighth-millennium BP. The large number of pot sherds that we recovered from these sites appears in a food-gathering context, where a broad range of resources were used. The most dominant were aquatic-fish resources, with hunting of large and small animals as well as exploitation of plant food such as wild sorghum. The ceramics consist of large pots, probably used for storing and of smaller pots (Figure 2.3), which could have been used for cooking or possibly serving. Beer may already have been in use at this early period. I will discuss this in more detail later. This ceramic tradition was maintained for several thousand years until the sixth-millennium BP, with some changes in surface treatment and decoration. It was probably during this time that people in the Khartoum area started to cultivate the local wild sorghum (Haaland 1987, 1995, 1999, 2013). This was the beginning of a long history and remarkable diversity in different forms of sorghum-based foodstuffs, porridge and beer (Edwards 2003); this will be discussed further later in the chapter.

I have argued earlier that the archaeological material during the Khartoum Neolithic period indicates that cultivation of sorghum was practised, as plant imprints on pottery show morphologically wild sorghum (Stemler 1990). However, the arguments that the inhabitants were cultivating sorghum are also based on the high number of grinders

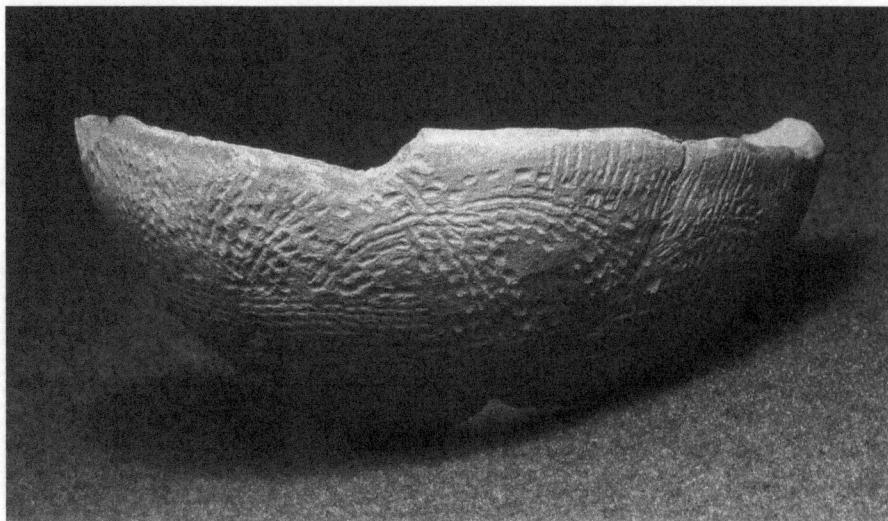

Figure 2.3. Small pot probably used for cooking-serving, with dotted wavy line decoration from the site of Aneibis (© Anne Marie Olsen).

(on one site 30,000 fragments were recovered from 140 m^2) and large sites (from 10 to 30,000 m^2) suggesting a sedentary lifestyle (Haaland 1987, 1999, Haaland & Haaland 2013). This was the beginning of the cultivation of the summer-growing African cereal.

Neumann (2003) does not agree with the perspectives presented above. She argues for the importance of pastoralism, which lead to a high degree of mobility. This was necessary for exploitation of the savannah resources, which were unequally distributed in space and time. The same argument was presented by Marshall and Hildebrand (2002). However, my argument is that we have large sedentary sites based on a broad spectrum of resources, especially aquatic ones, dated between 9–6000 BP along riverine areas such as the Nile and Atbara (Figure 2.4). It is within this river context that we find material indicating cultivation dated to the sixth-millennium BP. The heavy reliance on abundant fish resources made it possible to maintain a sedentary way of life. The population build-up in bigger and more sedentary communities along the Nile would have increased pressure on plant resources, which became scarcer. To be able to maintain the plant diet the inhabitants could adopt new activities in the exploitation of plants. Activities included cultivation, which in the long run led to domestic plants (for further arguments see Haaland 1987, Haaland 1999). I thus see the Nile Valley with remains of large sedentary sites as quite different from the savannah region with its dispersed resources. The inhabitants were exploiting a broad set of resources, the rich aquatic ones, hunting, gathering, as well as cultivating activities.

Figure 2.4. The site of Abu Darbein, located along the Atbara River. This is a typical location of the multi-resource aquatic focused sites (© Randi Haaland).

CATTLE AND ELITE FORMATION

We also find that the inhabitants over time were adopting new resources spreading from the north: domestic animals (sheep, goat and cattle) – these domestic animals ultimately originating from the Levant. Kadero is a major site and here it can be observed that already during the sixth-millennium BP a large number of animal bones were recovered of which 88 per cent were from domestic animals such as cattle (Krzyzaniak 1991, 2004). This supports the rapid importance of cattle over time in the local economy.

Cattle are wealth that can be accumulated and serve as a source of social differentiation. Furthermore cattle are not only economic wealth serving consumption needs, they are a form of wealth that can be converted into political support through clientships or redistribution. In addition they may be used as symbols of power – the cow and the bull have natural attributes that lend themselves to metaphoric and metonymic association from motherly nurturance and succour to manly political dominance and power. The introduction of cattle is thus a possible source of incipient elite formation, particularly in communities where it can be connected to differential control over goods flowing in exchange networks. It is important to keep in mind that the emergence of elites takes place in political contexts where acceptance of differences is related to the elites' ability (real or imagined) to defend group members' interest in competition with other groups. An important step in institutionalization of elite differentiation occurs with the concentration of military means in the hands of elites. Differential access to valued goods and services within groups is generally cemented by the development of ideologies that legitimize the elite positions, and the development of chiefdoms. We can see this when comparing graves from rich and poor people. Some graves contain a few funerary objects such as simple utility pottery or no funerary objects at all, while in another, smaller cemetery graves are richly furnished. Krzyzaniak (1991, 2004) interprets the rich graves as burials of a social elite.

During the Khartoum Neolihic phase, the graves along the Middle Nile Valley consisted of small pit-graves with bodies buried in a contracted foetal position and the graves were similarily built. However, there were striking differences in the types of funerary goods deposited in the grave as seen in the site of Kadero. This is reflected in funerary goods such as bucrania and human scarifies. Some graves contain large numbers of items used as body decoration such as beads, amulets, bracelets, carnelian, zeolite, marine shells from the Red Sea, and lip-nose plugs made from ivory. Palettes with remains of red ochre and malachite were probably used to decorate the body, indicating a strong focus on bodily decoration throughout the Nile Valley from northern Egypt to Khartoum. There was an emphasis on a mortuary practice related to the decoration of the body and personal objects in contrast to the Near East where the emphasis was on burial ceremonies and rituals connected to the house (Wengrow 2001, 2006).

We also find in the graves an increased occurrence from the fifth-millennium BP of different types of drinking vessels such as small cups, beakers (Caneva 1994) and elaborately decorated caliciform beakers (Caneva 1994, Reinold 2001, Krzyzaniak 2004) (Figure 2.5). Dirar (1993) has argued that fermented foods such as sorghum beer have an extensive history in Sudan, with 30 to 50 different varieties recognized. They may well have been drunk from vessels of this kind, and it seems likely that it was during the fifth-millennium BP or even earlier that the typical African 'pot and porridge/beer' cuisine emerged (Haaland 2007). Residue analyses of pottery from the Blue Nile provide some support for their use in such beer brewing (Fernandez & Tresseras 2000). Beer – and other kinds of alcoholic beverage – are now widely recognized to have played important roles in cultural history in many parts of the world (See Edwards 1996, Dietler 2001) and the occurrence of both simple and elaborated drinking vessels in the Sudanese Neolithic indicates socially mediated differences in how it was consumed there. Indeed, it is difficult to see that the elaborately shaped ceramic vessels were used for anything

Figure 2.5. A caliciform beaker from the site of Kadero. Courtesy Poznan Archaeological Museum (© Maciej Jordecka).

else than serving drinks like beer on festive occasions. Occasions for such conspicuous consumption may express differences internal to specific political units, but also relations of competition, as well as alliances between the leaders of different units in inter-group elite feasting where beer and the material items involved in its consumption served as signs validating rank.

FISH TABOO

I have argued (1992) that domestic animals were only exploited as primary products; however, recent research has shown that pastoralism which specialized to produce secondary products such as milk and blood appear to have become important in the late sixth-millennium BP. Inside pottery fat remains from milk shows that dairying was adopted by prehistoric Saharan African people in the fifth-millennium BP (Dunne et al. 2012). These findings confirm the importance of milk and provide an evolutionary context for the emergence of lactase persistence in Africa. Over time it appears that specialized pastoralism became more important. Domestic animals were valuable capital, which could be individually accumulated. As well as being items of wealth, they served as a source of social differentiation and cattle also served as a symbol of power. With the use of secondary products, animal husbandry could be exploited more efficiently, and this played a major factor in the emergence of specialized pastoralists. In the eastern Sudan ceramics from sites in the Kassala area with remains from Mokram groups show close similarities with the pottery used by Pan-Grave population in Upper Egypt and Nubia. This suggests connections among a broad range of pastoralist communities in and west of the Red Sea hills (Fattovich 1991, Sadr 1991: 47, Hafsaas 2006: 5–6).

Interestingly, these sites have been attributed to the Medjay in Egyptian texts (Sadr 1991: 47). The Medjay were cattle herders and are assumed to have been the ancestors of the present-day Beja, Cushtic speakers who inhabit the Red Sea hills (Arkell 1961, Haaland 1992). The Beja observe a strict fish-taboo. We might thus have evidence that indicates the presence of people practising fish-taboo from the fifth-millennium BP and that this became an important marker of identity with the emergence of specialized pastoralism. Material from two sites along the Nile Valley dated to the fifth-millennium shows remains of cattle camps, and these appear not to have exploited aquatic resources (Caneva 1991), which is in support of the argument that fish taboo was practised (Haaland 1992). These nomadic camps indicate that pastoralists were embedded with other groups along the Nile. Certain Cushitic-speaking groups in eastern Sudan, such as stated above, the Beja, still maintain a strong fish-taboo tradition today.

ETHNOGRAPHIC CASES OF FOOD SYMBOLISM IN AFRICA

In Africa the staple food is primarily porridge and beer made from sorghum and millet. The importance of food as a medium for initiating and maintaining social relations is well recognized in both anthropological and archaeological literature (Appadurai 1981, Douglas 1984, Harris & Ross 1987, Gosden & Hather 1999, Dietler & Hayden 2001). However, it is beer that characteristically binds people together and serves to reinforce social hospitality during ceremonial and communality during ritual and everyday life. An example of practical and symbolical uses of cereal food is provided by Karp's study of the Iteso of Kenya where beer is both considered as food (an alcoholic nourishing gruel) and a ritual substance (Karp 1980). Beer drinking is a daily activity for most people and is a pervasive feature of social life. The Iteso use the beer partly to organize work groups and the willingness to participate in reciprocal beer drinking is a fundamental part of the definition of the social person. It is part of life-cycle rituals such as complex mortuary ceremonies and rituals connected with birth. Consumption of beer generally takes place among a group of people sitting round a big pot and sipping beer through a straw. Similar ways of consumption are widespread in East Africa like among the Fipa in Tanzania (Figure 2.6). Joint consumption of beer symbolically expresses and fosters communal solidarity. There is, furthermore, a close association

Figure 2.6. The Fipa of Tanzania party-drinking beer through straws (© Randi Haaland).

of beer with mother's milk. This is manifested in rituals performed after the birth of a child. By these examples I just want to draw attention to a familiar theme in African ethnography, namely that the two food items (porridge and beer), and the items and activities involved in their preparation serve as important sources for symbolic elaboration in a wide range of African communities.

AGRICULTURE AND FOOD PREPARATION IN THE NEAR EAST

Material evidence from the Near East suggests that production and consumption of cereal (wheat and barley) was quite different from the African system. During the first period of cultivation of these winter-growing cereals, no pottery remains are found. This is within the so-called Pre-Pottery Neolithic period between the eleventh and the eighth-millennium BP (Kujit & Goring-Morris 2002)

The tools for grinding cereal and preparing food are very numerous on most settlement sites, and consisted of a wide variety of ground stone tools; the cereal that people cultivated was obviously made into flour. The problem is what kind of food people were preparing from this flour when they did not use pottery. From the site Tell Sabi Abayad II in Syria, excavations revealed several circular bread ovens in the courtyard surrounded by square rooms. The ovens are classified as tannours (Akkermans & Schwartz 2002: 64). These ancient ovens resemble the modern tannour, which is widespread in the Near East and parts of North Africa (Figure 2.7). The tradition of baking bread in the oven does not appear to be a technique used before the Pre-Pottery Neolithic B period during the ninth-millennium BP, while the tradition of making bread/cakes in the hearth seem to have preceded the use of the oven by more than a thousand years. With the emphasis that we see from the earliest period on bread, there was not the same need to use pottery as containers for cooking as it was for porridge making in Africa.

HEARTH AND HOUSE IN THE MIDDLE EAST

During the time when ovens seem to become quite numerous one can also see important changes taking place in domestic arrangements, such as the shape of complex buildings. Very striking changes are seen in the decoration of the houses and rooms. There is the elaborate use of plaster floors, described as lime plaster, which are often highly burnished. The temperature required to make lime plaster is 7–800°C (Gourdin & Kingery 1975). So obviously they had the technology to make ceramics, but they did not use it! Special care seems to have been taken when making the hearths. They were often surrounded by a rim made of clay and stones, or

Figure 2.7. Reconstruction of a tannour oven from the site of Catal Huyuk, Turkey, dated to the eighth-millennium BP (© Randi Haaland).

special decorations as seen at such sites as Ain Ghazal and Jericho (Kujit & Goring-Morris 2002). The emphasis in the Near East was not like Africa on pots, but it was on the hearth, the oven and the house. This is apparent in the richly decorated houses (Hodder 1990) and, as discussed by David Wengrow (2001), the importance played by the house in Near Eastern cultures that are quite different from cultures along the African Nile Valley.

CEREAL AND BEER

I have up to now focused on bread but I will now address beer as a food item, which is and was important in the diet not only in the Near East but also, as we discussed earlier, for Africa. It has been suggested that the first cultivated cereals were used for making beer (Katz & Voigt 1986). There are several advantages other than nutrition that come with the production and consumption of fermented food.

It is crucial here to understand how the process of fermentation could have started since this usually requires containers for heating in the pre-pottery period before pottery was not in use. However, Katz and Voigt (1986) do not see this as a problem since the daily temperatures in the Near East can reach 40°C or more and there would thus have been little need to heat the brew. They suggest that the use of beer could go back to the early phase of manipulation of grain, i.e., in 11–12,000 BP. The importance of beer is very well documented from later periods, especially dynastic time.

TEXTUAL AND ETHNOGRAPHIC CASES OF FOOD SYMBOLISM IN THE NEAR EAST

There is also plenty of iconographic evidence and religious texts that show the importance of bread, these food items being used as 'food for thoughts' expressing important ideas about human and cosmological relations. This comes across very strongly in the Bible where the sharing of bread is used as a metaphor for the unity of those who partake in eating it, as we see in the Last Supper: 'And as they were eating, Jesus took bread, blessed and broke *it*, and gave *it* to the disciples and said, "Take, eat; this is My body"' (St Matthew 26:26).

As discussed by Bottero (2000), in the Sumerian language there are 200 names for different types of bread. In the epic of Gilgamesh it is described how Enkidu was not a beast but became human by eating bread and drinking beer. These items were inseparable. The early Sumerian sign for eating was a sign for the mouth and bread. Gods appear to have been eating and drinking similar to humans, only better.

The earliest textual evidence of beer consumption is dated to 4000 BC (Katz & Voigt 1986: 31, Joffe 1998). Beer features prominently in many later Sumerian and Mesopotamian texts between 2600 and 2350 BC. Some texts include a song celebrating the Sumerian beer house and the beer goddess Ninkasi, to whom a hymn was written. In the third and second millennia BP one sometimes encounters beer as a general metaphor for drinks. In general beer seemed to have been an influential food item that was integrated into the mythology, religion and economy of the Sumerians (Katz & Voigt 1986).

I will here refer to some ethnographic studies, which show the significance of bread and the rich symbolism surrounding it. Delaney (1991) describes how bread represents the major part of most people's diet and bread is the generic name for food. Symbolically bread making is looked upon as analogous to the process of procreation. The rising of the dough is a mysterious and creative process similar to pregnancy.

EGYPT

As I have stated earlier there does not seem to have been any contact between the Levant and the Nile Valley during the pre-pottery Neolithic period, the links having been between the northern Levant and Anatolia. It was probably during the early pottery Neolithic, during the late eighth-millennium BP, that contacts between the Nile Valley and the Levant were initiated, involving the diffusion of a package of domesticated plants and animals. Pottery material recovered from Nabta Playa dated to 7000 BP shows similarities to the Levantine material (Wendorf & Schild 1998, 2002).

The excavators assume this new pottery to have been introduced with domesticates from the Near East. With the contact now established between Egypt and the Levant, winter-growing cereals (emmer wheat and barley) were introduced and became the vital food products in Egypt.

These cereals were probably made into bread and beer, and were to be the main staples in the Pharaonic diet (Samuel 2000). Egypt became part of the Near Eastern bread-eating world (Edwards 2003) culturally distinct from the region to the south with summer rain where sorghum-millet was cultivated, and used for porridge and beer.

Egypt in the north was part of the Near Eastern cuisine based on bread and beer from winter-growing wheat/barley crops (Edwards 2003, Haaland 2007). Bread was prepared in the oven and in bread moulds. Bread and beer were used by elite and commoners alike. Bread moulds and beer jars found together in burials suggest that they were also linked together in ritual offerings. This is evident from paintings on the tomb walls and models illustrating the complete beer-making process. The central role of beer is seen in the recovering of several large beer breweries from Tell el-Farkha dated to the fourth millennium BC; according to the excavator these would be the earliest breweries in the world (Cialowicz 2007). The consumption of beer and bread was of no less importance to the Pharaohs than it was to the workers. Workers' villages with bakeries and breweries show that both bread and beer were provided as rations for workers as part of the payment system (Kemp 1986, 1989). The importance of bread for the Egyptians comes across quite clearly during the New Kingdom when there were more than 40 words for different types of bread (Smith 2003).

Beer today is of fundamental importance in large parts of Africa. Based on material from Nubia dated to AD 350–550, Armelagos (2010) has argued that consumption of beer has important health benefits because it contained antibiotic tetracycline. Beer was made from grain kept in mud stores where it had been contaminated by the bacteria streptomycedes, which produces tetracycline. Armelagos found that 90 per cent of the human bones he studied showed traces of tetracycline even in two-year-old children. He believes that tetracycline protected the Nubians from bone infections, since all the bones were infection free (see Roach 2009 for further discussion). Beer was thus important both nutritionally and medicinally. Furthermore it is commonly used on occasions expressing and fostering social solidarity that ties people together and reinforces hospitality and communality in everyday life. In a wide range of African communities one sees that the two food items, porridge and beer, and the activities involved in their preparation serve as important sources for symbolic elaborations. Beer provides the occasion for an important expression of sociability (Figure 2.3), especially beer drinking connected with communal work, and rituals (see Karp 1980 for extensive references).

Nubia was on the crossroads between the bread-beer consuming cuisine in the north and the porridge/beer-consuming world to the south. Egyptian epigraphic inscriptions on temples from the Middle Nile dated to the first-millennium BP describe people of the south as porridge eaters. Food is here used to recognize social identity and this is expressed in writing suggesting contrast between the sub-Saharan and Mediterranean traditions (Pope 2013).

REFERENCES

Amiran, R. 1965. 'The beginning of pottery-making in the Near East'. In Matson, F.R. (ed.), *Ceramics and Man*, Chicago: Aldine, pp. 240–7.

Arkell, A.J. 1961. *The History of the Sudan. From the Earliest Times to 1821*. Second edition. London: Athole Press.

Akkermans, P.M.G., Schwartz, G.M. 2003. *The Archaeology of Syria: From Complex Hunter-Gatherers to Early Urban Societies.* Cambridge: Cambridge University Press.

Appadurai, A. 1981. 'Gastropolitics in Hindu South Asia'. *American Ethnologist*: 494–511.

Armelagos, G. 2010. NationalGeographicNews28.10.http://news.nationalgeographic. com/news/2005/05/0516_050516_ancientbeer_2.html.

Barley, N. 1994. *Smashing Pots*. London: British Museum Press.

Beldados, A., Constantini, L. 2011. 'Sorghum exploitation at Kassala and its environs, North Eastern Sudan in the second and first millennia BC'. *Nyame Akuma* 75: 33–9.

Bottero, J. 2000. *Ancestors of the West. Writing, Reasoning, and Religion in Mesopotamia, Elam and Greece.* Chicago: Chicago University Press.

Caneva, I. 1983. 'Pottery using gatherers and hunters at Saggai (Sudan). Preconditions for food production'. *Origini* 1: 7–278.

Caneva 1991. "Prehistoric hunters, herders and tradesmen in Central Sudan: data from the Geili region", In W.V. Davies (ed.). *Egypt and Africa*. London: British Museum Press, pp. 6–15.

Caneva , I. 1994. 'Recipienti per liquidi nelle pastorali dell alto Nilo'. In Caneva, I. (ed.), *Drinking in ancient societies*. Padua: Sargon, pp. 209–226.

Carsten, J., Hugh-Jones, S. 1990. 'Introduction'. In Carsten, J., Hugh-Jones, S. (eds), *About the House*. Cambridge: Cambridge University Press.

Cialowicz, K.M. 2007. 'From residence to early temple: the case of Tel el-Farkha'. In Kroeper, K. (ed.), *Archaeology of Early Northeastern Africa*. Poznan: Poznan Archaeological Museum, pp. 916–34.

Close, A.E. 1995. 'Few and far between: early ceramics in North Africa'. In Barnett, W.K., Hoopes, J.W. (eds), *The Emergence of Pottery*. Washington: Smithsonian Institution Press, pp. 23–37.

Counihan, C., van Esterik, P. (1997). *Food and Culture*. New York: Routledge.

Dale, D., C. Z. Ashley 2010. 'Holocene hunter-fisher-gatherers of Western Kenya'. *Azania* 45: 24–28.

Douglas, M. 1984. *Food in the Social Order*. New York: Russel Sage.

Delaney, C. 1991. *The Seed and the Soil. Gender and Cosmology in Turkish Village Society*. Los Angeles: University of California Press.

Dietler, M. 1990. 'Driven by drink: the role of drinking in the political economy and the case of the early Iron Age France'. *Journal of Anthropological Archaeology* 9: 352–406.

Dietler, M., Hayden, B. 2001. *Feasts: Archaeological and Ethnographic Perspectives on Food, Politics and Power*. Washington DC: Smithsonian Institute.

Dirar, H.A. 1993. *The Indigenous Fermented Food of the Sudan: A study of African Food and Nutrition*. Wallingford: CAB International.

Dunne, J. et al. 2012. 'First dairying in green Saharan Africa in the fifth millennium BC'. *Nature* 486: 390–4.

Edwards, D.N. 1996. 'Sorghum, beer and Kushite society'. *Norwegian Archaeological Review* 29: 65–77.

——— 2003. 'Ancient Egypt in the Sudanese Middle Nile: a case of mistaken identity'. In O'Connor, D., Reid, A. (eds), *Ancient Egypt in Africa*. London: UCL University Press, pp. 137–50.

——— 2004. *Nubian Past. An Archaeology of Sudan*. London: Routledge.

Fattovich, R. 1991. 'At the periphery of the empire: the Gash Delta (eastern Sudan)'. In Davies, W.V. (ed.), *Egypt and Nubia*. London: British Museum, pp. 40–8.

Fernandez, V., Tresseras, J.J. 2000. 'New Data on intensive plant processing and beer brewing in the Mesolithic and Neolithic periods in Central Sudan'. *Nyame Akuma*, 54.

Garfinkel, Y. 1999. *Neolithic and Chalcolithic Pottery of the Southern Levant*. Jerusalem: Hebrew University Press, Institute of Archaeology.

Gosden, C., Hather, J. 1999. *The Prehistory of Food*. London: Routledge.

Gourdin, W.H., Kingery, W.D. 1975. 'The beginnings of pyrotechnology: Neolithic and Egyptian lime plaster'. *Journal of Field Archaeology* 2: 133–56.

Hafsaas, H. 2006. *Cattle Pastoralists in a Multicultural Setting. The C-Group People in Lower Nubia. 2500–1500 BCE*. Ramallah: Birzeit University, Bergen University.

Haaland, R. 1987. *Socio-Economic Differentiation in the Neolithic Sudan*. Oxford: British Archaeological Reports.

——— 1992. 'Fish pots and grain: early and mid-Holocene adaptations in the Central Sudan'. *African Archaeological Review* 10: 43–64.

——— 1995. 'Sedentism, cultivation, and plant domestication. In the Holocene Middle Nile region'. *Journal of Field Archaeology* 22: 157–73.

——— 1997. 'Emergence of sedentism: new ways of living, new ways of symbolizing'. *Antiquity* 71(272): 374–85.

——— 1999. 'The puzzle of the late domestication of sorghum'. In Gosden, C., Hather, J. (eds), *The Prehistory of Food*. London: Routledge.

——— 2007. 'Porridge and pot, bread and oven: foodways and symbolism in Africa and the Near East from the Neolithic to the present'. *Cambridge Archaeological Journal* 17: 167–83.

——— 2013. 'Early Farming Societis along the Nile'. In Mitchell, P., Lane. P., *Handbook of African Archaeology*. Oxford University Press: Oxford, pp. 541–553.

Haaland, R., Magid, A.M. 1995. *Aqualithic Sites along the Rivers Atbara and the Nile*. Bergen: Alma Mater.

Handwerker, W. 1983. 'The first demographic transition: an analysis of subsistence reproductive consequences'. *American Anthropologist* 85: 5–27.

Harris, M., Ross, E. 1987. *Food and Evolution, Towards a Theory of Human Food Habits*. Philadelphia: Temple.

Hodder, I. 1990. *The Domestication of Europe. Structure and Contingency in Neolithic Societies*. Oxford: Blackwell.

Holthoer, R. 1987. 'New kingdom Pharaonic sites'. *Scandinavian Joint Expedition to Nubia* 5. Oslo: Norwegian University Press.

Jesse, F. 2010. 'Early pottery in northern Africa – an overview'. *Journal of African Archaeology* 8(2): 219–38.

Joffe, A.J. 1998. 'Alcohol and social complexity in ancient western Asia'. *Current Anthropology* 39: 297–322.

Karp, I. 1980. 'Beer drinking and social experience in an African society. An essay in formal sociology'. In Karp, I., Bird, C.S. (eds), *Exploration in African Systems of Thoughts*, Washington: Smithsonian Institution Press.

Katz, S.H., Voigt, M.M. 1986. 'Bread and beer: the early use of cereals in the human diet'. *Expedition* 28: 23–35.

Kemp, B. 1986. 'Amarna Reports III'. *Occasional Publications* 4. London: Egyptian Exploration Society.

——— 1989. *Ancient Egypt: Anatomy of a Civilization*. London: Routledge.

Kuper, R., Kröpelin, S. 2006. 'Climate-controlled Holocene occupation in the Sahara: motor of Africa's evolution'. *Science* 313: 803–7.

Krzyzaniak, L. 1991. 'Early farming in the Middle Nile Basin. Recent discoveries at Kadero (central Sudan)'. *Antiquity* 65: 159–72.

——— 2004. 'Kadero'. In Welsby, D.A., Anderson, J.R. (eds), *Sudan Ancient Treasures*. London: British Museum.

Kujit, I., Goring-Morris, A.N. 2002. 'Foraging, farming, and social complexity in the pre-pottery Neolithic of the southern Levant: a review and synthesis'. *Journal of World Prehistory* 16: 361–440.

Lupton, D. 1996. *Food, Body and the Self*. London: Sage Publications.

Magid, A.M. 1989. *Plant Domestication in the Middle Nile Basin. An Archaeoethnobotanical Case Study*. Oxford: British Archaeological Reports International series 523.

——— 1995. 'Plant remains from the sites of Aneibis, Abu Darbein and El Damer and their implications'. In Haaland R., Magid, A.A. (eds), *Aqualitic Sites along the Rivers Nile and Atbara Sudan*. Bergen: Alma Mater Press.

Marshall, F., Hildebrand, E. 2002. 'Cattle before crops: the beginnings of food production in Africa.' *Journal of World Prehistory* 16: 99–143.

Miracle, P., Milner, N. 2002. *Consuming Passions and Patterns of Consumption*. Cambridge: Cambridge University Press.

Moore, M.A.T. 1995. 'The inception of potting in western Asia and its impact on economy'. In Barnett, W.K., Hoopes, J.W. (eds), *The Emergence of Pottery*. Washington: Smithsonian Institution Press, pp. 39–53.

Neuman, K. 2003. 'The late emergence of agriculture in sub-Saharan Africa: archaeobotanical evidence and ecological considerations'. In Neumann, K., Butler, A., Kahleber, S. (eds), *Food, Fuels and Fields. Progress in African Archaeology*, Köln: Heinrich Barth-Institute.

Pope, J. 2013. 'Epigraphic evidence for a "Pot-and-Porridge" tradition on the ancient Middle Nile'. *Azania* 48: 473–497.

Reinold, J. 2001. 'Kadruka and the Neolithic of the northern Dongola reach'. *Sudan and Nubia* 5: 2–10.

Roach, J. 2009. '"Antibiotic" beer gave ancient Africans health buzz'. *National Geographic News*. Accessed on 13.04.09: http://news.nationalgeographic.com/news/2005/05/0516_050516_ancientbeer.html.

Sadr, K. 1991. *The Development of Nomadism in Ancient Northeast Africa*. Philadelphia: University of Pennsylvania Press.

Samuel, D. 2000. 'Brewing-Baking'. In Nicholson P., Shaw, I. (eds), *Ancient Egyptian Materials and Technology*. Cambridge: Cambridge University Press, pp. 537–76.

Smith, S.T. 2003. 'Pharaohs, feasts and foreigners: cooking, foodways, and agency on ancient Egypt's southern frontier'. In Bray, T.L. (ed.), *The Archaeology and Politics of Food and Feasting in Early Empires*. New York: Kluwer Plenum Press, pp. 39–65.

Stemler, A.B.L. 1990. 'A scanning electron microscopic analysis of plant impressions in pottery from the sites of Kadero. El Zakiab Um Direiwa and el Kadada'. *Archaéologie du Nil Moyen* 4: 87–105.

Sutton, J.E.G. 1974. 'The aquatic civilization of Middle Africa'. *Journal of African History* 15: 527–46.

Veen, van der, R. 2003. 'When is food luxury?' *World Archaeology* 34: 405–27.

Wasylikowa, K., Mitka, J., Wendorf, F., Schild, R. 1997. 'Exploitation of wild plants by the Early Neolithic hunter-gatherers of the Western Desert, Egypt: Nabta Playa as a case study'. *Antiquity* 71: 992–41.

Wendorf, F., Schild, R. 1998. 'Nabta Playa and its role in northeastern African Prehistory'. *Journal of Anthropological Archaeology* 17: 97–123.

—— 2002. 'Implications of incipient social complexity in the late Neolithic in the Egyptian Sahara'. In Friedman, R. (ed.), *Egypt and Nubia. Gifts of the Desert*. London: The British Museum Press.

Wengrow, D. 2001. 'Rethinking "cattle cults" in early Egypt: towards a prehistoric perspective on the Narmer Palette'. *Cambridge Archaeological Journal* 11: 91–104.

—— 2006. *The Archaeology of Early Egypt. Social Transformations in North-East Africa, 10000 to 2650 BC*. Cambridge: Cambridge University Press.

Wright, R.P. 1991. 'Women's labour and pottery production in prehistory', in Gero, J.M., Conkey, W.C. (eds), *Engendering Archaeology: Women and Prehistory*. Oxford: Basil Blackwell.

3   How did the Nile Water System Impact Swine Husbandry Practices in Ancient Egypt?

Louise Bertini

INTRODUCTION

The flourishing landscape of the Nile valley provided the necessary environment for the ancient Egyptians to transition from a hunter-gatherer people to a more sedentary agrarian society. More than simply providing a source of water for this burgeoning economy, the Nile's annual flood and its resultant silt deposits created an environment ideal for subsistence farming that would not have otherwise been possible. In the fifth millennium BC, permanent settlements along the Nile valley began to appear, leading to the emergence of early civilization by the fourth millennium BC. Attempting to understand ancient Egypt without considering the Nile and its profound impact on both plant and animal life only paints a partial picture of everyday life at this time. The Nile floodplain and its fertile soil proved essential for both the prosperity of the agrarian economy but also for the animals in this environment. Generally speaking, one such way of estimating the full impact of the Nile, and its broader ecological effects, is by taking into account the bio-archaeological record of excavated animal remains. In particular, this chapter explores swine husbandry practices along the Nile valley as a means of evaluating the distinctive ecology of this central river system.

It is the generally believed view (Zeuner 1963; Reed 1969; Epstein 1971; Clark 1971) that domesticated animals in Egypt derived from south-west Asia slightly before 5000 BC, replacing an existing hunting-gatherer economy. This replacement is partially supported by the faunal record as seen in Figure 3.1, where in the Late Paleolithic – c.18000–13000 BC (Wadi Kubbaniya) – there is primarily reliance on fishing and wild game. In the Neolithic period – c.5500–4400 BC (Sais I levels) – there still is a reliance on fishing, but domestic animals begin to make an appearance, accounting for less than ten per cent of the total faunal assemblage. By the Predynastic period – c.4200–3000 BC (Merimde, Maadi, Buto, Sais II levels) – domestic animals make up about 50 per cent of the total faunal assemblage. Approaching the unification of Egypt c.3000 BC and start of the Early Dynastic period (Sais III levels), domestic animals

Figure 3.1. Relative importance (%) of different food strategies from selected settlements from the Late Paleolithic-New Kingdom Egypt (© Louise Bertini).

account for well over 50 per cent of the faunal assemblages. This trend continues throughout the rest of dynastic Egypt at sites such as South Abydos, Tell el-Daba and Amarna. Although the faunal record does seem to support that Egyptian foragers were gradually adopting the south-west Asian domesticates as new resources to exploit, it is important to note that this view is largely due to the complete absence of domestic animals from Late Paleolithic assemblages. Furthermore, all of the known Paleolithic assemblages come from the Western desert (Gautier, 1976, Gautier & Van Neer, 1989, Van Neer, 2000, Linseele & Van Neer 2009), which is not the ideal habitat for domestic animals. This lack of domesticates could also be attributed to the absence of Paleolithic sites from the Nile Delta (Bertini & Cruz-Rivera 2014). Recent evidence has found that at least three mammals may indeed have been domesticated in Africa including the donkey (Beja-Pereira et al. 2004, Rossel et al. 2008, Blench 2000), cat (Driscoll et al. 2007, Linseele, Van Neer & Hendrickx 2007, 2008), and North African cattle (Dunne et al. 2012, Marshall & Hildebrand 2002, Cunningham 2000, Bradley & Magee 2006). This new evidence has now prompted a re-visiting of both arguments for animals of indigenous origin and animals for importation.

There have been many studies on the general husbandry of animals in ancient Egypt (Brewer, Redford & Redford 1994, Germond & Livet 2001, Houlihan 1996), transition from hunter/gatherer economies to a reliance on domesticated animal (Wetterstrom 1995, MacDonald 2000) and meat production (Ikram 1995), with particular reference to the highly valued cattle (Ghoneim 1977). However, little to no work has been done on the herd structure and possible changes in the husbandry regime of the pig (*Sus scrofa*). Understanding the role of the pig in both the ecology and ancient Egyptian husbandry regime is particularly important. Despite the lack of artistic and textual sources (Redding 1991: 23), it is one of the most commonly identified species in faunal assemblages from settlement sites (Table 3.1), and it is the animal that is most affected by the environment in which it lives (Bertini 2011: 90).

The role of the pig in ancient Egypt was primarily for subsistence. Though the pig was slaughtered for meat and useful for trampling grain into the ground (Lobban 1998: 141), pig rearing in ancient Egypt did not yield a surplus that could be renumerated by the state. As pig rearing was maintained by individuals to supplement other sources of animal protein, it was exempt from taxation, which may also explain its infrequency in texts and tomb/temple reliefs (Redding 1991: 23). The relationship of pig husbandry to the rearing of other domestic animals, most notably cattle and sheep/goat, is in part dependent on the type of agriculture being practiced. This is especially relevant when considering the degree of grain cultivation (which would put the pig in direct competition with humans), and the degree of cattle reliance, which may depress pig husbandry (Redding 1991: 23-5).

While this relationship scheme may hold up for ancient Egypt, it is also important to take into account the environment in which a

Table 3.1. Number of pig elements and their percentages of total faunal assemblage from selected settlement sites.

Site	Time Period	Date	Number of Pig Elements	Pig % of Total Assemblage	Reference
Merimde-Benisalame	Neolithic	c. 5000–4000 BC	6568	41	Von den Driesch and Boessneck 1985
Saïs	Neolithic	c. 5000–4000 BC	353	54	Bertini and Ikram 2014
Ma'adi	Predynastic	c. 4000–3400 BC	21045	15	Boessneck and Von den Driesch 1989
		c. 4000–3400 BC	38	21	Hecker 1982
Kom el-Hisn	Old Kingdom	c. 2663–2160 BC	397	51	Wenke et al. 1988
Giza	Old Kingdom	c. 2663–2160 BC	386	1.67	Redding 2010
South Abydos	Middle Kingdom	c. 1870–1831 BC	360	7	Rossel 2007
Tell el-Dab'a	Second Intermediate Period	c. 1650–1549 BC	63	5	Boessneck 1976
Amarna (Workmen's Village)	New Kingdom (Dyn 18)	c. 1352–1336 BC	302	47.30	Hecker 1982
Kom Firin	New Kingdom	c. 1550–1069 BC	950	53.98	Bertini 2014
Saïs	Third Intermediate Period	c. 1295–664 BC	271	39.30	Bertini and Linseele 2011
Malkata	New Kingdom (Dyn 18)	c. 1388–1348 BC	74	9.50	Ikram 1995

particular animal is being utilized. Some (Harris 1974: 42–3, Diener 1978: 496) argue that pigs cannot survive in temperatures over 98F/36°C. However, pigs are known to adapt to high temperatures. Such acclimatization is not only seen in modern Egypt (Bertini 2011: 99) where summer temperatures can reach over 100F/38°C, but also in south-western Iran where summer temperatures reach over 120F/49°C (Redding 1991: 23). Although the wet marshes of the Nile Delta and the Fayum may have been the most ideal for maintaining pigs, archaeological remains have been found as far south as Amarna and Elephantine (Figure 3.2), as long as the animals maintained access to wallows and shelter. It has thus been suggested that there is evidence for two types of swine management: free range (Miller 1990: 126) and pens (Shaw 1984, Kemp 1991: 256). This chapter will thus explore the husbandry regime of the pig and the impact of the Nile water system through the presence and frequency of the dental defect, linear enamel hypoplasia (LEH).

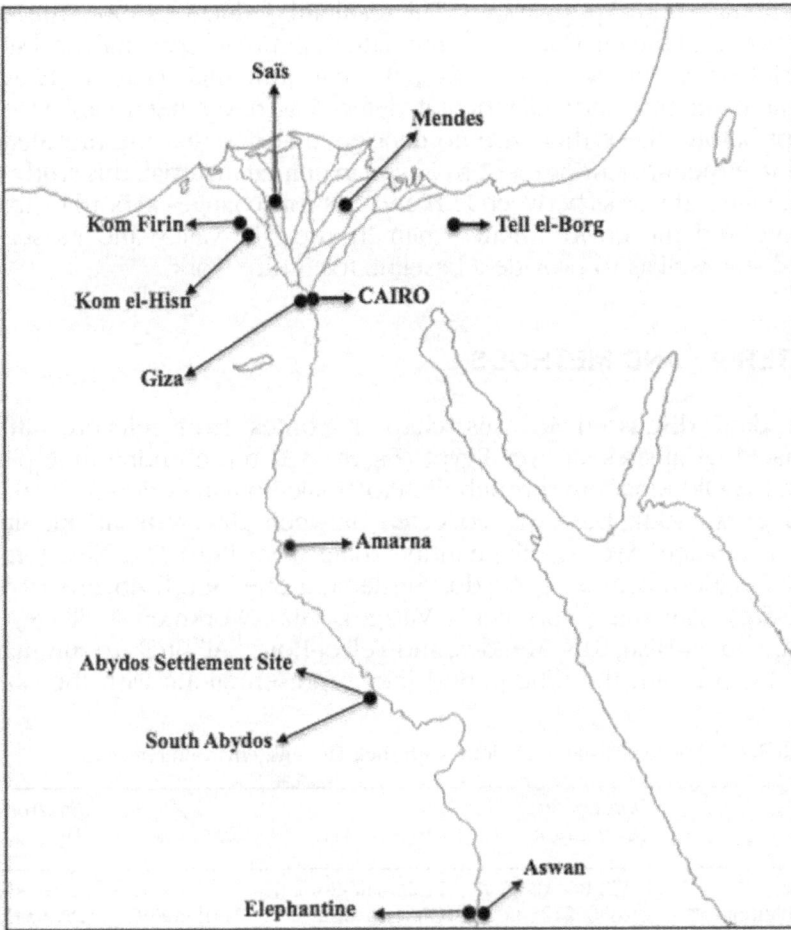

Figure 3.2. The location of each of the eleven archaeological sites where samples were collected (© Louise Bertini).

Recent zooarchaeological studies on European, Near Eastern, and East Asian material have shown how mammal teeth can provide information on husbandry practices, environment, health status, and the diet of an individual through the recording of the developmental defect LEH (Dobney & Ervynck 1998, 2000, Dobney et al. 2002, 2004, Ervynck & Dobney 1999, 2002, Ervynck et al. 2001). LEH is defined as a deficiency in enamel thickness that occurs during tooth crown formation. This defect is caused by physiological stresses such as disease, poor nutrition, and environmental factors. Such stresses cause a disruption in enamel secretion and manifest themselves in a variety of ways on the tooth's surface, such as lines, depressions, or pits (Goodman & Rose 1990). Thus, it serves as a means to study the health status of both ancient and modern human and non-human animal populations. The resultant deductions about husbandry practices and the environment can relate to penning, seasonal food supply, possible scavenging, and seasonal environmental factors. However, research on this particular dental defect has never been carried out in Egypt before. Since there are no datasets that describe the prevalence of LEH in either ancient or modern Egyptian animal material, this study seeks to highlight the links between LEH and possible changes in both husbandry regime and the environment – namely the Nile valley and its seasonal flood – as well as to provide a baseline for future work.

MATERIAL AND METHODS

The data discussed in this chapter comes from eleven different archaeological sites all over Egypt (Figure 3.2) representing time periods from the Old Kingdom through the Ptolemaic-Roman period (c. 2686 BC up through AD 400). Data was collected between 2006–9 from the sites of Aswan (ancient *Syene*), Elephantine (data from both Old Kingdom and New Kingdom contexts), Abydos Settlement Site, South Abydos (town of *Wah-Sut*), Amarna (Workmen's Village), Giza Workmen's Village, Kom Firin, Kom el-Hisn, Saïs, Mendes, and Tell el-Borg.[1] All sites are summarized in Table 3.2 with the time period they represent along with the site and

Table 3.2.   Total sites analysed along with their site and environment type.

Site	Dates of Site Occupation	Time Period	Site Type	Environment Type
Aswan	c. 332 BC–AD 395	Ptolemaic-Roman	Settlement	Riverside
Elephantine	c. 2686–2125 BC	Old and New Kingdom	Settlement	Riverside
	c. 1550– 1069 BC			
Abydos Settlement Site	c. 2160–2055 BC	First Intermediate Period	Settlement	Floodplain
South Abydos	c. 1870–1831 BC	Middle Kingdom	Settlement	Desert Edge
Amarna	c. 1352–1336 BC	New Kingdom	Settlement	Desert Edge
Giza	c. 2686–2125 BC	Old Kingdom	Settlement	Desert Edge
Kom Firin	c. 1550– 1069 BC	New Kingdom	Settlement/ Possible Fortress Settlement	Floodplain Edge
Kom el-Hisn	c. 2686–2125 BC	Old Kingdom	Settlement	Floodplain Edge
Saïs	c. 1295–664 BC	Third Intermediate Period	Settlement	Floodplain
Mendes	c. 2686–2125 BC	Old Kingdom	Settlement	Floodplain
Tell el-Borg	c. 1550–1069 BC	New Kingdom	Fortress Settlement	Marshy land

environment type. A modern comparative sample of domestic pig mandibles from Cairo was also collected to act as a control group against which to compare the archaeological material to.

The selection of sites was based on the availability of materials, the generosity of excavation directors, and permission from the Egyptian Supreme Council of Antiquities (current Ministry of Antiquities). As is often the case with earlier excavations in Egypt, animal bones were either not collected or not properly curated and recorded. Therefore, the selection of teeth in this study comes from sites that have been excavated within the last 30 years, which rather restricts the sample.

The following protocol was used to record all teeth based on the methodologies developed by Dobney and Ervynck (1998) and Dobney et al. (2002): tooth type (as first molar [M_1], second molar [M_2], or third molar [M_3]), side (right or left), and wear stage of each tooth following Grant (1982). The presence of LEH was recorded and measured on the buccal surface of all separate cusps of the permanent mandibular molars (M_1, M_2 and M_3). The LEH position on the crown is recorded by measuring the distance between the hypoplasia and the root-enamel junction (REJ) on each cusp, along a perpendicular axis. The measurement (Figure 3.3) is taken from the midpoint of the hypoplasia line, depression etc. down to the REJ (Dobney & Ervynck 1998: 267) using digital calipers with measurements being recorded to the nearest 0.1 mm. This measurement technique, however, is only used when the tooth crown is complete. Occurrences of hypoplasia can easily be observed under a strong light, and frequently multiple occurrences of LEH can be observed on the same tooth (Figure 3.3), with each occurrence noted separately. Additionally, the crown height of all teeth was recorded, which could influence the relative position of LEH on the tooth crown. As teeth wear down, occurrences of LEH can be obliterated, thus prompting the need for the measurement to which height distribution for each tooth can be established. The distributions are then smoothed using the running mean in order to reveal underlying trends more clearly and to allow easier comparisons between sites and other established samples.

All archaeological data in this chapter will be compared with a sample of modern domestic pigs collected from the Shobra Pork Store, which provide a control group with a known life history. This sample consists of nineteen domestic mandibles from pigs that were raised in pens on a private farm outside Mohandessein, Cairo and fed a diet primarily of potatoes, tomatoes, and other various grains and vegetables (Bertini & Cruz-Rivera 2014).

RESULTS

A total of 562 recordable suid teeth were analysed from all of the archaeological sites, along with the modern sample of 57 teeth, totaling

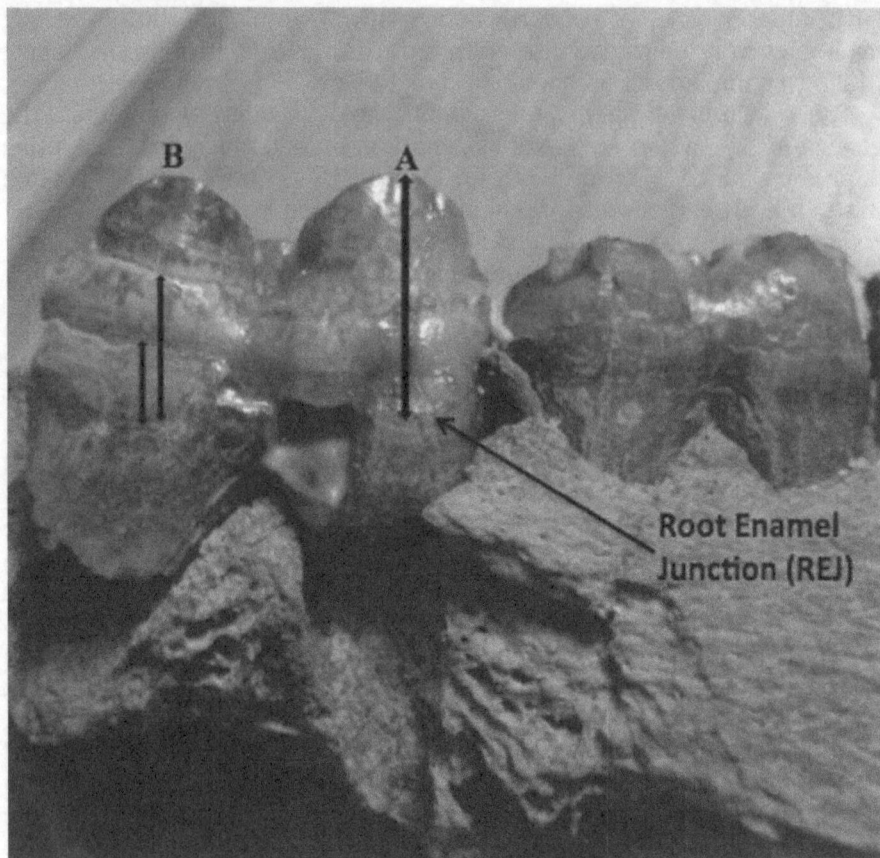

Figure 3.3. Pig mandible from Tell el-Borg showing the lingual surface of the second molar (M_2), left side (A). The measurement of the maximum height of the tooth crown; (B) the measurement(s) of hypoplastic defects on a complete tooth crown from the root enamel junction (REJ) (© Louise Bertini).

619 teeth. An inventory of the archaeological material studied is presented in Table 3.3. All tooth types are sufficiently abundant in the study collection and from most sites a large number of teeth were available.

Interpreting swine husbandry practices

Table 3.3 summarizes both the archaeological and modern pig samples in terms of the frequency of LEH, of which 408 teeth (66 per cent) have hypoplasia present. Of note, when simultaneously compared, there are no significant differences in the frequency of hypoplasia between any of the archaeological or modern pig samples (P = 0.338, Chi square test).

The sample of Old Kingdom teeth from Elephantine has the highest frequency of LEH (88 per cent) as compared to both the modern and

Table 3.3.   Total teeth and frequency of enamel hypoplasia for archaeological and modern pig samples.

Site	Time Period	M1	M2	M3	Total Number of Teeth	Number of Teeth with LEH	Teeth with LEH (%)
Elephantine	Old Kingdom	5	3	0	8	7	88
Giza	Old Kingdom	5	4	5	14	11	79
Kom el-Hisn	Old Kingdom	11	9	12	32	22	69
Mendes	Old Kingdom	13	24	33	70	49	70
Abydos Settlement Site	First Intermediate Period	9	13	13	35	24	69
South Abydos	Middle Kingdom	13	20	19	52	39	75
Elephantine	New Kingdom	16	13	10	39	28	72
Amarna	New Kingdom	45	38	33	116	66	57
Kom Firin	New Kingdom	48	38	27	113	72	64
Saïs	Third Intermediate Period	10	14	5	29	20	69
Tell el-Borg	New Kingdom	1	3	5	9	7	78
Aswan	Ptolemaic-Roman	16	17	12	45	30	67
Shobra	Modern	19	19	19	57	32	56
TOTAL	**ALL SITES**	**211**	**215**	**193**	**619**	**407**	**66**

other archaeological samples. Amarna, has the lowest frequency of LEH (57 per cent), which is almost the same frequency as the modern sample. The percentage of LEH for each of the three molars per site is provided in Table 3.4.

Understanding seasonality and physiological impacts

The archaeological and modern sample distributions of hypoplasia on the anterior and posterior cusp of the M_1 and M_2 and the anterior and middle cups of the M_3 are shown in Figure 3.4. Calculations are based on the running means of the values of the individual tooth height classes recorded. The use of running means was chosen for the analysis in order to smooth variation and allow underlying trends to become more clearly visible (Dobney & Ervynck 2000: 601, Dobney et al. 2002: 37, Dobney et al. 2004: 200).

For the modern sample, defects of the M_1 are located in the middle of the tooth crown, with the majority of defects occurring between 3.5 and 6.5 mm above the REJ. The archaeological samples are slightly different with the majority of defects located between 2.5 and 3.5 mm above the REJ. There are a few sites, however, that have a second peak at about 6.5 mm above the REJ. A different pattern, however, is seen on the M_2 and M_3, where defects are located in the cervical 7.5 mm of

Table 3.4.   Percentage of teeth affected by enamel hypoplasia for each site.

Site	Time Period	% of M1	% of M2	% of M3
Elephantine	Old Kingdom	100	67	0
Giza	Old Kingdom	40	100	100
Kom el-Hisn	Old Kingdom	36	89	83
Mendes	Old Kingdom	38	79	76
Abydos Settlement Site	First Intermediate Period	11	85	92
South Abydos	Middle Kingdom	62	85	74
Elephantine	New Kingdom	38	92	100
Amarna	New Kingdom	33	71	73
Kom Firin	New Kingdom	35	84	85
Saïs	Third Intermediate Period	50	86	60
Tell el-Borg	New Kingdom	0	67	100
Aswan	Ptolemaic-Roman	38	76	92
Shobra	Modern	11	79	79
TOTAL	**All Sites**	**36**	**81**	**81**

the tooth crown on both teeth, with the majority of defects located about 3.5 mm of the tooth crown above the REJ on the M_2 and about 4.5 mm of the tooth crown above the REJ on the M_3.

Hypoplasia distribution of the M_2 and M_3 are exclusively located in the cervical 8.5 mm of the tooth crown, with the majority of defects located in the 3.5 mm of the crown above the REJ. Similar to the M_1, there are a few sites that have a second peak of hypoplasia at 7.5 mm above the REJ.

There are a few sites that do not have LEH frequency distributions calculated, due to either a lack of defects present, or too small a sample size. The exempted sites include the M_1 from both the Abydos Settlement Site and Tell el-Borg and the M_3 from Old Kingdom Elephantine.

At this point, the results from the modern material only were plotted on to a schematic dental developmental graph (Figure 3.5) based on the developmental rates of pigs as described in McCance et al. (1961), and the known time of birth. An increase in the frequency of enamel hypoplasia is observed around the time of birth. On the M_2, there is a peak of enamel hypoplasia that occurs at about eight to nine months after birth and then drops sharply. A similar picture is seen with the M_3, as the number of defects peak between 19 to 20 months after birth and again sharply drops.

INTERPRETING SWINE HUSBANDRY PRACTICES

In ancient Egypt, there is evidence for two types of swine management strategies: free-range and penning (Miller 1990: 126, Shaw 1984, Kemp 1991: 256). Management strategies along with the environment seem to be

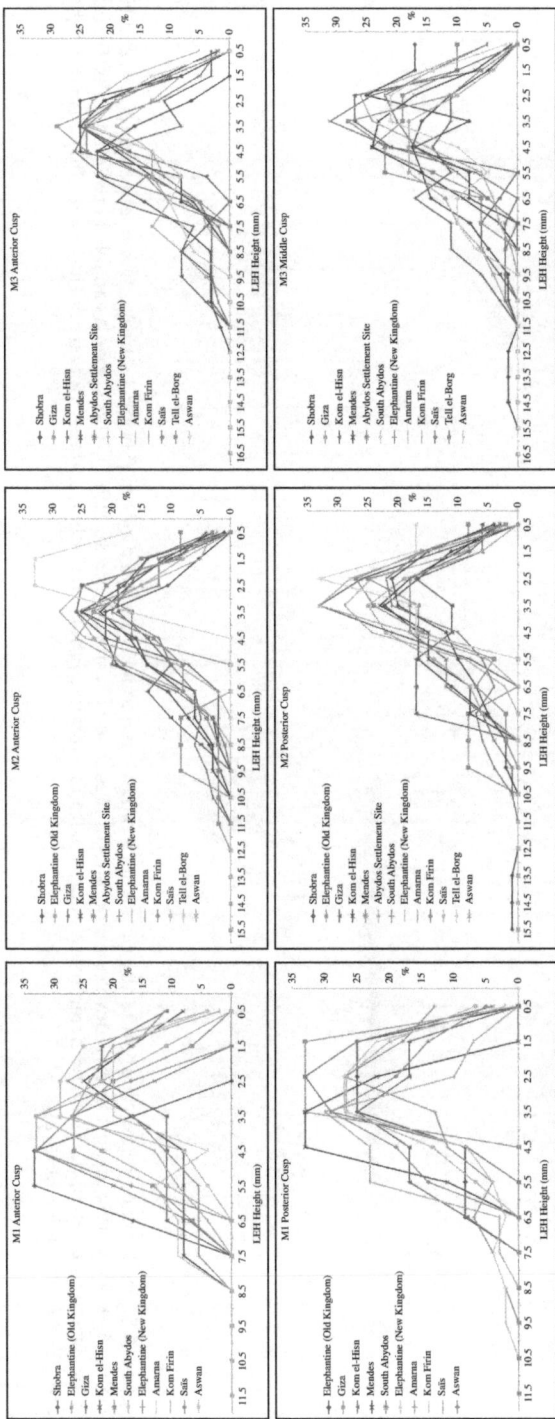

Figure 3.4. Frequency distribution shown as percentages of LEH heights (*Y-axis*) for each sample, per individual tooth and cusp (M_1 = first molar; M_2 = second molar; M_3 = third molar). Root enamel junction (REJ) is at 0 with maximum crown height on the left of the *X-axis* (© Louise Bertini).

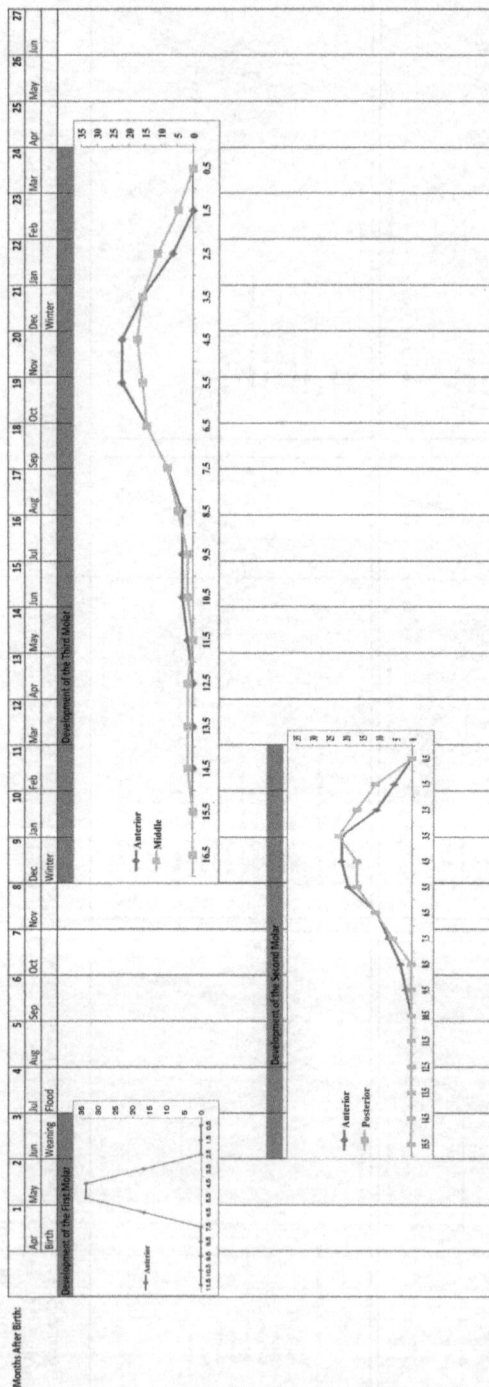

Figure 3.5. Schematic representation of the occurrence of LEH frequencies for the modern (Shobra) sample shown as percentages of LEH heights (*Y-axis*) plotted against known development rates of pigs based on McCance et al. (1961). The *X-axis* is variable between graphs, based on the average size of tooth crown with minimum wear. Maximum crown height is on the left, and the root enamel junction (REJ) is on the right of the *X-axis* (© Louise Bertini).

the two major factors in determining the frequency of enamel hypoplasia present.

The New Kingdom site of Amarna stands out from all the other sites as it has the lowest frequency (57 per cent) of enamel hypoplasia. The only site that has a similar frequency is the modern sample (Shobra), of which 56 per cent of teeth show evidence of hypoplasia. This result is quite surprising as compared to all the sites, Amarna quite possibly has the harshest environment. It is located on the desert edge, in an area that is particularly exposed to the sun and the hot summer winds from the south with no natural shade (Shaw 1984: 49). Amarna also stands out as it is the only known ancient Egyptian site to date that has evidence for pig pens (Shaw 1984, Kemp 1991: 256).

Modern studies have demonstrated that pigs in confined spaces are more likely to be affected by various health-related conditions (such as respiratory diseases) as compared to free-range animals. Penning can cause a stress on the animals, be it a respiratory disease or even stress from overcrowding (McCosker 2011: 24). However, the Amarna assemblage displays a considerably lower frequency of LEH. Although puzzling, it is highly possible that Amarna swine husbandry was very good, resulting in fewer pigs having physiological stresses. What stresses that would have been encountered were reduced, and as a result, there was a lower LEH frequency. Furthermore, the percentage of M_1's from Amarna display the lowest frequency (33 per cent) of enamel hypoplasia (Table 3.4). Again, similar to the modern sample (11 per cent of M_1's have LEH), this may indicate that both Amarna and the modern sample (which were also kept in pens) were not as heavily affected by early life stresses as compared to the other archaeological populations. This seems to suggest that penning provides a buffer for the pigs from the environmental factors that might otherwise produce enamel hypoplasia.

Although Amarna is the only known site with specific pig penning, Aswan also seems to have evidence of penning based on various pathologies identified on non-cranial bones (Johanna Sigl, personal communication) along with the Old Kingdom site of Kom el-Hisn, although the type of animal they serve is unknown (Richard Redding, personal communication). Like Amarna, Kom el-Hisn, displays both a low frequency of teeth with LEH at 69 per cent (Table 3.3) and a low percentage of M_1's with LEH, at 36 per cent (Table 3.4). Aswan is also similar at 67 per cent of teeth with LEH, with 38 per cent LEH on the M_1. Other sites that also show a similar pattern of low overall percentage of teeth with LEH along with a low percentage of M_1's with LEH are: the Abydos Settlement Site (69 per cent, with 11 per cent LEH on M_1) and Kom Firin (64 per cent, with 35 per cent LEH on M_1). Although it is unknown if any of these sites may have had penning as well, it is highly possible that these sites at least had more intensive involvement in pig husbandry as compared to other archaeological sites such as

Elephantine, Giza, Mendes, and Tell el-Borg, indicating a possible free-range management strategy.

Conversely, the Old Kingdom sample from Elephantine displays the highest frequency (88 per cent) of LEH (Table 3.3). As Elephantine is a riverside settlement in an island, it is highly possible that the pig management strategy here was free-ranging, leaving the pigs to forage for themselves. Furthermore, it is possible that the impact of the Nile flood may have affected on these pigs adversely, as when the river flooded, it may have limited the pigs' normal environment and foraging range.

Elephantine also displays the highest percentage of M_1's (at 100 per cent) that have evidence of LEH (Table 3.4). The Old Kingdom Elephantine sample, however, is very small, possibly skewing the data and creating a curve that may not be representative of the whole sample. The only other site that displays a similar pattern is that of South Abydos, which also has a high percentage of teeth with LEH (75 per cent), and the second highest percentage of M_1's displaying evidence of LEH (62 per cent). A similarity between these two sites is that here pigs are of fairly low status as compared to other animals (Angela von den Driesch, personal communication in 2008 on Elephantine material; Rossel 2007: 191–2 on South Abydos) that are more predominantly consumed, such as cattle, caprines, and even fish. Thus, it is unlikely that there was a heavy emphasis on pig husbandry and care for these animals, thereby exposing pigs to increased physiological stresses, and a resulting increase in the presence of LEH.

Other sites that display a high percentage of teeth with LEH include Giza (79 per cent) and Tell el-Borg (78 per cent). Both these sites, however, have significantly lower percentages of M_1's with LEH (Table 3.4), which is due to the small sample size. These two sites not only represent the smallest samples, but are also similar in that pig remains are not prominently represented in the overall faunal assemblage, indicating that there were not as commonly consumed as other animals (Redding 2010: 72).

UNDERSTANDING SEASONALITY AND PHYSIOLOGICAL IMPACTS: PHYSIOLOGY AND ANIMAL HUSBANDRY

Birth. The dental development graph for the modern Egyptian pigs (Figure 3.5) shows a peak of enamel hypoplasia between 3.5 and 6.5 mm, which according to the chart, is around the time of birth. Because of the moderate sample size (N=19), only 2 M_1's (11 per cent – Table 3.4) have any evidence of LEH, which is not large enough to be statistically valid. It is also possible that since many the M_1's were in more advanced wear stages (Bertini & Cruz-Rivera 2014: 94), evidence of enamel hypoplasia may have been obliterated. Alternatively, the modern sample may not have experienced much early life stress. Looking at the archaeological samples, there are a number of sites that also have defects occurring on the M_1 in the range between 3.5 and 6.5 mm (Figure 3.4) including: Old Kingdom Elephantine, Giza, Kom el-Hisn, and Amarna. Thus it is possible that Egyptian pigs, both ancient and modern, may have encountered some physiological stress from birth, despite the low frequency of its presence in the modern sample.

Weaning. With the exception of the modern sample, all the archaeological samples (Figure 3.4) show a significant peak of enamel hypoplasia on the cervical half of the M_1 (between 2.5 and 3.5 mm). When this data is related to the chronology of dental development, this peak of LEH must be related to a stress occurring after birth, but before the animal has reached four months of age, when the development of the M_1 is completed. Such a conclusion is based on a normal, well-nourished population, although development of the M_1 can take up to seven or eight months in an under-nourished population (McCance et al. 1961: 220). Thus, it is most likely that weaning and the associated physiological stresses that are connected with this period are responsible for this peak of hypoplasia. However, this peak may also indicate a difference in the timing of weaning at different sites.

The vast majority of the sites that were discussed in the previous section as having intensive pig rearing (Amarna, Kom Firin, Abydos Settlement Site, Kom el-Hisn, and Aswan) include some (Kom el-Hisn and Amanra) that peak at 2.5 mm on the M_1, while others (Kom Firin and Aswan) peak at 3.5 mm. Sites that may not have had as intensive a pig rearing program all have similar peaks, for example South Abydos at 2.5 mm, while the Old Kingdom Elephantine sample has a peak at 3.5 mm. Thus, pig husbandry does not seem to have been responsible for this particular hypoplasia peak. Geographical context does not seem to reveal any patterns either, as most of the desert edge sites such as Amarna and South Abydos peak at 2.5 mm, while the riverside New Kingdom Elephantine settlement has a peak at the same height as well. The other floodplain sites (Kom el-Hisn, Kom Firin, and Saïs) along with the riverside settlement of Aswan instead peak at 3.5 mm. Although this is a very small discrepancy, suggesting that weaning occurred naturally, this does support that weaning is the most common stress encountered during the development of the M_1. Hypoplasia peaks on the M_2 and M_3, however, have more variation.

UNDERSTANDING SEASONALITY AND PHYSIOLOGICAL IMPACTS: SEASONAL ENVIRONMENTAL FACTORS

Upex (2012: 264) states that if some stress episodes are connected to the seasonal deterioration of environmental conditions, then a distribution of hypoplastic defects that reflect that particular seasonal cycle should be expected. It is further noted that in order for seasonal patterning to be identified, there needs to be a known season of birth to ensure that seasonal events affect all animals when they are at a similar age. There is no exact season of pig birth in modern Egypt, as the current practice is to have continuous breeding that results in five litters every two years. The data from modern sample, however, has a known season of birth and time of death, which will provide the base for comparison of the archaeological material.

The modern and ancient seasonality in Egypt and its impact on birthing seasons

The single most important seasonal factor in Egypt is the annual flooding of the Nile, which is the country's lifeline. The flood typically lasted from mid July–October, winter from mid November to February, and summer, from mid March to June made up the three seasons that Egypt would experience on a yearly basis. However, the construction of the Aswan Dam in the 1960s completely stopped the annual flooding north of the dam. Traditionally, inundation was the time when farmers prepare their fields, followed by winter when crops are planted, and summer when they are harvested and the temperature is at or reaching its peak.

The absence of the flood, however, made it easier on both the animals and their herdsman, allowing for the continuous breeding of pigs, as they do not have to deal with possible displacement and, depending on the level of the flood, parasitic infection. In antiquity, the annual flood had an impact not only on the preparation of planting crops, but on the animals as well, causing seasonal stress. However, it is unknown at this point if it was the actual flood, or its repercussions, such as waterlogging of the soils, that had a greater impact on the animals. The flood levels consistently varied throughout dynastic Egypt, with some years being uncontrollably high, while others were dangerously low, ultimately resulting in drought (Said 1993: 133–52).

Modern seasonality

With the absence of the flood, modern day Egypt has only two seasons: winter and summer. Winter lasts typically from November–March followed by a brief *khamsin* hot wind in April (Wilcocks & Craig 1913: 304) before the start of summer lasting through October.

The modern pig sample (Figure 3.5) shows hypoplasia peaks relating to birth that occur at roughly 4.5 mm on the M_1. On the M_2, there is a single peak that occurs at 3.5 mm. This peak falls at about eight to nine months of age, or at about the beginning of winter. The M_3 also has only one peak that occurs at 4.5 mm. This peak occurs between 19 to 20 months of

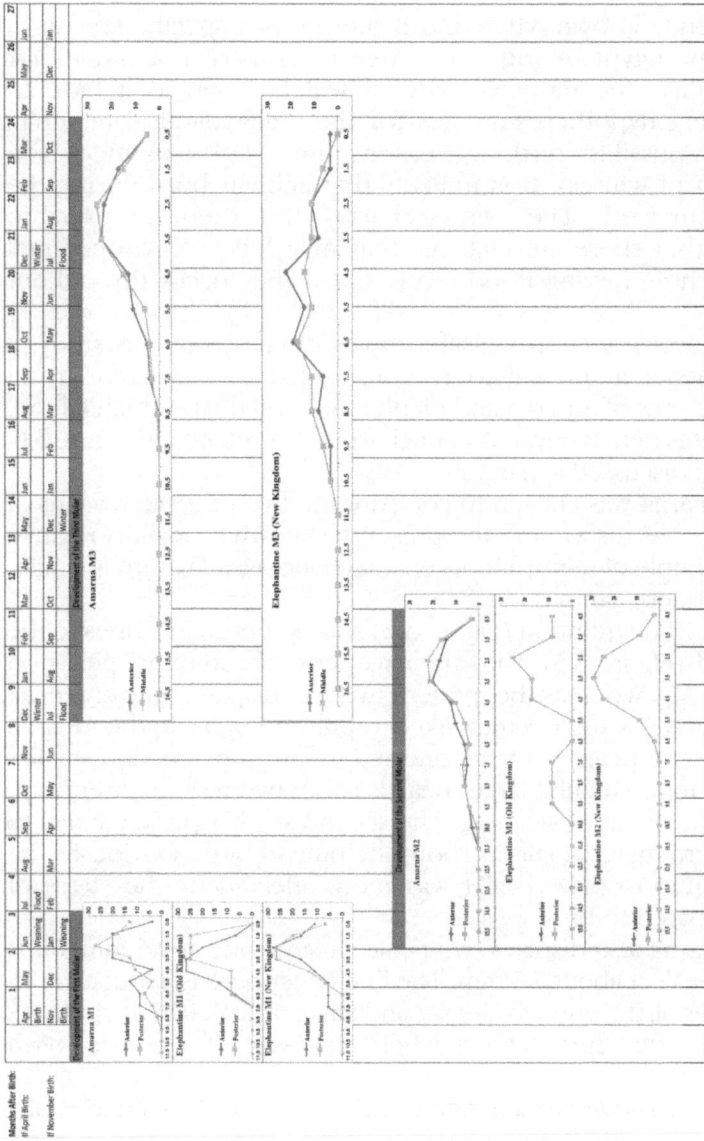

Figure 3.6. Schematic representation of the occurrence of LEH frequencies for the Amarna, Old Kingdom Elephantine, and New Kingdom Elephantine samples. Compared against two possible birthing seasons (summer vs winter), each sample is shown as percentages of LEH heights (*Y-axis*) plotted against known development rates of pigs based on McCance et al. (1961). The *X-axis* is variable between graphs, based on the average size of tooth crown with minimum wear. Maximum crown height is on the left, and the root enamel junction (REJ) is on the right of the *X-axis* (© Louise Bertini).

age, also at about the beginning of the second winter in the animal's life, at a time when temperatures in Cairo decrease dramatically.

Archaeological seasonality

It is not completely known when and if there was a specific season of birth for ancient Egyptian pigs. In order to construct a model for ancient seasonality, the modern material will be used as a base to construct working hypotheses/models for the archaeological material. The two most possible birthing seasons are April (summer) and November (start of winter). It is unlikely that animals born during the flood season survived. The two scenarios will then be used to investigate whether there are changes that may reflect environmental occurrences such as varying flood levels and if they occur throughout dynastic Egypt.

As there are twelve archaeological samples discussed in this study, a dental developmental graph will not be constructed for each site to avoid redundancy. The sites of Amarna and Elephantine will thus be highlighted. Although both sites dental development charts are combined in Figure 3.6, they each will be discussed separately.

The site of Amarna was chosen to compare the two possible seasons of birth (i.e., summer versus winter). It represents a known husbandry regime, has the largest sample of teeth, and its enamel hypoplasia pattern is similar to the known modern sample.

The Amarna LEH distribution (Figure 3.6) shows some similarities to the modern material (Figure 3.5), with the exception of additional peaks on both the M_1 and M_2. Whereas the modern material displays only one peak on the M_1 between 3.5 and 6.5 mm, likely representing birth, the Amarna material displays two peaks – a large one at 2.5 mm, and a second smaller one at about 5.5 mm. The first larger peak is about two months after birth, most probably the result of weaning. The second smaller peak, parallel to that in the modern material falls at about the time of birth. Obviously, the pig population that displayed LEH was more affected by the stress of weaning as compared to birth.

The M_2 at Amarna also displays two peaks: a large one at 2.5 mm and a second smaller peak at about 7.5 mm. The first larger peak is about eight to nine months after birth, whereas the second smaller peak falls at about six to seven months after birth. This is where the issue of season of birth becomes important.

Figure 3.6 shows the Amarna material with the two birth scenarios of a summer and winter birth. If the pigs experienced a summer birth in April, the larger peak at about eight to nine months would fall at the time of winter, same as the modern material. The second possibility of a winter birth in November instead places this second peak at the time of the flood. Both scenarios are possible, as the flood and winter have the potential to cause stress for the animals. However, when considering the geographical context of the site on the desert edge, the likelihood that the flood is the cause of the peak becomes slim. The site is quite a long distance away from the floodplain, so the flood being the contributing factor becomes very small, especially in terms of animal displacement.

As the flood is known to affect food production by flooding the fields, it is conceivable that the flood could affect the animals through the food they consumed – or lack thereof. However, considering the intensive rearing regime that was in place at Amarna, the pigs experienced a combination of being fed, as evidenced by the stone troughs that were excavated in the pens, along with possible foraging at the nearby garbage heaps for organic refuse (Shaw 1984: 47). Thus, a lack of food as a result of the flood was less likely to have been a problem.

For these reasons, the possibility of winter being the causing factor of the LEH peaks is more plausible. This is further supported when considering that the temperatures can drop dramatically during winter months at this site (modern temperatures are known to drop as low as 49F/9.3°C in the winter), in addition to the decrease in food availability, as this is the planting season when food sources would have been low. The factors of food lack and temperature stress could therefore have resulted in the LEH peak. Another possibility is that these peaks could but do not necessarily represent the same individual. Although the double peaks on the M_2 could represent two different physiological events, another possibility is that they represent two different birthing seasons. As Amarna is known to have an intensive pig breeding, this is a strong possibility, especially when considering the fact that having pigs born more than once a year would create a greater turnover of pigs all year round, so supplies are not interrupted. Although there are some individual M_2's that display two occurrences of LEH, most display only one. This indicates at least two pig populations born at different times throughout the year.

The pattern seen on the M_3 displays only one large peak at 2.5 mm, about 21 to 22 months after birth. Following the summer birth scenario this also would fall during winter, whereas following the winter birth scenario, this would occur during flood season. Again, if the pigs were born in the winter, it is possible that the decrease in food that results from the flood is a possible cause. However, considering the intensive rearing regime in place at Amarna, and the significant distance of the site away from the flood plain, this proves to be a slim possibility. Rather, the dramatic decrease in temperature that occurs during the winter months combined with the possibility of even fewer food resources available further supports a spring season of birth.

Despite the fact that at Amarna, the site context seems to suggest a summer birth, the possibility still remains that pigs could have been born in the winter, and the flood could have a strong effect on the formation of LEH. To test this possibility, the site of Elephantine was chosen, as it is a riverside settlement at the southernmost border of Egypt that is known to have higher flood levels as compared to the other sites. Two data sets were available from Elephantine (Old and New Kingdom), and are also seen in Figure 3.6.

Both the Old and New Kingdom samples at Elephantine seem to have a peak on the M_1 at about 3.5 mm, about two months after birth, probably the result of weaning. Although both samples also show evidence of LEH as the result of birth, the Old Kingdom sample is slightly more affected by birth than the New Kingdom sample, shown in the slightly larger peak at about 6.5 mm. The M_2, however, has a bit more variation.

Looking at the distribution of LEH on the second molar, the Old Kingdom sample shows two discrete peaks: one at 2.5 mm at about nine months after birth, and a second smaller peak at 8.5 mm at about six months after birth. The New Kingdom sample only shows one peak at 3.5 mm at about eight months after birth. If these two populations were born in the summer, then the larger peak seen in both samples occurs during winter, with the second smaller peak seen on the second molar occurring during the flood. But why is there a second peak on the Old Kingdom sample and not on the New Kingdom sample?

Before the First Intermediate Period (c.2160 BC), the island of Elephantine would have split into two parts (Franke 2000: 465) during the annual flood (Baines & Malek 2002: 15). However, during the First Intermediate Period, the river level sank, and the swamp between the two islands was filled in, forming one larger island (Baines & Malek 2002: 15, Moeller 2005). If there were indeed a summer season of birth, then the flood covering the island, splitting it in two would explain the second smaller peak on the second molar as a result of stress from displacement. However, in the New Kingdom when the island was unified and not as drastically affected by the flood, the pigs would not have been as adversely affected, as this second peak on the M_2 disappears. This could not be tested on the M_3, as none were available for analysis from the Old Kingdom sample. However, if the pigs were born in the winter, then another possibility arises.

If the Elephantine pigs were born in November then the larger peak on the M_2 for both Old and New Kingdom populations occurs at the time of the flood. The second smaller peak on the Old Kingdom samples would then occur right at the beginning of summer, when Aswan can reach temperatures of over 120F/49°C), thus the extreme heat could be a factor for this second peak. However, aside from the fluctuating Nile flood levels, the general climate did not change that much between the Old and New Kingdom in Egypt. Thus, the possibility of a summer birth is more plausible than a winter birth, as if extreme heat is the cause of the second smaller peak, it should show up in the New Kingdom material as well.

Although there is no M_3 data for the Old Kingdom sample, the possibility of a summer birth fits in better with the LEH distribution seen in the New Kingdom Elephantine sample, as the larger peak at about 4.5 mm occurs during winter. If the pigs were born in the winter, then this larger peak would occur before the flood. As previously discussed with reference to the M_2's in the Amarna sample, these peaks could, but do not necessarily, represent the same individual. In the case of the Old Kingdom Elephantine material, all the second molars have LEH representing both peaks, so this does in fact represent two different physiological impacts which are likely the result of nutritional stresses from winter and the annual flood.

There are other causes of hypoplasia that could contribute to some of the peaks observed in the different pig samples such as infection, short-term nutritional deficiency, or even a sporadic metabolic dysfunction (Dobney et al. 2002: 40). Infections from soil parasites could have been rather common, especially during periods of increased Nile flood levels, such as the Middle Kingdom (c. 1870 BC) when long-term water logging of soil took place. The extra water would have increased soil parasites that were likely to have been consumed by animals along with endangering crops through rot and vermin (Butzer 2000: 550, 1984: 105, Seidlmayer 2001: 73–80, Said 1993: 145–6, Bell 1975: 238–45). This is an area that would require future study with larger sample sizes, particularly from the Middle Kingdom, as the one sample from South Abydos is not enough to see if high flood levels are the cause of LEH or not. The data would also need to be tested against environmental sampling of the soils and sediments from drill augering.

CONCLUSION

Overall the Nile water system and its resultant ecology had a profound impact on swine husbandry practices in ancient Egypt. The type of agriculture and degree of grain cultivation being practiced could very well determine the degree to which pig husbandry is practiced (Redding 1991). However, it is rather the environment, and to an extent the reach of the annual flood of the Nile, that also needs to be taken into consideration especially when determining the choice of swine husbandry regime. Riverside, floodplain, or marshy land settlement sites such as Elephantine, Mendes, and Tell el-Borg were located in wet, marshy areas ideal for pig access to wallows and food. However, such locations can result in pigs experiencing high degrees of physiological stress (and thus LEH) as a result of the annual inundation of the Nile, especially in times of high flood levels like those experienced in the Old Kingdom. These stresses can be minimized with more intense pig rearing such as penning, which does seem to occur by the New Kingdom as seen by a lower frequency of LEH at Elephantine, and certainly by the Ptolemaic-Roman period at Aswan.

Alternatively, desert and floodplain-edge settlements such as Amarna, Abydos Settlement Site, and Kom Firin were more likely involved in an intensive regime of penning due to the slightly drier environment as compared to settlements in closer proximity to the river. This is especially noted at Amarna, where not only have the actual pens been excavated, but a low frequency of LEH is observed.

This study also demonstrates that LEH is a common occurrence both in ancient and modern Egyptian pig samples, and that the occurrence of the condition can be explained by the same events within the animal's life (birth, weaning, winter starvation) as has been previously suggested for archaeological domestic pig samples (Ervynck & Dobney 1999, Dobney et al. 2002, Dobney et al. 2004), in additional to the seasonal stress of the annual flooding of the Nile. While most of the LEH frequencies are relatively high, sites that seem to have more intense pig rearing (i.e., penning) have a slightly lower frequency of LEH in comparison to sites that were more likely free-range. This high frequency of LEH seems to be indicative of the harsher, desert environment of Egypt. The lack of year round food resources in addition to the possible displacement that some animals may have experienced at sites such as Elephantine during the Old Kingdom may have been particular problems. The environmental conditions seem to be counteracted at sites where intensive rearing took place such as at Amarna, where LEH frequency is comparable to the modern pig sample. This suggests that Egyptian swineherds were aware of the effects of the environment on their pigs, and took measures to rear them more effectively. LEH thus proves to be a valuable tool for understanding both swine husbandry and the effects of seasonality, namely the annual flood of the Nile.

NOTES

1 The research in this chapter was made possible by the financial support of the United States Department of Education and Cultural Affairs and the American Research Center in Egypt. I would also like to thank the excavation directors who made their material available and kindly helped with the logistics of its recording. The directors include: Matthew D. Adams, Matthew J. Adams, James Hoffmeier, Salima Ikram, Barry Kemp, Mark Lehner, Cornelius von Pilgrim, Dietrich Raue, Richard Redding, Donald Redford, Janet Richards, Steven Snape, Neal Spencer, Penelope Wilson, and Josef Wegner. A special word of thanks is also extended to the Egyptian Supreme Council of Antiquities (current Ministry of State for Antiquities) for allowing access to all of the sites where data were collected.

REFERENCES

Baines, J., Malek J. 2002. *Atlas of Ancient Egypt*. Cairo: American University in Cairo Press.

Beja-Pereira, A. et al. 2004. 'African origins of the domestic donkey'. *Science* 304 (5676): 1781.

Bell, B. 1975. 'Climate and history of Egypt: the Middle Kingdom'. *American Journal of Archaeology* 79(3): 223–69.

Bertini, L. 2011. 'Changes in Suid and Caprine Husbandry Practices throughout Dynastic Egypt using Linear Enamel Hypoplasia (LEH)'. Unpublished PhD dissertation, Durham University.

Bertini, L. 2014. 'Faunal Remains at Kom Firin'. In Spencer, N. (ed.), *Kom Firin II: The Urban Fabric and Landscape*. London: The British Museum Press, pp. 306–11.

Bertini, L., Cruz-Rivera, E. 2014. 'The size of ancient Egyptian pigs. A biometrical analysis using molar width'. *Bioarchaeology of the Near East* 8: 83–107.

Bertini, L., Ikram, S. 2014. 'Faunal Analyses'. In Wilson, P., Gilbert, G., Tassie G. (eds), *Sais II: The Prehistoric Period at Sa el-Hagar*. London: Egypt Exploration Society, pp. 133–9.

Bertini, L., Linseele, V. 2011. 'Appendix 7: Faunal Report'. In Wilson P. (ed.), *Sais I: The Ramesside-Third Intermediate Period at Kom Rebwa*. London: Egypt Exploration Society, pp. 277–85.

Blench, R. 2000. 'A history of donkeys, wild asses, and mules in Africa'. In Blench, R., MacDonald M. (eds), *The Origins and Development of African Livestock: Archaeology, Genetics, Linguistics, and Ethnography*. London: Routledge, pp. 339–54.

Bradley, D.G., Magee, D.A. 2006. 'Genetics and the Origins of Domestic Cattle'. In Zeder, M.A., Bradley, D.G., Emshwiller, E., Smith, B.D. (eds), *Documenting Domestication New Genetic and Archaeological Paradigms*, Ewing: University of California Press, pp. 317–28.

Brewer, D.J., Redford, D.B., Redford, S. (1994). *Domestic Plants and Animals: The Egyptian Origins*, Warminister: Aris and Phillips.

Boessneck, J. 1976. *Tell el-Dab'a III' Die Tierknochenfunde 1966–1969*. Vienna: Österreichischen Akademie der Wissenschaften.

Boessneck, J., Driesch, von den A. 1989. 'Tierknochen-und Molluskenfunde'. In Brink, E.C.M ven den et al. (eds), *A Transitional Late Predynastic-Early Dynastic Settlement in the Northern Nile Delta, Egypt. Mitteilungen des Deutschen Archäologischen Institutes, Abteilung Kairo* 45: 94–102.

―――― 1990. 'Tierreste aus der vorgeschichtlichen Siedling El Omari bei Heluan/ Unterägypten'. In Debono, F., Mortensen, B. (eds), *El Omari. A Neolithic Settlement and Other Sites in the Vicinity of Wadi Hof*. Mainz: Philipp von Zabern, pp. 99–108.

―――― 1992. *Tell el-Daba VII*. Wein: Verlag der Österreichischen Akademie der Wissenschaften.

Butzer, K. 1976. *Early Hydraulic Civilization in Egypt. A Study in Cultural Ecology*. Chicago: University of Chicago Press.

―――― 1984. 'Long term Nile flood variation and political discontinuities in Pharaonic Egypt'. In Clark, J.D., Brandt, S.A. (eds), *From Hunters to Farmers:*

The Causes and Consequences of Food Production in Africa. Berkeley: University of California Press, pp. 102–12.

—— 2000. 'Nile'. In Redford, D.B. (ed.), *The Oxford Encyclopedia of Ancient Egypt*. Oxford: Oxford University Press, pp. 543–51.

Clark, J.D. 1971. 'A re-examination of the evidence for agricultural origins in the Nile valley', *Proceedings of the Prehistoric Society* 37: 34–79.

Cunningham, P. 2000. 'Genetics and origins of African cattle'. In Blench, R., MacDonald, K. (eds), *The Origins and Development of African Livestock: Archaeology, Genetics, Linguistics, and Ethnography*. London: Routledge, pp. 240–3.

Diener, P. et al. 1978. 'Ecology, evolution, and the search for cultural origins: the question of Islamic pig prohibition'. *Current Anthropology* 19: 493–540.

Dobney, K., Ervynck, A. 1998. 'A protocol for recording linear enamel hypoplasia on archaeological pig teeth'. *International Journal of Osteoarchaeology* 8: 263–73.

—— 2000. 'Interpreting developmental stress in archaeological pigs: the chronology of linear enamel hypoplasia'. *Journal of Archaeological Science* 27: 597–607.

Dobney, K., Ervynck, A., La Ferla, B. 2002. 'Assessment and further development of the recording and interpretation of linear enamel hypoplasia in archaeological pig populations', *Environmental Archaeology* 7: 35–46.

Dobney, K., Ervynck, A. Albarella, U., Rowley-Conwey, P. 2004. 'Assessment and further development of the recording and interpretation of linear enamel hypoplasia in archaeological pig populations', *Environmental Archaeology Journal of Zoololology* 264: 197–208.

Driesch, von den A. 1997. 'Tierreste aus Buto im Nildelta', *Archaeofauna* 6: 23–39.

Driesch, von den A., Boessneck, J. 1985. *Die Tierknochenfunde aus der neolithischen Siedlung von Merimde Benisalâme am westlichen Nildelta*. Munich: Institut für Paläoanatomie, Domestikationsforschung und Geschichte der Tiermedizin der Universität München und Deutsches Archäologisches Institut Abteilung Kairo.

Driscoll, C.A., et al. 2007. 'The Near Eastern origin of cat domestication', *Science* 317(5837): 519–23.

Dunne, J. et al. 2012. 'First dairying in green Saharan Africa in the fifth millennium BC', *Nature* 486: 390–4.

Epstein, H. 1971. *The Origins of the Domestic Animals of Africa*. New York: Africana Publishing Corporation.

Ervynck, A., Dobney, K. 1999. 'Lining up on the M_1: a tooth defect as a bio-indicator for environment and husbandry in ancient pigs', *Environmental Archaeology* 4: 1–8.

—— 2002. 'A Pig for all seasons? approaches to the assessment of second farrowing in archaeological pig populations'. *Archaeofauna* 1: 7–22.

Ervynck, A., Dobney, K., Hongo, H., Meadow, R. 2001. 'Born free? new evidence of the status of pigs at Neolithic çayönü Tepesi Southeastern Anatolia, Turkey'. *Paléorient* 27: 47–73.

Franke, D. 2000. 'Elephantine'. In Redford, D.B. (ed.), *The Oxford Encyclopedia of Ancient Egypt*. Oxford: Oxford University Press, pp. 465–7.

Gautier, A. (1976). 'Freshwater mollusks and mammals from Upper Paleolithic sites near Idfu and Isna'. In Wendorf, F., Schild, R. (eds), *Prehistory of the Nile valley*. New York: Academic Press, pp. 349–61.

Gautier A., Van Neer W. (1989). 'Animal remains from the Late Paleolithic sequence at Wadi Kubbaniya'. In Wendorf, F., Schild, R., Close, A.E. (eds), *The Prehistory of Wadi Kubbaniya*. Dallas: Southern Methodist University Press, pp. 119–61.

Germond, P., Livet, J. 2001. *An Egyptian Bestiary*. London: Thames and Hudson.

Ghoneim, W. 1977. *Die Okonomische Bedeutung des Rindes im Alten Agypten*. Bonn: Rudolf Habelt.

Goodman, A.H., Rose, J.C. 1990. 'Assessment of systemic physiological perturbations from dental enamel hypoplasias and associated histological structures', *Yearbook of Physical Anthropology* 33: 59–110.

Grant, A. 1982. 'The use of tooth wear as a guide to the age of domestic ungulates'. In Wilson, B., Grigson, C., Payne, S. (eds), *Ageing and Sexing Animal Bones from Archaeological Sites*. Oxford: BAR, pp. 91–108.

Harris, M. 1974. *Cows, Pigs, Wars, and Witches. The Riddles of Culture*. New York: Vintage Press.

Hecker, H. 1982. 'A Zooarchaeological Inquiry into Pork Consumption in Egypt From Prehistoric to New Kingdom Times', *Journal of the American Research Center in Egypt*, 19, pp. 59–71.

—— 1984. 'Preliminary report on the faunal remains from the workmen's village'. In Kemp, B. (ed.), *Amarna Reports I*. London: Egypt Exploration Society, pp. 154–64.

Houlihan, P.F. 1996. *The Animal World of the Pharaohs*. Cairo: American University in Cairo Press.

Ikram, S. 1995. *Choice Cuts: Meat Production in Ancient Egypt*. Leuven: Uitgeverij Peeters en Department Oosterse Studies.

Kemp, B. 1991. *Ancient Egypt: Anatomy of a Civilization*. London: Routledge.

Linseele V., Van Neer W. 2009. 'Exploitation of desert and other wild game in ancient Egypt: The archaeozoological evidence from the Nile valley'. In Riemer, H., Förster, F., Herb, M., Pöllath, N. (eds), *Desert Animals in the Eastern Sahara: Status, Economic Significance, and Cultural Reflection in Antiquity: Proceedings of an Interdisciplinary ACACIA Workshop held at the University of Cologne December 14–15, 2007*. Köln: Heinrich Barth Institut, pp. 47–78.

Linseele, V., Van Neer, W., Hendrickx, S. 2007. 'Evidence for early cat taming in Egypt'. *Journal of Archaeological Science* 34: 2081–90.

—— 2008. 'Early cat taming in Egypt: a correction'. *Journal of Archaeological Science* 35: 2672–3.

Lobban, R. 1998. 'Pigs in Ancient Egypt'. In Nelson, S. (ed.) *Ancestors for the Pigs: Pigs in Prehistory*. Philadelphia: MASCA Research Papers in Science and Archaeology, pp. 137–48.

MacDonald, K. 2000. 'The origins of African livestock: indigenous or imported?'. In Blench, R., MacDonald, K. (eds), *The Origins and Development of African Livestock: Archaeology, Genetics, Linguistics, and Ethnography*. London: Routledge, pp. 2–17.

Marshall, F., Hildebrand, E. 2002. 'Cattle before crops: the beginnings of food production in Africa'. *Journal of World Prehistory* 16(2): 99–143.

Reed, C.A. 1969. 'The pattern of animal domestication in the Prehistoric Near East'. In Ucko, P.J., Dimbleby, G.W. (eds), *The Domestication and*

Exploitation of Plants and Animals. London: Gerald Duckworth and Co. Ltd, pp. 361–80.

McCance, R.A., Ford, E.H.R., Brown, W.A.B. 1961. 'Severe under nutrition in growing and adult animals. Development of the skull, jaw, and teeth in pigs'. *British Journal of Nutrition* 15: 213–24.

McCosker, L. 2011. *Free Range Pig Farming-Starting Out*. Raleigh: Lulu Enterprises.

Miller, R.L. 1990. 'Hogs and hygiene'. *Journal of Egyptian Archaeology* 76: 125–40.

Moeller, N. 2005. 'The first intermediate period: a time of famine and climate change?', *Ägypten und Levante* 15: 153–67.

Redding, R.W. 1991. 'The role of the pig in the subsistence system of ancient Egypt: a parable on the potential of faunal data'. In Crabtree, P., Ryan, K. (eds), *Animal Use and Culture Change*. Philadelphia: MASCA Research Papers in Science and Archaeology, pp. 20–38.

——— 2010. 'Status and diet at the Workers' Town, Giza, Egypt'. In Campana, D., Crabtree, P., deFrance S.D., Lev-Tov J., Choyke A. (eds), *Anthropological Approaches to Zooarchaeology. Complexity, Colonialism, and Animal Transformations*. Oxford: Oxbow Books, pp. 65–75.

Rossel, S. 2007. 'The Development of Productive Subsistence Economies in the Nile valley: Zooarchaeological Analysis at El-Mahâsna and South Abydos, Upper Egypt'. PhD dissertation, Harvard University, Ann Arbor, MI: UMI Microfilms.

Rossel, S. et al. 2008. 'Domestication of the donkey: timing, process, and indicators', *PNAS*: 105(10): 3715–20.

Said, R. 1993. *The River Nile. Geology, Hydrology, and Utilization*. New York: Pergamon Press.

Seidlmayer, S.J. 2001. 'Historische und moderne Nilstände. Untersuchungen zu den Pegelablesungen des Nils von der Frühzeit bis in die Gegenwart'. *Bulletin de la Société de Géographie de l'Egypte* 2: 71–101.

Shaw, I.M.E. 1984. 'The animal pens (Building 400)'. In Kemp, B. (ed.), *Amarna Reports I*. London: Egypt Exploration Society, pp. 40–59.

Upex, B., Dobney, K. 2012. 'Dental enamel hypoplasia as indicators of seasonal environmental and physiological impacts in modern sheep populations: a model for interpreting the zooarchaeological record'. *Journal of Zoology* 287: 259–68.

Van Neer W. 2000. 'Faunal remains from Shuwikhat 1'. In Vermeersch, P.M. (ed.), *Paleolithic Living Sites in Upper and Middle Egypt*. Leuven: Leuven University Press, pp. 153–4.

Wenke, R.J. et al. 1988. 'Kom El-Hisn: excavation of an Old Kingdom settlement in the Egyptian Delta', *Journal of the American Research Center in Egypt* 25: 5–34.

Wetterstrom, W. 1995. 'Foraging and farming in Egypt: the transition from hunting and gathering to horticulture in the Nile valley'. In Shaw, T. Sinclair, P., Andah, B., Okpoko, A. (eds), *The Archaeology of Africa. Food, metals, and towns*. London: Routledge, pp. 165–226.

Wilcocks, W., Craig, J.I. 1913. *Egyptian Irrigation*, London: E. and F.N. Spon, Ltd.

Zeuner, F.E. 1963. *A History of Domesticated Animals*, London: Hutchinson of London.

4 A Breadbasket, *Mais Encore*? The Socio-Economics of Food Production in the Nile Delta from Antiquity Onwards

Katherine Blouin

Now, indeed, there are no men, neither in the rest of Egypt, nor in the whole world, who gain from the soil with so little labour; they have not the toil of breaking up the land with the plough, nor of hoeing, nor of any other work which other men do to get them a crop; the river rises of itself, waters the fields, and then sinks back again; thereupon each man sows his fields and sends swine into it to tread down the seed, and waits for the harvest; then he makes the swine to thresh his grain, and so garners it.

Herodotus on the Nile Delta, *Histories* II, 14, Godley 1920 transl.

INTRODUCTION

By focusing on the socio-economics of food production in an area of the Roman Nile Delta called the Mendesian Nome, this chapter asks the following question: To what extent does the image of 'breadbasket'[1] apply to the ancient as well as to modern Nile Delta? Such a question might sound surprising if we take today's Egypt as a vantage point. As a matter of fact, since the country became largely dependent on food imports in the 1970s, its agricultural sector has been facing increasing challenges. In addition to the multifaceted negative effects that accompanied the national reforms encouraged by – and in many ways beneficial to – the USAID, the World Bank, and the IMF, Egyptian farmers have had to cope with the unwanted effects of the High Aswan Dam (i.e., the fact that the rich sediments carried by the flood waters are now trapped by the dam, forcing peasants to resort to chemical fertilizers and exposing the Nile Delta to increasing littoral erosion and salinization) (see notably on the matter Ayeb 2010, Ayeb & Bush 2014, Bethemont 2003, Bush & Sabri 2000, Mitchell 1991). Compounded with the gradual rise of the Mediterranean sea-level brought about by global warming, which contributes further to the 'destruction phase'[2] the Nile Delta has entered, and to the rise of tensions between Egypt and its upstream neighbours over the sharing of the Nile waters, the image one gets of today's Egypt is more often than not that of an unstable and overpopulated country whose powerless population faces the imminent threat of severe water and food shortages.[3] While the issues faced by many Egyptians are real, as T. Mitchell already showed in 1991, such alarmist images are a convenient way for many western organs to evacuate from the discussion the role played by global markets, neo-imperial dynamics, and local alimentary risk management strategies:

> The image of a vast, overbreeding population packed within a limited agricultural area is therefore quite misleading. Egypt's food problem is the result not of too many people occupying too little land, but of the power of a certain part of that population, supported by the prevailing domestic and international regime, to shift the country's resources from staple foods to more expensive items of consumption (Mitchell 1991).

From a historical perspective, Egypt's current vicissitudes and the representations associated with it stand in sharp contrast with the reputation the country had in earlier periods. Indeed, from Antiquity to modern times, it was celebrated as one of the most fertile and productive agricultural zones of the Mediterranean, as testifies Herodotus' passage quoted at the beginning of this chapter. Although one could reproach to the fifth-century BC Greek author his propensity to simplify Egypt's agricultural life – notably by taking away all merit to the work and empirical knowledge of the Egyptians themselves – the fact is, Egypt's agricultural yield did indeed generate surpluses for most of its history. Egyptian grain exports – in the form of tributes, taxes, and private stocks – are already attested in the fourth century BC,[4] and this trade grew in importance after the Macedonian conquest (Buraselis 2013, Rathbone 1983). In Roman times, in addition to supplying Alexandria and many cities of the eastern Mediterranean in grain, the province played – together with North Africa – a central role in the grain supply of Rome[5] and, starting from the fourth century AD, of Constantinople (Sirks 1991). Overall, the successive Arab and Ottoman conquests seem to have led to comparable patterns of public and private redistribution within imperial redistribution networks (see Mikhail 2011 for Ottoman Egypt, Udovitch 1999 for eleventh-century AD Egypt, Sijpesteijn 2014 for early-Arab Egypt.) It is not to say that everything was perfect. Indeed, beyond the sheer benefits associated with the Nile flood, bad inundations, changes in political regimes, demanding agro-fiscal policies, and corruption and abuses did give rise to tensions (over-indebtedness, endemic movements of fugitive peasants) and, at times, crises (fiscal revolts, poor harvests and food shortages). Yet a look at the general features of agricultural production Egypt in the *longue durée* allows us to identify two broad socio-economic continuities: the importance of local patterns of water management and farming that relied on the empirical knowledge and work of farmers, and the will of all successive political powers to manage an integrated, State-supervised system of taxation, storage, and transportation facilities.[6]

In the absence of any specific quantitative data, it is generally assumed that, because of its size (*c.*26,000 km^2), which is far greater than that of the Nile valley (*c.*9,900 km^2),[7] the agricultural land of the Nile Delta was the most productive of Egypt and, therefore, the source of most of its exports. Moreover, the ability of the Delta's – and more generally of Egypt's – agricultural production to generate important grain surpluses is the source of a series of Orientalizing *lieux communs* that have survived to this day: One might think of Herodotus' 'Gift of the Nile' (which, like the quote above, applies in fact to the Delta), as well as of formulas such as 'hydraulic civilization',[8] 'eternal Egypt'(Mitchell 1991), or, and this is the one this chapter focuses on, 'breadbasket of the Mediterranean'.[9] Although based on some truths, these *topoi* offer a monolithic, dehumanized, and romantic vision of Egypt that, just like today's dramatic images denounced by Mitchell, evacuates spatio-temporal specificities and brushes aside the complex interplay of local, regional, and international socio-economic and

Figure 4.1. Partially irrigated agricultural landscape in the vicinity of ancient Thmuis (© Katherine Blouin).

political interests at play.[10] Besides the fact that the Delta was a fertile, productive agricultural region where a great deal of Egypt's grain came from, what *exactly* do we know about the *specifics* of agricultural production in the different areas of this ecologically complex region from antiquity onwards (types of food produced, relative importance of each activity/crop, regional and microecological peculiarities, relationships between domestic, communal, State, and other economic agents' interests)? To answer this question, we need to submit the 'breadbasket' image to the test of locally based evidence. Starting from an analysis of administrative papyri related to the Roman Mendesian Nome, I will show what these documents tell us regarding local practices of agricultural diversification in that region in the second century AD. I will then turn my attention to a fourth-century cadastral register to reflect on the relationship between diversification and cash crop-oriented specialization. But before-hand, it is necessary to dedicate a few words to the primary sources and historico-environmental context this paper focuses on.

HISTORIANS AND THE NILE DELTA: SOURCES AND CONTEXT

What food was produced in the Nile Delta? In what quantity? And for whom? The most promising way for us to address these questions is to adopt a case study approach. For as mentioned in the introduction, apart from the general deduction based on the comparative surface of arable land in the Nile Delta and Valley, no data allows us to quantify and qualify the overall agricultural productivity of ancient Lower Egypt (neither *per se* nor compared to the Nile valley and Fayum Oasis), nor to establish the exact proportion that was dedicated to exports. Other variables to consider – for instance demography, domestic, rural, and urban food needs, local hydric conditions and irrigation systems, their maintenance, as well as agricultural storage and transportation facilities – are better documented for Egypt in general,[11] but the data are sparse and, accordingly, any generalization stemming from them provides us at best with general overviews and coarse orders of possibility. The same goes, as far as I know, for the Arab and Ottoman periods (see Cuno 1999, Udovitch 1999, who both stress regional differences).

Another setback to take into account when it comes to the early history of the Nile Delta is that of the evidence. Indeed, the damp climatic conditions and high population density that prevail in the region are detrimental to the conservation of ancient remains. For this reason, scholars have traditionally neglected the region, focusing instead on the better documented – and often more spectacular – sites and material of the Nile valley and Fayum oasis. This led to an unfortunate historiographical paradox, whereby our limited knowledge of the region's history does not match its socio-economic, political, and agrarian importance, to which Egypt's founding myths and

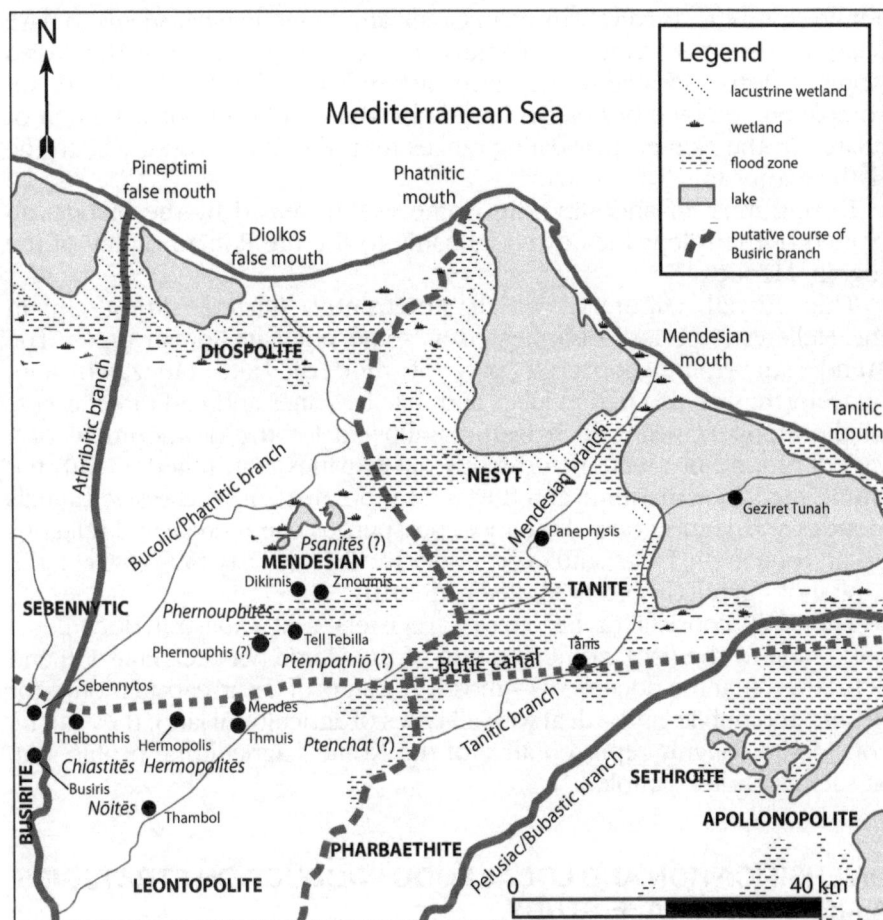

Figure 4.2.   Map of the Mendesian Nome in the Roman period (© Katherine Blouin).

Egyptian, Greek, and Latin sources testify (Watrin 2003, Yoyotte 1958, Zivie 1975: 15). While the situation has been slowly improving since the 1980s,[12] some major discrepancies still remain. It is particularly the case with papyri. In comparison with Upper Egypt, the Fayum, and oasian/desertic sites, where tens of thousands of papyri and ostraca (texts written on stone or sherds) have been found, we still possess very few papyri from the Nile Delta,[13] and absolutely none were found in Alexandria. Most of them have, alas for us, decomposed in the region's mostly rich, humid soil. This is not to say that no data is available. As I hope to show in this chapter, one of the three groups of papyri so far found in the Delta – the 'carbonized archives from Thmuis' (CAT) – give us very precious information regarding agricultural productivity in the region under the Roman Principate. These second-century AD administrative texts were written in Greek and found in what seems to have been the district's archives of Thmuis, the Roman capital of the Mendesian Nome, a district located in the north-eastern Nile Delta. These archives are complemented by other papyri found elsewhere in Egypt but written in or related to the Nome; their dating ranges from the fourth century BC to the sixth century AD.

During most of antiquity, the Nome was traversed by the Mendesian branch of the Nile, which owed its name to the pre-Roman capital of the region, Mendes.[14]

This fluvial tributary, which apparently started silting up in the Hellenistic period, disappeared at some point in Roman times. The Mendesian Nome bordered part of modern Lake Menzaleh and, henceforth, was also rich in lakes and marshy zones and had direct access to the Mediterranean. Such features allowed for the development of a variety of food production and industrial activities that, together with the Nome's strategic maritime and fluvial location, made its successive capitals Mendes and Thmuis one of the most prosperous commercial and religious urban zone in the Delta, and even, under the 29th dynasty (399–380 BC), the capital of Egypt (Grimal 1988, Redford 2010).

The CAT consists of tax arrears registers and cadastral documents pertaining to the fiscal administration of the Nome between AD *c.* 150 and 200. Given that the documents – mostly registers of tax arrears in money for private and public land – deal with all types of agricultural land, they can be considered roughly representative of the Nome's agricultural profile and, as such, a choice sample.

DIVERSIFICATION AND LOCAL FOOD PRODUCTION STRATEGIES: A MENDESIAN CASE STUDY

Generally speaking, papyri from Hellenistic and Roman Egypt show how agricultural diversification remained a central strategy throughout this period despite the prevalence of cereal culture. As regional studies focusing on other areas have revealed, food production was characterized by a prevalence of wheat land, followed by a variety of production (barley and other cereal, legumes, fodder, fruits and vegetables, poultry, cattle, game, fish, etc.) (Rathbone 1991 (Fayum), Sharp 1999 (Fayum), Rowlandson 1996 (Oxyrhynchite), Bagnall 1997 (Kellis), Bousquet 1996 (Douch). See also Schnebel 1925 (Ptolemaic period)). The relative importance of each activity changed according to local contexts and resulted from the conjunction of multi-scaled interests like the quests for subsistence, autarky, and profit (Rathbone 1991: 212–64, Rowlandson 1996: 19–26). What do the CAT tell us regarding the Roman Nile Delta? Both similarities and peculiarities with other Egyptian regions can be observed.

First, just as in Roman Egypt in general, the Mendesian Nome's agricultural territory was divided into three main 'functional' groups of land categories: grain land, vineyards and gardens, and pastures (on sheep and goat pastures in Roman Egypt, see Langellotti 2012). In the CAT, the prevalence of grain-land categories over vineyards (no gardens are specifically attested) and pastures is striking. Of 42 different land categories preserved, 35 correspond to grain land (83 per cent), four to vineyards (10 per cent), and three to pastures (7 per cent). The discrepancy is such that the quantitative preponderance of grain-land over gardens and pastures on the Nome's agrarian territory seems unequivocal.

This phenomenon, which agrees with Egyptian evidence from other regions and periods,[15] is beautifully complemented by fiscal terminology. Indeed, because of their specific and often descriptive nature, the names of taxes preserved in the CAT document a wider variety of food production activities. Mendesian papyri refer to more than a hundred taxes and fees, mostly in money, but also in kind, many of which deal with agricultural land or activities. These pertain essentially to specific types of agricultural land (*P. Thmouis* 1= dry: *P. Ryl.* II 216= vineyards, orchards, and gardens). Consequently, some activities (cattle breeding, domestic gardening, fishing) that were certainly much more widespread than the papyri show are barely or not represented. Work organization within villages and practices such as fallow intervals, crop rotation, and mixed cropping are not documented either, and there is no reference to some crops attested in third-century BC papyri found in the Fayum but mentioning agricultural products from the Nome (i.e., papyrus, lotus, and sesame). Finally, the territorial subdivisions of the Nome known as 'toparchies' are very unevenly represented. Yet, as Table 4.1 shows, the richness of the terminology gives unique insights into the agricultural landscape of this portion of the Delta under Roman rule.

The CAT mention nine different taxes on grain levied in twelve of the Nome's 15, maybe 16 known toparchies. In reality, it was most probably practised in all of them. This seems all the more probable since wheat farming is attested on plots whose name ('limnitic', from the Greek *limnē*, 'marsh') indicates that they were originally damp or submerged. Given that wheat does not tolerate damp soil, these parcels must have been subject to a geomorphologically or humanly induced drainage, or both (on limnitic land, see Blouin 2012b.) Taxes on barley, bean, and lentil farming are also documented. Until the Hellenistic period, when it was superseded by free-threshing durum wheat, barley was, with emmer, the main cereal grown and consumed (in the form of food or beer) in Egypt (Murray 2000a, Samuel 2000).[16] Since this cereal is more tolerant to drought and salinity than wheat, it appears to have been privileged on dryer or more saline soils. Legumes are rich in proteins and in nutrients that are lacking in cereals, more tolerant to drought than wheat (Bousquet 1996: 250), and movable over long distances. As such, they complement cereals and are the perfect

Table 4.1.   Diversification in the Mendesian Nome according to papyri.

Sector	Activity (only attested in Ptolemaic documents)
Agriculture	Grain farming (wheat and barley)
	Bean farming
	Lentil farming
	Viticulture
	Gardens (general)
	Olive growing
	Flax farming
	Castor-oil tree (*croton*) farming
	Reeds farming
	Sesame farming
Husbandry	Livestock breeding (general)
	Pig breeding
	Goat and sheep breeding
	Goat breeding (religious context)
	Calf breeding (religious context)
	Donkey breeding
	Poultry breeding (general)
	Geese and hen breeding
	Pigeon breeding
Hunting and fishing	Hunting and fishing
Picking?	*Lotus picking and/or farming*

substitute for meat, which is much more expensive to produce and acquire. No wonder, then, that they were – and still are – a key staple in ancient Mediterranean diets (Garnsey 1998: 214–25, Murray 2000b: 637–42, Garnsey 1999: 12–21).

Numerous taxes document the cultivation of fruits and vegetables in the Nome. Unfortunately, apart from those on vineyards and olive tree orchards, they are general in nature, so that we do not know which specific crops were grown on the plots in question. This could result from the fact that fruits and vegetables were mostly produced on a small scale and through mixed cropping for local and domestic consumption (Bagnall 1993: 25–7, Crawford 1971: 130–1, Garnsey 1996, Horden & Purcell 2000: 203, Murray 2000b, Rathbone 1991: 381, Scheidel 2001: 237, Thanheiser 1992: 118). Garden products were nevertheless a great way to vary one's diet, and an interesting investment for farmers. The close ties that universally link rural peripheries and urban centres no doubt explain the great profitability of these products.[17]

Vineyards are attested in at least nine toparchies.[18] Vines require very well drained soils. Given the annual flood of the Nile, they had to be planted above the major riverbed and artificially irrigated (Murray, Boulton & Heron 2000: 582–3). Like other fruit plantations,[19] vineyards were hence essentially located outside the areas suited for grain cultivation (on higher ground, along shores, on dykes) and required considerable care and expertise. The time and effort invested by wine growers was however compensated by the generous profits that this commercial activity could generate (Brun 2004: 144, Rathbone 1991: 212–3, Schnebel 1925: 239–92).[20]

Tax records also document the farming of olive trees, ricin and flax. Only two taxes on olive oil production are mentioned in Mendesian papyri; such meagre data can be seen as another illustration of the more general marginality of olive growing in Egypt. For olive trees grow on well-drained soil and, consequently, are not suited for the floodable land of the Nile valley and Delta. Accordingly, olive cultivation in Egypt, whose origins remain uncertain, could only take place in oases and above the major riverbed, like viticulture (Serpico & White 2000: 398–401).[21] Apart from towns and villages themselves, such land was scarce in the Mendesian Nome, as in the Delta as a whole (except for the desert margins). The oil obtained from the fruit of the castor-oil plant, known as *kiki* in Egyptian and *krotōn* in Greek, was toxic and, for this reason, reserved for lighting (Pliny the Elder, *HN* XV, 7). Flax – essentially farmed for textile purposes – remained, throughout Antiquity and the Medieval period, a major cash crop in Egypt, especially so in the northern Delta (Blouin 2012c, Udovitch 1999). Finally, third-century BC papyri document the export of sesame seeds from Mendes to the Fayum; this seems to imply that sesame was produced in the Nome at the time. The silence of Roman papyri could result from their fragmentary nature, or from the inclusion of sesame land into general taxes on fruit and vegetable growing.[22]

The very numerous references to taxes on husbandry indicate that this activity was conspicuous in the Nome, and that it consisted of a combination of pigs, sheep, goats, and poultry (See *P.Ryl.* II 213, intro., Wallace 1969 (1938). See also *PSI* III 233, which seems to record the confiscation and auction of non-registered sheep.) Poultry breeding is attested in about ten toparchies, and we possess a specific reference to a tax on pigeon levied in one. Porcine and caprine breeding are documented in three toparchies, and most probably more if we take into account the 'general' taxes. In reality though, it seems almost certain that husbandry was practiced, though in different proportion and intensity, throughout the Nome, at least in a domestic context (Rowlandson 1996: 22). Interestingly, apart from a religious tax and, probably too, general taxes, no tax deals specifically with cattle breeding (Wallace 1969 (1938): 242). Yet, literary sources on the *Boukoloi* show that this activity was widespread in the northern Delta, and notably in the Lake Menzaleh area; it is still the case today.[23] Finally, several occurrences of payments for pasture royalties indicate the presence of public grazing land, and we note two mentions of royalties on donkeys; these animals were (and are still) largely used in Egypt for the transportation of goods and people (Bagnall 1985: 1–6).

Figure 4.3. Vegetable growing and ploughing of an irrigated plot at the north-eastern edge of ancient Thmuis (© Katherine Blouin).

To sum up, according to the CAT, the Mendesian rural economy was characterized by a pre-eminence of cereal culture – mostly wheat, but also barley – accompanied by other crops (legumes, fruits and vegetables, including viticulture and oleaginous plants, fodder) and husbandry. In any case, the prevalence of wheat land agrees with data from elsewhere in Egypt (Bagnall 1993: 24–5, Rathbone 1991: 213–4, Rowlandson 1996: 19), and seems to have been directly linked to the high demand for Egyptian wheat within imperial, private and public networks of grain supply (see notably Erdkamp 2005). To that effect, the Nome's land seems, indeed, to fit with the idea of the Delta being a breadbasket. Yet there was more to the Mendesian foodbasket than bread.

In particular, certain practices typically associated with damp environments seem to have played a major economic role in the Nome, and one might suspect that it was the same in other areas of the northern Delta. This is the case with flax (and possibly papyrus) farming, water plant picking, cattle breeding and, more generally, pastoralism. Hunting, fishing, and fish farming also seem to have been widespread, if not predominant, in three toparchies located in the vicinity of the Menzaleh Lake, and these activities were taxed by the State. This is what three pieces of information contained in the CAT show: mentions of renting offers for submerged land for hunting, fishing, and fish farming; the collection of taxes named after a category of originally wet land (*limnitika*); the presence of a community of fishermen in a village located in the area of Lake Menzaleh (*P. Thmouis* 1, 82, 10–91, 12,

Figure 4.4.   Sheep grazing on a clover parcel the day after harvest in the vicinity of ancient Thmuis (© Katherine Blouin).

115, 21–116, 18 and ref.). More generally, literary sources testify to the presence of communities of fishermen and herders in the vast lacustral areas that border the Delta's littoral (Blouin 2011 & 2014, Chapters 7 and 9). Such communities survived to this day, although they are facing increasing environmental and economic pressures (Bush & Sabri 2000, Henein 2010). Overall, the Mendesian data nuances the typicality of wheat predominance in Mendesian agro-fiscal data on a micro-regional level and poses the question of the relative importance of each food production activity within the Nome. On this matter, another Mendesian document proves particularly helpful.

BETWEEN DIVERSIFICATION AND SPECIALIZATION

Our only source for Mendesian land typology in late antiquity is *P. Oxy.* XLIV 3205 (Świderek 1971). This papyrus, which was found in Oxyrhynchos but originally written in the Mendesian Nome between AD 297–301 and 308, is one of the only two known sowing registers in Egypt (the only other register of this type from the same period is *P. Ryl.* IV 655 (origin unknown), which is very mutilated (Świderek 1971: 32)) that date from the fourth century AD.

Figure 4.5. Fish farming area, Izbat Burj Rashid, close to the Rosetta mouth of the Nile (© Katherine Blouin).

As such, it is a crucial source of information on the agro-fiscal reforms implemented in the province under the emperor Diocletian in AD 297 (ibid.: 31–2).[24] The register is made of two sections. The first is a land register of a toparchy called Phernouphitēs; the second, a topographical register of a village from that toparchy whose name starts with Psen- (the rest of the name is lost). The information compiled in the document was most probably collected during the general land *census* of AD 297–301, on the basis of land declarations made by landowners under the supervision of an official (*P. Oxy.* XLIV 3205, 3; see Świderek 1971: 31).

Interestingly for us, *P. Oxy.* XLIV 3205 includes the official (i.e., we may assume, optimistic) proportion of land dedicated to the main types of agricultural productions in the toparchy at the time. Although grazing land and areas open for hunting and fishing are not mentioned, the document allows us to grasp the unequivocal predominance of grain cultivation on the Phernouphite landscape.

The new agrarian typology implemented under Diocletian documents the persistent prevalence of grain cultivation in the Nome in the early Dominate. In addition to grain land, the toparchy's agricultural territory also included a small amount of vineyards, gardens, as well as bean and, probably too, reed land (see *P. Ryl.* II 427, fr. 19). When sorting the cadastral data converted into percentages by type of culture, the preponderance of grain farming appears even more striking. According to the first part of the register, c.77 per cent of the toparchy's agrarian territory was dedicated to cereal cultivation. Viticulture occupied c.17.5 per cent of the land, the rest corresponding to sandy and bushy land, and very small areas of orchards, beans, and reed land. Interestingly, c.81 per cent of the toparchy's orchards were dry, and this percentage rises to c.85 per cent in the case of vineyards. This phenomenon most probably results from both the constraints inherent to vine and olive cultivation and the priority given to wheat on fertile parcels. In fact, when looking only at the productive land, the proportion of grain land climbs to c.95.5 per cent; vineyards, as well as orchards and bean land, follow far behind, occupying respectively c.3.5 per cent and 0.5 per cent of the land. In spite of the presumed optimism of these official statistics, the discrepancy is such that no ambiguity persists as to the territorial dominance of wheat cultivation.

The second section of the document, which deals with the agricultural land of the village of Psen-, lists even more polarized totals. Indeed, c.94 per cent of the village's tillable land was grain land. If we include dry land, the total rises to c.98 per cent, the remaining c.2 per cent including orchards and bean land. This hybrid category attests to the generality of the Greek term for garden, *paradeisos*, in the context of this document, and allows us to suspect that in addition to fruit trees and beans, other crops (vegetables, other legumes, flowers) were grown, most probably through mixed cropping procedures.

As for vineyards, two categories present in the toparchic survey are absent from the survey of Psen-. Does this mean that no viticulture was practised in the village? This is possible, and could result from the fact that Psen-'s territory was not suited for this activity. The comparison between the composition of dry land in Psen- and in the toparchy as a whole reinforces this hypothesis.[25] These statistics illustrate how agrarian choices relied on an opportunistic and adaptive relationship to local agrarian potentialities.

The persistence of beans and orchards on less than 2 per cent of Psen-'s territory matches the toparchic data. Overall, the very small amount of land dedicated to these products should not keep us from acknowledging their essential role for local dietary supply, as well as their profitability as cash crops. It could also be interpreted as a sign of the essentially local, small scale (including domestic), but potentially very intensive nature of gardening.

CONCLUSION

The agrarian typology and the fiscal terminology preserved in the Mendesian papyri show that the overwhelming part of the

Nome's territory was dedicated to cereal, and especially wheat, cultivation. Yet as Mendesian data show, in addition to grain, one notes the omnipresence – in areas not suited for grain farming – of a variety of complementary activities: gardening (including viticulture and some olive tree growing), legume farming, as well as cattle and poultry breeding and the exploitation of wetlands for flax farming, hunting, fishing, fish farming and picking. Just as the quantitative prevalence of grain farming, diversification, too, appears as an *essential* part of the Nome's agricultural economy, which both enhanced the protection of landholders and tenants from alimentary risk and allowed for a maximized exploitation of the complex, diverse, and complementary local environments, including agriculturally marginal land.

How much initiative and decisional power was in the hands of the different economic agents is difficult to know with precision. What the evidence shows, however, is that agrarian choices involved a series of environmental, socio-economic, and fiscal factors: the potential and the limits of the land itself (which was largely dependent on its proper access to Nile water); the relationship between costs and potential yield regarding the farming of a plot and the types of activities practiced on it; market dynamics, and fiscal constraints (proceeding from the state's power to measure, categorize, and tax the land and its yields). If the very modes of land categorization and taxation, as well as the incentives and coercive measures aimed at the cultivation of marginal plots show a certain degree of state control beyond the realm of public land (see Blouin 2014: Chapter 5), the decisional role of landholders, whose interests must often have been similar to those of the fisc, and the empirical *savoir-faire* of the farmers themselves should not be underestimated.

Let us now come back to the question this chapter deals with: to what extent does the image of 'breadbasket' apply to the ancient as well as to the modern Nile Delta? Overall, the evidence discussed in this paper give the impression that agricultural diversification was a practice reserved for land that was either not suited for wheat cultivation or in need of fallowing. The configuration of the Mendesian landscape therefore appears as the fruit of a multi-scaled economic rationalism that relied on the complementarity of agricultural crops and aimed at maximizing the dietary, commercial, and fiscal output of *all* (potentially) productive land. In this context, the overwhelming predominance of wheat cultivation can be seen as resulting from a series of interconnecting factors: the predisposition of the land to grain farming, the advantages of wheat from the point of view of alimentary risk management, the great demand and concomitant speculative profitability of grain on local, provincial, and imperial markets, and its consequent appeal to landholders, including the State. The same would apply, for later periods, to other cash crops such as flax, rice, cotton, tobacco, and sugar cane.[26] In other words, specialization and diversification *were not* necessarily incompatible.

Given its size and the ecological similarities between the Mendesian Nome and other deltaic areas, we can safely assume (for want of definitive proof) that, from antiquity until the modern period, the Nile Delta was, indeed, a regional and imperial breadbasket. Yet as I hope to have shown in this chapter, the socio-economics of food production in this environmentally complex and fluctuating region was way more diversified than what that eye-catching term might allow us to see at first. Furthermore, the ability of the region to generate substantial agricultural surpluses for most of its history owes much more to locally based *savoirs* and *savoir-faire*, than to any top-down interventionism. This, I believe, could serve as a constructive starting point for whomever wishes to properly address the agrarian challenges Egypt and its Delta are currently facing.

NOTES

1 I define this word as 'A part of a region that produces cereals for the rest of it'; Oxford Dictionaries Online, http://www.oxforddictionaries.com/definition/english/breadbasket (accessed 15 May 2016).

2 I am referring to the article 'Nile Delta in its destruction phase' (Stanley and Warne 1998).

3 Testify to this the title of recent newspaper articles: 'Nile Delta Disappearing Beneath the Sea' (McGrath 2014), 'We are going underwater. The sea will conquer our lands' (Shenker 2009), 'Amid political instability, farmers' plight goes from bad to worse' (Viney 2013).

4 According to Diodorus Siculus (*Hist.* 14, 79, 4), in 396 BC, the Spartans received a substantial shipment of grain as a gift from the pharaoh Nepherites (Buraselis 2013: 97–9), with discussion of other fourth-century BC evidence. See also Erdkamp 2005: 181–5, who discusses an oration of Demosthenes (382–322 BC) documenting Athenian and Rhodian imports of Egyptian wheat during his lifetime.

5 The bibliography is abundant. See notably Erdkamp 2005: 225–37, Garnsey 1983, Rickman 1980, Sirks 1991.

6 For a general description of these processes in Roman Egypt, see Lewis 1999. See also Blouin 2014, Chapters 8–9 and, for later periods, Michel 2000 and 2001, Mikhail 2011: Chapter 2, Sijpesteijn 2014.

7 These general estimates are based on 1990 data regarding agricultural land use in Egypt published by the FAO: http://www.fao.org/docrep/w4347e/w4347e0k.htm (Table 26; accessed 15 May 2016).

8 Butzer 1976 was already critical of this top-down model. For more recent discussions, see Bowman and Rogan 1999, Moreno Garcí 2005: 45–50, Manning 2002: 612–3, 2010: Chapter 2.

9 See for instance Mikhail 2011: 96.

10 For similar view regarding more specifically clichés associated with water, see Mikhail 2011: 1.

11 The bibliography is rich. See notably Adams 2007, Bagnall 1993, Bagnall and Frier 1994 (demography), Blouin 2014, Bonneau 1971 and 1993, Bowman

2013, Bowman and Rogan 1999, Moreno García 2005, Parsons 2007, Rathbone 1991, Rowlandson 1996, Scheidel 2001.

12 For an overview, see Bietak 2009 and Blouin 2014: 2–3.

13 These, which are all carbonized, have been found in three sites, all of which are located in the eastern Delta: Bubastis (Frösén and Hagedorn 1989, Hagedorn and Maresch 1998), Tanis (Chauveau and Devauchelle 1996), and Thmuis (Kambitsis 1976 and 1985).

14 On the transfer of the capital from Mendes to the nearby urban zone of Thmuis, see Blouin 2012a.

15 See notably Sijpesteijn 2014 and Mikhail 2011. I will address later the fact that vineyards and pastures are only marginally represented; as for the absence of references to gardens, it could partially result from the fact that the main published papyrus from the archives (*P.Thmouis* 1) deals only with dry plots (hence unsuitable for gardening).

16 Generally speaking, barley seems to have been considerably less cultivated than wheat, of which it was worth half the value (Bagnall 1993: 25, Rathbone 1991: 214, Rowlandson 1996: 20).

17 See Cuno 1999: 315, regarding the Saturday market of Badaway (a village located in an area of the Delta corresponding roughly to the ancient Mendesian Nome) in the nineteenth century.

18 Their presence in the Nile Delta is not surprising. Indeed, to date, the oldest traces of vine growing in Egypt come from the deltaic sites of Tell Ibrahim Awad and Tell el-Faraïn (Buto) and date from the Predynastic period, while viticulture in the region goes back to at least the first dynasty. See Murray, Boulton, and Heron 2000: 577 and ref.

19 Vegetables, just like cereals and grain, offer a much faster yield in exchange of a lesser time investment (Murray 2000b: 616).

20 On a 'white Mendesian wine' (*oinos leukos Mendēsios/Mendaios*) attested by Hippocrates in the fifth century BC (VII 200, 206, 208, 212, 228 *Littré*) and otherwise unattested, see Redford 2010: 176.

21 Rathbone (1991: 244–7), Rowlandson (1996: 24), and Sharp (1999) underline the scarcity of sources on olive tree growing in Theadelphia and the Oxyrhynchite.

22 On the relative importance of sesame farming in nineteenth-century Egypt, see Batou 1991: 408, Table 4, which shows that in 1844, it was, after cotton and rice, the third most important irrigated cash crop culture in Egypt.

23 On the *Boukoloi* (lit. cattlemen) who are attested by Greco-Roman literary sources as well as, indirectly, in papyri, see Blouin 2011. See also Henein 2010: 18, who specifies that the economy of the current inhabitants of the Lake Menzaleh islands relies essentially on husbandry and, on available dried up land, clover (*rebba*) cultivation.

24 On Diocletian's fiscal and monetary reforms, whose exact dates are difficult to establish with certainty, see notably Adams 2004, Bonneau 1971: 198–207, Carrié and Rousselle 1999: 190–5, 593–615.

25 In fact, while all of Psen-'s dry land was dedicated to grain farming, at the scale of the toparchy, 57 per cent of the dry land was composed of vineyards, followed by grain land (c.25 per cent), sandy and bushy land (16 per cent) and, in a significantly lesser proportion, orchards (2 per cent).

26 The period of the so-called 'monopoly system' implemented by Muhammad Ali at the beginning of his reign, and which lasted until 1840, needs to be considered as an exception rather than a rule, see Batou 1991.

REFERENCES

Adams, C. 2004. 'Transition and change in Diocletian's Egypt: province and empire in the late third century'. In Swain, S., Edwards M (eds). *Approaching Late Antiquity: The Transformation from Early to Late Empire*. Oxford: Oxford University Press, pp. 82–108.

———— 2007. *Land Transport in Roman Egypt: A Study of Economics and Administration in a Roman Province*. Oxford: Oxford University Press.

Ayeb, H. 2010. Crise de la société rurale en *Égypte: la fin du fellah?* Paris: Karthala.

Ayeb, H., Bush, R. 2014. 'Small farmer uprisings and rural neglect in Egypt and Tunisia'. *Middlea East Report* 272, http://www.merip.org/mer/mer272/small-farmer-uprisings-rural-neglect-egypt-tunisia (accessed 14 October 2014).

Bagnall, R.S. 1985. 'The camel, the wagon, and the donkey in Later Roman Egypt'. *Bulletin of the American Society of Papyrologists* 22: 1–6.

———— 1993. *Egypt in Late Antiquity*. Princeton: Princeton University Press.

———— (ed.) 1997. *The Kellis Agricultural Account Book (P.Kell. IV Gr.96)*. Oxford: Oxbow Monograph.

Bagnall, R.S., and Frier, B.W. 1994. *The Demography of Roman Egypt* (Cambridge Studies in Population, Economy and Society in Past Times 23). Cambridge: Cambridge University Press.

Batou, J. 1991. 'L'Égypte de Muhammad-'Ali. Pouvoir politique et développement économique, 1805–1848'. *Annales, économies, sociétés, civilisations* 46(2): 401–28.

Bethemont, J. 2003, 'Le Nil, l'Égypte et les autres', VertigO – La revue électronique en sciences de l'environnement, 4(3), http://vertigo.revues.org/3727 (accessed 14 October 2014).

Bietak, M. 2009. 'Archaeology in the Nile Delta'. *Egyptian Archaeology*, 35: 11.

Blouin, K. 2011. 'La révolte des *Boukoloi* (delta du Nil, Égypte, ~166–172 de notre ère): regard socio-environnemental sur la violence'. *Phoenix* 64 (3–4): 386–422.

———— 2012a. 'Les jumelles non identiques: Mendès et Thmouis aux époques hellénistique et romaine'. In Subias, E., Azara, P., Carruesco, J., Fiz, I., Cuesta, R. (eds), *The Space of the City in Graeco-Roman Egypt. Image and Reality*. Tarragona: Institut Català d'Arqueologia Clàssica, pp. 57–67.

———— 2012b. 'Régionalisme fiscal dans l'Égypte romaine: le cas des terres limnitiques mendésienne'. In De Angelis, F. (ed.), *Proceedings of the Conference Regionalism and Globalism in Antiquity*. Leuven: Peeters.

———— 2012c. '*Minimum firmitatis, plurimum lucri*: le cas du « lin mendésien »', in Schubert, P. (ed.), *Actes du 26e Congrès international de papyrologie: Genève, 16–21 août 2010*. Geneva: Librairie Droz, pp. 83–9.

———— 2014. *Triangular Landscapes. Environment, Society, and the State in the Nile Delta under Roman Rule*. Oxford: Oxford University Press.

Bonneau, D. 1971. *Le fisc et le Nil: incidences des irrégularités de la crue du Nil sur la fiscalité foncière dans l'Égypte grecque et romaine*. Paris: Éditions Cujas.

———— 1993. *Le régime administratif de l'eau du Nil dans l'Égypte grecque, romaine et byzantine*. Leiden: Brill.

Bousquet, B. 1996. *Tell-Douch et sa région: géographie d'une limite de milieu à une frontière d'Empire*. Cairo: Ifao.

Bowman, A.K. 2013. 'Agricultural production in Egypt'. In Bowmand, A.K., Wilson, A. (eds), *The Roman Agricultural Economy: Organisation, Investment and Production*. Oxford: Oxford University Press, pp. 219–53.

Bowman, A.K., Rogan, E. (eds) 1999. *Agriculture in Egypt from Pharaonic to Modern Times*, Oxford: Oxford University Press.

Brun, J.-P. 2004. *Archéologie du vin et de l'huile: de la préhistoire à l'époque hellénisque*. Paris: Éditions Errance.

Busarelis, K. 2013. 'Ptolemaic grain, seaways, and power'. In Buraselis, K., Stefanou, M., Thompson, D.J. (eds), *The Ptolemies, the Sea and the Nile*. Cambridge, Cambridge University Press, pp. 97–107.

Bush, R., Sabri, A. 2000. 'Mining for fish. Privatization of the "commons" along Egypt's northern coastline'. *Middle East Report* 126, http://www.merip.org/mer/mer216/mining-fish (accessed 14 October 2014).

Butzer, K.W. 1976. *Early Hydraulic Civilization in Egypt: A Study in Cultural Ecology*. Chicago: Chicago University Press.

Carrié, J.-M., Rousselle, A. 1999. *L'Empire romain en mutation: des Sévères à Constantin, 192–337* (Nouvelle histoire de l'Antiquité 10), Paris: Seuil.

Chauveau, M., Devauchelle, D. 1996. 'Rapport sur les papyrus carbonisés de Tanis'. *Bulletin de la Société des fouilles françaises de Tanis* 10(3): 107–11.

Crawford, D.J. 1971. *Kerkeosiris: An Egyptian Village in the Ptolemaic Period*. Cambridge: Cambridge University Press.

Cuno, K.M. 1999. 'A tale of two villages: family, property, and economic activity in rural Egypt in the 1840s'. In Bowman, A.K., Rogan, E. (eds), *Agriculture in Egypt from Pharaonic to Modern Times*. Oxford: Oxford University Press, pp. 301–29.

Erdkamp, P. 2005. *The Grain Market in the Roman Empire: A Social, Political and Economic Study*. Cambridge: Cambridge University Press.

Frösen, J., Hagedorn, D. (eds) 1989. *Die verkohlten Papyri aus Bubastos*, vol. 1, Opladen: Westdeutscher Verlag.

Garnsey, P. 1983. 'Famine in Rome'. In Garnsey, P., Whittaker, C.R. (eds), *Trade and Famine in Classical Antiquity*. Cambridge: Cambridge University Press, pp. 56–65.

———— 1996. *Famine et approvisionnement dans le monde gréco-romain: réactions aux risques et aux crises*. Paris: Les Belles Lettres.

———— 1998. 'The bean: substance and symbol'. In Scheidel, W. (ed.), *Cities, Peasants, and Food in Classical Antiquity*. Cambridge: Cambridge University Press, pp. 214–25.

———— 1999. *Food and Society in Classical Antiquity*. Cambridge: Cambridge University Press.

Godley, A.D. (transl.) 1920. *Herodotus. The Persian Wars. Books 1–2*. Harvard: Harvard University Press.

Grimal, N. 1988. *Histoire de l'Égypte ancienne*. Paris: Fayard.

Hagedorn, D., Maresch, K. (eds) 1998. *Die verkohlten Papyri aus Bubastos*, vol. 2, Opladen: Westdeutscher Verlag.

Henein, N. 2010. *Pêche et chasse au lac Manzala*. Cairo: Ifao.

Horden, P., Purcell, N. 2000. *The Corrupting Sea: A Study of Mediterranean History*. Malden: Blackwell.

Kambitsis, S. 1976. 'Un nouveau texte sur le dépeuplement du nome mendésien: *P.Thmouis* 1, col. 104–105'. *Chronique d'Égypte* 51(101): 130–40.

—— (ed.) 1985. *Le papyrus Thmouis 1, colonnes 68–160*. Paris: Publications de la Sorbonne.

Langellotti, M. 2012. *L'allevamento di pecore e capre nell'Egitto romano: aspetti economici e sociali*. Bari: Edipuglia.

Lewis, N. 1999. *Life in Egypt under Roman Rule*. Oakville: American Society of Papyrologists.

McGrath, C. 2014. 'Nile Delta Disappearing Beneath the Sea', *Inter Press Service*, 29 January 2014, http://www.ipsnews.net/2014/01/nile-delta-disappearing-beneath-sea/ (accessed 15 May 2016).

Manning, J.G. 2002. 'Irrigation et état en Égypte antique'. *Annales histoire, sciences sociales: politiques et contrôle de l'eau dans le Moyen-Orient ancien* 3: 611–23.

—— 2010. *The Last Pharaohs. Egypt under the Ptolemies, 305–30 BC*. Princeton: Princeton University Press.

Michel, N. (2000). 'Devoirs fiscaux et droits fonciers: la condition des *fellahs* égyptiens (13e–16e s.)'. *Journal of the Economic and Social History of the Orient* 4: 521–78.

—— (2001). 'Migrations de paysans dans le Delta du Nil au début de l'époque ottomane'. *Annales islamologiques* 35: 241–90.

Mikhail, A. 2011. *Nature and Empire in Ottoman Egypt*. Cambridge: Cambridge University Press.

Mitchell, T. 1991. 'America's Egypt. Discourse of the development industry'. *Middle East Report* 169, http://www.merip.org/mer/mer169/americas-egypt (accessed 15 May 2016).

Moreno García, J.C. (ed.) 2005. *L'agriculture institutionnelle en Égypte ancienne: état de la question et perspectives interdisciplinaires*. Lille: CRIPEL.

Murray, M.A. 2000a. 'Cereal production and processing'. In Nicholson, P.T., Shaw, I. (eds), *Ancient Egyptian Materials and Technology*. Cambridge: Cambridge University Press, pp. 505–36.

—— 2000b. 'Fruits, vegetables, pulses and condiments'. In Nicholson, P.T., Shaw, I. (eds), *Ancient Egyptian Materials and Technology*. Cambridge: Cambridge University Press, pp. 609–55.

Murray, M.A., Boulton, N., Heron, C. 2000. 'Viticulture and wine production'. In Nicholson, P.T., Shaw, I. (eds), *Ancient Egyptian Materials and Technology*. Cambridge: Cambridge University Press, pp. 577–608.

Parsons, P. 2007. *City of the Sharp-Nosed Fish*. London: Orion Books.

Rathbone, D.W. 1983. 'The grain trade and grain shortages in the Hellenistic East'. In Garnsey, P., Whittaker, C.R. (eds), *Trade and Famine in Classical Antiquity*. Cambridge: Cambridge University Press, pp. 45–55.

—— 1991. *Economic Rationalism and Rural Society in Third-Century A.D. Egypt: The Heroninos Archive and the Appianus Estate*. Cambridge: Cambridge University Press.

Redford, D.B. 2010. *City of the Ram-Man*. Princeton: Princeton University Press.

Rickman, G. 1980. *The Corn Supply of Ancient Rome*. Oxford: Oxford University Press.

Rowlandson, J. 1996. *Landowners and Tenants in Roman Egypt*. Oxford: Oxford University Press.

Samuel, D. 2000. 'Brewing and baking'. In Nicholson, P.T., Shaw, I. (eds), *Ancient Egyptian Materials and Technology*. Cambridge: Cambridge University Press, pp. 537–76.

Scheidel, W. 2001. *Death on the Nile: Disease and the Demography of Roman Egypt*. Leiden: Brill.

Schnebel, M. 1925. *Die Landwirtschaft im hellenistischen Aegypten*. Munich: Beck.

Serpico, M., White, R. 2000. 'Oil, fat and wax'. In Nicholson, P.T., Shaw, I. (eds), *Ancient Egyptian Materials and Technology*. Cambridge: Cambridge University Press, pp. 390–429.

Sharp, M. 1999. 'The village of Philadelphia in the Fayyum: land and population in the second century', in Bowman, A.K., Rogan E. (eds), *Agriculture in Egypt from Pharaonic to Modern Times*. Oxford: Oxford University Press, pp. 159–92.

Shenker, J. 2009. 'We are going underwater. The sea will conquer our lands', *Guardian*, 21 August 2009, http://www.theguardian.com/environment/2009/aug/21/climate-change-nile-flooding-farming (accessed 14 October 2014).

Sijpesteijn, P. 2014. *Shaping a Muslim State. The World of a Mid-Eighth-Century Egyptian Official*. Oxford: Oxford University Press.

Sirks, A.J.B. 1991. 'The size of the grain distribution in imperial Rome and Constantinople', *Athenaeum* 79: 215–37.

Stanley, J.D., Warne, A.J. 1998. 'Nile Delta in its destruction phase', *Journal of Coastal Research* 14(3): 795–825.

Świderek, A. 1971. 'The land-register of the ΦΕΡΝΟΥΦΙΤΟΥ toparchy in the Mendesian Nome', *Journal of Juristic Papyrology* 17: 31–44.

Thanheiser, U. 1992. 'Plant-food at Tell Ibrahim Awad: preliminary report'. In van den Brink, E.C.M. (ed.), *The Nile Delta in Transition: 4th.–3rd. Millennium B.C.* Tel Aviv: E.C.M. van den Brink, pp. 117–21.

Udovitch, A.L. 1999. 'International trade and the Medieval countryside'. In Bowman, A.K., Rogan, E. (eds). *Agriculture in Egypt from Pharaonic to Modern Times*. Oxford: Oxford University Press, pp. 267–85.

Viney, S. 2013. 'Amid political instability, farmers' plight goes from bad to worse', *Egypt Independent*, 7 April 2013, Amid political instability, farmers' plight goes from bad to worse' http://www.egyptindependent.com//news/amid-political-instability-farmers-plight-goes-bad-worse (accessed 14 October 2014).

Wallace, S.L. 1969. *Taxation in Egypt from Augustus to Diocletian*. New York: Greenwood.

Watrin, L. 2003. 'Les enjeux de l'archéologie prédynastique en Basse–Égypte', *Bulletin du GREPAL* 1: 6–9.

Yoyotte, J. 1958. 'Promenade à travers les sites anciens du Delta', *Bulletin Trimestriel de la Société Française d'Égyptologie* 25: 13–24.

Zivie, A.-P. 1975. *Hermopolis et le nome de l'ibis: recherches sur la province du dieu Thot en Basse Égypte*, vol. 1: *introduction et chronologie des sources*. Cairo: Ifao.

5   Water, Migration and Settlement in the Southern African Iron Age

Johann W.N. Tempelhoff

INTRODUCTION: THE 'DRYING UP' OF SOUTHERN AFRICA

The availability of water has always been a vital consideration for human settlement in southern Africa. In the 1860s J.C. Brown (1808–95), the Cape colonial botanist, was of the opinion that the subcontinent was in the process of drying out. He was aware of severe water shortages in many parts (Cape of Good Hope 1865: iv, 23) and later published a work dealing with what he considered to be a major hydrological problem in the region (Brown 1875). Brown was influenced by J.F. Wilson (1865) a traveller and explorer, who after exploring the arid interior of the Cape Colony, was of the view that the 'drying out' of southern-Africa was of recent origin and measures should be taken to literally 'green' the subcontinent by planting trees (Grove 1995: 468–9). In motivating their case both authors took note of the distribution of the peoples resident in the region. Ever since the seventeenth century, successive colonial administrations, first the Dutch, and by the end of the eighteenth century the British, were aware that the subcontinent was populated by a variety of distinct ethnic communities who inhabited regions where the available water resources were highly variable.

By the nineteenth century the philologist, Wilhelm Bleek (1827–75), who had done considerable work on the language and culture of the hunter-gatherer San (Bushmen), was aware that these people were probably the earliest residents of the subcontinent and that they tended to live in the arid areas. Bleek also noted that the most populous ethnic communities of the region, primarily living in the north-eastern parts of the subcontinent, were Negroid-like and spoke a variety of regional interrelated vernacular languages. He coined the term 'Bantu' languages for the way these people communicated (Parkington & Hall 2012: 69).

In their exposition of the imminent 'drying out' of the subcontinent Brown and Wilson categorized the Bantu-speaking peoples in relation to their lighter skinned neighbours, the San hunter and gatherers and the latter's Khoi livestock-farming (or 'herder') relatives. The 'Black' Africans, as they termed them, were resident in the eastern parts of the subcontinent where the landscape tended to be fertile and largely covered with green succulent trees and forests, that is, in parts adjacent to the Drakensberg mountain range from the northeast of South Africa, down to the Eastern Cape. The central area of the subcontinent was the territory of the 'Bechuana', a landscape notable for rolling plains and 'arid prairies' with few fountains and hardly any rivers or forests. The third and most westerly region was the territory of the 'Namaqua and Bushman'. This was an arid region, considered to be sterile and barren in the extreme, with sporadic thunderstorms responsible for flash floods in dry riverbeds (Brown 1875: 109–10) – an area better known as the Namib desert (Seddon 1968: 493). By implication this description seemed to suggest that the 'Bechuana' were Africans who had partially integrated with the San, of which the almost Bantu-like Khoi pastoralists, were representative. The western region was predominantly inhabited by San people, with a presence of Khoi pastoralists; they were designated the 'Namaqua'.

By the twentieth century, a more sophisticated interpretation of the distribution of the Bantu-speaking peoples suggested that they tended to reside in areas with an average rainfall above 200 mm, and to the east of the winter rainfall area. Their Khoisan neighbours seemed to prefer the western and northern arid regions of the subcontinent (Figure 5.1).

The Bantu-speaking peoples of southern Africa are said to have originated in the border area of modern Cameroon and Nigeria. One group, the proto-Western Bantu speakers hunted, fished, kept goats and also cultivated yams; later they resorted to working with iron. Traces of their presence in Gabon and Congo-Brazzaville have been dated to 3000–2300 BP. The Eastern Bantu-speakers moved across the continent to the East African Lakes region and became acquainted with the cultivation of cereals and keeping sheep, goats and cattle. The Uruwe ceramic culture, generally associated with both Western and Eastern Bantu-speaking groups, was evident in East Africa by 2500 BP (Mitchell, 2002: 260–1).

In the discussion to follow, attention will be given specifically to Bantu-speaking Iron Age (c. 200–1850 CE) farming communities of southern Africa and the way their livelihoods and mobility were influenced by water.

THE IMPACT OF THE INTER-TROPICAL CONVERGENCE ZONE (ITCZ)

Sub-continental Africa, situated between the Atlantic Ocean to the west and the Indian Ocean in the east, is prone to dry conditions and uneven rainfall (Anon., *sine die*). Periods of drought are frequently interrupted by large-scale flood events. In South Africa floods are a fairly regular occurrence along the Indian Ocean coastline. On the Atlantic side of the

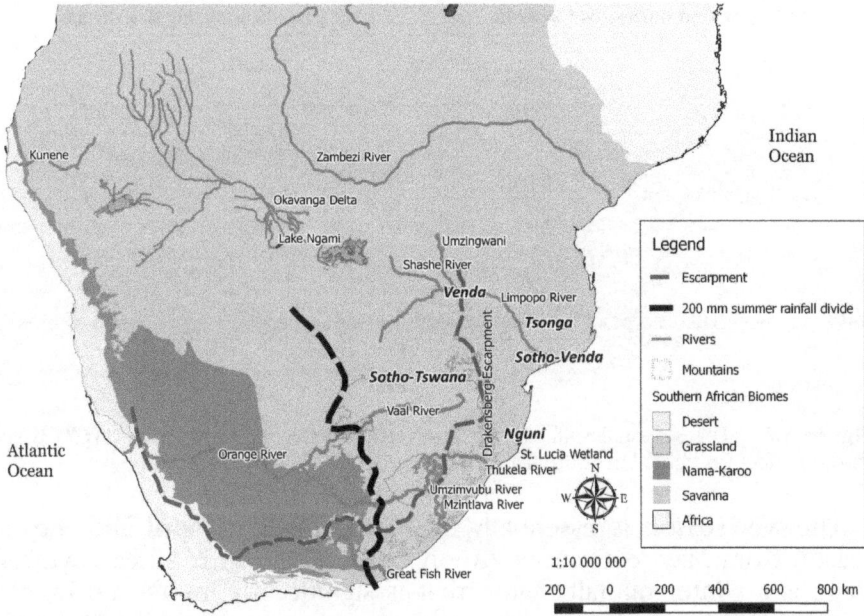

Figure 5.1. By the eighteenth century the settlement of Bantu-speaking people of southern Africa was in many respects the result of environmental factors. The Early Iron Age people settled in the savannah regions. They also tended to settle in the summer rainfall region, east of the 200mm divide (© Emile Hoffmann).[1]

subcontinent floods in the more arid parts of the region are less common (Rowntree 2000: 403). Rainfall in southern Africa is the direct result of the air circulation in the northern parts of the region where warm air rises into the atmosphere of the inter-tropical convergence zone (ITCZ) and creates conditions conducive for rain (Figure 5.2). The ITCZ follows the position of the sun shining on to the equator. It moves to the south in the summer months and to the north of the equator in the winter months. Because of the ITCZ the rain falls more heavily near the equator and decreases progressively in a southward direction (Pallet 1997: 13).

There are also other movements of air to the south of the ITCZ, which shape the rainfall pattern. Descending air is responsible for a high-pressure cell over the Indian Ocean. It blows for a long distance over the sea. The air picks up moisture and then moves into the subcontinent. This brings rain. A similar high-pressure cell is located on the western side of the subcontinent in the Atlantic Ocean and blows across the colder waters over a relative short distance. The result is that the west coast of southern Africa receives little rain. Moisture comes over the continent from the east and rainfall decreases progressively towards the west (Pallet 1997: 13).

ITCZ position during wet season   **ITCZ position during dry season**

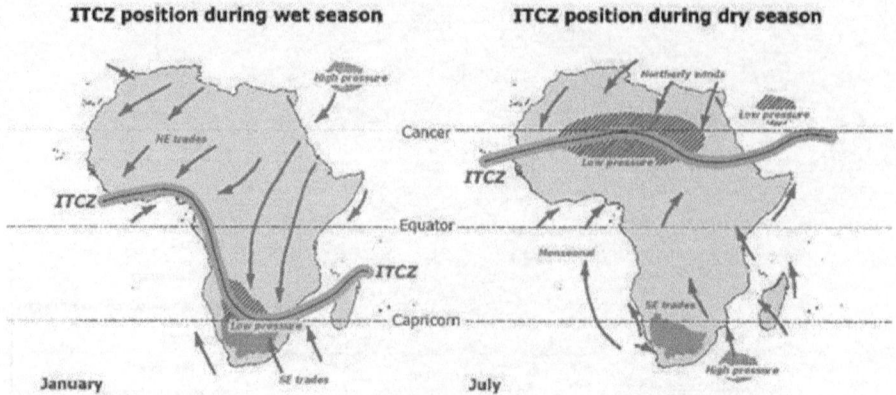

January   July

Figure 5.2. The seasonal shifts in the Inter-Tropical Convergence Zone (ITCZ) have a marked effect on rainfall in southern Africa (source: Anon., *sine die*).

The wet season is essentially from November to April and the dry season from May to October (Anon., *sine die*). South Africa's Western Cape is a winter rainfall region that is significantly influenced by cold fronts that originate in the south Atlantic and move eastwards. They pass over the mountains of the southern Cape and bring with them cold and wet weather (Pallet 1997: 13).

The mean annual rainfall estimate for southern Africa stands at 497 mm/a (Rowntree 2000: 394). South Africa's is even lower at 450 mm/a – both figures well below the world average of 860 mm/a (DWAF 2004: 15). Annual rainfall is highly variable. In semi-arid and arid regions the rainfall can be considerably lower. Evaporation rates are high (ibid.).

WATER, HUMANS AND MIGRATION

Water, but specifically rainfall, has determined the lives of humankind in Africa since time immemorial (Collins & Burns 2013: 160). Its availability is highly variable. In the north-eastern areas of South Africa the annual rainfall reaches a maximum of 1500 mm. It is here where a formidable human footprint in the region has been most profuse in the past two millennia. In the south western parts of southern Africa, such as the Namib desert, the annual rainfall can be as low as 25 mm (Pallet, 1997: 1) making it one of the more inhospitable areas for humans on earth (Carruthers 2003: 256). With the exception of the period 2000–1400 BP there has not been a significant human presence in the region (Avery 1995: 350).

According to McCann (1999: 262), rainfall, more than extreme temperature, has been the most limiting climatic factor in respect of food production for human consumption in Africa. His observation also holds for southern Africa and its Iron Age farming communities who settled in the region two millennia ago:

> The timing of rainfall and its particular relationship to the constellation of labor, cropping patterns, and capital requirements of a specific farming system is a critical but neglected aspect of both the historical and contemporary development of rural society and economy. It is, after all, the pattern of seasonal rainfall that trigger social and economic processes of labor, renewal of resources (food, seed cash crops, and forage) and the shortage or abundance of harvests (McCann 1999: 262–3).

Understandably the importance of rain is deeply seated in the culture of the subcontinent's peoples and this is evident in the way they engage with each other (Etherington 2010: 361–75). In Lesotho, for example, where there is a high rainfall and plentiful water supplies, when people greet each other, they say 'Pula!' (meaning: May it rain!) (Berger 2009: 2). In Botswana, one of the more arid parts of the subcontinent, the country's currency is the Pula (rain) – a symbolic cultural pointer suggesting that rain is highly valued.

At the time of their arrival the Iron Age farming communities engaged with people of the Late Stone Age (40,000–200 BP) (Mitchell 2002: 135–6) representative of the aboriginal population of southern Africa (Lombard, Schlebusch, et al. 2013: 27–34). The San and Khoikhoi (collectively known as the Khoisan) have a long history of engagement, especially with the arid regions of southern Africa. They were able to adapt continuously to changing social and ecological conditions. Apart from having mastered natural water resources management strategies, and according to tradition, also the spiritual powers of making rain (Alcock 2010: 199–203) they were responsible for basic, but important technologies of water storage in ostrich eggshells (Klein 2001: 12) and preparing food on fire by boiling broths in potted water (Alcock 2014: 12, 24). They were resilient and whenever local water resources dried up in the arid regions they proved themselves able to thrive on a variety of endemic water-rich flora (Barnard 1992: 43–4, Van Wyk & Gericke 2007, Chapters 4 and 5).

In southern African history little mention is made of a Neolithic period marking the transition of Stone Age society to more complex cultural and technological innovations. In recent times some archaeologists have hinted at the likelihood of hunters who herded sheep in northern Botswana some two millennia ago who could perhaps have been representative of the region's Neolithic era (Sadr 2003: 195–209, Robbins, Campbell, et al. 2005: 671–7). However, mainstream thinking of pre-colonial southern Africa has not yet absorbed the idea. The historical record seems to suggest that Iron Age farming communities drifting into the subcontinent some 2,000 years ago were at the cusp of a major phase of cultural innovation and change that transformed social, political and economic lifestyles. Their impact on the region was so powerful that the significance of the Iron Age simply overshadows whatever might have resembled a Neolithic period.

REASONS FOR THE SOUTHWARD MIGRATIONS

There are various explanations for the southward movements of Early Iron Age farming communities. A strong case has been made out for climate change. Between 200 and 600 CE a number of scattered Early Iron Age communities in South Africa's North West Province (at Broederstroom) and in Botswana's Kalahari region suggest that the local climate was warmer and wetter (Huffman 1996: 55–60). Perhaps there was an abundance of food and water in southern Africa, resources that could accommodate a consistent inflow of new settlers from the north (Parsons 2008: 44, Seddon 1968: 489).

According to Phillipson (1994: 188), at the time of the southward migration, there were anthropogenic climatic conditions in East Africa. East Africa's Chifumbaze complex (a ceramic culture) was representative of communities that started settling in villages, practising agriculture and livestock farming and metallurgy. As a result of charcoal burning for iron-smelting activities and accompanying forest-clearing processes, local climate conditions began to change. As the sustainability of large settlements diminished many people started moving away.

Another reason for the southern migratory process was rainfall. Isichei explains:

> [R]ainfall [...] determined the pattern of settlement in Africa south of the Limpopo [...] Communities defined their rights to water precisely, but away from it boundaries were vague and ill defined (Isichei 1997: 141).

The availability of water was key to the intrinsic value of land. Ruling elites tended to determine the territorial area of a community and were also in many communities responsible either for procuring rain or at least for using authoritative intermediaries who had close ties with the spirits responsible for providing rain (Alcock 2010: 198). Climate refugees coming from drought-ridden regions sought places with sufficient and consistent rainfall, but claiming ownership of the land was of secondary importance. People could easily relocate and move to where rains were more plentiful and we know that new settlers intuitively steered clear of low rainfall regions. Low rainfall posed a barrier to basic subsistence livelihoods (Turton, Meissner, et al. 2004: 27). A pronounced territorial demarcation of land for settlement would, in the early years, have meant an impediment to mobility in times of hardship.

Another reason for the southward shift was the growing contact along the coastline of East Africa between African communities and foreign traders and sailors, from as far afield as Malagasy, Arabia and Persia. In time to come cultural exchange and the dissemination of new ideas accompanied the economic activity between local African communities and foreigners (Parsons 2008: 44). Coastal urban settlements of trade, bustling with people from distant parts of the Indian Ocean world, were part of a transformation process driven by the supply and demand for goods and services. As the demand for ivory, gold and slaves increased, there was a constant need for extending ties with areas that had not yet been infiltrated by trade. Some coastal urban trading centres, for security reasons, were on off-shore islands (Freund 2007: 25), but this did not prevent them from touching the lives of indigenous communities on the mainland. Trade settlements were usually in the proximity of a coastal bay area with nearby rivers that could provide water and straddling land routes into the interior.

TEMPORAL AND SPATIAL CLASSIFICATION OF THE IRON AGE

The Iron Age in southern African has been classified into three phases:

1. The Early Iron Age (200–900 CE): initial settlement of farming communities in southern Africa;
2. The Middle Iron Age (900–1300 CE): the beginning and expansion of trade relations between Iron Age communities and the Indian Ocean trade; and
3. The Late Iron Age (1300–1820 CE): the emergence of new states on the Zimbabwean Plateau and further south in South Africa (Huffman 2007: Chapters 19–21; Badenhorst 2010: 89).

There were three dominant spatial directions of settlement (Silva & Steele 2012). The first of these, known as the eastern coastal stream, was used by settlers from about 200–300 CE to seek out extensive wetlands and riverine areas along coastal Maputoland and KwaZulu-Natal where there were substantial food producing opportunities (Hall 1987: 36–8).

The second group, known as the eastern highland stream (also known as the central stream) first settled in the Limpopo River Valley. They are dealt with in more detail below, but it is important to note that some were pastoral farmers (Poland, Hammond-Tooke & Voigt 2004: 14). Their southward migration, in early times, tended to be in the interior of southern Africa through Zimbabwe (Smith 1992: 134). They preferred the drier areas of southern Africa to the west of the Drakensberg and areas on the edges of the Kalahari desert (Parsons 2008: 41–3).

A third grouping, known as the western stream, entered southern Africa from Angola and/or Uganda and then moved into the northern parts of South Africa's Limpopo valley where they were presumably resident by about 350 CE (Denbow 1990: 139–75, Parsons 2008: 45). They were highly mobile and moved as far south as the coastline of KwaZulu-Natal and the Eastern Cape by 800 CE (Parsons 2008: 45). The Okavango Delta region of Northern Botswana became home to settled communities of agro-pastoralists who shared the delta region, one of the largest desert wetland areas in southern Africa, with local hunting and gathering communities by the seventh century (Turner 1987: 25–40, Denbow and Wilmsen 1990: 1509–15, Miller and Van der Merwe 1994: 101–15). There still seems to be uncertainty about the third stream, but it is apparent that the availability of sufficient water resources for watering livestock and satisfactory grazing in semi- to extremely arid regions would have been an important consideration when they chose settlement sites (Figure 5.3).

THE EARLY IRON AGE

The Early Iron Age settlement period, between 250 and 400 CE, appears to have taken place gradually over an extended period of time. There were also phases of rapid intrusion and settlement. Moreover, there were significant movements of communities and the diversity of the regional populations augured well for a significant cultural transition of southern Africa (Parkingston & Hall 2012: 70).

In the coastal region of modern day KwaZulu-Natal where there were riverine floodplains in the lowlands, people were soon planting crops. Typical sites would have been in the region of the Lake St Lucia estuary and the Mdlanzi swamplands bordering on the Mkuze River (Hall 1987: 36–8, Turner & Plater 2004: 220–9). Some groups veered into the interior lowlands and the fringes of the Drakensberg mountain range. The escarpment of the mountain range extending from the northern parts of South Africa down to the southern Cape appears to have been a barrier to settlement to the west. However, there was a consistent move southwards, sometimes close to the Indian Ocean; at other times the farming communities moved more to the interior.

In the northeastern parts of South Africa evidence of the eastern stream of Early Iron Age settlement was notable for crop farming activities. At first, cattle did not comprise a large part of the subsistence strategies of communities (Huffman 2001: 19–35). Agriculture, in the form of hoe culture, seemed to be dominant (Poland, Hammond-Tooke & Voigt, 2004: 15). The crops they planted were similar to those of farming communities elsewhere in the southern Africa interior and included sorghum (*Sorghum* sp.), pearl millet (*Pennisetum typhoides*), finger millet (*Eleusine caracasna*), cowpeas (*Vigna unguiculata*) and

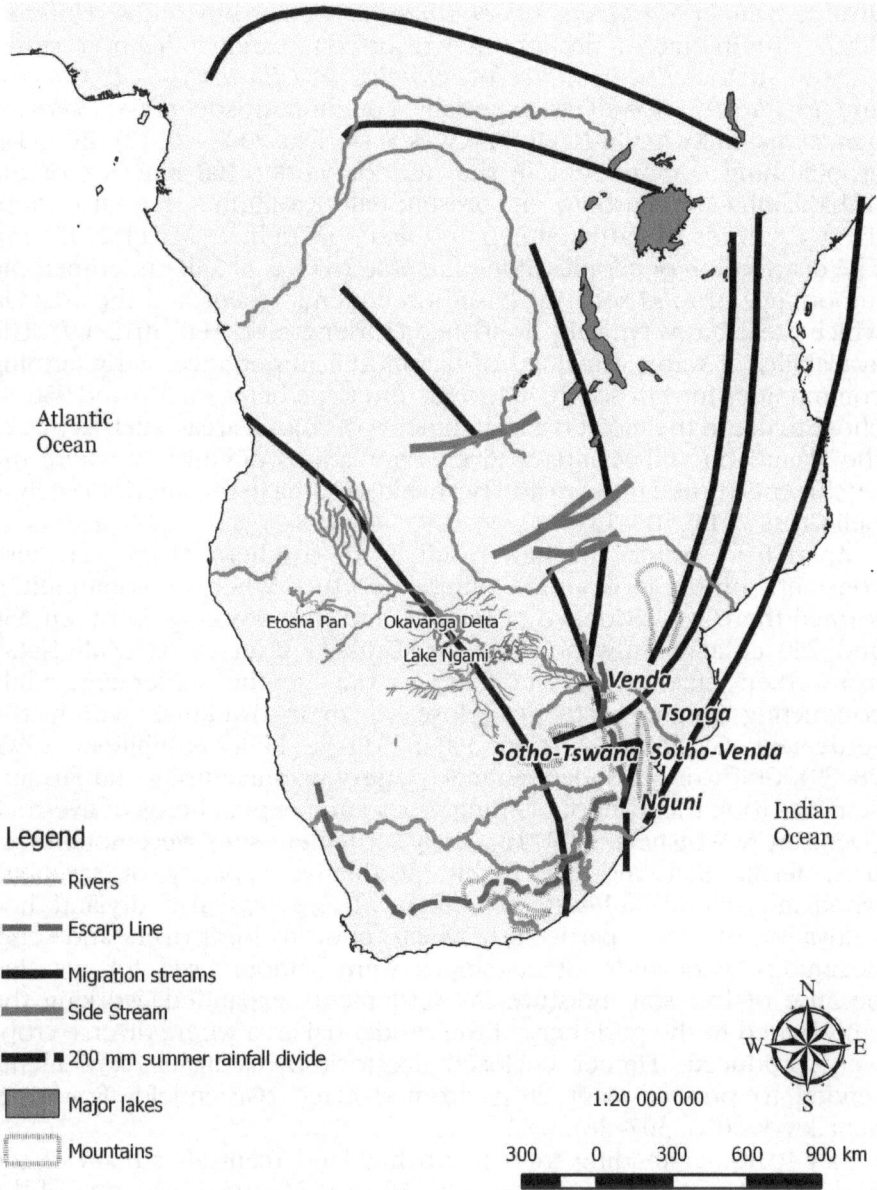

Figure 5.3. The three streams of Early Iron Age farming communities that moved into southern Africa two millennia ago (© Emile Hoffmann).[2]

ground beans (*Voandzeia subterranea*) as well as phytoliths (Huffman 2007: 338). In times of drought they resorted to eating indigenous cereal grasses, such as *Eragrostic chlorommelas, E. Ciliannensis, E. Curvela* and *E. Plana*, as well as especially common bristle grass (*Setarias phacelata*) in KwaZulu-Natal (Van Wyk & Gericke 2007: 10, 12). By 550 CE crop-farming communities in the interior, about 100 km east of the uKhahlamba-Drakensberg in present-day KwaZulu-Natal, also kept livestock herds of cattle, sheep and goats (Wright & Mazel 2007: 46). The distribution of peoples along the eastern coastal belt underlined the importance of food security, based on endemic grains and the areas in which these crops typically flourished (Parkington & Hall 2012: 69). The availability of water remained of paramount importance. Early farming communities further south, in the Eastern Cape between 650 and 950 CE, chose to live in the moist river catchments of coastal areas, such as that of the Mzimvubu and Mzintlava River. Populations diminished where the catchments turned in more arid bushveld savanna in the interior (Feely & Bell-Cross 2011: 105–12).

Apart from water, the new residents of southern Africa were also constantly on the lookout for mineral deposits. Wherever communities formed there was evidence of mining, and iron working. Between 650 and 750 CE at Nondwane near the Thukela River in KwaZulu-Natal, ironworkers smelted iron in furnaces outside the settlement, while conducting forging activities close to their dwellings within the settlement (Greenfield & Miller 3004: 1511–32, Miller & Whitelaw 1995: 78–89). Craftwork included ceramic pottery manufacturing and farming activities took the form of planting crops and keeping herds of livestock (Denison & Wotshela 2009: 10). Early settlement sites were notable for their fertile soil and the highly productive capacity of savannah woodlands along well-watered valleys. There was also dryland hoe cultivation of crops, particularly wheat, close to local rivers and semi-deciduous woodlands. Steep slopes were seldom used for planting because of low soil moisture. As settlements expanded, working the land shifted to the periphery of the residential area where diverse crops were produced. Timber enclosed livestock byres inside settlements tended to protect plant crops from damage (Greenfield, Fowler & Schalkwyk 2005: 307–28).

Key to understanding the eastern highland (central) stream of the Early Iron Age presence in southern Africa is Mapungubwe, one of the earliest large human settlements in the subcontinent, situated at the confluence of the Shashe and Limpopo, close to the borders of present-day South Africa, Botswana and Zimbabwe. Rivers and streams have historically been the sources of water that humans exploited most comprehensively in the settlement process. Although lakes contain far more water than rivers, as a rule humans have settled them less densely, probably because their accessible perimeters are smaller than those of riverbanks (Solomon 2010: 12–3). The first residents of Mapungubwe began their farming between 350 and 450 CE (Huffman 2000: 16). Further downstream in the middle and lower Limpopo River valley, farmers also lived in close proximity to the river. In the hinterland they tended to favour open grassland areas, such as small lakes where there were edible grasses. In turn they tended to avoid forests near the rivers and woodlands (Ekblom, Gillson & Notelid 2011: 11). Livestock used in the early farming industry primarily comprised sheep and goats (Badenhorst 2010: 89). Although there were also cattle, they were not as numerous as the caprines (Huffman 2010: 164–74). Apart from growing the usual millet and sorghum crops, they busied themselves, as was the case with other contemporary communities in southern Africa, with ironwork; carving utensils from bone and ivory; manufacturing ceramic pots, dishes and bowls from soapstone; and carving bones and ivory. They extracted salt from saline mineral springs and also created platforms of potential trade (Hammond-Tooke 1993: 27).

Early Iron Age communities tended to settle and freely integrate with local aboriginal hunter-gatherers. The groups tended to live in relative harmony (Robertson & Bradley 2000: 312). Although hunter-gatherers reserved for themselves the right and freedom to move elsewhere, many tended to adapt and become integrated in a transformational open frontier communal system (Parkington & Hall 1987: 1–25). There is evidence of technology transfer and the development of inter-group relations. Hunter-gatherers adapted with significant ease to the culture and technologies of the new incumbents (Schoeman 2006: 6). Some embraced pastoralism and only periodically reverted to old ways in times of environmental crises. The new farming communities also brought with them deep-seated traditions of working with water and managing available resources to promote farming activities. Many were descendants of communities resident in the East African lakes region where they had been part of a thriving Iron Age culture. Apart from farming, they also relied on fishing (Whitelaw 2009: 195–212) and knew how to interact with large waterscapes. At the same time the new settlers needed to learn the knowledge of the terrain from the indigenous people who had experience of local climatic conditions and could make predictions of expected rainfall; the locals could also purportedly intervene by means of rainmaking ceremonies, all of which were crucial to the planning of crop planting activity.

One of the notable Early Iron Age transformations was in the characteristics of domestic dwellings. For hunter-gatherer communities temporary grass structures, rock shelters and caves were customary living spaces (Schapera 1960: 86–7). With the onset of new community formations, settlements emerged that comprised clay and wattle-and-daub type structures (Robertson & Bradley 2000: 313). Early communal subsistence activities tended to focus on cultivating land, keeping livestock, working with metals and living in rudimentary self-built dwellings in valleys alongside rivers (Badenhorst 2010: 87–106).

Gradually there was a shift away from the valleys and nearby water sources to settle on higher elevations (Robertson & Bradley 2000: 315–6). Security considerations played a role here (Orpen 1964: 73, Quin 1959: 134), as did the threat of disease. Early Iron Age farming community formations represent the beginnings of proto-urban settlements that later became more complex. Where populations of people lived in close proximity to each other the potential for disease is also higher. In hot and dry, as well as humid conditions close to water supplies, there was relatively more disease. In the low-lying regions of the southern African Lowveld, waterborne diseases were endemic. Bilharzia, known to have been prolific in Ancient Egypt and East Africa (Cox 2002: 601–2, Morgan, De Jong Adeoye et al. 2005: 3898–9) was a major threat in southern African waters (Farley 2003: 5–6). Beside the rivers, malarial fever, carried by the *Anopheles* mosquito that breeds in stagnant water pools, posed an added danger. Not all Iron Age communities were immune to the fever although immunity did eventually occur over time (Packard 2001: 591–612).

Tsetse flies have a long history in Africa where they are responsible for the transmission of the Trypanosoma parasite of *nagana* in livestock and certain species of wildlife, and sleeping sickness in humans (Steverding 2008, Selous 1909: 113–29, Livingstone 1857: 353–66, 1858: 94–7). There is reason to believe that the tsetse fly (*Glossina* sp.) prevented Early Iron Age pastoralists from penetrating southern Africa (Clark 1963: 219–22). There is evidence that in the northern parts of present-day Botswana, pastoralists were present two millennia ago, living just south of a zone notorious for animal diseases, such as tsetse (Robbins, Campbell, Murphy et al. 2005: 671–7). Huffman explains that a migration from the west of people belonging to the Uruwe tradition of Bantu-speaking West Africa might well have taken place two thousand years ago, prompted by climate fluctuations which created a tsetse fly-free route into southern Africa (Huffman 2007: 359).

The transition from the Early Iron Age was related to changing climatic conditions. From about 600 ce, for perhaps more than three centuries, there were colder conditions in the Limpopo-Shashe area, which could have had an impact on farming in the Mapungubwe region. According to Huffman hardly any farming took place in this period (Huffman 2000: 16). Climatic conditions had changed in the Limpopo valley by 800 ce. People began clearing lands for planting seeds by means of slash-and-burn methods also known as swidden agriculture (Ekblom, Gillson & Notelid 2011: 12) although this practice was not commonly practised (Maggs 2010: 216–7). In other parts of southern Africa's eastern low-lying regions, Iron Age people resident in settlements in KwaZulu-Natal that had been inhabited for several centuries tended to refrain from using swidden strategies (ibid. 216). Even in parts of the Limpopo valley local wooded areas in the region did not necessarily decrease. Instead, local farmers preferred to make use of forest gaps for planting crops (Ekblom, Gillson & Notelid 2011: 12). Huffman interprets traces of charcoal as the residue of the burning of grain storage sites and even the dwellings of people responsible for rainmaking rituals (Huffman 2009a: 991–1005). The prominence of rainmaking in the archaeological record suggests that the need for local water supplies for farming were at a premium and required of local leaders to provide interventions to bring rain and mitigate the looming threat of drought.

MIDDLE IRON AGE

By 900 CE when the cold, dry conditions had come to an end in the Mapungubwe area, a new group of farmers (of the Zhizo ceramic culture) moved in and started productive agricultural pursuits (Huffman 2000: 16). This marked the onset of the Middle Iron Age of pre-colonial southern Africa. At the Chixuludzi Pan in the Lower Limpopo valley there was an overall decrease in the amount of charcoal after 950 CE, considered to be an indication of increased rainfall – associated with the medieval warm period (MWP) in Europe (Ekblom, Gillson & Notelid 2011: 13). The best and most productive lands at Mapungubwe were made available for agricultural production while local élites usually resided on second-choice farming land (Huffman 2000: 25). At this stage there was little incentive to secure surplus crops and the local community resorted to *zunde* farming, meaning that everyone helped with the cultivation of the fields as a form of tribute. Town residents were entitled to agricultural land to feed their own families. More fields would then have been required as the population increased (Huffman 2000: 25–6).

Politically the agricultural towns pointed to the emergence of strong leaders. To the west of Mapungubwe emergent Tswana chiefdoms agglomerated in large populations in open country as a deterrent to potential attacks from bands of plunderers. Good water supplies were essential and edible endemic veld plants were a boon to foraging community members. At Mapungubwe there emerged a similar trend, but on a smaller scale. The settlement was smaller than the extensive, populous urban areas of the Tswana. The Mapungubwe leadership resorted to living on a hill in relative opulence with remarkable works of art and handcraft that hinted at the onset of trade with the East coast (Freund 2007: 4–5).

From as early as the eighth century there were indications of urban development in southern Africa and at Mapungubwe this phenomenon started playing itself out (Manyanga, Pikirayi & Chirikure 2010: 574).

By the tenth century the availability of ivory, gold and food production in the Shashe-Limpopo basin created ideal circumstances for trade between the urban Mapungubwe and traders operating from the Indian Ocean coast (ibid. 574–5). Whereas earlier human settlement patterns tended to be in the low-lying regions, now the emphasis was on growth of the urban agglomerations on the foothills overlooking the Limpopo and tributary river plains. Various other communities in the Limpopo catchment similarly settled in the river's foothills (Manyanga, Pikirayi & Chirikure 2010: 576). A notable feature of these later Iron Age settlements was the increasing number of cattle (Badenhorst 2010: 89, Huffman 2010: 25–6).

Climatic conditions had clearly improved. This was evident from the local population. Between 900 and 1000 CE in the Zhizho urban area of Mapungubwe there may well have been as many as 1,900 residents (Huffman 2000: 23, Huffman 1996: 55–60, Huffman 2008: 2034–5). A substantial population of townspeople now required food. At the time there was a notable absence of many large livestock populations (Badenhorst 2010: 92). To meet the growing demand for food farmers resorted primarily to floodplain agriculture (Huffman 2008: 2032). The flood plains of the Shashe-Limpopo confluence held considerable water and for longer periods of time than they do today. Conditions were favourable for cereal crops of sorghum and millet. It was the floodplains below the confluence that were the most productive. A short narrow gorge downstream of the confluence Shashe-Limpopo created a dam effect and acted against the strong stream of the Limpopo, forcing water back into the tributaries where farming settlements flourished at a small delta, known as Kolope, on the edge of the floodplain (Huffman 2007: 382). According to Huffman, in the area there were many farming settlements in the basin on natural terraces above the floodplains. The years 900 to 1300 CE were notable for high rainfall and conditions that favoured agriculture. There were abundant water supplies and warm weather during the growing season (Huffman 2000: 26).

In about 1000 CE new settlers arrived at Mapungubwe. They formed part of the Shona culture from Leopard's Kopje in present-day Zimbabwe. They took control and established a new capital at a settlement that archaeologists have named K2 (Huffman 2000: 17). The presence of the Leopard's Kopje community at K2 between 1000 and 1230 is said to have been as a result of improved local crop production opportunities (Figure 5.4). These people were also linked to communities resident in the Lydenburg region in present-day Mpumalanga Province (Huffman 2008: 2037–8, Whitelaw 1996: 75–83). From this point there appears to have been a strong dispersion of new cultural traditions in the Shashi-Limpopo confluence region (Huffman 2000: 17).

Politically, there were also changes that directly affected important rituals of rainmaking. The control of rainmaking was central to political

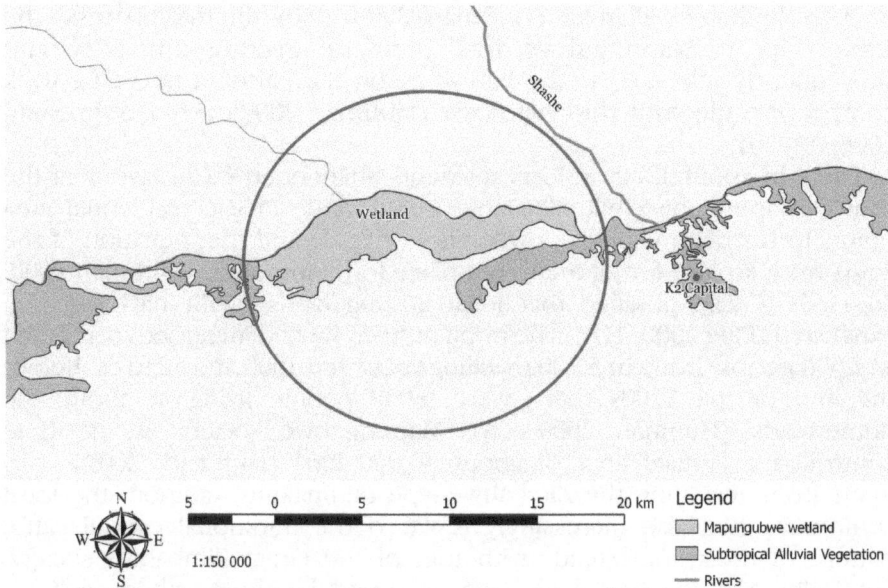

Figure 5.4. The floodplain below the Limpopo used by Middle Iron Age crop farmers in the area of K2 in the Mapungubwe era. The highlighted strips on the present-day South African side of the Limpopo point to the sub-tropical alluvial vegetation in the immediate catchment of the river, indicating a meaningful choice of location and potential sites for planting crops. The K2 settlement is situated in the centre of a subtropical, alluvial site (© Emile Hoffmann).[3]

power. At Mapungubwe rainmaking ceremonies were conducted in different localities. Sometimes ceremonies took place outside the settlement (in the natural surroundings) or on hillsides. Hunter-gatherers initially participated in these activities (Schoeman 2006: 12). In later times the responsibility of rainmaking came under the control of the ruling élite (Schoeman 2006a: 152–65). In times of severe drought, local residents resorted to desperate measures when rainmakers were unable to produce the desired effect (Huffman 2009a: 40–1). One such change manifested after 1000 CE when Mapungubwe rainmaking practices increasingly came under the influence of a tradition developed on the Zimbabwean Plateau. Shona traders and farmers who had become residents of Mapungubwe were instrumental in the implementation of such changes (Schoeman 2006: 12). Politically they had evidently secured a position of authority that their approach to dealing with issues of rainmaking was acceptable to the local community.

Ultimately it was trade that counted. The demand on the east coast for gold induced the Mapungubwe leadership to expand their activities further northwards by the 1200s to take control of some of the goldfields on the Zimbabwe Plateau (Huffman 2009a: 50). In the process, the leadership at Mapungubwe lost their rainmaking authority and subsequently relocated to Zimbabwe. Soon thereafter, a new elite took control of trade with the east coast (Huffman 2009a: 52–3, Schoeman 2006: 15–20).

Then, in about 1220 CE, local residents abandoned K2 in favour of the nearby Mapungubwe hill. The absence of cattle in the residential area points to restrictions on the ownership of cattle and the transition of the court from a place for all men, to a place for commoners (Huffman 2000: 20–1). It is also possible that local grazing lands might have become exhausted (Plug 2000: 118). The population at K2 had increased to as much as 5,300 people living in 3,800 dwellings between 1000 and 1230 CE. Before the end of the 1200s there were 9,000 people living in about 400 homesteads (Huffman 2000: 23). Mapungubwe society at the time controlled a surface area of about 30,000 km^2 (Huffman, 2009a: 44). Apart from accepting the Zimbabwe-type rainmaking tradition, the local community had also increasingly replaced the traditional central cattle pattern of their homesteads with that of the Great Zimbabwe society. Politically, Mapungubwe had been absorbed by the northern culture. Further change was in the offing; in 1300 CE there was an evacuation of the area (Mitchell, 2013: 663, Huffman, 2009a: 39–40).

THE LATE IRON AGE

Traditionally the Late Iron Age in southern Africa was said to have begun in about 1000 CE (Parkington & Hall 1987: 1–25, Phillipson 1975: 321–42, Plug 2000: 117–26). However, in the 1980s, in the light of more extensive research findings and the subsequent growing awareness of global climate change, a new categorization has emerged. The new periodic synchronization suggests that the Late Iron Age began in about 1300 CE with the abandonment of Mapungubwe – one of the pivotal Iron Age centres of southern Africa. Moreover, the new classification also coincided globally with the onset of the Little Ice Age (Huffman 2008: 2032–47, Denbow 1986: 3–28). In South Africa, during the Little Ice Age (1300 to c.1800 CE) average temperatures could have been 1°C colder. Nor was this all. There were also fluctuations. During the coinciding medieval warm period (MWP) in Europe, which in southern Africa tended to prevail between 1000 and 1400 CE, average temperatures could have been 3°C higher (Tyson, Karlén, Holmgren & Heiss 2000: 121–6). These significant changes in temperature might well have had the effect of stimulating extreme drought and/or flood conditions on the subcontinent. A clear indication of these changing climatic conditions was the tendency of Mapungubwe residents to start evacuating the Shashe-Limpopo confluence area. Some moved to the south into the

Soutpansberg region of present-day Venda in Limpopo Province. Others moved northwards into modern-day Zimbabwe (Huffman 2000: 24).

However, these changes of climate outlined by Huffman have since been subjected to contestation on the basis of research findings in Botswana. It has been suggested that politics and local economic matters probably played a more important role than climate change in the fourteenth century and that there was no shortage of water supplies at that time (Denbow, Smith, Ndobochani et al. 2008: 459–80). It is argued that it was precisely because of climate change that there were indications of political conflict. As climatic conditions changed these would naturally have affected the livelihoods of communities and conflict was bound to ensue.

In the arid parts of southern Africa hunter-gatherer communities, in times of drought, would typically have resorted to reciprocal altruism, across genetic kinship lines (Fukuyama 2011, 29–31, Fukuyama 2014: Locs, 1538–50, 8333–7, 8342, Christian 2011: 123–4, 146–8, 245–82) by means of exchanging gifts (*hxaro*) as part of a strategy aimed at mutually complementing existing resources. The action of exchange implied a greater sense of togetherness in the face of potentially disastrous ecological circumstances. This tended to open up communication channels between people over extended distances – sometimes 200 km and more (Mitchell 2002: 215–6). It also points to social and political systems aimed at gaining control over local natural resources. In areas such as the eastern parts of the subcontinent, where water resources would be ubiquitous, control of the resource was localized, whereas in more arid parts such as the Kalahari, the resource would potentially have been subjected to centralized regional control (Huffman 1986: 284).

Endemic drought conditions are known to have been the single greatest cause of climate-induced human mortality rates on the African continent (Tyson & Gatebe 2001: 106). In the Late Iron Age drought conditions were responsible for creating distrust between people of different cultures living together in specific localities. The historical record suggests that drought conditions were responsible for: the decline of peasantries in many parts of the subcontinent; which in turn caused migrations; and that this contributed to changing forms of production and distribution (Beinart & McGregor, 2003: 13). One example of major catastrophic circumstances that might well have followed in the wake of intermittent climate change was originally advanced by Omer-Cooper (1978) in his history of the *Mfecane/Difaqane* (*c.* 1760–1830). The era was notable for significant dispersals of people in many parts of southern Africa, as a result of political changes amongst the emergent Zulu peoples in KwaZulu-Natal. We do know that the *Madlathuli* drought that occurred in 1800–1803 in the region decimated crops and livestock herds in Zululand. This led to significant destitution (Alcock 2010: 192, Poland, Hammond-Tooke & Voigt 2004: 13).

Circumstances of this kind would typically have had a political spin-off in Late Iron Age farming communities.

Zimbabwean state formation

The apparently peaceful southward migration of the Sotho-Tswana groups coincided with the shift of regional dominance of Mapungubwe to Zimbabwe in 1300 CE. The leadership at the impressive urban capital of Great Zimbabwe seized control of the gold trade via Indian Ocean ports. This state, arising in about 1100 CE, was situated in close proximity to the Sabi River and held sway over a significant agro-pastoral population on the Zimbabwean Plateau. In the southeast the countryside was well watered with fertile soil and cultivated lands. The region formed part of the line of communication with the East African emporium of Kilwa (Collins & Burns 2014: 164–6). At its height this early African city complex spread out over a surface area of about 720 ha, with 25,000 m^2 of stone-built architecture and an estimated population of about 18,000 residents. There were even three access routes to the east coast. Great Zimbabwe is considered to have been one of the most comprehensive African polities of the southern African Iron Age (Mitchell 2002: 313, 322). Its area of influence covered more than 30,000 km^2 (ibid. 332). The empire of Great Zimbabwe thrived between about 1250 and 1450. Even after its apparent decline its influence remained in numerous satellite communities such as Mutapa on the northern edge of the Zimbabwe Plateau between 1450 and 1760 CE; and the Khami River settlement, near modern day Bulawayo, which flourished on the southern Zimbabwean Plateau from about 1450 to 1700 CE (Collins & Burns 2014: 169–70).

Great Zimbabwe was in an ideal locality. The capital city was centrally situated in a highly populated area and its rulers had control over numerous Tonga communities along the tsetse fly-free route between the Zimbabwean Plateau and the Mozambican coast. In addition there were valuable salt supplies, abundant quantities of ivory, supplies of gold and cereal crops from the plateau region (Mitchell 2002: 325–6). Maize, introduced to southern Africa by the Portuguese in 1550 could potentially have been included in Zimbabwean crop production cycles (Huffman 2007: 41, Burtt-Davy 1914: 12–4). Requiring less water under dry conditions than sorghum, substantial supplies of surplus maize might well have been sold to stimulate trade with the coast, if not consumed locally.

The influence of the Zimbabwean state, specifically the Shona people, in the region to the south of the Limpopo River basin in modern-day South Africa, did not come to an abrupt halt after the evacuation of Mapungubwe. Ties of friendship and trade remained with the Soutpansberg peoples who had previously moved from Mapungubwe. The Sotho-Tswana entered the Bushveld of the Soutpansberg range, straddling the Limpopo River catchment by about 1300 and in turn, moved into today's Limpopo Province, North West Province and the southeastern parts of Botswana (Hall 2012: 128–9). Both groups subscribed to the central cattle pattern and this strengthens the theory that they had both come from East Africa and that the reason for their southwards move was changing climatic conditions. The Sotho Tswana's arrival in the 1300s was probably a move during slightly wetter times (Hall 2012: 129–31). The Sotho settlements in the northern Bushveld were conducive to relations with Zimbabwe. In the 1700s there was a marked increase in the Venda population in the Limpopo Province region. Local communities developed strong ties in the Lowveld where, as was the case in the Limpopo Valley at Messina, ore deposits were exploited at Phalaborwa, Gravelotte and Leydsdorp – all areas situated to the east of the Drakensberg range that had its origins in the northern region where the legendary Balobedo Rain Queen, Modjadje (Jensen Krige and Krige, 1980) with strong ties to the Venda, built up formidable authority and prestige in a vast geographical region in southern Africa (Huffman 2000: 15).

Nguni movements

The first Nguni-speaking peoples in southern Africa were resident by about 1100 CE in the coastal regions of what is now KwaZulu-Natal (Hall 2012: 129–31). They had a distinctly different tradition of rainmaking to that of their neighbours in the west (Hammond-Tooke 1993: 81). The Nguni, who seemed ambivalent in their relations with the San people, relied on these 'free' people to make rain. Especially in the Drakensberg region the Nguni believed that the San and their descendants had mystical rainmaking powers (Parkington & Hall 2012: 97). The Nguni, comprising the Xhosa in the southeastern Cape and the Zulu of KwaZulu-Natal, inhabited the region originally settled by the eastern stream of Early Iron Age settlement. However, there were also dispersed people who were Nguni descendants of the Moor Park settlement phase in the Eastern Cape (c. 1300 CE) who were linked with parts of the Waterberg in the Limpopo Province. These communities are said to have settled in the region after 1500 CE (Hall 2012: 144–5).

The Nguni cultures also followed the tradition of the central cattle pattern – a cultural manifestation of spatial use and economic priorities (Badenhorst 2009: 149–53). Politically, it signified the importance of cattle, especially amongst Nguni people in the eastern parts of southern Africa. The byre was in the centre of the settlement and also appears to have been the area where the male members of the community deliberated on matters of authority, governance and the manner in which certain rules would be applied. Around the byre there was a residential area, with the senior male leader's location at the back of the settlement and the houses of the wives adjacent to the senior dwellings. The rest of the population lived on the edges of the settlement's perimeter, leading to the front entrance. The spread of livestock farming and the patterning of settlements increasingly pointed to the emergence of links of trade and communication between regional communities along the eastern coastal lands, as well as to the north and south of the Shashe-Limpopo basin, and also further west into Botswana's Tsodilo Hills and the Okavango. The regional exchange of cattle, hides, ivory, fish, salt and other mineral resources point to more complex networks of integration in southern Africa. In many cases the central cattle pattern was representative of an emergent architectural culture of organization (Parkington & Hall 2012: 80–91).

The Tswana and Khoisan in the Kalahari

The apparent dominance of a cattle culture amongst farming communities in southern Africa in the emergent Late Iron Age seems to overshadow crop farming and by implication also the issue of water availability and the manner in which communities managed the resource. Competing demands for scarce water resources by different user communities caused indigenous communities to resort to specific strategies of appropriating these resources. From about 1300 CE the formation of culturally distinct communities with unique languages and lifestyles became a feature of regional development. However, there was also considerable mobility. Sotho-Tswana communities moving south-ward in search of favourable warmer climates in colder times came into contact with San and Khoi communities. These groups were either integrated or moved away. In the western and southern Kalahari region some San lived menial lives under the Tswana communities who had started forming centralized entities by the eighteenth century (Hall 2012: 162).

There is every reason to believe that Khoisan pastoralists and hunter-gatherers in the arid parts of the southern African interior did manage to maintain reasonable control of the environment in which they resided, despite the growing presence of Bantu-speaking communities. They appear to have had a reliable knowledge of the available water resources and how these valuable areas needed to be managed. In southern Botswana the Khoisan people are known to have been associated as family groups with specific waterholes, which they exploited when good rains fell and filled the depressions. These waterholes usually occurred in pairs. The 'female' waterhole was said to have been the one that would hold water longer. The 'male' hole had the more feeble supply and would also allow for the water to evaporate faster. The waterholes were frequently cleared of external deposits in an effort to enlarge their capacity and to ensure that the water supply would be clean. The lines along which the waterholes occurred spoke, according to Traill, of the features of the landscape. It linked distant parts of the territory. These waterholes also determined the routes travellers used in the Thirstland (Traill 1998: 290).

There was a high rate of seasonal movement amongst the Khoi and San in the Kalahari. It is said that people frequently visited relatives and they would move for extensive periods of time to distant areas. One implication was that waterholes would constantly have a high rate of population turnover with a permanent population of about 13 per cent and 35 per cent sharing their time each season between two or more sites. In the case of permanent waterholes the times of residence were not of a random nature. People would have fixed schedules of staying at the waterholes (Parkington 2001: 5).

WORKING WITH WATER IN THE IRON AGE

Although pastoral industry tended to dominate human food production patterns in the era, there is reason to believe that the southward-migrating peoples were capable of switching between pastoral and agricultural pursuits over the medium to long-term (Mace 1993: 363–82). Changing climatic conditions, such as drought and famine, the abstention of traditions of migration, and population growth in some regions, had a definitive impact on agriculture. Lifestyles were also subject to change. In parts of modern-day Botswana, when livestock farming activities increased, the edible wild foods available for hunter-gatherers began to diminish. They then resorted to livestock farming and only seasonally returned to hunting activities. In areas such as Leopard's Kopje, in Zimbabwe the local water supply also affected other activities in settlements. The seasonal rising of the water table halted all mining (Hammel, White, Pfeiffer et al. 2000: 49–56) and at such times, at least temporarily, farming activities potentially flourished.

Wetland farming

In southern Africa, prior to the widespread settlement of Europeans, the farming of wetlands was commonplace. This type of farming, known as *dambos*, *mapani*, *matoro*, *amaxhapozi*, *shiramba* or *mashamba* and *vleis* featured prominently in the folklore of early farming communities (Wood, Dixon & McCartney 2013: 11). They are primarily situated in wetland environments that retain water close to the surface for the greater part of the year. This availability of surface water supports the vigorous growth of grasses and sedges, as well as high plant and animal diversity during the dry season. Traditionally, farmers planted sorghum, pumpkins and a variety of gourds in these areas (Shaw 1974: 92). Other crops included *Clues esculentus* (tzentsa), cucurbits, coco yams and a variety of vegetables. In later times these crops were increasingly replaced by the cultivation of maize. In the case of well-flooded *dambos*, rice crops were produced in parts of Mozambique by the nineteenth century (Van Wyk & Gericke 2007: 10). Water was near the surface and this meant that shallow wells could be used to water vegetable gardens throughout the year (Whitlow 1990: 197). Zimbabwean agriculturalists used the central watershed plateau in the country extensively by the mid nineteenth century (ibid. 197). The colonial authorities halted the system in the twentieth century. Farmers were told that these activities promoted erosion (Whitlow 1990: 197, Wood, Dixon & McCartney 2013: 12–3).

New technologies also shifted the interest in *dambo* farming. African farmers who had made use of *dambo* agricultural activities began exploiting the technology of ploughs drawn by oxen. Larger portions of land could now be ploughed (Whitlow 1977: 197). How extensive the use of the *dambo* system might have been in pre-colonial Zimbabwe is evident when one realizes that in the late twentieth century some 1.3 million ha of the country's surface area was identified as *dambo* lands (Shoniwa 1998: 6).

Floodplain agriculture

As pointed out earlier, in the case of Mapungubwe, floodplain agriculture played a significant role. A floodplain is different to a wetland, in that its water tends to periodically dry up. East of Mapungubwe where the valleys of the Luvhuvhu and Limpopo converge there are floodplains. Although the average rainfall is about 430 mm the floor is flat with an elevation of between 200 and 250 mm. In the summer this region is frequently prone to flooding as a result of the poor drainage (Plug 2000: 117). One site, Thulamela, flourished in the 1300s when the area of the Limpopo-Shashe confluence went into decline and also later when Great Zimbabwe began to decline in about 1450. What then happened was the Khami culture in the south-western parts of Zimbabwe started flourishing. These conditions favoured Thulamela in that it attracted elites who had significant numbers of cattle (Ekblom, Gillson & Notelid 2011: 14–5).

There were also indications of extensive agricultural activity. Tsonga communities in the northern Lowveld of South Africa first drained the wetlands and the women cultivated the wet hollows near lakes between the dunes. According to Junod, in the vicinity of Rikatla there was a natural depression, the centre of which was filled with papyrus and slender typha. The roots and stems of these plants were collected and piled into heaps about one metre high and then left to dry. They were not burnt but left to decay; these were then used as a growing medium for pumpkin seeds which were planted in the heap (Junod 1913: 22). Further south, in the Bokoni region near Carolina in what is now the Mpumalanga

Province, there are also indications of pre-colonial floodplain irrigation on the farm Rietfontein 95. Physical remains of early farming can be seen, including a number of canals on a fairly level tract of land. It appears that a dam of sorts was built to take water from the Gemsbokspruit to form a floodplain so that water could siphon through onto the farming lands (Myburgh 1956: 44).

There is relatively little in the literature on the Iron Age history of floodplain farming in the Okavango Delta (Mitchell 2013a: 107–21, Gumbo 2010: 14–5, 74–5), or even on South Africa's Pongola Floodplain (Stump 2013: 678–9). We do know that intermittent floodplain agriculture known as *molapo* farming was practised in Botswana prior to the region becoming a protectorate in the 1880s. *Molapo* lands proliferated, depending on the quantity of water emanating from the seasonal flooding of the delta and lake lands (Wood, Dixon & McCartney 2013: Chapter 3, Mackenzie 1946: 5, Kgathi, Mmopelwa & Mosepele 2005: 70–81, Bendsen & Meyer 2002).

STONE TERRACING AND IRRIGATION

Although there have been many attempts to identify water canals and furrows representative of a sophisticated irrigation farming culture in southern Africa in pre-colonial times, descriptions have been ephemeral, largely because irrigation is not always considered as 'indigenous' to Africa. Denison and Wotshela provide one reason for this opinion. They argue that the development of stone tools in North Africa, the Middle East and Southern Europe was instrumental in the subsequent evolution of settlements notable for water-diversion technologies that became evident in Egypt, Tunisia and the Middle East. In Africa south of the Sahara the same technology was not considered quintessential for communities to flourish. Furthermore, there were substantial potential sources of food in the environments in which early farmers settled. Consequently, irrigation and related water harvesting strategies appear not to have been as sophisticated as those that evolved in the Northern Hemisphere (Denison & Wotshela 2009: 40). This implies that a Neolithic collective consciousness was absent in southern Africa in the transition from the Stone to the Iron Age some two millennia ago.

Stone wall architecture and terracing

The use of stone appears to be one of the important prerequisites for pre-colonial irrigation. A variety of ways in which stone has been used as building material in human architectural structures suggests that attention was also given to water. In southern Africa there are typically two types of stone walling structures that were utilized in the Iron Age period. They are defined as those belonging to the Zimbabwean pattern and the emergent style of the central cattle pattern of settlement (Sadr 2012). Related to these structures are elements of terracing that are thought to date back to early forms of agricultural irrigation in southern Africa.

Basic practices such as the levelling of the land, or even the construction of terraces, are not necessarily original in one part of the world or another. Terracing has a pronounced and long history in Africa. However, it seems to have been transmitted from the Middle East into Africa, and not Asia as some authorities argued earlier (Widgren 2007: 65–6). In Africa terracing occurs in areas where farmers plan to conduct the planting of crops on the slopes of hills and steep slopes. The terraces serve a number of purposes, i.e. for the simple conservation of soil; clearing the area of stones; increasing the soil depth; and creating a suitable surface supply for gradient cultivation of the land (Soper 1997: 227–8). Terracing is also described as a form of landesque capital investment, with builders contemplating the work performed as an investment made in land with the understanding that there would be an afterlife in the land once the crop had been harvested (Soper 1997: 227–8). Terracing also creates an awareness of the anticipated life of the land well beyond the present crop.

The construction of terraces on hillsides is a prolific tradition in West and East Africa (Netting 1974: 36, Summers 1971). Similar structures have also been found on the island of Madagascar (Dick-Read 2005: 102), built primarily for the purpose of agriculture. These terraces tended to slant downwards to ensure natural irrigation by gravitation, the water reaching different levels along the hillside. This was a popular form of land-use in southern Africa until the end of the nineteenth century. Elements of terracing have also been found in Angola (Summers 1971: 183–4). In southern Africa terracing appears to have been developed primarily for the purposes of clearing lands of rocks and controlling water-flow (Denison & Wotshela 2009: 23). They are fairly commonplace on escarpments such as those in the eastern Highlands of Zimbabwe (Trevor 1930: 391, Soper 1999).

Terracing in southern Africa, according to Stump, points to two essential sites of early irrigation in southern Africa. The one was at Nyanga in Zimbabwe and the other at Bokoni in South Africa's Mpumalanga Province.

Nyanga

Perhaps the best regional example we have today of pre-colonial Late Iron Age agriculture in southern Africa is found in the Nyanga area in the north-eastern Highlands of Zimbabwe. It is here that there is definitive evidence that a pre-colonial farming community made use of what can be classified as irrigation, albeit not on an expansive scale (Soper 2006:

49, 56). The site of Nyanga extends over a surface area of about 7,000 km^2 in the north-eastern part of Zimbabwe. An outstanding feature of the landscape is the evidence of agricultural terracing and cultivation, water furrows, homesteads and defensive fortification. The stone terraces at Nyanga occur on the escarpments and slopes of foothills as well as on nearby detached hills. In some areas the terraces number as many as 100 structures on slopes while their altitudinal heights vary between 900 and 1,700 m on the escarpment and the highland. Soper has identified four classes of terracing: multipurpose for domestic, livestock and homestead garden watering purposes; purely irrigation purposes; directing water to prevent excessive flooding; gently sloped terraced fields terraces, not used for irrigation (Soper 2006: 56). For him the third category is a positive indication of irrigation farming activities.

Nyanga is thought to have been established in about 1200 CE (Soper 2005: 1164) which is considerably older than an earlier estimate of 1500 CE (Soper 1994: 19–20). In the 1970s, on the evidence of structures in the Manicaland region of Zimbabwe, Summers suggested that the construction of irrigation furrows could have been the work of the Portuguese who lived in trading settlements that operated in the hinterland of Mozambique (Summers 1971: 85–6). This is a view that Huffman (in some of his earlier studies) also seems to subscribe to, while Hall tends to be cynical about Nyanga, maintaining that it was primarily a settlement of 'losers' (Hall 1987: 136). Interestingly, in recent years there has been a lively debate in which a strong case has been made out for Nyanga as the site of a rich gold mining area. The extensive terraces and peculiar architecture of many local structures, in the view of Kritzinger, are related to intensive mining activities in the region (Kritzinger 2008: 1–44, 2010: 10–16, 2012: 58–67, 2012a, 2012b: 60–84). Her views are considered controversial by most archaeologists working in the field.

The outside world first became familiar with Nyanga in the late nineteenth century (Chirawu 1999: 1–2). In some respects the Zimbabwean culture of working with water is clearly reflected at Nyanga. Stream headwaters in pre-colonial times were seasonally dammed up as an important element of agricultural activity. The water would then be used for watering fields (Stump 2013: 679). Summers, who began his investigations of the site at the end of the 1940s, compiled a comprehensive report on Nyanga (Summers 1958). He was convinced that the settlers responsible for these structures were of the Iron Age and claims that the people of the sixteenth-century Ziwa culture, an early Iron Age community (they were presumably also pastoralists with goats and sheep) were responsible for the first phase of development. They were the founders of the *Uplands culture* of the meta site. The corresponding *Lowland culture* was a continuation of the Upland culture's architecture, but there was evidence of more refined activity in the form of pottery. It is presumed that this culture started flourishing in the seventeenth century (ibid. 467: 8). The new settlers were people of a second phase in the Iron Age and are said to have been cattle herders. Technologically they were also more advanced than their predecessors.

Soper sees terracing in a social context, viewing it as the organization of labour and an integral part of the social system and settlement pattern of the community (Soper 1997: 228). In his study, Davies points out that Late Iron Age communities, such as those at Engakura in northern Tanzania and Nyanga, have been studied as agricultural landscapes in an effort to understand their agronomies (Davies 2010: 280). Terracing is said to have served as a strategy for clearing the soil of stones for cultivation and also as a method of protecting the soil from erosion. In the case of Nyanga the terraces are not all levelled along the contours; this makes provision for longitudinal drainage and flood irrigation (Soper 2005: 1164). In some areas there are stone-lined drains on the downslope to carry excess run-off, while some of the walls are pierced to let siphoned water through. It appears that the majority of terraces were not irrigated (Soper 1994: 20, 2005: 1164). Stump argues that the structures were not necessarily the result of intensive labour activity but were built gradually, making use of the many stones in the area and were part of an effort to create clearings for planting crops in a particularly stony terrain (Stump 2013: 679).

The networks of cultivation ridges at Nyanga are situated on the less stony slopes below the escarpment; they are found over a distance of some 60 km and are far more extensive than the terraced area. These are parallel and have sub-parallel linear banks; they are about 10 m wide, and up to a metre deep. These formations occur in impeded drainage wetlands and on the valley sides. One piece of agricultural land is said to extend over a surface area of more than 1,000 ha. The water furrows of Nyanga have primarily remained in a fairly good state in the upper highlands areas. They seemed to have served as domestic water supply sources, but also for livestock and gardens (Soper 2005: 1165). In an earlier observation, Soper was of the opinion that they were very old and primarily served as a means of providing water to pit structures (1994: 20). Although it is known that Nyanga was a cattle-farming region, there are few signs of dung. Consequently, there is general acceptance that dung must have been used as fertiliser. Crops planted in the area of Nyanga include millet, sorghum, cowpeas, ground beans, as well as a traditional root crop *Plectranthus esculentus* and *Colacasia* (Soper 2005: 1165).

BOKONI

Some time ago the author conducted research on pre-colonial irrigation in southern Africa in an effort to create an awareness of the potential for research in the field of indigenous water-related technology (Tempelhoff 2008). Working from a vast array of randomly selected secondary materials at face value has opened up a number of questions about why so little attention has been given to our understanding of indigenous practices of water and land-use. One of the case studies of the article dealt with a settlement near the town of Carolina in the Mpumalanga Province on the Drakensberg escarpment. At the time a group of historians had started working on what has been termed the Five Hundred Year Initiative (FYI) a comprehensive interdisciplinary foray into the field of South Africa's pre-colonial history. The project has made significant progress. By 2010, working from a rich variety of advanced contemporary archaeological and anthropological material those involved in the project had succeeded in opening up highly relevant issues on the indigenous and material memory landscape of South African history. One of the flagship initiatives has been the investigation into the history of the Bokoni structures in Mpumalanga. The reconsideration of Bokoni, a complex of walled settlements and terraces with roughly circular homesteads extending over a distance of about 150 km between Ohrigstad in the north and Carolina in the south, has generated significant interest among historians and archaeologists alike. Comparisons have also been drawn with the terraced sites of Nyanga in the north-eastern highlands of Zimbabwe (Delius, Maggs & Schoeman 2012: 399–414). However, there are no indications of ethnic linkages between the two societies (Maggs 2007: 3).

In the 1970s there was considerable work undertaken on the Iron Age in Mpumalanga. The work of Evers (Evers 1975: 71–83, Maggs & Schoeman 2012: 402–3), followed by that of Maggs (Maggs 1976) and later also Schoeman kept the discourse open. However, until the 2000s, little seemed to emanate in terms of broader perspectives (Delius, Maggs & Schoeman 2012: 403–4). We do know that the stone-built settlements in an area extending over a distance of about 150 km along the Drakensberg escarpment between the urban settlements of Carolina in the south and Ohrigstad in the north were signs of a flourishing regional farming society for a period of about 500 years (Maggs 2007: 2).

An outstanding feature of the Bokoni sites is the complexity of the landscape. There are traces of dense settlement (as many as 200 homesteads in a single built-up area) and road networks linking homesteads and agricultural terraces. The features that stand out in the case of Bokoni include: terracing of hillside fields; the specialized manipulation of soils, levelling of plots, mounding and ranging; irrigation; cattle (fed in stalls); manuring of fields; composting/mulching fields; specific crop rotations; and the introduction of new crops (Schoeman 2013: 5–6). The residents apparently produced finger and pearl millet (*Eleusine coracana* and *E. glaucum*) (Hattingh, Schoeman & Bamford 2014). Sorghum was also produced in the area, but against all expectations there have not been traces of maize production (Maggs 2007: 1–4). Apart from cattle the Bokoni also had sheep and presumably goats, but they also busied themselves with agriculture. Several iron hoes have also been found during excavations (Delius & Schoeman 2008: 139).

The Koni settlers of the region appear in Pedi traditions dating back to the 1600s. It seems they were active in trade with Delagoa Bay and Inhambane on the east coast. Their first settlement site was at Moxomatsi near Belfast on the Highveld, but they were harassed by Nguni groups and as a result moved to other areas, one of which was the valley of the Dorps River at Lydenburg, which at one stage may have accommodated as many as 57,000 people (Delius, Maggs & Schoeman 2012: 404–5). A later phase of settlement appears to have been operational in the proximity of Ohrigstad at the time, under the control of the Pedi (Sotho), but this community came under threat from internecine clashes amongst the Pedi, forcing some of the Bokoni to join up with the local Ndzundza Ndebele. Others sought more secure refuge in the mountains where they continued to build their stone walled terraces (ibid. 406–7).

The decline of Bokoni, according to Delius and Schoeman, is thought to have been the result of the conflict between the Pedi's following their defeat by the Ndwandwe who had settled in the Steelpoort River valley in the mid 1820s. Because the Koni sided with the Pedi they were subject to attacks from marauding groups (among others, those of the *Difaqane*) that extended from the Mpumalanga Lowveld as far as the region's section of the Drakensberg range (Delius and Schoeman 2008: 150–1) and into regions where farming Koni settlements had been entrenched for a considerable period of time. Delius and Schirmer depict the Koni as being caught up in the disputes of predatory polities such as the Ndwandwe, Swazi and the Ndebele (Delius & Schirmer 2014: 52). Eventually, in the final phase, there were remnants of Koni groups in the settlement run by the Berlin Mission Society at Botšhabelo, near the later town of Middelburg, but these appeared to have been largely absorbed into Pedi society, and were certainly not as influential as they had been in former times (Delius, Maggs & Schoeman 2012: 407).

The terracing structures of the Bokoni settlements in Mpumalanga are of significant importance in that structures resembling those at Nyanga in Zimbabwe, or at Cherengani in East Africa, have not previously been identified in South Africa (ibid. 409). Understandably then, there are potentially fertile areas for comparative regional studies on the stone structure.

The trade of Bokoni appears to have been with the east coast in the form of ivory, metals and beads. One school of thought, influenced by economic historical perspectives, is that the Koni were middle-dealers between specialist iron producing communities and ivory hunters (after

1500 CE) and that they exchanged these products in barter arrangements that included cattle and also goods brought from the coast (Delius & Schirmer 2014: 44). The Koni are also said to have acted as intermediaries in trade between local elephant hunting communities and traders operating from Delagoa Bay. The valleys in which the Koni were resident, such as the Komati, Crocodile, Sabie and the tributaries of the Steelpoort River, all flow into the Rio Incomati which, in turn, flows into the Indian Ocean at Maputo Bay. The Koni, although they are thought to have been ironworkers, did not have direct access to iron. They presumably acquired their supplies from the Pedi and other communities who mined in the Steelpoort Valley (Delius & Schoeman 2008: 161–3).

In terms of related economic activity it appears that the Koni owned extensive cattle herds and also practised agriculture. However, as a result of high magnesium levels in the water in the river valleys in Mpumalanga it is not considered to be suitable for irrigation. The river streams are situated in sweet veld grasslands. Consequently these proved ideal for cattle farming. Hill slopes in the valleys were preferred for agricultural activity and it was here that the Koni built sloped, terraced agricultural landscapes. The agricultural land was fertilized with the cattle dung (ibid. 162–3, Delius & Schirmer 2014: 43). The organization of Bokoni seems to have been relatively decentralized; political and military systems and settlement patterns tended to be influenced by intensive agriculture instead of strong defences (Delius, Maggs & Schoeman 2012: 414, Delius & Schirmer 2014: 38).

Stump (2010: 255–78) after visiting several sites in South Africa and Zimbabwe, suggested that more attention be given to a regional understanding of terracing and field systems in southern Africa, and potentially also East Africa, to inform archaeological and historical research. Davies is critical of studies done on southern Africa. In respect of the Bokoni landscapes of Mpumalanga, he noted that too much attention has been given to the people who were responsible for the construction of the sites. For him, too much emphasis is given to 'what was', and too little on 'what is' (Davies 2010: 291–2). By 2013 the focus of the project in terms of the FYI had changed significantly, with Schoeman pointing out that the research group intended to work more comprehensively on the collection of oral tradition and historical sources to throw more light on the sites which formed part of the Bokoni settlements (Schoeman 2013: 5).

CONCLUSION

It is intriguing to think that Brown and Wilson's views on southern Africa 'drying out' in the 1860s could well be linked to the first traces of anthropogenic climate change in the south, changes that had their origin in the eighteenth-century industrial revolution in the Northern Hemisphere. They were not the last to take note of what they considered to be marked changes in the region's climate. In the first-half of the twentieth century, drought conditions and the prospect of diminishing water supplies formed part of the strategic planning of senior government engineers (U of SA 1926, Kokot 1948). The discourse on global climate change has contributed much to a more comprehensive understanding of climate change and how landscapes are bound to change over the long term. Brown and Wilson's views on the ethnic demographics of human settlement patterns have changed slightly in the sense that we currently have a more complex subcontinental population of peoples, from many parts of the globe resident in the region. What seems to have remained consistent is the incidence of drought conditions, intermittent floods and frequent, but futile attempts, even in the era of advanced science and technology, to accurately predict trends of the inter-tropical convergence zone and the much-needed supply of rainfall.

Interestingly the settlement cycles of successive Iron Age communities in southern Africa do not necessarily point to a rise-and-fall discourse. Instead, they can be interpreted as complex panarchical cycles of emergence, conservation, collapse and then gradual restoration before new societies, with new horizons, emerge (Gunderson & Holling 2002, Weeks, Rodriguez & Blakeslee 2004, Folke, Carpenter, Walker et al. 2010, Van der Leeuw, Costanza, Aulenbach et al. 2011, Allen, Angeler et al. 2014, 578–589). For example, the gradual collapse of Mapungubwe in the thirteenth century was a creative destruction process leading to the emergence of the Great Zimbabwe Empire. However, Great Zimbabwe also transformed itself in cycles of creative destruction. Much the same can be said of the waves of settlement by the Nguni Xhosa and Zulu societies in the eastern parts of South Africa during the Iron Age, as well as the societies of the Sotho and Tswana. Regional and state-forming societies formed part of the creative transformation processes. Under these circumstances of ongoing change, water appears to be an important natural resource without which societies are unable to thrive.

Socially, it appears as if we are still in an explorative phase of reaching an understanding of the dynamics of the subcontinent's water history. A more holistic view of the human engagement with water over the past two millennia is necessary. Bokoni, in South Africa's Mpumalanga Province and Nyanga, on the Highlands of Zimbabwe, form part of a larger network of aquatic systems shaped by African communities for more than two millennia. Moreover, our understanding of the historical process should extend from the relatively deep past, up to the present. Currently, our understanding of the way African communities interacted with the hydrosphere remains a discourse in need of more attention.

NOTES

1 Quantum GIS hydrographic mapping of southern Africa (based on Hammond-Tooke 1993: 1).
2 Quantum GIS hydrographic mapping synthesis of maps (Huffman 2007: 336, Silva and Steele 2012).
3 Map data of the South African Biodiversity GIS Map (2006) (based on Huffman 2007: 382).

REFERENCES

Anon., 'The regional climate of southern Africa', Orange-Senqu: River awareness kit, http://www.orangesenqurak.com/river/climate/basin/regional.aspx (accessed 15 October 2014).
Alcock, P.G. 2010. *Rainbows in the Mist: Indigenous Knowledge, Beliefs and Folklore in South Africa*. Pretoria: South African Weather Service.
———— 2014. *Venus Rising: South African Astronomical Beliefs, Customs and Observations*. First edition. Pietermaritzburg: Private publication.
Allen, C.R., Angeler, D.G., Garmestani, A.S., Gunderson, L.H. and Holling, C.S. 2014. 'Panarchy: Theory and Application' in *Ecosystems* 17(4): 578–589.
Avery, D.M. 1995. 'Physical environment and site choice in South Africa'. In *Journal of Archaeological Science* 22: 343–53.
Badenhorst, S. 'The descent of Iron Age farmers in southern Africa during the last 2000 years'. In *African Archaeological Review* 27(2), June 2010: 87–106.
Barnard, A. 1992. *Hunters and Herders of Southern Africa: A Comparative Ethnography of the Khoisan Peoples*. Cambridge: Cambridge University Press.
Beinart, W., McGreggor, J. 2003. 'Introduction'. In Beinart, W., McGreggor, J. (eds), *Social History and African Environments*. Oxford: James Currey, pp. 1–24.
Bendsen, H., Meyer, T. 2002. 'The dynamics of the land use system in Ngamiland, Botswana: changing livelihood options and strategies'. Paper presented at an international conference on wetlands, held at the Okavango Swamps.
Berger, I. 2009. *South Africa in World History*. Oxford: Oxford University Press.
Brown, J.C. 1875. *Hydrology of South Africa: Or Details of the Former Hydrographic Conditions of the Cape of Good Hope, and of Causes of its Present Aridity, with Suggestions of Appropriate Remedies for this Aridity*. London: Henry S. King and Co.
Burtt-Davy, J. 1914. *Maize: Its History, Cultivation, Handling, and Uses: With Special Reference to South Africa*. London: Longmans, Green and Co.
Cape of Good Hope, 1865. 'Report of the select committee appointed by the legislative council to consider the colonial botanist's report'. Cape Town: Cape Legislative Council, Paul Solomon and Co., Steam Printing Office.
Carruthers, J. 2003. 'Past and future landscape ideology: the Kalahari Gemsbok National Park'. In Beinart, W., McGreggor, J. (eds), *Social History and African Environments*. Oxford: James Currey, pp. 255–66.

Chirawu, C. 1999. 'Ancient terrace farming in north eastern Zimbabwe'. Paper presented at *The archaeology of farming communities at the World Archaeological Congress 4, 10–14 January*, Cape Town: University of Cape Town.

Christian, D. 2011. *Maps of Time: An Introduction to Big History*. Berkeley: University of California Press.

Clark, J.D. 1962. 'The spread of food production in sub-Saharan Africa'. *Journal of African History* 3(2): 211–28.

Collins, R.O., Burns, J.M. 2014. *A History of Sub-Saharan Africa*. Second edition. Cambridge: Cambridge University Press.

Cox, F.E.G. 2002. 'History of human parasitology'. *Clinical Microbiology Reviews* 15(2): 595–612.

Davies, M. 2010. 'A view from the east: An interdisciplinary "historical ecology" approach to a contemporary agricultural landscape in Northwest Kenya'. *African Studies* 69(2): 279–97.

Delius, P., Maggs, T., Schoeman, M. 2012. 'Bokoni: old structures, new paradigms? Rethinking precolonial society from the perspective of the stone-walled sites in Mpumalanga'. *Journal of Southern African Studies* 38(2): 399–414.

Delius, P., Schirmer, S. 2014. 'Order, openness, and economic change in precolonial southern Africa: a perspective from the Bokoni terraces'. *The Journal of African History* 55(1): 313–22.

Delius P., Schoeman, M.H. 2008. 'Revisiting Bokoni: populating the stone ruins of the Mpumalanga escarpment'. In Swanepoel, N., Esterhuysen, A., Bonner, P. (eds), *Five Hundred Years Rediscovered: Southern African Precedents and Prospects*. Johannesburg: Wits University Press, pp. 135–68.

Denbow, J. 1986. 'A new look at the later prehistory of the Kalahari'. *The Journal of African History* 27(1): 3–28.

—— 1990. Congo to Kalahari: Data and hypothesis about the political economy of the western stream if the early Iron Age. In *The African Archaeological Review* 8: 139–75.

Denbow, J.R., Smith, J. Ndobochani, N.M., Atwood, K., Miller, D. 2008. 'Archaeological excavations at Bosutswe, Botswana: cultural chronology, paleo-ecology and economy'. *Journal of Archaeological Science* 35(2): 459–80.

Denbow, J.R., Wilmsen, E.N. 1990. 'Advent and course of pastoralism in the Kalahari'. *Science* 234(4783): 1509–15.

Denison, J., Wotshela, L. 2009. *Indigenous Water Harvesting and Conservation Practices: Historical Context, Cases and Implications*. Report TT 392/09, Pretoria: Water Research Commission.

Department of Water Affairs and Forestry (DWAF), 2004. National Water Resource Strategy (NWRS). First edition. Pretoria: DWAF.

Dick-Read, R. 2005. *The Phantom Voyagers: Evidence of Indonesian Settlement in Africa in Ancient Times*. Winchester: Thurlton Press.

Ekblom, A., Gillson L., Notelid, M. 2011. 'A historical ecology of the Limpopo and Kruger National Parks and lower Limpopo Valley'. *Journal of Archaeology and Ancient History* 1: 2–29.

Etherington, N. 2010. 'Historians, archaeologists and the legacy of the discredited short Iron-Age chronology'. *African Studies* 69(2): 361–75.

Evers, T.M. 1975. 'Recent Iron Age research in the Eastern Transvaal, South Africa'. *The South African Archaeological Bulletin* 30(119–20): 71–83.

Farley, J. 2003. *Bilharzia: A History of Imperial Tropical Medicine*. Cambridge: Cambridge University Press.

Feely, J.M., Bell-Cross, S.M. 2011. 'The distribution of Early Iron Age settlement in the Eastern Cape: some historical and ecological implications'. *South African Archaeological Bulletin* 66(194): 105–12.

Folke, C., Carpenter, S., Walker, B., Scheffer, M., Chapin, V., Rockström, J. 2010. 'Resilience thinking: integrating resilience, adaptability and transformability'. *Ecology and Society* 15(4), at http: //www.ecologyandsociety.org/vol15/iss4/art20/ (accessed 7 February 2013).

Freund, B. 2007. *The African city: A history*. Cambridge: Cambridge University Press.

Fukuyama, F. 2011. *The Origins of Political Order: From Prehuman Times to the French Revolution*. London: Profile Books.

—————— 2014. *Political Order and Political Decay: From the Industrial Revolution to Globalisation and Democracy*. Kindle version, London: Profile Books.

Greenfield, H.J., Fowler, K.D., Van Schalkwyk, L.O. 2005. 'Where are the gardens? Early Iron Age horticulture in the Thukela River basin of South Africa'. *World Archaeology* 37(2): 307–28.

Greenfield H.J., Miller, D. 2004. 'Spatial patterning of Early Iron Age metal production at Ndondondwane, South Africa: the question of cultural continuity between the Early and Late Iron Ages'. *Journal of Archaeological Science* 31(11): 1511–32.

Grove, R.H. 1995. *Green Imperialism: Colonial Expansion, Tropical Island Edens and the Origins of Environmentalism, 1600–1860*. Cambridge: Cambridge University Press.

Gunderson, L.H., Holling, C.S. (eds). 2002. *Panarchy: Understanding Transformations in Human and Natural Systems*. Washington, D.C.: Island Press.

Hall, M. 1987. *The Changing Past: Farmers, Kings and Traders in Southern Africa, 200–1986*. Cape Town: David Philip.

Hall, S. 2012. 'Farming communities in the second millennium'. In Hamilton, C., Mbenga, B.K., Ross, R. (eds). 2012. *The Cambridge History of South Africa. Volume 1: From Earliest Times to 1885*. Cambridge: Cambridge University Press, pp. 112–67.

Hamilton, C., Mbenga, B.K., Ross, R. (eds). 2012. *The Cambridge History of South Africa. Volume 1: From Early Times to 1885*. Cambridge: Cambridge University Press.

Hammel, A., White, C., Pfeiffer, S., Miller, D. 2000. 'Pre-colonial mining in southern Africa'. *The Journal of the South African Institute of Mining and Metallurgy* 100(1): 49–56.

Hammond-Tooke, D. 1993. *The roots of black South Africa*. Johannesburg: Jonathan Ball Publishers.

Hattingh, T., Schoeman, A., Bamford, M. 2014. 'A phytolith analysis of Bokoni soils'. Paper presented at the 14th Congress of the Pan African Archaeological Association, 14–18 July. Johannesburg: University of the Witwatersrand.

Hitchcock, R.K. 1987. 'Socioeconomic change among the Basarwa in Botswana: an ethnographic analysis'. *Ethnohistory* 34(3): 219–55.

Hornberg, A., McNeill, J.R., Martinez-Alier, J. (eds). 2007. *World-System History and Global Environmental Change*. Lanham: AltaMira Press.

Huffman, T.N. 1986. 'Archaeological evidence and conventional explanations of southern Bantu settlement patterns'. *Africa* 56(3): 280–98.

———— 1996. 'Archaeological evidence for climatic change during the last 2000 years in Southern Africa'. *Quaternary International* 33: 55–60.

———— 2000. 'Mapungubwe and the origins of the Zimbabwe Culture'. *Goodwin Series*, Vol. 8, African Naissance: The Limpopo Valley 1000 Years Ago: 14–29.

———— 2007. *Handbook to the Iron Age: The Archaeology of Pre-Colonial Societies in Southern Africa*. Pietermaritzburg: University of KwaZulu-Natal.

———— 2008. 'Climate change during the Iron Age in the Shashe-Limpopo basin, southern Africa'. *Journal of Archaeological Science* 35: 2032–47.

———— 2009. 'A cultural proxy for drought: ritual burning in the Iron Age of southern Africa'. *Journal of Archaeological Science* 36: 991–1005.

———— 2009a. 'Mapungubwe and Great Zimbabwe: The origin and spread of social complexity'. *Journal of Archaeological Anthropology* 28(1): 37–54.

———— 2010. 'Debating the Central Cattle Pattern: A reply to Badenhorst'. *South African Archaeological Bulletin* 65(192): 164–74.

———— 2010a. 'Mapungubwe and the origins of the Zimbabwe culture'. *Goodwin Series* 8: 14–29.

Isichei, E. 1997. *A History of African Societies to 1870*. Cambridge: Cambridge University Press.

Jensen Krige, E., Krige, J.D. 1980. *The Realm of the Rain-Queen: A Study of the Pattern of Lovedu Society*. Cape Town: Juta and Company.

Junod, H.A. 1913. *The Life of a South African Tribe: II. The Psychic Life*. Neuchatel: Imprimerie Attinger Freres.

Kgathi, D.L., Mmopelwa, G., Mosepele, K. 2005. 'Natural resources assessment in the Okavango Delta, Botswana: Case studies of some key resources'. *Natural Resources Forum* 29(1): 70–81.

Klein, R.G. 2001. 'Southern Africa and modern human origins'. *Journal of Anthropological Research* 57(1): 1–16.

Kokot, D.F. 1948. 'An investigation into the evidence bearing on recent climatic change over southern Africa' *Irrigation Department Memoir*. Pretoria: Government Printer.

Kritzinger, A. 2012. '*Guyo* and *huyo*: science proves the stone tools used in liberating gold fines from waste in Nyanga district, Zimbabwe'. *Nyame Akuma* 78: 58–67.

———— 2012a. 'Laboratory analysis reveals direct evidence of precolonial gold recovery in the archaeology of Zimbabwe's Eastern Highlands'. *Proceedings of 9th International Mining History Congress*, Johannesburg, 17–21 April (paper no longer available at listed website www.imhc.co.za (accessed 10 October 2014).

———— 2012b. 'Location of Zimbabwe's sixteenth-century mines identified from a Portuguese document – with particular reference to Manyika in the Eastern Highlands'. *Zimbabwea* 10: 60–84.

———— 2010. 'Gradient and soil analysis identify the function of stone-built tunnels in the archaeology of the Eastern Highlands, Zimbabwe'. *Nyame Akuma* 73: 10–16.

———— 2008. 'Gold not grain – precolonial harvest in the terraced Hills of Zimbabwe's Eastern Highlands'. *Cookeia Journal* 13: 1–44.

Livingstone, D. 1858. *Missionary Travels and Researches in South Africa; Including a Sketch of Sixteen Years' Residence in the Interior of Africa, and a Journey from the Cape of Good Hope to Loanda on the West Coast: Thence Across the Continent, Down the River Zambezi, to the Eastern Ocean*. New York: Harper and Brothers.

——— 1857. 'Explorations into the interior of Africa'. *Journal of the Royal Geographical Society of London* 27: 349–87.

Lombard, M., Schlebusch, C., Soodyall, H. 2013. 'Bridging disciplines to better elucidate the evolution of early *Homo sapiens* in southern Africa'. *South African Journal of Science,* 109(11/12): 27–34.

McCann, J. 1999. 'Climate and causation in African history'. *The International Journal of African Historical Studies* 32(2–3): 261–79.

Mace, R. 1993. Transitions between cultivation and pastoralism in sub-Saharan Africa. In *Current Anthropology,* 34(4), August–October: 363–82.

Mackenzie, L.A. 1946. Report on the Kalahari expedition. Being a further investigation into the resources of the Kalahari and their relationship to the climate of South Africa. Government Printer, Pretoria.

Maggs, T.M. O'C. 2009. 'The FYI Workshop and excursion: valuable lessons from, East Africa'. *African Studies* 69(2): 213–17.

——— 2007. 'Iron Age settlements of the Mpumalanga escarpment: some answers but many questions'. *The Digging Stick* 24(2): 1–4.

——— 1976. *Iron Age Communities of the Southern Highveld.* Pietermaritzburg: Council of the Natal Museum.

Manyanga, M., Pikirayi, I., Chirikure, S. 2010. 'Reconceptualising the urban mind in pre-European southern Africa: Rethinking Mapungubwe and Great Zimbabwe'. In Sinclair, P.J.J., Nordquist, G., Herscend, F., Isenahl, S. (eds), *The Urban Mind: Cultural and Environmental Dynamics.* Uppsala: Uppsala University, pp. 573–90.

Menotti F., O' Sullivan, A., (eds). 2013. *The Oxford Handbook of Wetlands Archaeology*. Oxford: Oxford University Press.

Miller, D., Whitelaw, G. 1995. 'Early Iron Age metal working from the site of Kwagandaganda, Natal, South Africa'. *The South African Archaeological Bulletin* 49(160): 79–89.

Miller, D.E., Van der Merwe, N.J. 1994. 'Early Iron Age metal working at the Tsodilo Hills, Northwestern Botswana'. *Journal of Archaeological Science* 21(1): 101–15.

Mitchell, P. 2013. 'Early farming communities of southern and South-central Africa'. In Mitchell P., Lane, P. (eds), *The Oxford Handbook of African Archaeology.* Oxford: Oxford University Press, pp. 657–70.

——— 2013a. 'People and wetlands in Africa'. In Menotti, F., O'Sullivan, A. (eds), *The Oxford Handbook of Wetlands Archaeology.* Oxford: Oxford University Press, pp. 107–21.

——— 2002. *The Archaeology of Southern Africa.* Cambridge: Cambridge University Press.

Mitchell P., Lane, P. (eds). 2013. *The Oxford Handbook of African Archaeology.* Oxford: Oxford University Press.

Morgan, J.A.T., De Jong R.J., Adeoye, G.O., Ansa, E.D.O., et al. 2005. 'Origins and diversification of the human parasite *Schistosoma mansoni*'. *Molecular Ecology* 14: 3889–902.

Myburgh, A.C. 1956. *Die stamme van die distrik Carolina*. Pretoria: Government Printer.

Netting, R.M. 1974. 'Agrarian ecology'. *Annual Review of Anthropology* 3: 21–56.

Omer-Cooper, J.D. 1978. *The Zulu Aftermath: A Nineteenth-Century Revolution in Bantu Africa*. London: Longman.

Orpen, J.M. 1964. *Reminiscences of Life in South Africa from 1846 to the Present Day, Vols I and II*. Cape Town: C Struik.

Packard, R. 2001. 'Malaria blocks development' revisited: the role of disease in the history of the agricultural development in the Eastern and Northern Transvaal Lowveld, 1890–1960'. *Journal of Southern African Studies* 27(3): 591–612.

Pallet, J. (ed.). 1997. *Sharing Water in Southern Africa*. Windhoek: Desert Research Foundation of Namibia.

Parkington, J. 2001. 'Presidential address: seasonality and Southern African hunter-gatherers'. *The South African Archaeological Bulletin* 56 (173/4): 1–7.

Parkington, J. and Hall, M. 1987. 'Patterning in recent radiocarbon dates from southern Africa as a reflection of prehistoric settlement and interaction'. *The Journal of African History* 28(1): 1–25.

Parkington, J. and Hall, S. 2012. 'The appearance of food production in southern Africa'. In Hamilton, C., Mbenga, B.K., Ross, R. (eds), *The Cambridge History of South Africa. Volume 1: From Early Times to 1885*. Cambridge: Cambridge University Press, pp. 63–111.

Parsons, N. 2008. 'South Africa in Africa more than five hundred years ago: some questions'. In Swanepoel, N., Esterhuysen, A., Bonner, P. (eds), *Five Hundred Years Rediscovered: Southern African Precedents and Prospects*. Johannesburg: Wits University Press, pp. 41–54.

Phillipson, D.W. 1975. 'The chronology of the Iron Age in Bantu Africa'. *The Journal of African History* 16(3): 321–42.

———— 1994. *African Archaeology*. Second edition. Cambridge: Cambridge University Press.

Plug, I. 2000. 'Overview of Iron Age fauna from the Limpopo Valley'. *Goodwin Series* 8: 117–26.

Poland, M., Hammond-Tooke, D., Voigt, L. 2004. *The Abundant Herds: A Celebration of the Cattle of the Zulu*. Vlaeberg: Fernwood Press.

Quin, P.J. 1959. 'Foods and feeding habits of the Pedi with special reference to identification, classification, preparation and nutritive value of the respective food'. Johannesburg: Witwatersrand University Press.

Robbins, L.H., Campbell, A.C., Murphy, M.L., Brook, G.A., Srivastava, P., Badenhorst, S. 2005. 'The advent of herding in southern Africa: Early AMS dates on domestic livestock from the Kalahari Desert'. *Current Anthropology* 46(4): 671–7.

Robertson, H., Bradley, A. 2000. 'A new paradigm: The African Early Iron Age without Bantu migrations'. *History in Africa* 27: 287–323.

Rowntree, K. 2000. 'Geography of drainage basins: hydrology, geomorphology, and ecosystems management'. In Fox, R., Rowntree, E. (eds), *The Geography of South Africa in a Changing World*. Cape Town: Oxford University Press, pp. 393–416.

Sadr, K. 2003. 'The Neolithic of South Africa'. *Journal of African History* 44(2): 195–209.

———— 2012. 'The origins and spread of pre-colonial stone-walled architecture in southern Africa'. Invited lecture at the Centre for African Studies, University of Leiden, 5 April.

Schapera, I. 1960. *The Khoisan People of South Africa: Bushmen and Hottentots*. London: Routledge and Kegan Paul.

Schoeman, M.H. 2006. 'Clouding power? Rain-control, landscapes and ideology in Shashe-Limpopo state formation'. PhD thesis, Johannesburg: University of the Witwatersrand.

———— 2006a. 'Imagining rain-places: rain-control and changing ritual landscapes in the Shashe-Limpopo confluence areas, South Africa'. *The South African Archaeological Bulletin* 61(184): 152–65.

———— 2013. 'The archaeology of Komati Gorge: forming part of the broader project exploring precolonial agriculture and intensification: the case of Bokoni South Africa'. Johannesburg: University of the Witwatersrand.

Seddon, D. 1968. 'The origins and development of agriculture in East and Southern Africa'. *Current Anthropology* 9(5), Part 2: 489–509.

Selous, F.C. 1909. 'Big game in South Africa and its relation to the tse-tse fly'. *Journal of the Royal African Society* 8(30): 113–29.

Shaw, M. 1974. 'Material culture'. In Hammond-Tooke W.D. (ed.), *The Bantu-Speaking Peoples of Southern Africa*. London: Routledge and Kegan Paul, pp. 85–134.

Shoniwa, F.F. 1998.The effects of land-use history on plant species diversity and abundance in dambo wetlands of Zimbabwe. Morgantown: MSc. West Virginia University.

Silva F., Steele, J. 2012. 'Modelling boundaries between converging fronts in prehistory'. *Advances in Complex Systems* 15(1 and 2).

Sinclair, P.J.J., Nordquist, G., Herscend, F., Isenahl, S. (eds). 2010. *The Urban Mind: Cultural and Environmental Dynamics*. Uppsala: Uppsala University.

Smith, A.B. 1992. 'Origins and spread of pastoralism in Africa'. *Annual Review of Anthropology* 21: 125–41.

Solomon, S. 2010. *Water: The Epic Struggle for Wealth, Power and Civilization*. New York: Harper.

Soper, R. 1994. 'Zimbabwe: Ancient fields and agricultural systems: new work on the Nyanga terrace complex'. *Nyama Akuma* 42: 18–21.

———— 1997. 'East African terraced-irrigation systems'. In Vogel, J.O., Vogel, J. (eds), *Encyclopedia of Precolonial Africa*. Walnut Creek: Altramira, pp. 228–30.

———— 1999. 'The agricultural landscape of the Nyanga area of Zimbabwe'. Paper presented at *The archaeology of farming communities at the World Archaeological Congress 4, 10–14 January*, University of Cape Town.

———— 2005. 'Nyanga Hills'. In Shillington, K. (ed.), *Encyclopedia of African History: Volume 2 H-O*. New York: Fitzroy Dearborn, pp. 1164–5.

———— 2006. *The Terrace Builders of Nyanga*. Harare: Weaver Press.

Steverding, D. 2008. 'The history of African trypanosomiasis'. *Parasites and Vectors* 1: 3.

Stump, D. 2010. 'Intensification in context: Archaeological approaches to precolonial field systems in Eastern and southern Africa'. *African Studies* 69(2): 255–78.

———— 2013. 'The archaeology of agricultural intensification in Africa'. In Mitchell, P., Lane, P. (eds), *The Oxford Handbook of African Archaeology*. Oxford: Oxford University Press: pp. 671–88.

Summers, R. 1958. *Inyanga: Prehistoric Settlements in Southern Rhodesia*. Cambridge: Cambridge University Press, for the Inyanga Research Fund.

———— 1971. *Ancient Ruins and Vanished Civilizations of Southern Africa*. Cape Town: TV Bulpin.

Swanepoel, N., Esterhuysen, A., Bonner, P. (eds). 2008. *Five Hundred Years Rediscovered: Southern African Precedents and Prospects*. Johannesburg: Wits University Press.

Tempelhoff, J.W.N. 2008. 'Historical perspectives on pre-colonial irrigation in Southern Africa'. *African Historical Review* 40(1): 121–60.

Terblanche, D.E., Pegram, C.G.S., Mittermaier, M.P. 2001. 'The development of weather radar as a research and operational tool for hydrology in South Africa'. *Journal of Hydrology* 241(1–2): 3–25.

Trevor, T.G. 1930. 'Some observations on the relics of pre-European culture in Rhodesia and South Africa'. *The Journal of the Royal Anthropological Institute of Great Britain and Ireland*, 60: 389–99.

Turner, G. 1987. 'Hunters and herders of the Okavango Delta, Northern Botswana'. *Botswana Notes and Records* 19: 25–40.

Turner, S., Plater, A. 2004. Palynological evidence for the origin and development of late Holocene wetland sediments: Mdlanzi Swamp, KwaZulu-Natal, South Africa'. *South African Journal of Science* 100: 220–9.

Turton, A.R., Meissner, R., Mampane, P.M., Seremo, O. 2004. 'A Hydropolitical History of South Africa's International River Basins'. WRC Report No. 1220/1/04. Pretoria: Water Research Commission.

Tyson, P.D., Gatebe, C.K. 2001. 'The atmosphere, aerosols, trace gases and biogeochemical change in suthern Africa: a regional integration'. *South African Journal of Science* 97: 106–16.

Tyson, P.D., Karlén, W., Holmgren, K., Heiss, G.A. 2000. 'The Little Ice Age and Medieval warming in South Africa'. *South African Journal of Science* 96: 121–6.

Union of South Africa (U of SA) 1926. Department of Agriculture, 'The great drought problem of South Africa'. Pretoria: Government Printing and Stationary office.

Van der Leeuw, S., Costanza, R., Aulenbach, S., Brewer, S., et al. 2011. 'Toward an integrated history to guide the future'. *Ecology and Society* 16(4): 2. At http://dx.doi.org/10.5751/ES-04341-160402 (accessed 8 February 2013).

Van Wyk, B.-E., Gericke, N. 2007. *People's Plants: A Guide to the Useful Plants of Southern Africa*. Pretoria: Briza Publications.

Weeks, B., Rodriguez, M.A., Blakeslee, J.H. 2004. 'Panarchy: complexity and regime change in human societies'. In *Proceedings of the Santa Fe Institute Complex Systems Summer School*, Santa Fe: Santa Fe Institute.

Whitelaw, G. 1990. 'An Iron Age fishing tale'. *Southern African Humanities* 21: 195–212.

———— 1996. 'Lydenburg revisited: another look at the Mpumalanga Early Iron Age sequence'. *South African Archaeological Bulletin* 51: 75–83.

Whitlow, R. 1990. 'Conservation status of wetlands in Zimbabwe: past and present'. *GeoJournal* 20(3): 191–202.

Widgren, M. 2007. 'Pre-colonial landesque capital: a global perspective'. In Hornberg, A., McNeill, J.R., Martinez-Alier, J. (eds), *World-System History and Global Environmental Change*. Lanham: AltaMira Press, pp. 61–77.

Wilson, J.F. 1865. 'Water supply in the basin of the River orange, or "Gariep", South Africa'. *Journal of the Royal Geographic Society of London* 35: 106–29.

Wood, A., Dixon, A., McCartney, M. (eds). 2013. *Wetlands Management and Sustainable Livelihoods in Africa*. Abingdon: Routledge.

Wright, J., Mazel, A. 2007. *Tracks in a Mountain Range: Exploring the History of the uKhalamba-Drakensberg*, Johannesburg: Wits University Press.

Part II   Water Control and Irrigation

6   The Nile and Food in the Early Modern Ottoman Empire

Alan Mikhail

INTRODUCTION[1]

No land has ever depended on water management more than Egypt. Its irrigation system is among the world's oldest, its soils some of the richest in Africa. After its conquest by Ottoman armies in 1517, Egypt – as it had been for the Roman Empire – immediately became the most lucrative and important province of the Ottoman Empire.[2] It was the Empire's largest producer of agricultural goods, it generated more revenue for the state than any other province, and its capital was the second largest city in the Empire after Istanbul. It was the gateway to the Red Sea and Indian Ocean and to North and sub-Saharan Africa and was a crucial hub for the management of the pilgrimage sites of Mecca and Medina. The basis for Ottoman Egypt's wealth, population, and power was its land and water.

THE OTTOMAN CONTEXT

The Ottoman Empire was the longest lasting and geographically largest Empire to rule in the Mediterranean basin since antiquity.[3] Its rule for over 600 years across the Middle East, North Africa, and south-eastern Europe both continued and created precedents for nearly all modern states in these regions. The Empire first emerged around the turn of the fourteenth century in the context of centuries of steady migrations by nomadic Turkic peoples across Central Asia, the Mongol invasions of Anatolia, and internal crises in the Byzantine Empire. Through a series of military victories, Osman, the leader of one of the many Turkish tribal groups that came to settle in what was then still Byzantine north-western Anatolia, was able to carve out for himself an area of autonomy from which to extend his power. Osman's son Orhan captured the city of Bursa from the Byzantines in 1326, making it the first capital of the rising polity that would eventually come to be known as the Ottoman Empire.

Ottoman military conquests in western Anatolia and around the Sea of Marmara continued throughout the rest of the fourteenth century and the beginning of the fifteenth. The final blow to waning Byzantine power in Anatolia and south-eastern Europe and the greatest conquest of the early Ottoman state was the capture of Constantinople in 1453. With this new strategic and symbolic capital – Istanbul – the Empire was fully in place to strengthen and extend its rule throughout the second half of the fifteenth century in the Morea, the southern Black Sea coast, the Crimea, and areas further south and west. The reigns of Sultans Selim and Süleyman in the first half of the sixteenth century saw the Empire gain Egypt and most of the Arab world from the Mamlūks, including the religious centres of Mecca and Medina. In what one scholar has termed 'a sixteenth-century world war', Ottoman forces fought not only the Mamlūks in the Arab world and the Safavids in Iran on their eastern frontiers, but also the Spanish and the Habsburgs in the western Mediterranean; the Portuguese in the Indian Ocean; the Hungarians, Serbs, and Bulgarians in the Balkans; and the Venetians and Genoese in the central Mediterranean and the Aegean (Quataert 2000: 21).

These sixteenth-century military successes came in addition to and themselves necessitated new modes of imperial rule (one of which I will discuss shortly) in these vast newly acquired territories. Numerous law codes were promulgated during this period; the practices of various kinds of legal courts and other venues were adapted to changing circumstances throughout the Empire; new forms of administrative regulations, taxation regimes, and means of revenue collection were instituted and refined; commercial relations were further strengthened and extended. As the central imperial bureaucracy expanded in the sixteenth century, the day-to-day administration of the Empire slowly moved away from the person of the sultan to his surrounding retinue in the palace – to the mothers and wives of sultans, to the Empire's grand viziers (chief administrators), and to others in the dynasty's ruling elite. Political power thus became decentralized and more diffuse from the end of the sixteenth century through the seventeenth as new powerbrokers emerged. Chief among these were the large households of viziers and other elites throughout the Empire that, even as they challenged central imperial authority, came to mirror the internal workings of the royal family in their hierarchies, abilities to accumulate wealth and followers, and eventually also in their power in various urban and rural locales.

The Ottoman eighteenth century has mostly been noted for the Empire's military defeats, territorial losses, internal urban and rural rebellions, and its increasing inability to compete economically with Europe on both global and local scales. Scholars have more recently, however, offered important reassessments of this period, as a time when, among other things, new imperial actors were emerging to negotiate novel governing arrangements

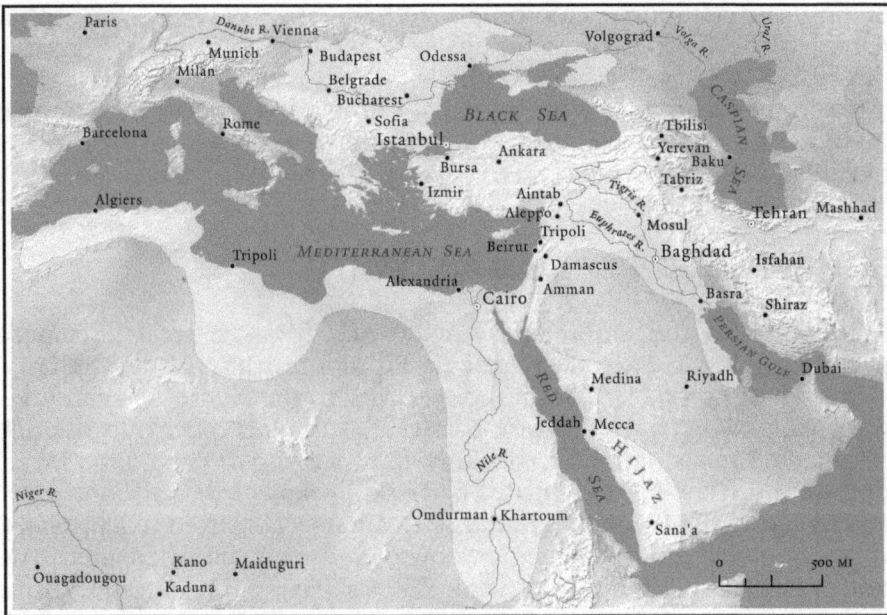

Figure 6.1. The Ottoman Empire, c. 1650 (modified from Maples, 2014).

with the central bureaucracy and new economic mechanisms worked to extend property ownership throughout the Empire.

The Empire's territorial losses continued into the nineteenth century, however, now under the guise of what was termed in European capitals 'the Eastern question' – namely, how were European powers to deal geopolitically and strategically with the contracting Ottoman Empire. For the Russians, the British, and the French (and eventually the Germans as well), the answer was to keep a weakened Empire fledging along so as to check the encroachment of any one power over what remained of Ottoman lands. This was, however, mostly only a European conversation.

For their part, the Ottomans responded to nineteenth-century ideas and political reforms like most other states in the period. Their governmental bureaucracy greatly expanded with the establishment of new ministries, schools, and legislative bodies. New industries, military units, medical institutions, and social clubs were either founded or greatly expanded during this period. As with most other areas of the world, the Ottomans also had to deal with the specter of various rising nationalisms during the nineteenth century. Given the multi-ethnic, multi-confessional, and multi-linguistic makeup of the Empire's populations, there were all sorts of competing and overlapping nationalisms, interests, and desires at work as the century wore on. Sometimes encouraged by outside forces interested in weakening or ending the Ottoman Empire altogether, Arab, Balkan, and other nationalists pushed their claims against the Ottoman state. The crucible of World War I brought these conflicting passions and political agendas to a violent climax that saw the eventual dismemberment of the Ottoman Empire after over six centuries of rule and its replacement by over 30 twentieth-century nation states.

REGULATING EGYPT'S LAND AND WATER

As outlined in the Ottoman law code of Egypt – a legal document promulgated in 1525 formally making Egypt a province of the Empire – almost all of Egypt's land from the sixteenth to the eighteenth centuries was legally owned by the Ottoman state.[4] For administrative purposes, all rural land was divided into plots known as *muqāṭa'āt* (sing. *muqāṭa'a*) that were further divided into 24 parts (known as *qīrāṭ*, with one *qīrāṭ* equaling roughly 175 m^2). A *muqāṭa'a* generally consisted of a principal village, its surrounding villages and towns, and their cultivated areas. The holders of these plots were responsible for paying the state treasury a basic yearly tax and for maintaining irrigation works and agricultural fields in areas under their control. The incentive for someone to take on this responsibility of delivering to the state a set amount of revenue each and every year was the right to raise additional amounts of profit for himself. Thus, the Ottoman state devolved authority over the day-to-day maintenance of agriculture to these local leaders who, in turn, guaranteed the state's revenues and were also able to make a profit for themselves. Each year the provincial governor of Egypt collected and then paid to the Ottoman state the cash remittances from all the province's plots of land.[5] The profit garnered by the Empire from this payment – more than that from any other single province – was the ultimate reason for the Ottoman state's control and maintenance of irrigation, land, and agriculture in Egypt (Hathaway 1997: 6, Shaw 1968b).

The annual cycle of agricultural cultivation in Egypt was of course timed to the Nile's flood. Summer rains in the Ethiopian highlands swelled the river causing it to rise in Aswān in Upper Egypt by June and in Cairo by early July. Water continued to rise through the summer until its peak in Cairo in late August or early September. From then it began to fall steadily, reaching half of its flood height by the middle of November and its minimum by May before the cycle began anew. The onset of the flood in the late summer was designated as the start of the agricultural year in Egypt. Lands in Upper and Lower Egypt watered at the beginning of this year in September or October produced the major harvest of the year consisting of wheat, barley, lentils, clover, flax, chickpeas, onions, and garlic. This was known as the winter crop (*al-shitwī*). Lands were also planted and harvested from January through May with stored water from basins and canals, producing a second major yield for the agricultural year known as the summer crop (*al-ṣayfī*) consisting mainly of wheat, barley, cotton, melons, sugarcane, and sesame. There was of course wide regional variation in the kinds and amounts of crops grown. Rice cultivation, for example, was concentrated in northern Lower Egypt, tobacco and sugarcane in Upper Egypt, cotton in Middle and Lower Egypt, and flax in the interior of the Delta and in the Fayyum oasis. Wheat was grown most everywhere.

The legal status of the water of Ottoman Egypt that grew these foodstuffs and other crops was far more nebulous than the status of land. No single entity 'owned' the waters of the Nile or a canal. At the same time, however, the equitable use of this water was a priority maintained by the Ottoman administration at all costs.[6] Thus though water was owned by no one, it was in many ways owned by all the users of a particular water source or conduit. In the Ottoman law code of 1525 we find a clear statement of the interests of the Ottoman Empire in maintaining Egypt's irrigation network. The very first duty enumerated by this document as incumbent upon each subprovincial leader was 'the proper and timely repair of canals and the work of dredging them' (Mutawallī 1986: 29–30, Barkan 1943: 360). Likewise, this law code also stipulated that village elders and the peasantry of Egypt were to maintain canals that ran through their villages on a regular basis so that no potentially-cultivatable land was unwatered. Similarly, in the season of the flood, peasants were required to plant irrigated land so that no land with access to water was left uncultivated.

Much of the Ottoman law code and especially these sections dealing with irrigation in Egypt were based on earlier laws and precedents. For example, throughout the text of the law code we find references to what was done during the reign of Qaitbay, the Mamlūk ruler of Egypt at the end of the fifteenth century only decades before the Ottoman conquest (Mutawallī 1986: 32–3, Barkan 1943: 360–2). One of the precedents established during his reign and referenced in the Ottoman document was the following. If during the time of the flood a village's canal embankments were overwhelmed or otherwise destroyed and that village's tax revenues were not enough to cover the repair of the canal, then the peasants of that village were to pay for the repairs themselves. If these peasants' funds were insufficient, then – and only then – was money to be supplemented from the province's annual tribute.

While not unexpected, the inclusion of irrigation in this foundational document of Ottoman rule in Egypt illustrates just how vital irrigation was to the entire project of the Ottoman Empire in what would prove to be its richest province. Canals and the water they carried were given as much attention as the various military cadres of Egypt, religious institutions, taxes, and trade. Again, irrigation – and the food it produced – were of the utmost concern not only to those peasants who lived along canals in the countryside, but also to the sultan in Istanbul. The point is clear in this document: water made everything else possible in Ottoman Egypt.

MOVING THE FIELDS

The most important thing it made possible was food. Egypt was the breadbasket of the Ottoman Empire producing more foodstuffs than any other single province. The rich soils of rural Egypt renewed every year by the sediment-rich waters of the Nile flood provided a steady supply of nutrients for the cultivation of all sorts of grains and vegetables in the Egyptian countryside. Rains in the Ethiopian highlands collected nutrients from the volcanic rock of the mountains on which they fell and from the Sudd swamps in the Sudan through which these flood waters eventually travelled every year to Egypt. Once these waters covered the soil of Egypt, seemingly endless supplies of sunshine for photosynthesis produced vast quantities of chemical energy stored in the calories of food grown throughout southern Egypt and the Delta. A large peasant and animal population throughout the province provided the labour needed to harvest, pack, and move this food that would eventually make its way to other parts of Egypt, to Istanbul, to the cities of Mecca and Medina, and to many other areas of the Empire and beyond. Egypt's surplus energy thus provided the caloric power needed to maintain the political power of the Ottoman Empire.

The Ottoman administration of Egypt and the province's peasant and urban populations created, maintained, and harnessed an established web of storage facilities, labour, and shipping networks that provided the infrastructure necessary for the movement, trade, and consumption of massive quantities of Egyptian foodstuffs and goods throughout the Empire. The story of food in Ottoman Egypt is thus really a story of imperial connections and relationships stretching across the geographical expanse of the Empire – a story of early modern natural resource management. The need to move food out of Egypt created and relied upon a complex and shifting constellation of linkages maintained by both the Ottoman bureaucracy of Egypt's massive expenditures of resources and by peasant energies put toward the cultivation of foodstuffs in rural Egypt. Thus, both the Ottoman state and Egyptian peasants worked in tandem not only to irrigate Egypt but also to grow and move food from the province to the rest of the Empire.

As important nodes in this imperial system, grain storage facilities (*wakālas*) in cities like Cairo, Rosetta, Alexandria, and Suez housed large amounts of grain and, perhaps more importantly, provided the commercial environments in which these foodstuffs could be bought and sold. The strategic location of each of these major cities was meant to serve a specific area of the Ottoman Empire: grains in Suez were destined for the Hijaz while food in Rosetta and Alexandria served Istanbul and other cities in North Africa, the Levant, and Anatolia. Linking all of these places and providing the means to transport this food was a network of shipping and trade that served as the vital strings connecting these various locales to one another. Moving the proper amount of food from areas of excess like Egypt to other areas of need in the Ottoman Empire was a fundamental interest of the imperial state as it sought to balance various natural resources around the Ottoman Empire. Food served as a vital commodity for trade and commerce, and sustained urban and rural populations throughout the Empire. Egyptian food moreover also sustained the Ottoman army on campaign and was thus a strategic good.

To plow the fields that would grow these grains, farmers used large wooden tools pulled by animals. Wood, animals, humans, and the labour of the latter two thus came together to cultivate food. This production of food and its movement and trade brought the countryside, urban

Figure 6.2. Threshing Grain in Late Eighteenth-Century Rural Egypt. Commission des sciences et arts d'Égypte, *État moderne*, vol. 2, pt. 2 of *Description de l'Égypte, ou, recueil de observations et des recherches qui ont été faites en Égypte pendant l'éxpédition de l'armée française, publié par les ordres de Sa Majesté l'empereur Napoléon le Grand* (Paris: Imprimerie impériale, 1809–28), Arts et métiers, pl. 8. Beinecke Rare Book and Manuscript Library, Yale University.

hinterlands, and cities throughout Egypt much closer together as peasants took their goods to towns and then to cities so that they could then be transported even further away from their original points of growth and harvest. Lentils, wheat, maize, and above all rice were the primary foodstuffs grown and shipped from Egypt (McNeill 1992: 89).[7] Each of the organisms that produced these consumable goods required a specialized amount of knowledge and a specific kind and amount of labour to turn its products into commodities that could then be moved and traded. In the genetic and physical makeup of each of these organic structures was a specific history of human interaction with the natural world. The improvement of each of these organisms through the human technology of selective breeding ensured that the history of these grains, the lives of the animals who helped to cultivate them, and the lives of humans who harvested them, moved them, and consumed them would remain forever intertwined. The patchwork of fields in which this biota grew dictated the natural and built environment of much of rural Egypt. The countryside was a quilt of different colours and hues as rice fields bordered those of wheat, corn, and various other foodstuffs. Fences and concerns over property were largely tied up in this need to maximize food production for a demanding market. In short, the push to grow grains in the Egyptian countryside was perhaps the largest factor dictating the lives and futures of Egyptian peasants and their relationships to urban centres, the natural world, and the Ottoman administration of Egypt.

That both Egyptian peasants and Ottoman bureaucrats charged with the administration of the province expended so much caloric energy, time, and resources on the maintenance, repair, and improvement of Egypt's irrigation infrastructure is proof enough of the importance of food cultivation and production in Ottoman Egypt. After all, 'food supply was the Achilles heel of the early modern state' (Scott 1998: 29). Indeed, food and grain were the very reasons the Ottoman administration of Egypt and Egyptian peasants cared so deeply about the irrigation system of the province in the first place. Without water, the fields of Egypt would not have been able to supply the quantities and kinds of grains needed for the trading and transport that was so central to the longevity and durability of the Ottoman Empire. Irrigation and its maintenance demanded much from both the Ottoman administration of Egypt and from those peasants under its administration. Indeed, the Nile's flood and the management of this water was central to the production of food and to the general wellbeing of Egypt – a theme that runs throughout much of the correspondence between Egypt and Istanbul.

When food was properly grown and harvested in the Egyptian countryside, it was immediately entered into a network of shipping and transport that allowed grain and other foodstuffs to move from villages to towns, to cities, and then to capitals. This network relied on peasant and animal labour, a web of roads and canals, proper storage facilities, money, credit, and shipping to ensure that Egyptian crops arrived in Mecca, Medina, Dubrovnik, Istanbul, Tunis, Rhodes, and various other locales throughout the Ottoman Empire. Once foodstuffs were cultivated in villages, they were moved to a number of large regional cities – places like Asyūṭ, Jirja, al-Damanhūr, al-Manṣūra, and Manfalūṭ. These medium-size cities were usually the centres of subprovinces within Ottoman Egypt, the seats of major Islamic courts, and hubs of commerce. In terms of the transport of grains throughout the Empire, these cities were important regional nodes in a vast imperial network of commerce since it was to these cities that grains and other foodstuffs from Egypt's thousands of villages first went after harvest.

Some of this grain was bought and sold in the markets of each of these cities to be used by local merchants and others. The bulk of it, however, was shipped from these cities along the Nile to usually either Cairo, Alexandria, or Rosetta. Cairo was, of course, the largest single marketplace in Ottoman Egypt and had a thriving commercial life. Alexandria, Rosetta, and Suez were cities from where Egypt's grains and foodstuffs left by ship for other

Figure 6.3. Ottoman Cairo. Commission des sciences et arts d'Égypte, *État moderne*, vol. 1, pt. 2 of *Description de l'Égypte, Environs du Kaire* (Paris: Imprimerie impériale, 1809–28), Arts et métiers, pl. 15. Beinecke Rare Book and Manuscript Library, Yale University.

locations throughout the Empire. In each of these three Egyptian port cities, there were storage facilities for grain, sleeping quarters for merchants coming from all over the Ottoman Empire and from elsewhere, market weighers and overseers, and an entire infrastructure of commerce. Within this general schema of commerce and transport, the central means of getting grains from the villages of Egypt to subprovincial cities were roads. Dirt roads and waterways and seas were the primary conduits of goods, people, animals, and bureaucrats that connected all parts of the Ottoman world. The city of Manfalūṭ in southern Egypt, for example, was a central point from which emanated dozens of roads leading to smaller villages in its hinterland. Many of these smaller villages then served as nodes from which roads led to other villages. For its part, Manfalūṭ was then connected by roads to other large cities. Thus, cities like Manfalūṭ served as both destinations and points of origin for grain shipments moving on dendritic networks of roads and waterways throughout Egypt and between it and other regions of the Empire.

This network of roads that was so central to bringing grains from villages to towns and then to cities was dwarfed in scale and importance by Egypt's ultimate road – the Nile itself. The Nile and its many branches linked extremely distant areas of Egypt together into one system of collective responsibility, commerce, and transport. It was akin to a massive highway running the entire length of Egypt that connected top to bottom and sea to village (Sulaymān 2000). This colossal thoroughfare, along with a vast network of canals branching off of the river and roads connecting cities and villages, ensured that food was in constant motion within Egypt and that it was able to move quickly and efficiently from Egypt's hundreds of villages to its major port cities and on from there.

Landlocked cities like al-Damanhūr, Asyūṭ, Manfalūṭ, Jirja, al-Manṣūra, and Isnā – middle-size towns that were smaller than Rosetta, Suez, Alexandria, and Cairo but significantly larger than most villages – served as collection points for grains and other foodstuffs destined for Cairo, Egypt's ports, and other cities across the Ottoman Empire. These sources of caloric energy were transported by animals and people on the many roads of rural Egypt from the hundreds of villages that formed the hinterlands of these middling cities. In the southern Egyptian city of Asyūṭ, for example, there was a thriving state market (sūq sulṭānī) in which one could buy and sell grains and other goods from villages all over the Asyūṭ hinterland.[8] Peasants, Ottoman officials, merchants, weighers, people serving tea and coffee, and many others jostled their way through this crowded marketplace in one of Egypt's subprovincial centres to sample the wares on display and to acquire the food and provisions necessary to sustain their lives and livelihoods. In cities like Asyūṭ, the Ottoman administration of Egypt demanded that set amounts of money and grain be paid to the state every year.[9] This fixed amount was to be acquired from the harvests and taxes of all the villages in the orbit of Asyūṭ – in other words, those villages in the hinterland of the city that would use its markets, courts, and other resources. Each village was to contribute a part to this overall amount.

An Ottoman order from 1717 to the subprovincial head of the southern city of Asyūṭ makes clear the system by which state grain and money were to be acquired.[10] According to this decree, at the beginning of the Coptic month of Tūt (the first month of the Egyptian agricultural year), this official was to arrive in Asyūṭ. As was customary in years past, he was to make sure that each of the villages surrounding Asyūṭ sent its required annual share of grain to the city. Once all this grain was collected and inspected in Asyūṭ, it was then put onto ships in the port of Asyūṭ and sent to the imperial granaries of Cairo. This decree ended with a warning against any mistakes in this charge and with an entreat to defend the ships carrying the state's grain to Cairo against Bedouin attack. This order thus illustrates how grain moved from link to link in the Empire's food chain – from villages outside Asyūṭ to the city itself and on to Cairo before it was either sold there or sent further away to somewhere else in the Empire.

In other medium-size cities throughout Egypt, there was a similar administration in place to move grains and other foodstuffs from villages to cities and onto Cairo, Alexandria, and elsewhere. Various cases from the court of al-Baḥayra in the north-west Nile Delta, for example, sketch out a very robust trade in large quantities of wheat grown in the hinterland of large cities like al-Damanhūr to be shipped elsewhere in Egypt and the Ottoman Empire more generally.[11] Like Asyūṭ, the city of al-Damanhūr (the seat of the court of al-Baḥayra) was an important regional centre of trade and commerce. An Ottoman decree from 1771 about transport to and from al-Damanhūr confirms the importance of the city's markets to trade within and beyond Egypt.[12] On this point, we read in another Ottoman decree from 1777 about the buying and selling of cotton in al-Damanhūr that was destined to be shipped to Alexandria and from there to other cities in the Mediterranean.[13] To ensure the proper transport of grains from smaller Egyptian cities like Manfalūṭ or al-Damanhūr to the large Mediterranean port cities of Alexandria and Rosetta or to Cairo, the Ottoman state often appointed a grain official (*gılâl ağası*) to oversee the collection, inspection, and transport of grains from smaller cities to larger ones. In Manfalūṭ, for example, a man named Habahancı Başı was appointed as the official grain overseer for the years between 1765 and 1767.[14]

Like Manfalūṭ or al-Damanhūr, the north-eastern city of al-Manṣūra was also the centre of an extremely rich agricultural area. And the Ottoman administration of Egypt was keen to ensure the most expedient means possible to transport grain and other foodstuffs from this city to other areas of the Empire. As such, much of the grain sent from al-Manṣūra bypassed the larger coastal cities of Rosetta and Alexandria altogether on its way to Istanbul and elsewhere. This direct link between al-Manṣūra and the Ottoman capital of Istanbul is evidenced by numerous imperial

orders sent to the court of al-Manṣūra about the preparation of ships
to transport grain and about the appointment of an imperial functionary
to oversee this work.[15] It was also common for the Ottoman
administration – in al-Manṣūra and elsewhere – to keep and store excess
grains and foodstuffs in the possession of Egyptian merchants and
notables.[16] Shipments of grain from al-Manṣūra and from other Egyptian
cities were, moreover, under near constant threat of attack from bandits
and Bedouins, and numerous Ottoman decrees directed great caution to
fend off such raids.[17]

The transport of grain from smaller cities like al-Manṣūra, Manfalūṭ,
and Jirja to Cairo, Rosetta, Istanbul, or Alexandria was not organized
and carried out solely by the Ottoman state itself. Indeed, the imperial
bureaucracy only transported those grains and foodstuffs that would
eventually make it into its own coffers. Alongside this imperial system of
grain transportation, many wealthy individual merchants financed their
own shipments of grain and other items from smaller to larger cities. One
such individual – a Cairene merchant named al-Shaykh Humām Yūsif –
hired two brothers, Ḥusayn and Shāwīsh, who owned their own boat, to
acquire for him 249 *ardabb*s of wheat.[18] al-Shaykh Humām Yūsif did not
specify from where the wheat was to be obtained, but he promised to pay
the brothers one *ardabb* of wheat for every 100 *ardabb*s they brought him.
These two brother sailors recorded in the court of Isnā (situated at an
impressive distance of over 450 miles south of Cairo) that they were able to
get the 249 *ardabb*s requested by this merchant from two different sources
in Isnā (150 *ardabb*s from one unnamed source and 99 from another)
before returning with the wheat to Cairo.

This case is particularly striking because of the long distance traversed
by Ḥusayn and Shāwīsh in delivering wheat to Cairo. We are
unfortunately lacking several important pieces of information about
this case. Did the brothers solicit this work in Cairo and then sail up and
down the Nile looking for the best and cheapest source of wheat? Did
they know that wheat was much more easily obtained in Isnā than
anywhere else? And did they calculate that this better price of wheat in
Isnā was worth the long trip? Was there perhaps a shortage of wheat in
Egypt at this time, requiring the two brothers to sail very far away (again,
over 450 miles) from Cairo to find it? Or were they perhaps already in Isnā
when commissioned for this job? If so, how did they communicate with
al-Shaykh Humām Yūsif? Despite the lack of clear answers to these and
other questions, what is clear from this case is that merchants and those
they charged to acquire and transport their goods were part of a huge
network that stretched far beyond major cities like Cairo and that
included peasants, sailors, merchants, government officials, animals, ship
builders, lifters, and many others.

The large amount of wheat moved in this case is but one example of the
fact that enormous quantities of agricultural products were everywhere in

Ottoman Egypt and, more importantly, that they were probably the most common living organisms with which human Egyptians interacted. The lives of most Egyptian peasants were dictated by the responsibilities of growing and sustaining these organisms that would bring them both bodily and economic sustenance. They lived in fields and among their crops and interweaved their own histories in the lives of these plants. Through selective breeding and crop rotations, humans changed the course of Egyptian plants' histories at the same time that the dictates of agricultural life and economic necessity determined much of the course of Egyptian peasants' own histories.

Indeed, food often formed very large portions of people's inheritable estates.[19] This was especially true of a commodity like coffee, which was an important form of capital for many individuals.[20] Moreover, because the variety and amount of food consumed (and served) was a sign of status and wealth in Ottoman Egypt, it was quite common for those endowing *waqfs* (pious endowments) to make provisions in their endowment deeds for the gifting of food and water.[21] Food was also sometimes used to pay the salaries of those in the employ of the Ottoman state.[22] Food is, of course, at the core of human society. In Donald Worster's words, 'every group of people in history has had to identify such resources [foodstuffs] and create a mode of production to get them from the earth and into their bellies' (Worster 1990: 1091–2). In Egypt, this centrality of food to society was seen in part by the presence of various markets for different consumable foods (Raymond, 1973–4, 1: 307–72). Cities large and small had very robust marketplaces that were usually at the geographic and economic centres of cities.[23] As a hub of export, the city of Rosetta, for example, had many kinds of markets, one of the most important of which was its thriving rice market.[24] Not surprisingly, coastal Rosetta also had a fish market.[25] Like fish, there were also markets for pigeons, though both of these were considered luxury items that most Egyptians could not afford to eat save on very special occasions.[26]

FEEDING THE EMPIRE

Although Egypt's network of roads connecting villages to towns and towns to cities ensured a constant circulation of foodstuffs within the province, trading relationships that stretched beyond Egypt to other parts of the Ottoman Empire proved much more significant to the harvesting of grains in rural Egypt than did the domestic Egyptian market itself. Thus, the external demands and pressures produced by Egypt's and its countryside's intimate connections to other areas of the Ottoman Empire very much shaped the contours and histories of rural Egypt. Indeed, as the largest single area of agricultural production anywhere in the Ottoman Empire, Egypt was the breadbasket of the Empire that fed peoples very far away from its borders. Countless Ottoman orders were sent to various Egyptian cities over the centuries directing the shipment of food to places like Tunis,[27] Yemen,[28] Aleppo,[29] Morocco,[30] Izmir,[31] western Tripoli,[32] eastern Tripoli,[33] Crete,[34] Salonica,[35] Algeria,[36] and – most importantly – Istanbul [37] and the Ḥaramayn (Mecca and Medina).[38] Shipping food to these latter two places (Istanbul and the Ḥaramayn) was indeed the subject of considerable correspondence between Istanbul and Egypt.[39]

Ships were the key vehicles by which vast quantities of food emanated out from Egypt and by which Egyptians and Ottoman bureaucrats alike linked the economically, environmentally, and politically vital province of Egypt to the rest of the Ottoman world (Brummett 1994, Bostan 1992, Imber 1980: 211–82). If roads and canals from Egyptian villages to town and cities linked the rural world to the urban, ships linked towns and cities to each other and to places far beyond the reach of the distance a peasant or a donkey could walk. Roads were likely only used to traverse distances of 10 miles or less.[40] Beyond that, the energy expenditure of a donkey would cancel out too much of the energy value of the food it was carrying, rendering the venture an overall net loss of energy. Thus, water was used to cover distances of over ten miles since wind and water currents were free sources of energy. Without ships, Egypt would never have been conquered by the Ottomans and would have remained too distant a territory for their administration from Istanbul. Likewise, without ships, Egyptian merchants could not have benefited from opportunities to trade with a network much larger than Egypt itself. Egypt's robust and extensive Mediterranean and Red Sea shipping networks were the primary reason that Egypt's major urban centres – with the important exception of Cairo of course – were all port cities. Of these, Alexandria, Rosetta, and Suez were the most important for shipping and transport.[41]

The two most frequent destinations for Egyptian grain were first Istanbul and then the Ḥaramayn of Mecca and Medina. Istanbul was by far the most

Figure 6.4. Port of Rosetta at the End of the Eighteenth Century. Commission des sciences et arts d'Egypte, *État moderne*, vol. 1, pt. 2 of *Description de l'Égypte, Rosette et environs* (Paris: Imprimerie impériale, 1809–28), Arts et métiers, pl. 81. Beinecke Rare Book and Manuscript Library, Yale University.

common over the period of Ottoman rule in Egypt (Murphey 1988: 217–63). As the Empire's largest city, the imperial seat, and the home of the sultan and his ruling elites, it is no surprise that Istanbul acquired for itself the most and best the Empire had to offer. This need and demand for calories, however, was met by a relative dearth of vast quantities of agricultural land in the immediate hinterland around Istanbul. The closest expansive areas of agricultural land near Istanbul were in the Balkans and central Anatolia, both long and arduous overland journeys from the imperial capital. As such, it was much more expedient and easier for the Ottoman capital to bring food from a place like Egypt – despite its greater distance from Istanbul – since the majority of the journey could be carried out by ship. In addition to the relative ease of sea travel they enjoyed because of copious amounts of free energy from wind and water currents, boats allowed larger amounts of food to be shipped than could be achieved by any form of overland transport. Because food imports from Egypt were more common, more plentiful, and more important than shipments from anywhere else in the Empire, it is no surprise that the docking area for boats carrying food from across the Empire to the capital city came to be known as the Egyptian Bazaar (Mısır Çarşısı). Egypt was, in other words, renowned in Istanbul and in other parts of the Ottoman Empire for being the Empire's most important source of food and trade commodities. These shipments included not only food from Egypt, but also coffee and other commodities from Yemen and spices from India that came to Istanbul via Egypt (Sertoğlu 1986: 225).[42] Egypt was thus the largest and most vital conduit of food to Istanbul.

Just as Egypt served the Ottoman capital city of Istanbul to its north, it was also a major supplier of food to the Ḥaramayn of Mecca and Medina to its south-east. Indeed, this was the second most common destination for Egyptian grains during the Ottoman period. [43] As the location of al-Masjid al-Ḥaram, the holiest site in Sunni Islam, Mecca was one of the most important symbolic and spiritual cities in the Ottoman Empire. It and Medina – home of the Mosque of the prophet where Muḥammad, Abū Bakr, and 'Umar are buried – were the site of the annual pilgrimage (ḥajj), the last of the five pillars of Sunni Islam. Surely the largest annual single gathering of people anywhere in the early modern world, the preparations for and organization of the ḥajj were a major responsibility of those in charge of the wider Hijaz region in which were located Mecca and Medina, and in the early modern period the Ottoman Empire was the responsible polity (Faroqhi 1994, Shaw 1962: 239–71). There were two main caravan routes in the Empire that carried pilgrims to the Ḥaramayn – one through Damascus and the other through Cairo (Coşkun 2001: 307–22). The former brought pilgrims from Istanbul, Anatolia, Greater Syria, Iran, and regions of Central Asia to the Ḥaramayn, while the latter was the point from which pilgrims from Egypt, North Africa, and other parts of Africa traveled toward the Hijaz.

The organization of the annual Egyptian pilgrimage caravan that left from Cairo was a huge economic and bureaucratic affair. Animals, food, supplies, and protection were just some of the expenses incurred by the state in its organization of the caravan.[44] Despite these enormous costs, Egypt nevertheless always remained central to the organization and execution of the yearly *ḥajj*. The most important role Egypt played in the history of the *ḥajj* was ensuring that pilgrims going to the holy cities had sufficient amounts of food to sustain them while in the Hijaz. Egypt was the main supplier of food for the two cities of Mecca and Medina during the pilgrimage season and, indeed, throughout the remainder of the year as well.[45] As noted previously, boats from Egypt left for the Hijaz from the Egyptian Red Sea port city of Suez, and the construction of ships in this port was of central importance to the Ottoman administration of food transport from Egypt to the Ḥaramayn.[46] Any delay in the shipment of Egyptian grains to Mecca and Medina would cause the people of these cities great hardship and suffering, and thus the complete and proper transport of these grains was of the utmost importance to the Ottoman state.[47]

As an example of the Empire's great interest in ensuring that these affairs connected to the *ḥajj* were completed correctly and expediently, the state, for example, sent one of its most trusted and dependable functionaries – a man named Süleyman – to Egypt in 1732 to assist the Vali Muḥammad Paşa in overseeing this work. The necessary grains were to be obtained as quickly as possible from the appropriate sources in Egypt and were then to be sent to Suez to be loaded onto state-owned ships as soon as possible. If, however, these ships were not enough to transport the entire load, then private vessels were to be obtained as well to ship the remaining amount. This order also states that if any grain was missing from last year's shipment, it should be compensated by adding extra grain to this year's shipment. This imperial decree then warns against any amount of laziness, delay, or ignorance in this matter since even an 'iota's worth' (*mikdar-ı zerre*) of delay or mistake would lead to great harm for the people of the Ḥaramayn since they had no sources of food other than these grains from Egypt. The spoils and fortunes of the Nile flood and the ability of Egypt's irrigation system and of the peasants who worked on it to deal with the yearly inundation thus affected the lives of thousands of people in the Ḥaramayn and elsewhere far beyond the Nile valley.

CONCLUSION

Because of nutrients from Ethiopian volcanic rocks and the swamps of Sudan's Sudd, because of the reliable flood that brought these nutrients downstream to Egypt, and because of endless supplies of sunshine for photosynthesis, Egypt produced a huge surplus of chemical energy in the form of food. Its energy surplus powered the Ottoman Empire, supplying calories to Istanbul, to the Hijaz, to pious pilgrims, to farmers, and to the Ottoman military. Egypt was thus the caloric engine of the Empire whose surplus energy supplies fueled the political power of the Ottoman state – powering the brain of the palace and the capital, the religious heart of the Hijaz, and the Empire's military muscle.

As we have seen in this chapter, the Ottoman Empire regularly undertook massively complex operations to move grains within Egypt and between it and other areas of the Empire to get food into the stomachs of people and animals across the Ottoman world. Achieving caloric parity across the early modern Ottoman Empire was an exercise by the imperial administration of Egypt in natural resource management that necessitated the organization of irrigation, rural labour, traders, warehouses, and ships.

NOTES

1 For a fuller treatment of the themes developed in this chapter, see Mikhail 2011 from which much of this is adapted.
2 On the agricultural wealth of Roman Egypt, see Blouin 2012: 22–37.
3 For a very useful and accessible synthetic treatment of the full arc of Ottoman history, see Finkel 2006.
4 I consulted the following versions of the *Kanunname-i Mısır*: Kanunname-i Mısır 1845, Mutawallī 1986, Barkan 1943: 355–87. Mutawallī's Arabic translation contains a copy of Barkan's Turkish text. For discussions of the status of land under the Kanunname, see Cuno 1992: 25–7, Shaw 1968a: 91–103.
5 This annual tribute was known as the *irsāliyye-i ḫazīne*. For a detailed discussion of the *irsāliyye-i ḫazīne*, see Shaw 1962: 283–312 and 399–401.
6 For a discussion of the management of water resources and irrigation during the Mamlūk period, see Tsugitaka 1997: 220–33.
7 Maize, for example, was known as 'Egyptian grain' (*mısır*) amongst the Turks of Istanbul and Anatolia. And, suggesting the itinerary of its transmission throughout the Old World (from Egypt to Anatolia to Europe), it was known as 'Turkish grain' in the Balkans and Italy (*granturco* in Italian).
8 Dār al-Wathā'iq al-Qawmiyya (hereafter DWQ), Maḥkamat Asyūṭ 7, p. 20, case 42 (1 S 1208/7 Sep. 1793); DWQ, Maḥkamat Asyūṭ 7, p. 34, case 73 (28 R 1208/2 Dec. 1793); DWQ, Maḥkamat Asyūṭ 9, p. 85, case 196 (9 B 1219/13 Oct. 1804); DWQ, Maḥkamat Asyūṭ 9, p. 88, case 202 (20 Ş 1219/23 Nov. 1804); DWQ, Maḥkamat Asyūṭ 9, p. 178, case 402 (25 Ca 1219/31 Aug. 1804). For a description of the villages in Asyūṭ's hinterland, see Ramzī 1994, pt. 2(4):1–85.
9 In 1596, for instance, Asyūṭ paid the Ottoman treasury 1,542,166 *para* of the total amount of 66,080,476 *para* from Egypt for that year (Shaw 1968b: 21 and 92). The Egyptian purse equaled 25,000 *para*, the official Ottoman name given to the *niṣf fiḍḍa*. In Stanford J. Shaw's words, 'The silver coin in common use during Mamlûk and Ottoman times in Egypt was called

nısf fiḍḍe colloquially and *para* officially' (1962: 65, n. 169). For an example of an Ottoman firman ordering the sending of money and grain from Asyūṭ to the state, see DWQ, Maḥkamat Asyūṭ 3, p. 338, case 1001 (17 Za 1135/18 Aug. 1723).

10 DWQ, Maḥkamat Asyūṭ 2, p. 484, case 1213 (23 Ca 1129/5 May 1717).
11 On the wheat trade in al-Baḥayra, see for example: DWQ, Maḥkamat al-Baḥayra 30, p. 27, case 50 (Evail L 1219/2–11 Jan. 1805); DWQ, Maḥkamat al-Baḥayra 30, p. 34, case 60 (8 Za 1219/8 Feb. 1805).
12 DWQ, Maḥkamat al-Baḥayra 9, p. 206, no case no. (19 N 1185/26 Dec. 1771).
13 DWQ, Maḥkamat al-Baḥayra, 10, p. 316, case 733 (13 S 1191/23 Mar. 1777). Likewise, the following series of Ottoman orders sent to the court of Manfalūṭ in 1782 aimed at preparing a group of ships to carry barley and other items from that city to Cairo: DWQ, Maḥkamat Manfalūṭ 1, p. 316, case 788 (28 L 1196/6 Oct. 1782); DWQ, Maḥkamat Manfalūṭ 1, p. 316, case 789 (28 L 1196/6 Oct. 1782).
14 DWQ, Maḥkamat Manfalūṭ 2, p. 184, case 620 (1 R 1179/17 Sep. 1765). The following is the Ottoman firman instructing that this same man's tenure as *ġılâl ağası* be renewed: DWQ, Maḥkamat Manfalūṭ 2, p. 184, case 621 (8 R 1180/13 Sep. 1766).
15 Cases of this sort include the following: DWQ, Maḥkamat al-Manṣūra 12, p. 439, no case no. (14 L 1103/29 Jun. 1692); DWQ, Maḥkamat al-Manṣūra 12, p. 454, no case no. (18 B 1104/25 Mar. 1693); DWQ, Maḥkamat al-Manṣūra 17, p. 381, no case no. (18 R 1119/18 Jul. 1707).
16 For example, according to an order recorded in the court of al-Manṣūra in 1691, the Amīr Dhū al-Faqqār Çavuş, one 'Abd Allah al-Jindī, his brother Ḥijāzī Bishāra, and a man named Ghānim ibn Ismā'īl (all in the village of Ṣāfūr) were charged by the state with storing 300 *ardabb*s of grain, 100.5 *ardabb*s of wheat, 180 *ardabb*s of barley, 17 *ardabb*s of Egyptian clover, and 6 *ardabb*s of lentils. This decree also served to appoint a state functionary to oversee the recollection of these goods in al-Manṣūra and their shipment to Istanbul. DWQ, Maḥkamat al-Manṣūra 12, p. 419, no case no. (20 Ca 1102/19 Feb. 1691). The value of an *ardabb* varied greatly over the course of the eighteenth century from a minimum of 75 L in 1665 to 184 L in 1798 (Raymond 1973–4, 1: LVII, Hinz 1955: 39–40).
17 See, for example: DWQ, Maḥkamat al-Manṣūra 12, p. 441, no case no. (15 L 1103/30 Jun. 1692); DWQ, Maḥkamat al-Manṣūra 17, p. 379, no case no. (29 R 1119/29 Jul. 1707).
18 DWQ, Maḥkamat Isnā 7, p. 36, case 67 (16 B 1173/4 Mar. 1760).
19 See, for example: DWQ, Maḥkamat Manfalūṭ 1, pp. 50–51, case 100 (22 Z 1215/6 May 1801). In this case, grain constituted almost a third (32 per cent) of the value of an inheritable estate.
20 For an example of an instance in which rice formed the bulk of an individual's personal wealth, see DWQ, Maḥkamat Rashīd 132, p. 63, case 89 (1 Ş 1137/14 Apr. 1725).
21 For examples of this, see DWQ, Maḥkamat al-Manṣūra 51, p. 125, case 237 (n.d.); DWQ, Maḥkamat Isnā 3, pp. 15–16, case 29 (8 L 1171/14 Jun. 1758); DWQ, Maḥkamat Rashīd 137, pp. 196–98, case 296 (20 L 1146/26 Mar. 1734); DWQ, Maḥkamat Asyūṭ 1, p. 418, case 1160 (19 L 1069/9 Jul. 1659); DWQ, Maḥkamat Asyūṭ 3, p. 330, case 890 (23 L 1135/26 Jul. 1723). Generally on *waqf* in Ottoman

Egypt, see Muhammad 'Afīfī, *al-Awqāf wa al-Ḥayāh al-Iqtiṣādiyya fī Miṣr fī al-'Aṣr al-'Uthmānī* (Cairo: al-Hay'a al-Miṣriyya al-'Āmma lil-Kitāb, 1991).

22 For example, a group of surgeons (*cerrahīn*) in the Ottoman navy received ten double loaves of bread (*on çift nan*), four portions of sheep's meat (*dört kıt' lâhm-i ganim*), and ten rations of barley (*on yem şair*) on a daily basis as payment for their services. Başbakanlık Osmanlı Arşivi (hereafter BOA), Cevdet Sıhhiye 518 (14–25 Ş 1215/31 Dec. 1800–11 Jan. 1801).

23 On the important market of al-Manṣūra and the need to protect it from bandit raids, see DWQ, Maḥkamat al-Manṣūra 12, p. 441, no case no. (5 M 1104/16 Sep. 1692). On the market of al-Damanhūr, see DWQ, Maḥkamat al-Baḥayra 9, p. 206, no case no. (19 N 1185/26 Dec. 1771).

24 DWQ, Maḥkamat Rashīd 125, pp. 323–24, case 530 (12 M 1133/13 Nov. 1720); DWQ, Maḥkamat Rashīd 132, p. 198, case 305 (13 B 1137/28 Mar. 1725); DWQ, Maḥkamat Rashīd 132, p. 206, case 323 (5 M 1138/12 Sep. 1725).

25 DWQ, Maḥkamat Rashīd 134, p. 332, case 443 (30 B 1140/11 Mar. 1728).

26 This was not from a lack of fish or pigeons in the Delta; rather, rights to fishing and hunting were held as a state monopoly used to generate revenue (Shaw 1964: 19 and 47). On the keeping of pigeons, see DWQ, Maḥkamat al-Baḥayra 8, p. 24, case 37 (19 S 1176/8 Sep. 1762).

27 DWQ, Maḥkamat Rashīd 125, p. 328, case 540 (8 Za 1132/11 Sep. 1720); DWQ, Maḥkamat Rashīd 132, p. 196, case 298 (25 R 1137/10 Jan. 1725).

28 DWQ, Maḥkamat Rashīd 125, p. 333, case 548 (23 L 1132/28 Aug. 1720).

29 DWQ, Maḥkamat Rashīd 125, p. 287, case 452 (13 Ca 1132/22 Mar. 1720).

30 DWQ, Maḥkamat Rashīd 125, pp. 323–24, case 530 (12 M 1133/13 Nov. 1720).

31 DWQ, Maḥkamat Rashīd 125, p. 319, case 517 (28 Ra 1133/27 Jan. 1721); DWQ, Maḥkamat Rashīd 154, p. 3, case 5 (6 C 1159/25 Jun. 1746).

32 DWQ, Maḥkamat Rashīd 125, p. 147, case 257 (27 Z 1132/29 Oct. 1720).

33 DWQ, Maḥkamat Rashīd 125, p. 287, case 452 (13 Ca 1132/22 Mar. 1720).

34 DWQ, Maḥkamat Rashīd 154, p. 341, no case no. (22 M 1163/1 Jan. 1750).

35 DWQ, Maḥkamat Rashīd 125, p. 319, case 517 (28 Ra 1133/27 Jan. 1721); DWQ, Maḥkamat Rashīd 154, p. 341, no case no. (4 Ra 1163/11 Feb. 1750).

36 DWQ, Maḥkamat Rashīd 154, p. 10, no case no. (21 S 1161/21 Feb. 1748).

37 DWQ, Maḥkamat Rashīd 122, p. 67, case 113 (21 Ca 1131/11 Apr. 1719); DWQ, Maḥkamat Rashīd 123, p. 142, case 241 (25 B 1131/14 Jun. 1719); DWQ, Maḥkamat Rashīd 124, p. 253, case 352 (1 Ca 1132/10 Mar. 1720); DWQ, Maḥkamat Rashīd 154, p. 182, case 203 (25 Z 1162/6 Dec. 1749); DWQ, Maḥkamat Rashīd 157, p. 324, case 319 (15 R 1166/19 Feb. 1753).

38 DWQ, Maḥkamat Manfalūṭ 2, p. 189, case 631 (24 Ca 1179/8 Nov. 1765); DWQ, Maḥkamat Manfalūṭ 2, p. 190, case 632 (20 C 1179/4 Dec. 1765); DWQ, Maḥkamat Manfalūṭ 2, p. 190, case 633 (3 Z 1180/2 May 1767); DWQ, Maḥkamat Asyūṭ 2, p. 235, no case no. (23 Z 1107/23 Jul. 1696).

39 We should also note that there were instances of famine in Egypt during which time food was moved from elsewhere around the Empire to the province. In the following case, for example, shortages of rice in Egypt were met by shipments from Crete: Topkapı Sarayı Müzesi Arşivi (hereafter TSMA), E. 2444/107 (n.d.).

40 In reference to England and Wales, Fernand Braudel similarly observes that in 1600 grain was not transported overland for distances of 10 miles or more and that most often these overland trips were less than 5 miles (1982: 43).

41 For two cases dealing with the high volume of commerce that passed through the port of Rosetta, see DWQ, Maḥkamat Rashīd 125, p. 91, case 158 (12 L 1132/17 Aug. 1720); DWQ, Maḥkamat Rashīd 130, no page no., no case no. (A) (6 Ca 1136/2 Feb. 1724). A case about Alexandria from 1720 estimates that the daily income of the city from shipping and commerce was the impressive figure of 198 copper coins (*bakır*) and 482 *akçe*. DWQ, Maḥkamat Rashīd 125, p. 341, no case no. (16 L 1132/21 Aug. 1720).

42 After the rebuilding of the market after the great fire of 1689, it became known as the Yeni Çarşı.

43 On the general provisioning of Mecca and Medina from Egypt, see BOA, Mühimme-i Mısır, 3: 210 (Evail Ş 1133/27 May–5 Jun. 1721); BOA, Hatt-ı Hümayun 29/1358 (29 Z 1197/24 Nov. 1783); BOA, Hatt-ı Hümayun 28/1354 (7 Za 1198/22 Sep. 1784); BOA, Hatt-ı Hümayun 26/1256 (10 Za 1200/3 Sep. 1786). There is no internal evidence for the date of this final case. The date given is the one assigned by the BOA. TSMA, E. 3218 (n.d.); TSMA, E. 5657 (13 Ra 1204/1 Dec. 1789); TSMA, E. 664/40 (n.d.); TSMA, E. 5225/12 (Evahir S 1194/27 Feb.–7 Mar. 1780); TSMA, E. 664/51 (n.d.); TSMA, E. 2229/3 (n.d.). For a detailed study of Egyptian relations with the Hijaz during the Ottoman period, see Ḥusām Muḥammad ‘Abd al-Mu‘tī, *al-‘Alāqāt al-Miṣriyya al-Ḥijāziyya fī al-Qarn al-Thāmin ‘Ashar* (Cairo: al-Hay’a al-Miṣriyya al-‘Āmma lil-Kitāb, 1999).

44 For an example of the great financial and bureaucratic costs associated with the organization of the Egyptian pilgrimage caravan, see BOA, Hatt-ı Hümayun 177/4744 (12 Ra 1206/9 Nov. 1791).

45 On Egypt's provisioning of food to Mecca and Medina, see BOA, Hatt-ı Hümayun 29/1364 (28 N 1198/15 Aug. 1784).

46 For a series of Ottoman firmans about the construction of ships in the port of Suez, see DWQ, Maḥkamat Rashīd 132, p. 88, case 140 (17 Ş 1137/30 Apr. 1725); DWQ, Maḥkamat Rashīd 132, p. 199, case 308 (16 Ş 1137/29 Apr. 1725); DWQ, Maḥkamat Rashīd 132, p. 199, case 309 (17 Ş 1137/30 Apr. 1725); DWQ, Maḥkamat Rashīd 132, pp. 200–01, case 311 (3 N 1137/16 May 1725); DWQ, Maḥkamat Rashīd 132, p. 201, no case no. (15 N 1137/28 May 1725). Both the complete dependence of the Haramayn on food shipments from Egypt and the importance of this food to the preservation of these cities and to the people who lived and travelled there are outlined in the following firman from the Ottoman Sultan to the Vali of Egypt Muḥammad Paşa in 1732: BOA, Mühimme-i Mısır, 4: 426 (Evasıt B 1144/9–18 Jan. 1732). This document explicitly states that the daily sustenance and food (*kuvvet ve ma‘ishet*) of the people of Mecca and Medina was limited *only* to those grains that were customarily sent from Egypt.

47 These are points made consistently throughout the archival record of grain shipments from Egypt to the Hijaz. For a case that makes the point very adamantly that there must be continuous supplies of grain sent from Egypt to the Haramayn with no breaks in the flow of this food, see TSMA, E. 5225/9 (Evahir Ca 1191/27 Jun.–6 Jul. 1777).

REFERENCES

Barkan Ö.L. 1943. *Kanunlar*, vol. 1 of *XV ve XVIinci asırlarda Osmanlı İmparatorluğunda Ziraî Ekonominin Hukukî ve Malî Esasları*, İstanbul Üniversitesi Yayınlarından 256. Istanbul: Bürhaneddin Matbaası.

Blouin, K. 2012. 'Between water and sand: agriculture and husbandry'. In Riggs, C (ed.), *The Oxford Handbook of Roman Egypt*. Oxford: Oxford University Press, pp. 22–37.

Bostan, İ. 1992. *Osmanlı Bahriye Teşkilâtı: XVII. Yüzyılda Tersâne-i Âmire*. Ankara: Türk Tarih Kurumu Basımevi.

Braudel, F. 1982. *The Wheels of Commerce*, trans. Siân Reynolds, vol. 2 of *Civilization and Capitalism, 15th–18th Century*. London: Collins.

Brummett, P. 1994. *Ottoman Seapower and Levantine Diplomacy in the Age of Discovery*. Albany: State University of New York Press.

Coşkun, M. 2001 'Stations of the pilgrimage route from Istanbul to Mecca via Damascus on the basis of the *Menazilü't-Tarik İla Beyti'llahi'l-'Atik* by Kadri (17th Century)'. *Osmanlı Araştırmaları* 21: 307–22.

Cuno, K.M. 1992. *The Pasha's Peasants: Land, Society, and Economy in Lower Egypt, 1740–1858*. Cambridge: Cambridge University Press.

Faroqhi, S. 1994. *Pilgrims and Sultans: The Hajj under the Ottomans*. London: I.B.Tauris.

Finkel, C. 2006. *Osman's Dream: The History of the Ottoman Empire, 1300–1923*. New York: Basic Books.

Hathaway, J. 1997. *The Politics of Households in Ottoman Egypt: The Rise of the Qazdağlıs*. Cambridge: Cambridge University Press.

Hinz, W. 1955. *Islamische Masse und Gewichte umgerechnet ins metrische System*. Leiden: Brill.

Imber, C.H. 'The Navy of Süleiman the Magnificent,' *Archivum Ottomanicum* 6: 211–82.

Kanunname-i Mısır. 1845. *Ḳānūn-nāme-i Mıṣr*, Topkapı Sarayı Müzesi Kütüphanesi 1845 (E.H. 2063).

McNeill, J.R. 1992. *The Mountains of the Mediterranean World: An Environmental History*. Cambridge: Cambridge University Press.

Mikhail, A. 2011. *Nature and Empire in Ottoman Egypt: An Environmental History*. Cambridge: Cambridge University Press.

Murphey, R. 1988. 'Provisioning Istanbul: the state and subsistence in the early modern Middle East'. *Food and Foodways* 2: 217–63.

Mutawallī, A.F. 1986. *Qānūn Nāmah Miṣr, alladhī Aṣdarahu al-Sulṭān al-Qānūnī li-Ḥukm Miṣr*. Trans. and intro., Cairo: Maktabat al-Anjlū al-Miṣriyya.

Quataert, D. 2000. *The Ottoman Empire, 1700–1922*. Cambridge: Cambridge University Press.

Ramzī, M. 1994. *al-Qāmūs al-Jughrāfī lil-Bilād al-Miṣriyya min 'Ahd Qudamā' al-Miṣriyyīn ilā Sanat 1945*, 6 vols. in 2 pts. Cairo: al-Hay'a al-Miṣriyya al-'Āmma lil-Kitāb.

Raymond, A. 1973–4. *Artisans et commerçants au Caire au XVIIIe siècle*, 2 vols. Damascus: Institut français de Damas.

Scott, J.C. 1998. *Seeing Like a State: How Certain Schemes to Improve the Human Condition Have Failed*. New Haven: Yale University Press.

Sertoğlu, M. 1986. *Osmanlı Tarih Lûgatı*. Istanbul: Enderun Kitabevi.

Shaw, S.J. 1962. *The Financial and Administrative Organization and Development of Ottoman Egypt, 1517–1798*. Princeton: Princeton University Press.

——— (ed. and trans.) 1964. *Ottoman Egypt in the Eighteenth Century: The Nizâmnâme-i Mısır of Cezzâr Ahmed Pasha*. Cambridge: Center for Middle Eastern Studies of Harvard University.

——— 1968a. 'Landholding and land-tax revenues in Ottoman Egypt'. In Holt, P.M. (ed.), *Political and Social Change in Modern Egypt: Historical Studies from the Ottoman Conquest to the United Arab Republic*. London: Oxford University Press, pp. 91–103.

——— 1968b. *The Budget of Ottoman Egypt, 1005–1006/1596–1597*. The Hague: Mouton.

Sulaymān, A.H.H. 2000. *al-Milāḥa al-Nīliyya fī Miṣr al-'Uthmāniyya (1517–1798)*. Cairo: al-Hay'a al-Miṣriyya al-'Āmma lil-Kitāb.

Tsugitaka, S. 1997. *State and Rural Society in Medieval Islam: Sultans, Muqta's and Fallahun*. Leiden: Brill.

Worster, D. 1990. 'Transformations of the earth: toward an agroecological perspective in history'. *Journal of American History* 76: 1087–1106.

7 Do not Imagine that Every Cloud Will Bring Rain: A History of Irrigation on Kilimanjaro, Tanzania

Matthew V. Bender

INTRODUCTION

In the summer of 1884, English explorer Harry H. Johnston arrived at the foothills of Africa's tallest peak, Mount Kilimanjaro. In his memoir (1886), he describes much about his journey, the landscape and features of the mountain, and the characteristics of its peoples, a group he refers to as the Chagga. He describes with awe the mysterious glaciers that lay at its highest peak and the green, well-watered slopes that radiate from it, referring to the whole scene as an 'African Switzerland'. Johnston's praise, however, is not limited to the natural landscape. He notes with some surprise the presence of hundreds of artificial water canals:

> Along one side and then across and down the other side [...] flowed a tiny artificial canal of clear water brought from a tumbling stream higher up the mountain, and carried along this hill from above in a gently descending channel. Thus we had water at our very door, and needed not to seek it in the ravine a thousand feet below. It seemed so strange and quaint to find a placid brooklet flowing along high ground up in the clouds and at the edge of a precipice (Johnston 1886: 123).

These irrigation canals, known by locals as *mifongo* (sing. *mfongo*), captivated Johnston, as they did other Europeans who encountered Kilimanjaro in the nineteenth century. To these individuals, they represented a surprising feat of ingenuity that had fueled the productive agriculture of these mountainside peoples. By the 1930s, however, opinions on these irrigation works had shifted dramatically. Once considered a symbol of indigenous ingenuity, they came to be thought of as inefficient, wasteful, and unnecessary.

This chapter examines the *mifongo* irrigation works on Kilimanjaro from the middle of the nineteenth century to the present, and asks why outsiders to the mountain, in particular colonial agents and – after

1961 – the independent Tanzanian state, came to consider highland irrigation harmful and unnecessary. As one of the most prominent manmade features of the mountain's landscape, the *mifongo* have been the subject of much inquiry. In recent years, scholars such as Alison Grove, Frances Vavrus, Donald Musgrove and Mattias Tagseth have delved into the furrows, looking at issues related to how they were utilized, their development over the twentieth century, and their decline in many areas (Some examples of scholarship include Grove (1993), Mosgrove (1998), Tagseth (2008) and Vavrus (2003)). Less attention has been paid to how these systems became a point of contention between mountain peoples and those who came to influence Kilimanjaro from outside its borders. The shifting – and often inconsistent and contradictory – views of the colonial regimes, the independent state, and the development community toward indigenous irrigation schemes such as the *mifongo* are important to consider, given the history of external political domination of the African continent. They also offer us much in understanding the politics and economics of agriculture development and food production, as well as the complexity of controlling a natural resource that is inherently dynamic.

This chapter argues that furrow irrigation on the upper slopes of Kilimanjaro came to be a point of contention between Chagga mountain communities and outsiders as a result of two sets of factors – shifting perceptions of the waterscape and the development of new demands for water. These factors, in essence, 'created' water scarcity on the mountain. The discourse of scarcity led the colonial administration to criticize irrigation works it had previously considered ingenious. At the heart of these criticisms lay belief that water needed to be under careful government management, removed from the inherent local control of the furrows. The end of colonial rule in the 1960s did not alter this dynamic, but rather reinforced it. The independent state, first under a socialist program called *Ujamaa* and in subsequent years under a neoliberal regime, has considered water to be a 'national' resource subject to control by government agencies and regional water boards. In general, policies since independence have considered highland irrigation to be wasteful and largely unnecessary.

These changes, however, did not lead to an immediate decline in highland irrigation on Kilimanjaro. In fact, the use of irrigation was on the rise through the 1960s in spite of government criticism. In the minds of Chagga farmers, irrigation was not merely a part of their cultural heritage; it was vital to the success of local agriculture. Since the 1970s, many *mifongo* in the upper slopes of Kilimanjaro have fallen into decline. This decline, however, owes more to changing economic and social circumstances than it does invasive government policies. Furrows became expendable in the face of decline of key agricultural commodities such as coffee, booming population density, the rise of job opportunities beyond the mountain, the prevalence of alternative water sources such as pipelines, and the steady erosion of the furrows' cultural currency. What this tells us is that Chagga farmers successfully negotiated political pressures from the outside to shift away from irrigation, and only in the face of new economic and social realities have they been less able to sustain it. The conclusions of this paper are particularly significant at a time when many NGOs and governments are lending support to the development and renewal of indigenous irrigation schemes.

ORIGINS AND DEVELOPMENT OF THE MIFONGO

Mount Kilimanjaro is among the most defining geographic features of the African continent. Rising above the East African steppe to an altitude of 19,340 ft, it is the world's tallest volcano as well as the tallest freestanding mountain. Though often thought of as a single massif, it is actually comprised of three separate peaks: the venerable, glacier-capped *kibo*, its shorter counterpart to the east, *mawenzi* (16,893 ft), and the now-flattened peak to the west called *shira* (12,998 ft). The mountain's surroundings are formidable, consisting of hundreds of square miles of inhospitable, arid grassland. As it rises, however, its climate shifts quickly to well-watered temperate forest, and then to rainforest. In this way, the mountain is a 'Sky Island', featuring sharp distinction between its humid lower slopes and the surrounding arid steppe.[1] The relative abundance of rainfall on the lower slopes feeds a wealth of vegetation as well as numerous rivers and streams that flow through deep valleys toward the plains. These eventually converge into a single river, the Pangani, which flows more than 300 miles toward the Indian Ocean. The presence of high levels of precipitation, in particular on the south and east sides of the mountain and at elevations between 3,500 and 6,000 ft, creates optimal conditions for intensive agriculture.

These lush lower slopes of Kilimanjaro have long been home to a group of people that have come to be known as the Chagga. The ancestors of the current population began to arrive on the mountain an estimated 800 years ago, and are closely related to the Pare, Taita, and Meru, and more distantly to other Bantu groups such as the Swahili. These migrants were likely drawn to the mountain due to its moderate temperatures, fertile soils, ample rainfall and surface water, and the security provided by the mountain landscape. Over time they settled along the ridges of the mountain's southern and eastern slopes and developed a thriving agrarian economy based on small homesteads called *vihamba* (sing. *kihamba*). By the 1600s, bananas had come to be the chief staple, and they were so central to livelihood that many came to identify themselves as *wandu wa mbdeny*, or people of the banana groves (Wimmelbücker 2002: 51). Bananas were typically intercropped in the *vihamba* with other crops including yams, cassava, beans, taro, sugar cane, sweet potatoes, maize, pumpkins, papaya, and other vegetables (Abbott 1891: 394). Many farmers also cultivated crops such as maize and finger millet (*Eleusine coracana*) in more arid downhill areas outside of the vihamba. Aside from crops, most farmers also held some livestock, typically kept in stalls within the *vihamba*. On the whole, the Chagga-speaking mountain communities of Kilimanjaro developed a highly successful system of intensive farming, one that enabled the population to grow and prosper. Most people identified themselves first and foremost by their family and their clan, though by the late nineteenth century a strong system of chieftaincy had developed. By the turn of the century, an estimated 80,000 people called the mountain home.

The waters of Kilimanjaro, relatively abundant in comparison to much of Eastern Africa, were crucial to the success of these peoples. At times, though, they could be quite precarious. Rainfall on the mountain falls unevenly throughout the year, with much of it concentrated into two distinct rainy periods: the *fuli* (short rains) from late October to early December, and the *kisiye* (long rains) from April to early June. Outside of these periods, rainfall tends to be minimal. These patterns not only define the seasonal limitations for crop cultivation, but they also align optimal cultivation periods with the likelihood of flooding. Across the mountain, rainfall levels vary dramatically with elevation and the presence of microclimates. Rainfall also tends to be higher in areas of higher elevation, diminishing as one approaches the mountain's base. These patterns can be seen in Table 7.1.

On a longer time scale, the mountain endures cyclical periods of drought, with a period of below average precipitation occurring every six years or so. Adding to these difficulties is the fact that the mountain's numerous streams and rivers, most of which originate in the rainforest zone, flow past areas of settlement in deep valleys, making them difficult to access. In south-east Kilimanjaro, people developed a proverb to describe the unpredictable nature of water: *kipfilefile kirundu kechiwa mvuo kilawe* (Marealle 1965). Literally translated as 'a little rainy cloud that never became rain', it served as a reminder that rainfall and surface water could fail, leaving people to struggle to find sufficient water even in a place of seeming abundance.

In response to these challenges, the peoples of the mountain developed a system of furrow-based irrigation. In Chagga languages, the furrows are referred to as *mifongo*, though some sources use the Swahili term *mifereji*. They were excavated ditches that diverted water from rivers at points high up on the mountain, and then used the power of gravity to channel it away from the valleys and directly into areas of settlement. From that point people used the waters for irrigation as well as domestic uses. Furrows varied greatly in length, capacity, and number of branches, in many ways as

Table 7.1. Rainfall (in inches) across Kilimanjaro as measured in communities from west to east over the course of 12 months. Note the presence of the two rainy seasons, as well as the relationship between elevation and rainfall.

Location	Alt (ft)	Jan	Feb	Mar	Apr	May	Jun	Jul	Aug	Sep	Oct	Nov	Dec	Ave.
Machame	4,987	3.3	2.6	3.8	22.2	22.0	6.3	5.1	2.0	1.1	1.2	3.3	2.8	75.9
Lyamungu	4,101	1.7	2.6	4.3	21.1	18.0	4.4	2.4	1.8	1.2	1.5	3.8	3.1	65.5
Kibosho	4,790	3.0	2.8	6.3	23.4	25.3	8.0	4.6	2.2	1.5	1.6	3.7	3.6	86.0
Moshi Town	2,690	1.6	1.7	4.6	11.4	6.2	1.2	.5	.5	.5	1.3	2.2	1.9	33.5
Old Moshi	5,413	3.3	3.8	7.0	22.3	16.2	5.2	3.9	2.1	1.5	3.3	6.7	4.8	80.0
Kirua	5,512	3.5	5.0	12.9	24.3	13.5	3.6	4.7	2.9	1.7	6.5	11.9	6.6	96.9
Marangu	4,691	2.8	3.4	7.7	15.4	10.4	2.6	2.6	2.6	1.9	3.5	6.9	5.7	64.0
RomboMkuu	4,691	3.2	3.7	8.8	10.9	4.5	1.2	.9	1.7	1.5	3.4	10.7	8.0	58.4

Source: data is from Maro, 1975.

individual as the people who built and maintained them. By the middle of the nineteenth century, hundreds of furrows covered the ridges of the *vihamba* zone, with the highest concentration being in the south-east corner of the mountain (Figure 7.1).

It is difficult to estimate the earliest moment at which mountain farmers began to develop these systems, given the lack of written records and archeological data. Likely it was in the frame of 200–400 years ago. The idea of channeling water from streams using gravity was not unprecedented in East Africa. Several neighbouring areas, including Engaruka near the Ngorongoro Crater and the Kamba areas of central Kenya, had water furrow systems that are more ancient than those on Kilimanjaro (Sutton 1978, Sutton & Widgren 2004). It is possible that the technology behind the mountain's furrows originated in these places or elsewhere.

Furrows built before the twentieth century shared similar design characteristics. Each began with a dump, a swell within a river or stream that would provide a good, consistent supply of water. From here the water was diverted into the main channel of the furrow, which would turn away from the river and follow a less steep path toward its destination. The amount of water diverted was controlled by the use of a weir constructed from sections of banana trunk, leaves, and mud. From this point the waters flowed downhill at a gradual slope. Once they reached their destination, they would again be diverted, this time from the main channel into numerous secondary channels leading directly into individual farms. Water from the furrow itself could be gathered into clay pots for any number of domestic purposes (cooking, washing, bathing, brewing, construction of mud block, etc.), or diverted one last time to flood irrigate areas of crops. Beyond these, furrows could have any number of variations. Some featured multiple intakes to maximize the flow of water in areas with many users. Many had culverts or aqueducts, made from hollowed out trees or banana stems, to lead water over other streams or under heavily traveled paths. Some even had reservoirs, used to store the overnight water flow and in turn increase the amount of water available during daylight hours. Perhaps the biggest variables were length and number of branch furrows. Furrows could be as short as a quarter of a mile or as long as several miles, with a handful of branch furrows or over a dozen (Figure 7.2).

The development of a new furrow became a highly ritualized process. All activities related to furrows, from development to maintenance and usage, were considered masculine, and therefore only men participated in these efforts. The founder of the furrow, the *meni mfongo*, held a special status in terms of its development. He took lead in determining the point of extraction, the path of the main channel, and in organizing labour for construction. Often this individual was a specialist, a man with special knowledge of furrow construction

Figure 7.1. An irrigation furrow on Kilimanjaro (© Matthew Bender, 2004).

Figure 7.2.   An intake diverts water from a river into a furrow (on the right) (© Matthew Bender, 2004).

techniques (Interview with Joshua Maina, Foo, Machame, 2004.) All men interested in using water from the new furrow were required to aid in its construction. The *meni mfongo* surveyed the new furrow using one or more techniques and then commanded the men to dig the furrow to the needed specifications of width and depth. Once completed, those individuals who assisted with the building of the furrow formed the community of users whose families and their descendants had rights to use its waters. In addition to these acts, the *meni mfongo* set the timetable for the usage of water, adjudicated any disputes that arose between users, managed all required maintenance work to keep the furrows flowing, and ensured that the local laws regarding the protection of furrows were followed. They acted as spiritual managers as well, ensuring that offerings were made to the spirits to ensure that the waters would always flow.

Altogether, the *mifongo* contributed to the development of a highly successful, intensive system of agriculture on Kilimanjaro. They helped to provide security against variations in the water supply, while also permitting year-round cultivation. By providing easier access to domestic water, they eased the burden on women and in turn freed up more labour for farming. As time passed, their significance extended well beyond agriculture. They became a focal point for cooperation between members of different clans, especially those living uphill or downhill from one another. It is possible that they played a role in the development of chieftaincy on the mountain in the nineteenth century, a conclusion supported by the fact that chiefdoms arose between river valleys, following the same contours as the *mifongo*. Furrows were linked to notions of gender and age identity, as well as to connectivity with the spirit world. For young boys, learning the art of maintaining and using irrigation works was central to their initiation into adulthood (Raum 1940). Though other alpine societies in East Africa developed similar irrigation systems, the *mifongo* system of Kilimanjaro was by far the most extensive and integral to local culture.

CHANGE AND CONTINUITY IN THE COLONIAL ERA

The middle decades of the nineteenth century brought about a number of changes for the peoples of Kilimanjaro. The arrival of the first European explorers and missionaries was among the most significant of these. Beginning with its alleged 'discovery' by Johannes Rebmann in 1848, the mountain became a place of intrigue for a generation of adventurers and eventually a prize to be fought after in the Scramble for Africa. Explorers imagined it as a magnificent and lush landscape, standing in stark contrast to its arid surroundings, with many describing it in their writings as an African 'Switzerland', or an African 'Olympus.'[2] By the early 1890s, Kilimanjaro had been colonized by Germany and incorporated into a new colony, *Deutsch-Ostafrika* (German East Africa). This led to the introduction of new political, economic, and religious institutions that would transform the region and the lives of its peoples.

The Europeans who encountered Kilimanjaro from the 1840s until 1918 – a diverse group including Catholic and Protestant missionaries, German colonial administrators and military personnel, and settlers from over a half dozen countries – considered the mountain to be a place of tremendous environmental richness and endless potential. Yet they too came to understand that not every cloud would bring rain, and that success in this new locale required the careful management of water. While they viewed the Chagga population as primitive and superstitious – as they did most African communities in this era – they considered the *mifongo* to be incredibly ingenious. Charles New, a British missionary who explored the mountain in 1871, expressed his fascination with them in his travelogue:

> The watercourses traverse the sides of the hills everywhere [...] And now I understand what I had been told upon the coast...that the Wachagga make the water in their country run up hill! The almost level run of the watercourses, viewed in connection with the sharp descending outlines of the hill tops and the natural course of the stream below, deceives the eye (New 1873: 132–5).

New sees the *mifongo* as an example of the technical expertise of local agriculture. Other explorers echoed his sentiments. William Louis Abbott, an American naturalist who visited the mountain in the late 1880s, noted that the furrows were 'constructed with great ingenuity' (Abbott 1891: 392). Hans Meyer, the German mountaineer and first person to reach the summit of the mountain in 1889, remarked that the 'art of constructing these irrigation canals [...] is among the most wonderful to be found among tribes like the [Chagga]' (Meyer 1891: 103). Perhaps the most noteworthy praise came from Harry Johnston. A renowned explorer, he also played a role in the Scramble, attempting to claim parts of East Africa for the British by concluding treaties with the various chiefs he visited. As mentioned in the introduction to this essay, he developed a fascination for Kilimanjaro and its waterworks during his 1884 expedition to the mountain. He considered the furrows to be an example of Chagga 'handiwork' and a sign of their 'industrious' nature (Johnston 1886: 120). Johnston not only explored the region but also remained for almost six months and actually constructed a homestead of his own (Figure 7.3), complete with its own furrow, which like his Chagga neighbours enabled him to have 'water at his very door' (Johnston 1886: 122–3).

Throughout the first few decades of colonial rule, the *mifongo* were not merely praised by European settlers; they were also embraced and replicated. Catholic and Lutheran missions, which created settlements scattered among the *vihamba*, tapped into the very same rivers and streams as did their Chagga neighbours. They also relied upon local *meni mfongo* to help them develop their own furrows. For example, in 1908 Father August Gommenginger, the head priest at the Catholic mission in the chiefdom of Kilema, determined that the settlement needed a new water furrow to help provide water for its church and gardens.[3] For this task, he assigned Brother Cereus (Cere) Spiekerman, a former carpenter who had earned a reputation as a skilled handyman. Late in the year they sought out permission from the local chief, Mangi Kirita, who gave his approval and arranged for the services of a local *meni mfongo* named Matonga. Over several months, Cere and Matonga worked together to design and construct the new furrow. It tapped the Mwona (Himo) River nearly a mile uphill of the mission, at a point that ensured reliable year-round water. From there it flowed through the lands of several clans, past dozens of *vihamba*, before leading directly to the grounds of the mission. The finished furrow opened for use on 5 April 1909. The mission, in breaking with Chagga conventions regarding the naming of furrows, called it the *Mtakatifu* furrow, meaning Holy Spirit. They also held a special ceremony to open it in which Gommenginger led a prayer before allowing the waters to flow. By local standards the new *mfongo* was immense, measuring around 2.5 miles in total length. However, it adhered to the same design as most other furrows in the area.

Figure 7.3. A sketch of Johnston's settlement on Kilimanjaro (from Johnston, 1886).

Settler estates, for the most part, tended to be located further downhill of the Chagga *vihamba*, and as such faced drier growing conditions and were as a consequence even more dependent on irrigation. Settlers, in general, had less cordial relations with their Chagga counterparts than did missionaries, since they very quickly became economic competitors. Yet they also developed their own furrows based on the *mifongo* model, sometimes alone but often with the cooperation of local *meni mfongo* and the blessing of the local chief.

The furrows developed by missionaries and settlers were, for the most part, functionally identical in design to those long constructed by the mountain communities. There were important differences, however, in terms of usage, maintenance, and meaning. Europeans tended to use furrows more intensively, as they used more water per capita for both domestic uses and also to irrigate new crops they brought to the mountain, notably coffee. Maintenance tended to be provided not by users of the furrow, but rather by paid workers. Most notably, the new furrows were divorced of the spiritual and cultural meanings that Chagga ascribed to them. The *Mtakatifu* furrow, for example, was clearly linked to Catholic, rather than Chagga, notions of spirituality. For most settlers, furrows were named after themselves, and were explicitly utilitarian and entirely devoid of any spiritual or cultural connection.

These new European actors introduced a wealth of changes to the region in terms of politics, economics, and religion, and Chagga communities changed over time as they adapted to these new conditions and took advantage of new opportunities. Some of the most significant changes include the implementation of colonial rule, which transformed chiefs from sovereigns into low-level bureaucrats accountable to District Officers, the development of external markets for agricultural commodities, and the rising prevalence of Christianity. The use of irrigation changed as well. In the early decades of the century, there was a steady expansion in the number of irrigation canals as well as in the amount of water being used. Several factors account for this. One, the people of the mountain began to adopt coffee cultivation. By the start of World War I, coffee emerged as a lucrative cash crop. Water from furrows was used to irrigate small coffee trees, and also to process raw coffee cherries into export-quality parchment coffee. High rates of population growth likewise generated a need for more irrigation, as more people desired to open up more farmland for *vihamba* cultivation, often at lower elevations where more water was needed. Lastly, Chagga were beginning to use more water for domestic uses, the biggest being more frequent bathing and washing of clothes, as required for young children being sent to mission schools.

The development of German colonialism faced a challenge in 1914 with the outbreak of World War I. For nearly two years, the mountain was a focal point of fighting between the Germans and the British, who had invaded from the neighbouring Kenya Colony. By the end of the war, Kilimanjaro had come to be part of a British mandate called the Tanganyika Territory. British colonial rule brought a wide range of changes to the colony, including the introduction of 'indirect rule' institutions adapted from other colonies in the Empire as well as the expulsion of German settlers (some of whom would return in the 1920s). In terms of water, the British began to introduce policies similar to those in their other colonies. This was meant to address a key issue of German rule, which is that they had never developed a formal legal code for the control of water resources. The first piece of legislation developed by the British was the Water Ordinance of 1923, a law that vested formal control of water in the hands of the state, created a set of regional water boards to manage the resource, and placed control of African water use in the hands of chiefs accountable to District Officials.

Though the shift to British rule had led to some changes in policy regarding water by the early 1920s, these officials tended to view the waterscape of Kilimanjaro, the *mifongo*, and the water use practices of Chagga farmers, much as their German predecessors had. A good example of this can be found in the writings of Charles Dundas, a colonial officer and enthusiastic student of Chagga culture who served as the District Officer for Moshi from 1921 to 1925. In his 1924 book *Kilimanjaro and its People*, Dundas describes the mountain as having an 'abundance of water', and the furrows to be 'utterly unbelievable in their ingenuity, seemingly capable of carrying water right up the hill' (Dundas 1924). Like his predecessors, he viewed the Chagga to be progressive in their management of water, and possessing technologies that needed to be both praised and replicated.

By the 1930s, however, the situation had changed dramatically. *Mifongo* came to be criticized as inefficient and primitive, and the people themselves chastised for being wasteful, prodigal users of water. In particular, irrigation furrows came to be thought of as prone to wastage, susceptible to breakage and leakage, and promoting of both soil erosion and excessive evaporation. This dramatic change in viewpoint stemmed from several factors. For starters, a series of prolonged droughts (1907–9, 1913–15, 1922–4, 1929–30) strained the mountain's supply of surface water and generated fear that the mountain was becoming progressively more arid. As the upstream users, Chagga farmers came to be criticized by downstream Europeans for using too much water. The American Dust Bowl (1930–1936) likewise raised concerns, though more due to the potential of furrows to promote erosion through the soil wash associated with flooding sloping fields. British officers criticized the practice of flooding finger millet fields, since they considered millet to be an uneconomic crop with limited market potential and linked to excessive consumption of alcohol.[4] The high rate of population growth and rising rates of consumption per capita both created more demand for water amid a consistent or even dwindling supply. Lastly, the development of hydroelectric power and commercial agriculture in the Pangani River Valley created a new type of demand for water, one resonating from well beyond the mountain itself. Altogether, these created a sense of fear among colonial officials that the water supply on Kilimanjaro was not abundant, but rather dangerously limited, even more so due to human action. Chagga users, for the most part unfairly, came to be the focal point of these concerns. Criticism of their water practices ranged from the inherent flaws of the *mifongo*, to the uncontrolled development of new furrows, to blatant overuse of water, to the dirtying of water bound for downstream users. Thus the *mifongo*, which had a mere 20 years earlier been extolled as a symbol of ingenuity, came to be considered one of the biggest threats to the mountain's future.

To determine the scope of the problem, the colonial administration commissioned a series of hydrological studies in the 1930s. The published reports that came from them were intended to help develop more scientific tools and strategies for controlling the waters of Kilimanjaro. The most influential study was the *Report on the Investigation of the Proper Control of Water*, led by prominent colonial scientists Edmund Teale and Clement Gillman in 1934 (Teale & Gillman 1935). The so-called Teale-Gillman Report summarized the hydrological features of the mountain and then addressed pressing questions concerning the control, distribution, and use of water. Based on their analysis, Teale and Gillman found that across Kilimanjaro there was 'a very widespread haphazard use of the water' (Teale & Gillman 1935: para 71). This was partly due to excess consumption, but mostly due to wastage associated with furrows. They cited as a problem furrows being located in porous rock, being constructed too steep or too wide (thus resulting in either excess erosion or evaporation), and inadequately maintained, and thus subject to leakage and breakage (Teale & Gillman 1935: paras 89, 153–9). To solve these problems, they suggested that many furrows be relocated and reconstructed, using modern measurement and surveying tools, taking account of soil conditions, and drawing upon advancements such as reservoirs (to store water flow from the rainy seasons) and concrete intakes. Furthermore they suggested a careful prioritization of the uses of water, with domestic being given foremost attention, followed by industrial (coffee processing), irrigation, and finally hydropower. To their credit, Teale and Gillman placed blame on both Chagga farmers and European settlers. A second study, completed by Francis Kanthack in 1936, was not as generous. His published *Report on the Natural Waters of Tanganyika* chastised the *mifongo* for being 'primitive, wasteful, and inefficient' (Kanthack 1936: 6).

In both studies, a broader solution that needed to be addressed was control of water. The colonial system had, in essence, created two separate, yet interconnected, spheres of authority with regard to water: one for the European population and another for Chagga. Under the Water Ordinance created by the British administration in 1923, Chagga chiefs possessed oversight of the waterworks of their subjects, while local water boards oversaw those of Europeans. While this kept with the underlying premise of Indirect Rule, it also made overall management of the resource tricky, especially in a geography where Chagga users almost always lived upstream. Both the Teale and Gillman and the Kanthack Reports sought an immediate change in the legal framework for water in the region. They called for an end to the two-tier system for all future irrigation works and water extractions, with authority for water being vested in a governmental authority that would issue permits, regulate development and usage, and adjudicate disputes. Existing waterworks beyond the point of abstraction from a river or stream, now referred to as 'customary' furrows, would remain under local control. However, the government should make every effort to 'regularize' these furrows and promote improvements in design and use.

In the 1930s and 40s, the colonial administration did take some steps to transform the water regime on Kilimanjaro. In 1931, for example, it persuaded Chagga chiefs to implement a new policy, nicknamed the '50 Paces Tangazo' Order, that banned cultivation within fifty paces of a watercourse. This was intended both to reduce soil erosion and to conserve water (Native Authority Orders, 1931, Government of Tanganyika, Annual Reports, Northern Province, 1931. Tanzania National Archives (hereafter TNA) 11681). On a broader scale, the colonial government implemented a revised Water Ordinance in 1948. Notably, it called for the creation of regional Water Boards to manage the resource as well as independent Water Courts to adjudicate disputes. It also made an important change in terms of water use by African populations. All waterworks, such as furrows, that were deemed 'customary' would remain under the control of local specialists and chiefs. However, new ones could only be constructed with the permission of the appropriate water board. This meant that a new furrow project would require detailed designs, surveys, and usage estimates, as well as the payment of appropriate fees, to the district water board. In theory, this posed a significant threat to the existing practice of irrigation on Kilimanjaro.

In practice, however, the new scrutiny on Chagga mountainside irrigation resulted in little change. Most farmers utilized existing furrows in much the same manner as they had in previous years. In cases where they needed more water or wished to route water to new locations, they either expanded an existing furrow or 'revived' an allegedly abandoned furrow, thus allowing them to circumvent the new regulations.[5] *Mifongo* irrigation practices did experience significant changes in the early decades of the twentieth century, but not as a result of government regulations. Rather, changes resulted from social and economic changes within Chagga communities. For example, the spiritual attributes of irrigation took on more Christian overtones as people began to convert from their existing beliefs to Christianity. Rituals held to honour the spirits became less frequent, or at least less public, and came to be replaced by Christianized versions (such as holding novenas and praying the rosary). In a similar way, the role of the *meni mfongo* in local communities came to be less prestigious, as furrows became more secularized and other routes to status came into being (such as Church membership, formal education, and cash cropping). Perhaps the most significant change was intensification in the use of *mifongo*. This came in part from greater use of domestic water – for more frequent bathing and washing – but mostly due to the swift growth of the coffee industry. By the late 1940s, coffee had become the most lucrative crop on the mountain, with nearly 32,000 growers cultivating nearly 24,000 acres of export-quality Arabica (Von Clemm 1964: 112). Yielding £284 per ton in 1949, this crop had transformed the economic landscape of the mountain. Water from the furrows was crucial to irrigating trees in dry periods, fostering the growth of sapling trees, and processing of raw coffee into export-quality parchment beans. Thus by the 1950s, the *mifongo* of Kilimanjaro had weathered a flood brought about by colonial rule yet remained a vital part of local life.

CHALLENGES OF THE POST-INDEPENDENCE PERIOD

In the early 1960s, decades of colonial rule on Kilimanjaro came to an end with the emergence of the independent nation of Tanzania. The end of British rule and the rise of an elected government in Dar es Salaam precipitated numerous political and social changes. The new government's views on highland irrigation, however, bore striking similarity to those of its predecessor. This can be seen in a short essay published in 1965 by A. G. Pike, the Regional Hydrological Officer for the Pangani Basin and former Water Warden in Kilimanjaro (Pike 1965: 95–6). He remarks that 'there are now so many furrows, most houses today are within a very short distance of one.' While he expresses some admiration for the system, he points out that it is 'beginning to get wasteful':

> Many of the areas served by furrows ... need little irrigation water, being in a high rainfall belt. Also, when a furrow is divided and subdivided many times, wastage from seepage, leakage and evaporation is considerable. Moreover, with the development and settlement of the lower areas, water not essential for the requirements of the mountain dwellers is required down below, where many acres could be brought under cultivation given the water to irrigate.

Perhaps the most direct statement made by Pike follows just after, when he claims, 'it may well be that eventually the furrow system may have to give way to a more efficient method of water distribution, either by large shared communal furrows on each ridge, or by pipelines, or by both'. Pike's words clearly indicate a view by some in the government that, given the scarcity of water on the mountain amid growing demands, the days of highland irrigation were numbered.

Over the next decade, *mifongo* irrigation faced a direct challenge, ironically in the form of water development. In 1967, the Tanzanian government announced an extensive project called the *Rural Water Supply Programme for Kilimanjaro Region*. It called for a series of gravity flow water pipelines, valued at nearly £1.6 million, to provide clean, reliable water to the mountain's population, which by that point had reached an estimated 476,000.[6] The RWSP embodied the values of the government's recently adopted socialist program of economic development, *Ujamaa*, announced by President Julius Nyerere in his 1967 Arusha Declaration. Under *Ujamaa*, the government promised to improve water supplies across the country in an effort to promote the growth of the agricultural sector. Water was conceived as a national resource, one to be provided by the government to its people for free (Figure 7.4).

As I have argued in a separate article, the government extolled the virtues of pipeline development, seeing it as an optimal means of bringing 'more and better water' to the people of Kilimanjaro (Bender 2008: 851). Yet the project itself proposed a fundamentally different way of thinking about water. Unlike the furrows, which were locally controlled and managed, the pipes were designed, constructed, and managed by outsiders. Thus they challenged local control of water and the cooperative attitudes embodied by furrows. The pipes also served existing areas of settlement, with many running alongside existing furrows and tapping into the same rivers. This close proximity positioned pipelines as rivals to the furrows. Lastly, and most importantly, the RWSP made no provision for irrigation. In fact, people were expressly forbidden from using piped water for any form of irrigation, including watering plants by hand. This reflected belief on the part of government officials that irrigation was unnecessary, and that with the passage of time people would eventually give it up, along with the furrows.[7] Thus what we see with the RWSP is a thinly veiled attempt by government officials to undermine traditional forms of authority and local control of water on Kilimanjaro, with the goal of establishing a stronger presence for the national government and more centralized control of a vital natural resource.

By 1980, the pipeline project had come to fruition. Though it did not provide complete coverage to the mountain, it did succeed in providing around 60 per cent of the mountain's population with public taps within 1,000 ft of their homesteads (Grove 1993: 434). Aside from the RWSP, the government's other water initiatives focused not on the mountain slopes, but rather on the arid lands at the mountain's base. One project in this area was particularly notable: the development of the *Nyumba ya Mungu* (House of God) Dam and Reservoir in 1965 (Denny 1978: 5–28). Located just south of Moshi, this project aimed to provide more consistent water flows in the Pangani River basin to aid in hydroelectric development and also provide water for lowland irrigation and agricultural expansion.

Another set of changes began to take shape by the mid 1980s, as Tanzania faced economic turmoil amid the failure of *Ujamaa*. In 1985, newly elected President Ali Hassan Mwinyi formally abandoned the policy

Figure 7.4. Public tap on Kilimanjaro (© Matthew Bender, 2004).

as he embarked on a series of neoliberal economic reforms. These led to more shifts in government policy related to water. Instead of water being a national resource that government should provide for free, it became one for which people should be required to pay. Throughout Kilimanjaro and elsewhere, government agencies have begun to demand that users pay fees for the water they use from pipes and, in some cases, furrows. The justification for charging for water is two-fold: to provide money for developing and maintaining water systems, as well as to encourage more cautious consumption of water. On East Kilimanjaro the local agency Kiliwater has implemented a system of charging users for water based on volume consumed. In other areas, users are charged flat fees. Other changes since the start of structural adjustment include increased money and technical assistance from donor countries and NGOs (in particular Japanese, German, and Scandinavian agencies) to support water projects related to Kilimanjaro, as well as the development of basin-wide water planning boards. In July 1991, the government established the Pangani Basin Water Board, a regional board whose purpose is to involve stakeholders in the planning and management of water resources. An underlying principle of the new agency is Integrated Water Resource Management, which aims to manage the resource in a coordinated manner that that maximizes the welfare of all.

The participatory ethos inherent in these reforms, in stark contrast to the top-down planning strategies of both the colonial state and the Nyerere government, would seem to favour mountainside irrigators. Agencies frequently invoke a democratic, cooperative, participatory language in discussing the management of water. Yet the pitfalls are quickly apparent. The Pangani Basin Water Board, one of the most influential bodies when it comes to water affairs related to Kilimanjaro, does not include a single member who represents the interests of mountainside irrigators. Furthermore, the agency has cited these irrigators as contributing to the problem of inadequate water throughout the rest of the Pangani Basin (for more, see Ngana 2001). The shifts in government policy on water – and the dismissive attitudes toward mountain irrigators – would seem to portend the demise of *mifongo* irrigation in the upper areas of settlement. Indeed, since the 1970s there has been an overall decline in the number of irrigation furrows, and the number of people employing irrigation, on the mountain's upper slopes. According to Grove and others, the number of furrows declined from over 1,000 in 1960 to fewer than 500 by 1986 (Grove 1993: 431–48). This decline has been highly uneven, with some areas experiencing higher rates of decline than others. Mattias Tagseth has argued that this represents a 'restructuring' of irrigation, marked by decline in upper areas and expansion in lowland areas (Tagseth 2008). For those living in the older communities high up on the slopes, the loss of furrows is inevitable but deeply saddening (Figure 7.5). In an interview in 2004, a farmer from the Kilema area of Kilimanjaro named Hubert Kuwana

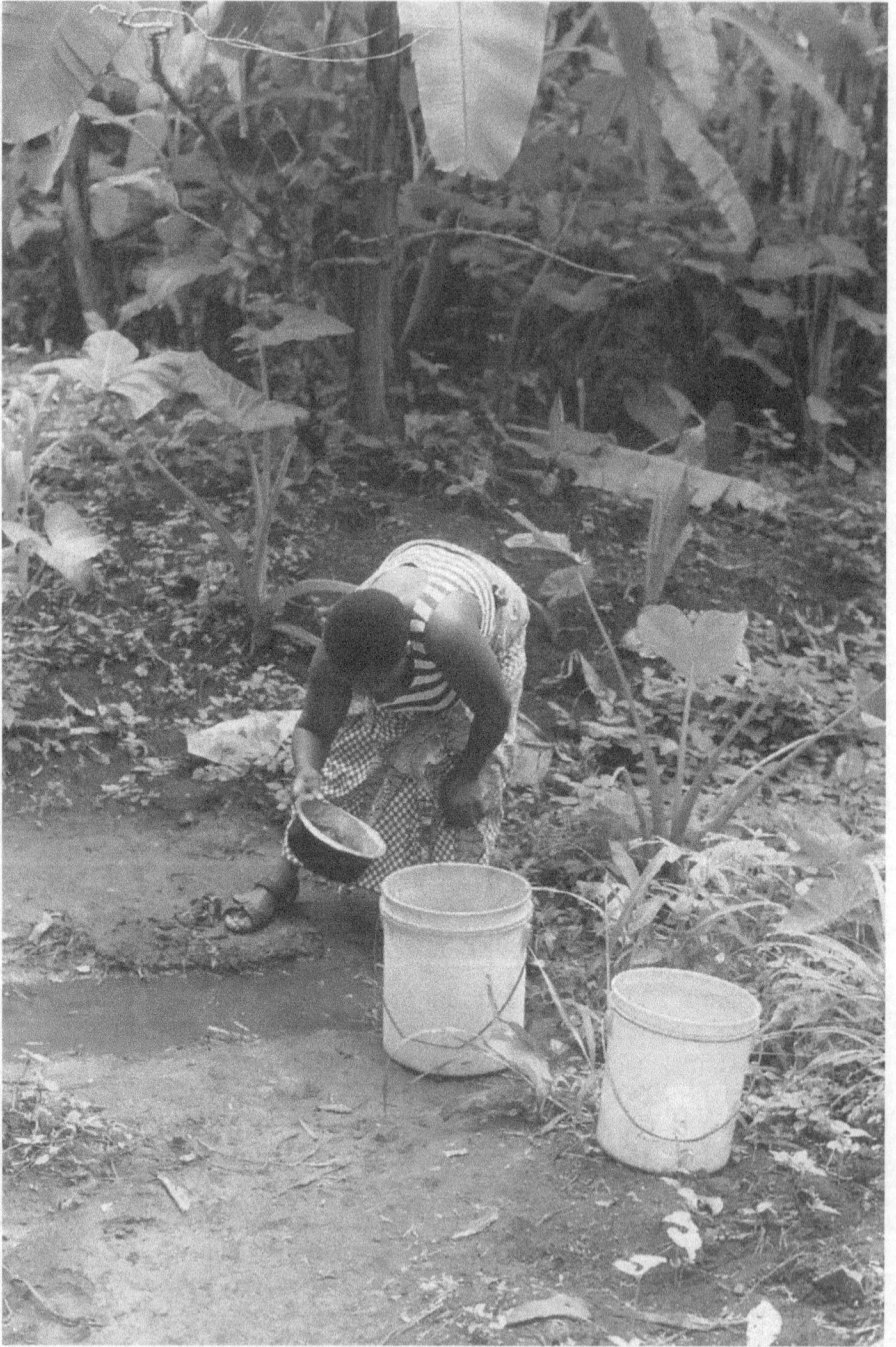

Figure 7.5. A woman takes water from a furrow to use for domestic purposes (© Matthew Bender, 2004).

expressed the feeling of many of his counterparts by saying that, 'the furrows are dying a natural death' (Interview of Hubert Buwana, Kilema Chini, conducted by the author in 2004.).

Why have the *mifongo* fallen into decline in the past few decades, especially given their resilience for much of the twentieth century? While it would be tempting to see government policies as the proximate cause of this decline, there is little direct evidence to support it. Rather, the process of decline owes to social and economic factors (Bender 2008: 856–7). At the heart of these is land scarcity. As the population of Kilimanjaro grew throughout the century, many families chose to subdivide existing *kihamba* as a means of providing land for their sons. The high value of coffee in the period, hitting a peak of £468 per ton in 1954, made it possible for smaller farms to be lucrative (Von Clemm 1964: 112). In the 1970s, however, the price of coffee collapsed, and has failed in recent years to recover significantly. The decline in the coffee industry, and the lack of an alternative crop of similar value, led to an overall decline in highland agriculture. As a result, young men began to seek new opportunities for employment away from the mountain. As Maro notes, by the late 1960s only 65 per cent of men claimed to work full-time in farming (Maro 1975: 34). Those who remained chose to embrace, in whole or in part, paid work in professions such as education or trades such as truck driving and carpentry. This created a 'commuter' lifestyle on the mountain that is clearly evident today, with the high number of *dala-dala* share taxis that connect the mountain communities with Moshi, Arusha, and other towns and cities.

These shifts in labour and employment made it increasingly difficult to sustain the furrows. Furrows depended on contributions of labour from those who made use of them. With so many young men working in whole or in part outside of the *vihamba*, there was less labour available in general. And with the decline of coffee, there was also less incentive to put what labour there was toward maintaining the waterworks. Time intensive, eroded of much of their social and cultural capital, and with less clear value, furrows became expendable. Many of the existing furrow societies began to collapse as elders passed away and younger people were less able to devote time to communal labour. To provide for domestic uses, people shifted to other sources of water, such as rivers, springs, and public taps. This process has been greeted with frustration in many areas, in particular among elders who lament the loss of furrows and the community dynamics they engendered. Albanus Mti, an elder from Kilema, expressed his frustration in a 2004 interview by saying that 'people became lazy and forgot the furrows' (Interview with Albanus Mti, Mkyashi, Kilema, conducted by the author in 2004).

In spite of the overall decline, furrows remain important in many areas of the mountain, in particular in the areas at the foot of the mountain that are more arid, and where plots of land are larger and more viable. Yet to call them *mifongo* is a bit misleading. In terms of technical attributes (such as the design of weirs and lining of the canals) and organizational structure, the furrows are very different from their ancestors. Nonetheless, they represent the perseverance of a system that is inherently small-scale, localized, and participatory. Though they may be a 'shadow of their former selves', in the words of Frances Vavrus, it is possible that these systems and the kind of intimate model they represent could have a strong future (see Vavrus 2003).

CONCLUSION

The *mifongo* system of irrigation on Mount Kilimanjaro has long been a prominent feature of the mountain landscape. For hundreds of years, it has provided mountain farmers with a means of bringing fresh water directly into their farms, for use in irrigation as well as domestic uses. It embodied a prevailing belief that 'not every cloud would bring rain', and that careful management of water was essential to the success of these agrarian communities. When European explorers first arrived, they considered these irrigation systems to be incredibly clever, even ingenious. As the region came to be colonized, European settlers embraced and replicated the systems to provide water to their farms. By the 1930s, however, *mifongo* had come to be thought of as primitive, and highland irrigation considered almost completely unnecessary. This precipitated several decades in which succeeding political regimes – the British colonial state, the independent Tanzanian socialist state, and most recently Tanzania in the era of neoliberal structural adjustment – all discouraged, either directly or passively, the use of irrigation canals in the settled slopes of the mountain.

This chapter illustrates how these changes were rooted in several beliefs on the part of people living beyond Kilimanjaro: that water was increasingly scarce, that irrigation was at best unnecessary and at worst destructive, and that control of water best lay in the hands of trained, skilled professionals living beyond the mountain's borders. For Chagga peoples, however, *mifongo* had long been central not just to local agriculture, but also to local culture and society. Furrows eventually fell into decline in many areas, but this decline owes more to changes within mountain communities than it does to government policy. To this day, people on Kilimanjaro claim the waters of the mountain to be rightfully theirs, and they resent the demands imposed by people living beyond its reaches.

The case of the *mifongo* on Kilimanjaro offers much to those interested in the challenges facing Africa in the twenty-first century. Now more than ever it is apparent that 'not every cloud will bring rain'. Water is increasingly scarce, a scarcity produced by population growth, increasing demand, and climate change. This promises to exacerbate conflicts between local communities and regional and national governments. This essay indicates the degree to which disconnects between the local communities and those who supposedly represent them can undermine successful water systems, while promoting solutions that may not be in the best interests of those affected.

NOTES

1 The term 'sky island' refers to mountains surrounded by lowlands that are drastically different in terms of climate, which in turn creates distinct, biologically isolated habitats. The term derives from the field of island biogeography and relates to the work of ecologists such as Robert MacArthur and E. O. Wilson.

2 Harry Johnston referred to Kilimanjaro as an 'African Switzerland' and remarked that it reminded him of the County Devonshire in England. Hans Meyer, who became the first European to reach the mountain's summit, referred to it as 'Olympus'. Other explorers and adventurers referred to the mountain in similar, otherworldly language.

3 These events are documented in the *Journal de la Communauté de Kilema*, the journal for the Kilema mission kept by the priests assigned there, in the years 1908 and 1909. Supporting information comes from an interview with Gustave Gerard, Kimaroroni, Kilema, 2004.

4 *Mbege*, a brew made from bananas and millet, had long been consumed widely on the mountain. It was used to mark important ceremonies and could also be used as a form of currency. By the 1920s, colonial officials and missionaries had come to think of the beverage as harmful and contributing to widespread laziness among men. Since millet was used to brew this beer, its cultivation came to be considered equally harmful. See Bender 2011: 191–214.

5 By the late 1940s, the number of so-called 'illegal' furrows was estimated to be as high as 50 per cent of the total number of legal ones. See TNA 26045. Letter from M. T. Avery, Hydrographic Surveyor, Water Development Department, Arusha, to the Provincial Commissioner, Northern Province, Arusha, 10 December 1947. See also related files in TNA 471/w. 2/8 v. 2.

6 Water Development and Irrigation Division, Ministry of Lands, Settlement, and Water Development, *A Rural Water Supply Programme for Kilimanjaro Region, 1967*. TNA 558/RCD/167/A.

7 This conclusion was supported by government's newly formed Water Development and Irrigation Division as well as the Kilimanjaro Region Water Advisory Board and the District and Regional Water Officers for Kilimanjaro. See Pike (1959); also *Kilimanjaro Region Water Advisory Board. Ministry of Lands, Settlement, and Water*, 1963. TNA 518/c. 50/3.

REFERENCES

Abbott, W. 1891. 'Descriptive Catalogue of the Abbott Collection of Ethnological Objects from Kilimanjaro, East Africa'. *Annual Report of the U.S. National Museum* 46(2).

Bender, M. 2006, 'Water Brings No Harm: Knowledge, Power, and Practice on Kilimanjaro, Tanzania, 1880–1980'. PhD Dissertation. Baltimore: The Johns Hopkins University.

───── 2008. 'For More and Better Water, Choose Pipes! Building Water and the Nation on Kilimanjaro'. *Journal of Southern African Studies* 34(4).

───── 2011. 'Millet is Gone! Considering the Demise of Eleusine Agriculture on Kilimanjaro'. *International Journal of African Historical Studies* 44(2): 191–214.

Denny, P. 1978. 'Nyumba ya Mungu Reservoir, Tanzania: The General Features'. *Biological Journal of the Linnean Society* 10(1): 5–28.

Dundas, C. 1924. *Kilimanjaro and its People*. London: Frank Cass.

Grove, A. 1993. 'Water Use by the Chagga on Kilimanjaro'. *African Affairs* 92: 431–48.

Johnston, H. 1886. *The Kilima-Njaro Expedition. A Record of Scientific Exploration in Eastern Equatorial Africa*. London: Kegan Paul, Trench, and Co.

Kanthack, F. 1936. *Report on the Control of the Natural Waters of Tanganyika and the Framework of a Water Law on which such Control should be based*. Dar es Salaam: Government of Tanganyika.

Marealle, P. 1965. 'Chagga Customs, Beliefs and Traditions'. *Tanganyika Notes and Records* 64(56).

Maro, P. 1975. 'Population Growth and Agricultural Change in Kilimanjaro, 1920–1970'. *BRALUP Research Paper No. 40*. Dar es Salaam: University of Dar es Salaam Press.

Meyer, H. 1891. *Across East African Glaciers*. Translated by E.H.S. Calder, London: George Philip & Son.

Mosgrove, D. 1998. 'Watering African Moons: Culture and History of Irrigation Design on Kilimanjaro and Beyond'. PhD Dissertation. Ithaca: Cornell University.

New, C. 1873. *Life, Wanderings, and Labours in Eastern Africa*. London: Hodder and Stoughton.

Ngana, J. 2001. *Water Resources Management in the Pangani River Basin*. Dar es Salaam: Dar es Salaam University Press.

Pike, A. 1959. *Water Resources, Moshi District*. Moshi: Water Development and Irrigation Division.

───── 1965. 'Kilimanjaro and the Furrow System'. *Tanganyika Notes and Records* 64: 95–6.

Raum, O. 1940. *Chagga Childhood*. London: Oxford University Press.

Sutton, J. 1978. 'Engaruka and its Waters'. *Azania* 13.

Sutton, J., Widgren, M. 2004. *Islands of Intensive Agriculture*. Athens: Ohio University Press.

Tagseth, M. 2008. 'The Expansion of Traditional Irrigation in Kilimanjaro, Tanzania'. *The International Journal of African Historical Studies* 41(3): 461–90.

Teale, E., Gillman, C. 1935. *Report on the Investigation of the Proper Control of Water and the Reorganization of Water Boards in the Northern Province of the Tanganyika Territory*. Dar es Salaam: Government Printer of Tanganyika.

Vavrus, F. 2003. 'A Shadow of the Real Thing: Furrow Societies, Water User Associations, and Democratic Practices in the Kilimanjaro Region of Tanzania'. *Journal of African American Studies* 41(3): 393–412.

Von Clemm, M. 1964. 'Agricultural Productivity and Sentiment on Kilimanjaro'. *Economic Botany* 18(2).

Wimmelbücker, L. 2002. *Kilimanjaro – A Regional History. Volume One: Production and Living Conditions, c. 1800–1920*. Munster: Lit Verlag.

8   Water and Rural-Urban Relations in the Maghreb

Brock Cutler

Algiers, like many other cities across the Maghreb, has a water problem. Or rather, certain areas of the city are provisioned with plentiful and accessible water, while others wring whatever drops they can from an uneven infrastructure. It will not be surprising to learn that the sectors better fed with water are those better connected to the conduits of power, economic and otherwise (Saïdi 2001). Neighbourhoods and regions of Algiers inhabited by the poor and marginal are rather left to desiccate as they will. And this they increasingly do, as the total water available to the region has been declining since the beginning of the twentieth century, a trend that will only increase in the near future due to changing climate. All of North Africa lacks rain, but this general situation has a more specific historical and geographical breakdown as economically- and politically-marginal regions suffer more than those with capital, social or otherwise.

Given the general dearth of water it is no small irony that three weeks after the Algiers Metro opened on 1 November 2011, it was underwater.[1] The flooding that submerged the metro came amid a month of remembrance for a tragedy of ten years prior. On 10 November 2001, after a mass day of prayer for rain, a cyclone docked at Algiers' harbour, drenching the Algerian coast for 36 hours.[2] The resultant flooding killed upwards of 800 people and left some 1,500 homeless. The worst of the flooding occurred in the poorer district of Bab el Oued, in the west of Algiers. Close to half of all the people killed in the floods were from this working-class neighbourhood, one which is usually left with only a few days of water per week, and well below the international 'poverty' level of 100 litres per person per day (Saïdi 2001: 306–8). After the waters subsided popular protests rocked the area, as citizens blamed the government for allowing a climatic event to become a human catastrophe. This is only an example of how a century and a half of centralized water works and large-scale infrastructural projects has in some cases only further marginalized the populations without access to state or economic power in North Africa.

Water and its progressive scarcity has become a point of focus in global environmental discourse. Yet it is not enough to claim that water has always been a scarce resource in the Maghreb. While water is in some areas abundant in North Africa, it is a scarce resource because of climatic, environmental, and social factors. The question is less one of the overall amount of water than the issue of who controls water resources and benefits from hydrologic infrastructure.[3] In this light the above anecdote about the flooding of Bab el Oued can help highlight one of the major themes in the history of water in North Africa over the last two centuries: water has become 'the friend of the powerful.'[4] In the last 200 years, population growth, capitalist agriculture, industrialization, an extractive colonial presence, autocratic independent regimes, and the high-modernism of international political economy have conspired to centralize hydrologic potential in relatively few hands. These hands have then directed water to a few select sectors. From a dispersed, localized, and decentralized system of water use (both urban and rural) has grown a modern system built on large-scale projects and centralized knowledge and control.

In this chapter we will explore this changing dynamic of water in North Africa through examples from the last 200 years. Our main aim will be to understand the way that water has conditioned rural-urban relations across the Maghreb. Over the course of the last two centuries, changes that have taken place in cities and the countryside have had broad similarities that allow us to consider the way that these seemingly-distinct areas are parts of one single social, economic, and political system. Changes in one part of the system thus occasioned changes in other parts. The chapter is organized in five sections. We will briefly cover the geographical and social situation of North Africa as a kind of background, with a view to understanding how the various environmental and climatic regions of the Maghreb relate. After this introduction, we will look at the example of water use in Algiers in the nineteenth century as a case study for changes that took place in urban areas across the region under the conditions of expanding French imperialism. We will move from the nineteenth to the twentieth century in the next section, and look at how agitation over maintaining traditional water rights went hand in hand with anti-colonial resistance. The story of the Meknes riots of September 1937 will help us understand how regular people responded to changes in resource control that were largely out of their hands. One of the lessons of the Meknes riots was that people did not only dislike colonial authorities, but rather any authorities that upset customary access to water. This lesson carries through to the next section, where we consider the rural sector of North Africa through the example of changes that came along in the 1970s to the *tabia* irrigation system of central Tunisia. Large-scale, centrally-directed projects replaced locally-constructed and maintained small-scale works, which had unforeseen and deleterious effects on the agricultural sector. These changes are emblematic of the overall trend in hydrologic infrastructure over the last centuries in rural and urban areas, and we will turn in the final section to a brief look at the post-independence and contemporary period to see how these changes continue to affect the countries of the Maghreb.

GEOGRAPHIC AND SOCIAL SITUATION

A few brief observations on the climate and both human and physical geography of North Africa are in order at the outset. There are a number of different geographic and climate regions that make up the Maghreb, ranging from forests to deserts, and from plains to mountains. For the water regime, the most significant of these formations is the Atlas Mountain range, which stretches from the Atlantic coast in southern Morocco to the Mediterranean coast in Tunisia. At least six distinct sub-ranges within the Atlas contain the northern sections of Morocco, Algeria, and Tunisia within a largely 'Mediterranean' climate by separating them from the Sahara. The Moroccan Atlas is divided into the Anti-Atlas (furthest southwest) and the Middle Atlas (in the north), with the High Atlas stretching west-to-east between them. Running east into Algeria, the *High Atlas* becomes the *Saharan Atlas*, the range that traces the northern border of the Sahara in north-western Africa. North of the Saharan Atlas are high, dry plains (the *Hauts plateaux*) that stretch north to the Tell Atlas, the smaller mountains that define the northern coast of Algeria and run parallel to that coast. The two ranges converge as they move east, coming together on the Algeria/Tunisia frontier as the Aurès range – the furthest east of the Atlas ranges. These mountains form a border between the desert and the sea, and as other mountain systems in the Mediterranean, generally trap moisture on their seaward faces, thus determining the zones of the Maghreb suitable for agriculture. While there are important elements of groundwater in the plateaus and desert areas of North Africa, seasonal rainfall plays the most important part in agriculture, and this rainfall is directed by the differential rise of the Atlas (Figure 8.1).

The other major feature of North African geography is the Saharan desert and pre-desert. Playing the part of a second sea, the Sahara effectively defined a northern and southern coast, with those people traversing the vast region more akin to sailors than, say, sedentary farmers. Oases and seasonal river beds provided the ability to pursue date agriculture and the raising of animals, although on a smaller scale than in the north. Populations living in the pre-Sahara tended to employ more terracing than those in the relatively more water-rich north, and had a water regime dissimilar from the typical Mediterranean arrangement that predominated in the north and west (Despois 1956). Importantly, the creation of underground aqueducts – *foggara* or *khettara* in North Africa, *qanats* elsewhere – allowed for the expansion of irrigated areas around oases and at a distance from sources of water (Beaumont, Bonine, McLachlan 1989). Systems like these allowed for a small but significant population and

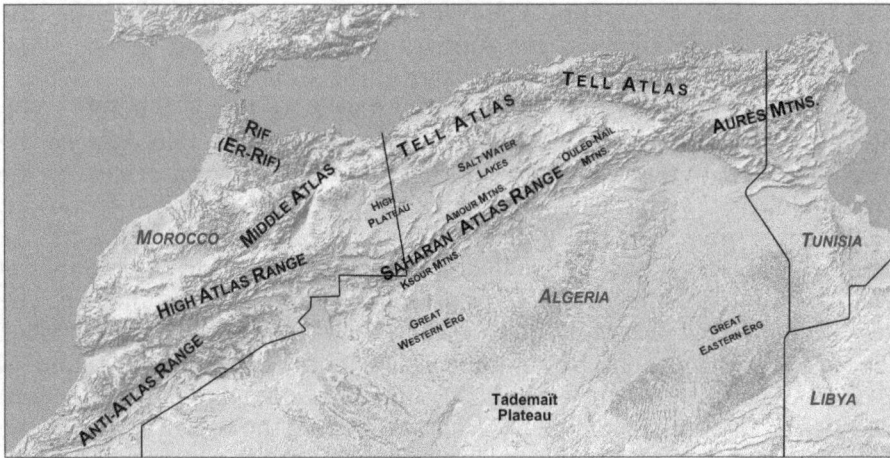

Figure 8.1. North Africa, Atlas Mountains (public domain).

production in the interior areas, and comprise an important area of local knowledge. Additionally, maintenance of the hydrologic system of oases and dry agriculture in the more arid zones of North Africa constituted an important element of community capital – a *foggara* demands significant labour outlays, and therefore is reliant upon the maintenance of local knowledge and the ability of the community to procure regular labour. We will see more about changes to this type of organization further on.

The northern mountains and plains are the areas of heaviest population in North Africa, shadowing the geography and climate of the region. There are multiple social regimes throughout North Africa, often badly fit into the binary of sedentary-nomad. Generally, there are groups that are permanent farming communities, those that farm during part of the year and drive herds during other parts, and then those groups committed to nomadic herding. The binary hides a more complicated social situation in which many populations moved on a spectrum from sedentary agriculture to pastoralism. Even those groups more firmly planted on either end of the spectrum depended on and had normal relations with those groups that had a different social-economic practice. In this way a highly-integrated and localized political economy defined much of the rural sector well into the colonial period.

The hydrologic regime in North Africa is of course varied from region to region, with highly differential rainfall patterns and human adaptations. In addition to the Atlas, rainfall in the Maghreb is also dependent on the North Atlantic Oscillation, one of the most important climate engines for Northwest Africa and Europe. These two elements conspire to produce a climate in which rain generally falls in the winter months, and more so in the north than the south. The various ranges do trap water in the mountains – sometimes in the form of snow – that then becomes the source of rivers and subterranean water in the spring.

The distribution of rain across the region is highly variable, with sometimes extreme local differences. Generally, more rain falls to the north-west of the Atlas ranges, leaving the south-eastern edges of the mountains drier and less densely populated. The high plateaus of Morocco north of Marrakesh, separated from the Algerian high plateaus by the Middle Atlas, receive around 500–600 mm of rain per year, double that of the Algerian highlands on the eastern (inland) side of the mountains. The northern edge of this range (called the Rif), also sees significant rainfall. The Tell, the most important agricultural and demographic region in Algeria, receives between 400 and 700 mm of rain annually, with the eastern reaches of that region generally being wetter than the western. Within the mountainous north lie many small plains and valleys, which serve as the agricultural base of Algeria. Further east along the coast lie the Kabylia mountain ranges, which receive the most rain out of any region in the Maghreb – some years near 1,600 mm. Rainfall in Tunisia follows these trends as well, with the eastern reaches of the Atlas range trapping water in the northern parts of the country, with less and less water falling as one moves south.

The trend over the last 200 years has been the centralization of control over water and food resources in these spheres, with modernizing state regimes attempting to rationalize the economic capacity of their territory. In 1800, the vast majority of North Africans lived in rural areas, a trend that has been changing slowly over the last 100 years. Depending on the geography of the area, people depended on small dams, dykes, and terracing for water control, as well as on wells and cisterns for drinking water. These sources had to be constantly maintained and watched over if they were to remain viable. In this way, engagement with the hydrologic system had a kind of continuity even as actors changed and populations shifted. While nineteenth-century French colonial archaeologists claimed that nearly all water works in the countryside dated to Roman times, in fact the populations of North Africa were highly engaged in hydrologic practice, as indeed they had to be if they were to continue to live under these climatic conditions. For instance, there is a wide and rich vocabulary touching on terrace agriculture and irrigation in the language spoken by the Seksaoua of the High-Atlas region – an indication of the importance of and engagement in hydrologic concerns for the population (Despois 1956: 48).

The introduction of colonial regimes across North Africa precipitated significant changes to agriculture and water use. Markets for agricultural products quickly shifted toward the northern coast, and the products of agriculture became those most desirable for the international market. The French colonial regime pursued these changes in what was the normal mode of imperial exploitation, using techniques such as demanding taxes in cash, legislating land and water into the public domain, and confiscating land and selling it to colonialist producers cheaply (or giving it away). Most importantly, colonial engineers took up the problem of scarce water in a series of large-scale projects meant to make the colonies economically profitable for France or European colonial populations. These projects ranged from the mundane (proposing expanded dams and aqueducts) to the spectacular (creating an 'inland lake' in the middle of the Sahara). When added together, however, the cumulative effect of these proposals was to denigrate 'traditional' indigenous uses and infrastructure in comparison to 'modern' or 'scientific' measures. These colonial projects, when they were carried out, were put to the benefit of large-scale capitalist agriculture or budding industry, further disrupting the small-scale local practices of the majority of North Africans. Water in the public domain and hydrologic/hydraulic projects directed by state engineers and metropolitan experts served to centralize decision making about water use even at the local level, as we will see in the following examples.

URBAN SPACES

In the urban zone we can see the beginnings of the process of centralization in the initial aftermath of the French occupation of Algiers.[5] The fall and occupation of Algiers had a profound impact on the water system of the area, and can serve as a model of some of the processes that were so transformative across North Africa over the next century. The hydrologic infrastructure of Algiers in 1500 was made up of wells and cisterns, most in multi-generational homes or for neighbourhood use. This system underwent major revision in the sixteenth century when people driven from Spain by the Reconquista brought aqueduct technology with them as they settled in North Africa.[6] In 1830, there were four main brick aqueducts that brought water to over 200 wells and fountains in Algiers – both public and private. Bab Azoun, fed by water from Hamma, created in 1662 and reworked in 1759 by Ali Pasha, was the most important; second came Telemly which led to the Porte Neuve; the third was from Aïn Zeboudja, the most recent and longest, which went to the Casbah; and the fourth was that of Birtraria, which entered in through Bab el Oued (Comité du Vieil Alger 2003: 53). All of this water came from sources in the Sahel (Lespès 1930: 447–58). In the decades leading up to the French occupation, Algiers' water infrastructure supplied about 77 litres of water per person per day (with seasonal fluctuations) (Goubet 2001: 365). The total also included collected rainwater and wells dug in private homes. As a point of comparison, in 1807 in Paris, the water system could only provide about 7.5 litres per person per day.[7] A French committee dedicated to uncovering the history of 'old Algiers' reported that the Ottoman rulers of Algiers (*Beys*: 1525–1830) were much more serious about their hydrologic infrastructure than the French who took the city in 1830. Strict laws protected the water supply, and anyone convicted of 'damaging a conduit of water' would be severely punished (Comité du Vieil Alger 2003: 54).[8]

One of the aspects that recommended Algiers as a colonial military establishment was its supposedly stable and plentiful supply of water, in comparison with other North African cities (Anon. 1830). However, when the French military occupied Algiers in 1830, they not only took over the city, but also damaged certain parts of it in an attempt to construct a European military centre in its place. The process of militarization destroyed buildings and infrastructure in order to create wide boulevards, for instance, and repurposed blocks of homes in order to build military institutions and barracks. The roster of destroyed urban infrastructure included the water system. The main aqueducts and secondary pipes had been constructed mainly from pottery, and as such were easily damaged by military equipment and troops moving through the narrow streets of the city (Lespès 1930: 448). Other parts of the infrastructure were rerouted by troops to bring water where they wanted it, or were accidentally destroyed by the workers as they were clearing space for the first boulevards, squares, and military buildings (Comité du Vieil Alger 2003: 45). In 1832, the

Figure 8.2. 'The Blue Fountain in Algiers' (from Herbert, 1881: 157).

neighbourhood of the Place de Chartres was razed to clear space for new streets and buildings. Because of these constructions, the water infrastructure of the area was damaged: after these destructions there was no fountain or other source for the whole neighbourhood to get water (Goubet 2001: 366).

When settlers finally started to populate the city, the aqueducts were reportedly in a bad state. The general decline of the quality and quantity of water available to the city due to military damage and new construction meant that in some parts of town the only water available was polluted or of poor quality (Lespès 1930: 448). The military was not entirely to blame. Parts of the system were old and not in perfect condition even before the arrival of French troops. Despite the poor state of constructions, as noted above, the system functioned well enough to bring about 77 litres of water per day per person. By 1855 the European inhabitants of Algiers only managed to squeak about 34 litres per day out of the system (Lespès 1930: 450). Part of the problem for water infrastructure in Algiers was the changing nature of who was in control of the municipal system (*Akhbar*, 23 August 1849). Attempting to maintain local control, water theft and pirating of water from larger networks ran rampant. Colonial authorities slowly clamped down on these types of interventions, punishing those who defied water laws or damaged water systems. Thus water slowly moved from the hands of neighbourhoods and small communities (and small cultivators in the surrounding countryside) into the control of larger institutions such as the military and capitalist colonization companies.

The French spent around a million francs on water projects by the early 1850s, and at least understood the importance of potable water for maintaining military advantage (Lespès 1930: 450). To this end they attempted to fix some aqueducts and wells, although the work was sporadic and of a questionable quality. Thanks to these interventions, however, the system worked well enough to hydrate the European settler and administrative sectors of the city. Other areas were left to fall into ruin. By the 1860s, there was a noticeable segregation between the European and Algerian areas of the Algiers, with the Europeans benefiting from the Turkish and new French infrastructure, while Algerians were left to only a few wells and the work of water carriers (Goubet 2001: 367). This is a part of the changes to the urban infrastructure that we have to consider: while water might have been available in town, changes to the urban geography of water distribution ended up privileging certain sectors of town over others. Some residents, notably those in the Casbah (the Algerian section of town), lost some of their infrastructure to French re-organization and were forced to traverse more city space to procure water. This did not just mean a slightly longer trip, for it often brought Algerians into contact with Europeans in a way that reinforced the narrative of Algerians as strangers in Algiers. People were forced to visit neighbourhoods provisioned with

Figure 8.3.   'Black Servant going to the fountain' (from Herbert, 1881: 119).

water, which were often European; their presence there then became part of the narrative of poor Algerians invading a civilized European space. Thus water in Algiers contributed to the discourses that alienated Algerians from the city.

Over the course of the first 50 years of colonial occupation in Algiers, then, the water system underwent changes not only in its physical composition but also in who it served. From a high-functioning and partly dispersed system that provisioned most of the city and was protected by *Beylical* mandate, the hydrologic infrastructure of Algiers consolidated into the hands of the military and colonial authorities who directed its flow to new areas of town occupied by settlers. This trend, apparent here in the 'queen of African cities', repeated in other towns and cities, and was generalized across colonial territory. Before formal colonialism came to Tunisia and Morocco some of these changes were taking place as well, as European merchants and government officials put pressure on the regimes to shape commercial sectors of important cities to their liking. So by 1900 hydrologists were drawing up plans for cities across the region, and implementing those plans when possible. We should not think of the process as inexorable or unproblematic, however. Former proprietors and benefactors of water systems did not just idly sit by while their water system was taken from them.

CHALLENGING WATER EMPIRE

As the French imperial presence in North Africa expanded, so did the colonial state's ability to control and centralize water. One of the first pieces of legislation introduced into each of the territories were laws putting water into the public domain (Algeria 1851, Tunisia 1885, Morocco 1914) (Perennes 1990: 18). This move set in motion efforts to 'rationalize' water use in order to orient it toward capitalist agriculture and international markets – meaning a change in the system of water management that had grown up over the centuries in customary and Muslim law. While it took decades to make these changes valid, the effect was to centralize hydrological infrastructure to the benefit of large landowners, industrial concerns, and state projects. These colonial changes were not only about a rational approach to water, of course. Large-scale projects also diverted water to new populations and thus created new relationships of power on the ground. It is easy to get the picture of an inexorable march of centralization without input from actors other than the state. This was far from the truth, and these centralizations were often the impetus for concerted resistances. For an example of the new relationships created by colonial water changes, we can look to Meknes in Morocco in the 1930s.

After becoming a Protectorate of France in 1912, a variety of resistance movements sprang up over the next forty years in Morocco. These movements have often been subsumed under the umbrella of 'nationalist' resistance, but as a recent article by Adam Guerin (2014) has shown, the initial impetus of many of these movements was not the desire for national self-determination but rather a return to local control over natural resources. The story of the Meknes water riots of 1–2 September 1937 can help us understand how colonial centralization of water resources involved resistance and negotiation on the part of local populations.

In Morocco as elsewhere in the Maghreb, water rights and customary usage was based on a system that combined official *shari'a* law rulings and local customary law. This malleable and dynamic process supported communal ownership of infrastructure and a locally-determined system of water allocation. We can see this in the *khettaras* that exist in Morocco and Algeria today, with their canalized distribution (see below). Similar dynamics appeared in urban spaces and agricultural zones, with the institution of local councils or religious brotherhoods often serving as arbiters and intermediaries. As Guerin states, 'while it would be overly romantic to claim that [this local] regulation always sufficed to mediate property disputes or resulted in a just distribution of resources, scholars tend to agree that the system was at least flexible enough to account for environmental fluctuations (drought, locusts, etc.) and political or economic developments from outside the region' (Guerin 2014: 9). Even before the introduction of formal colonialism with the 1912 protectorate, this situation was gradually replaced by one in which European agents and elite Moroccans were able to consolidate previously-inviolable lands (such as communal grazing land) and generate profits from world-market-oriented production. A similar process happened for water resources, which often went hand-in-hand with land resources. The colonial state took over water resources that were traditionally under the purview of the sultan and often left to local operators, and opened them to the same private property markets and logics that affected land across Morocco (Figure 8.4). 'By the mid-1930s, the bureaucratization of most pre-Protectorate institutions was complete and through a combination of French legal pressure and [Moroccan] military enforcement the sultan's government was able to make a powerful and enduring claim to the land and water of the realm' (Guerin 2014: 9).

In Meknes this trend came to a head in the 1930s. Due to loss of land and traditional water rights, many formerly-rural Moroccans began to move to Meknes (and other urban centres). The population of Meknes went from 36,500 in 1925 to 75,000 in 1936. With little agricultural work and scant manufacturing in the region, many of these people were confined to shanty towns on the edges of the city. Drought in the 1930s made grain scarce, yet grain exports continued during the period, leading to skyrocketing prices for basic foodstuffs. At the same time, the authorities in Meknes began a project to redistribute the water that came into town according to the growing European population. The project diverted water from the Bou Fekrane River that had gone to the pre-European city to the new colonial areas. These measures were not met with indifference, as people took to small-scale sabotage like diverting water from canals to the poor

Figure 8.4.   Agdal Reservoir, Meknes (2006). This reservoir was commissioned by Moulay Ismail, Moroccan Sultan from 1672–1727 (public domain).

neighbourhoods as well as filing formal petitions to authorities to have water rights returned.

Drought, the high price of grain, and a new lack of water combined to create a situation in which the local population had few options. This came to a head in September 1937. On the first of the month a nationalist rally brought around 400 people out into the streets. These people marched to the Hotel de Ville and chanted nationalist slogans. A large military and police presence met them there. There was little disorder. The next day, however, a much more explosive situation arose. Despite the petitions filed with authorities and small-scale resistances like water diversion, local authorities had not paid much attention to the mounting local concern over water rights. So when 2,000 people gathered in the streets on 2 September the authorities were not ready for them. Shouting 'Not a drop for the settlers!' the crowd swelled into the town square and forced the local police into a hasty retreat. Violence erupted as colonial reinforcements came into town to quell what the authorities saw as an unfocused riot.

23 Moroccans died, and another 80 were injured. Troops occupied town to keep any echoes of the riot from escaping to surrounding areas. People were sent back home and the water appropriation went on its way. The nationalist press grabbed the news of the riot and quickly turned it into a story about the nationalist aspirations of the Moroccan people. What went

un-discussed was the fact that the water riots targeted Moroccan authorities as much as the colonial French; the issue was maintaining local control of water against whatever centralizing outside force was changing the locally-adapted status quo. The Meknes water riots were not successful from the point of view of returning local control to the water system, but they can nonetheless help us understand the importance of the changes that were happening under colonialism to the people involved: Centralizing decision making about hydrologic infrastructure and water distribution was a contested process that occasioned impassioned resistance and negotiation by those people most affected by it.

RURAL SPACES

Arguably the most important element in the history of water and water use in North Africa is what took place in the rural sector. Despite changes in the political economy of the region during the colonial period, most people remained rural and agricultural up until the very recent past. Many of the changes to rural water systems consisted of processes like what we saw in Meknes – diversion of water resources to new colonial populations or commercial endeavours. This process started in Algeria in the nineteenth-century, with the damming of small rivers and modifying existing dams to bring water to colonial villages. These new projects often took the place of or 'improved' on previous works. As mentioned earlier, there were many hydrologic practices in the Maghreb, depending on the area. Most agricultural areas were home to multiple water structures, most often small dams and canals made from locally-available material. Some areas had larger works such as aqueducts or terracing, although the scale of these kinds of works determined their rarity. Built and maintained by generations of people interested in the long-term viability of the land, these works tended to be adapted to local conditions. In many cases those conditions included a small annual harvest or constrained planting so that groundwater could recharge and irrigation works could be locally maintained. With the introduction of world-market oriented agriculture, production imperatives changed and communities were forced to maximize their output in the short term, which meant changing the water regime. Often (but not always) this change was part of the constellation of processes that was French colonialism.

For instance in 1860 a new village was created near Philippeville in the Constantine province of Algeria.[9] Named Gastu, the village was originally placed on land taken from the local populations, including areas that accessed the main water sources of the region. Local water works were attributed to Romans, with the Algerian populations only 'maintaining' them. The village went in and settlers moved to take advantage of the cheap land provided by the colonial government. Later, in the late 1870s and early

1880s, the colonial regime enlarged the village, again taking land in a 'trade' from the local population – the Radjetta – in which those people were moved off land near town to drier areas further south. The creation of villages like Gastu involved the modification of local social practice with a view toward extending the reach of large-scale agriculture geared toward international markets – in other words, the type of agriculture practiced by colonists. The agricultural basis of the economy of North Africa combined with a modernizing colonialism to change the water infrastructure of the region in a way that helped direct resources toward export.

These types of changes were prevalent throughout the nineteenth century in Algeria, and late in the century in Tunisia, but most truly large-scale infrastructural works did not take off until after the first world war. The 1920s saw a rise in colonial attempts to build large dams (often by private companies paid with public funds). Dams such as Oued Fodda and Ghrib in Algeria diverted water to large-scale agriculture and managed to convince those remaining sceptical colonials that large-scale works would improve their position in the colony. Similar approaches then occurred in Morocco and Tunisia in the 1940s and 1950s, respectively (Perennes 1990: 15). These large-scale modernist enterprises continued in each of the countries after independence in the 1950s and 1960s, and have set the tone for hydrologic/hydraulic policy since then.

An example from rural Tunisia will help us better understand some of the problems with this shift in the water regime as well as allow us to see why large-scale works seemed so promising. The majority of Tunisia's usable surface water (90 per cent) is found in the north-west of the country, while the most important areas of rural usage are in the centre and south (Perennes 1990: 12). In order to cope with that insufficiency, populations in the centre and south of the territory have adopted a changing set of adaptations that have allowed them to maximize the water potential of the region. Jennifer Hill and Wendy Woodland (2003) have studied some of the varied hydrologic adoptions made in the Matmata Hills of the centre-south. In an area that gets relatively little rainfall, the local population has over the last few centuries developed small-scale, locally-managed water infrastructure that has allowed for sustainable agroforestry. Since rain usually comes in sporadic, high-intensity showers, trapping the water and allowing it to recharge groundwater is of the utmost importance. To this end, the farmers of the Matmata area have designed small dams (*tabias*) of a few meters high and about ten in length that trap material and sediment from washing down the hills, as well as slow rainwater, allowing it to recharge groundwater (Figure 8.5). The *tabias* have overflow systems developed over time and with porous materials that keep the water table from getting too high in a certain area, thereby increasing the potential for salinization of the soil. These porous overflow systems also ensure that upper fields do not deprive lower fields of water. The small scale of these works and their construction in locally-available material has allowed them to continue to

Figure 8.5.   View of Matmata Hills (2007) (public domain).

exist for generations with routine maintenance done by the local people
with local knowledge. 'Vernacular knowledge and craftsmanship, derived
from centuries of interaction with the local environment' have created a
system that is responsive to the extreme climate events that define the
region and manageable by people who live in the region and are dependent
on its continued viability (Hill & Woodland 2003: 347).

Since the late 1950s, however, and following the colonial model, the
Tunisian government has financed 'modernization' of some of these
overflow systems and *tabias*, both in Matmata and other areas around
Tunisia. The new constructions, made in cement and other materials not
produced locally and dependent on the knowledge of urban-based experts,
was not left in the hands of the local population. The rigidity of
construction lends the appearance of permanence and modernity and
allows for a higher level of initial rainwater capture, but that rigidity also
makes the overflows more likely to rupture under stress, as happened in
extreme rains in 1979. These modern systems were not the only large-scale
state projects in the area. After severe flooding in 1969, the Tunisian state
constructed the Sidi Saad dam to control water flow on the Kairouan plain
in the centre of the country. This plain had previously been irrigated by a
locally constructed and controlled set of barrages and dams, which had
allowed for the irrigation of 30,000 ha. The Sidi Saad dam, while able to
control flooding in later years, reduced the irrigated area of the Kairouan

plain to 4,000 ha. Local actors have also lost the ability to control their destiny as the water diverted by the Sidi Saad dam has gone mostly to large-scale cultivators. So while further severe flooding has been avoided for the most part, many people have lost cultivable land, which has in turn contributed to a declining rural population and urban overcrowding. As Hill and Woodland conclude: 'Modern large-scale developments have provided no more reliability over space and time than the earlier small-scale works ... The only difference is that modern dam developments can provide short-term yield maximization, but this requires greater volumes of water leading to insidious environmental degradation' (Hill & Woodland 2003: 352).

The trade-off that we see here is between immediate short-term gains and long-term environmental stability. In the rush to develop rural agricultural systems more amenable to the fast-changing needs of the international market, the countries in the Maghreb have largely followed the pattern laid down during the colonial period. There are benefits to this kind of approach, such as yield maximization and flood control, which help explain why these projects were – and continue to be – so popular among government officials and technological experts. The long-term viability of the agricultural sector, however, is largely left outside the purview of these projects. Hill and Woodland point out that one effect of the construction of the Sidi Saad dam was that lowland large-tract farmers benefited while higher-land farmers lost productive capacity. This then meant that more land ended up being either abandoned or consolidated into holdings by the wealthiest landowners. Land and work being then scarce, more people abandoned the land in search of jobs in cities. This trend threatens the long-term viability of the Tunisian economy as well as threatens the survival of specialized local agricultural and hydrological knowledge.

POLITICAL LIBERTY, HYDROLOGIC STAGNATION

As in the Tunisia example, the regimes that came to power with the wars of liberation in North Africa in the 1950s and 1960s have continued with the types of top-down water policies pursued during the colonial era. The ruling regimes of each new nation recruited nationals who had worked for and with the former colonial administrations, as well as technical experts from the former metropolis. This continuity of knowledge and expertise translated into administrations largely unable to think outside of the colonial paradigm of centralized and large-scale projects. Morocco created the National Irrigation Office, which helped create plans that would expand the reach of irrigated land beginning in 1968. Tunisia began its large dam construction in 1965 with the Nebhana dam, and continued these policies throughout the 1970s. Algeria was less dependent on the agricultural sector than its neighbours, and was less rushed to build new dams, constructing only three between 1962 and 1980. The next ten years saw 15 new large dams, however, as Algeria embraced the modernizing trends of the time. The aggressive modernization projects in each country attempted to reboot economies that had been kept dependent and undeveloped during the colonial period. These projects mandated the 'rationalization' of the agricultural sector, reorienting production to large-scale, industrial farming that drank more water more wastefully than had smaller-scale production. So while agriculture remained the largest consumer of water in each country (even Algeria), that water nourished fewer and fewer people.

All three countries have attempted to use 'modern' building materials and expertise to address the issue of water management. These projects have in some cases been successful in providing water to urban areas or diverting the water necessary for industrial productions, but recent research suggests that the long-term viability of these types of projects is less than that of smaller-scale, locally-managed infrastructure. The overwhelming problem, however, is the explosive growth of cities in the last 40 years. This growth has necessitated governments search further and further afield for sources of water. One result of these expanding networks of water infrastructure has been the directing of water from rural areas to urban centres, further impoverishing the local areas of groundwater. For instance, Algiers now gets water from the Mazafran region in the west as well as the Isser River in the east. This expanding reach puts more and more water into the hands of Algerois elite at the expense of those people living in the agricultural regions of the Mitidja (Perennes 1990: 14).

Another example of the problematic nature of these large-scale projects undertaken by the new regimes comes from Morocco. As discussed earlier, one important type of hydrologic work in the more arid zones of the country has been the *khettara*, or subterranean aqueduct. *Khettaras* tap the natural water table of an area to distribute water to fields, and the flow of the *khettara* is reliant on healthy ground water. In the 1970s the Moroccan state began its series of dam building and at the same time allowed for the building of diesel-generated pump wells around the country (Lightfoot 1996). These dams did not allow water to seep and replenish the water table, and the new wells actively pumped water from the ground. These modern technologies initially supplied more water to a larger region than the traditional *khettaras* could, allowing for the expansion of agriculture. By not recharging ground water, however, these new developments are proving unsustainable. Wells need to be dug deeper and deeper, and are progressively desiccating the agricultural region they originally irrigated. Reliant on non-local supplies and expertise, these wells and dams have also had the effect of drying up local sources of *khettara* knowledge. As these traditional wells are abandoned, so is the ability of local people to recreate them in the future. As Dale Lightfoot writes of the

Tafilalt oasis region in southern Morocco, 'landowners and farmers have mostly yielded to these changes, attracted to the short-term benefits – flood control and piped drinking water ... The result has been some loss of local control over water resources, the desiccation of the southern Tafilalt, and the loss of sustainable *khettara* irrigation in the north Tafilalt' (Lightfoot 1996: 267).

These large-scale, state-directed infrastructural works followed the trend laid down during the colonial period – a trend that consolidated control over water resources into the hands of government officials or large companies. While providing some immediate benefits to the domestic economy of these countries, the changes in water regimes ultimately contributed to dislocated local economies and environmental instability. With growing urban populations and increasingly precarious environmental situations, the three countries of the Maghreb will need to find ways to profitably combine the benefits of large-scale infrastructural works with locally-managed and small-scale operations if they are going to weather future climate changes.

CONTEMPORARY

If we remember back to the flooding of Bab el Oued and the Algiers metro from the beginning of the chapter, we can understand to what extent this problem persists today – and the continued difficulty for contemporary states in coming up with viable solutions. The ongoing crises of population growth and climate change will continue to demand comprehensive and creative approaches from governments and other actors in the Maghreb. Some of the most powerful of these actors are sometimes called 'states within states' – economic institutions that operate without regard to popular input or state oversight, such as the oil and gas sectors of Algeria – and are committed to maintaining the status quo (Perennes 1990: 14–15). However, it might not be too long before circumstance and changing environmental conditions require these actors to pursue or support innovative approaches to water infrastructure.

There might be hope for change in recent studies that have stressed the need for more localized control of resources. These studies are often state-funded and thus part of the same tradition of centralized control, but perhaps centralized decentralization will be the legacy of our contemporary period. Some traditionally conservative institutions such as the United Nations and the World Bank have also begun to support initiatives that look more to either decentralized or regional networks for knowledge and control. The *Projet Maghrébin sur les Changements Climatiques*, for instance, is working to develop resources across national lines. Part of the project is to allow for more local participation in regional water projects in the hopes of inculcating a new sense of shared purpose, necessary for the long-term viability of Maghrebi social systems. These inter-state and regional programs might turn out to be a way to facilitate the break-up of anti-democratic economic power within states, and thereby return some local control over water and other resources to the populations of the Maghreb. It is perhaps just as likely, however, that the decades-long (if not centuries-long) process of consolidation of decision making power will find new purchase at the regional scale.

The importance of mixing the scales of water use in the region is becoming more apparent. While large-scale damming and reservoirs allow producers to escape the limits imposed by the natural world, this escape often comes at the price of environmental and economic disequilibrium, contributing to rural exodus and stratified socio-economic conditions. This is the process that has happened in the Maghreb over the past few centuries. Small-scale and locally-operated hydrologic projects, on the other hand, are sustainable in the long term but limited by the dictates of nature. This makes them flexible, but that flexion can produce scarcity. Demanding that people give up short-term gains in the hopes of long-term stability is a difficult task in an era of international markets and immediate consumption. But the mixing of scales is necessary if the Maghreb is to confront the drying climate. The inflexibility of large-scale dams and reservoirs limits their ability to confront changing climate, something small-scale works are designed to do. An example of this mixing comes from Tunisia, where in the 1990s a series of hillside small lakes and catchments around large dams was able to reduce siltation rates and contribute to the dams' continued viability (Hill & Woodland 2003: 354). Local populations use and maintain the small catchments, which have also served to help replenish local aquifers and the ground table.

If there is hope for the kind of decentralization of power necessary to achieve this mixing of scales, we might see it in the recent upheavals across the region. The much-discussed 'Arab Spring' has not been without its problems, but the trend of more citizen involvement in decision making – if indeed it becomes a trend – is a kind of first step in breaking up the conglomeration of powers that wants to maintain the resource-use status quo. More citizen participation in local and national governance is a necessary first step. Whether this comes in the form of new constitutions or new governing structures, or is rather reflected in the willingness of current political and economic powers to take the desires of the population more seriously, it will be a welcome (and necessary) step in the direction of addressing the pressing need for new water regimes.

NOTES

1 See the websites for North African newspapers, such as: http://www.tsa-algerie. com/divers/des-stations-du-metro-d-alger-inondees-par-les-eaux-de-pluie_18278.

html; and http://www.emarrakech.info/Les-eaux-de-pluie-inondent-des-stations-du-metro-d-Alger-recemment-inaugure_a58818.html.

2 The floods were widely reported in the Algerian and international press; reports are still accessible online, such as the following: http://news.bbc.co.uk/2/hi/middle_east/1658078.stm; http://www.wsws.org/articles/2001/nov2001/alge-n16.shtml; http://www.merip.org/mero/mero121101; http://www.guardian.co.uk/world/2001/nov/12/.

3 Indeed, René Arrus claims that 'l'idée de rareté permettait-elle à la fois de masquer les carences passées et à venir des pouvoirs publics en la matière et de faire prendre aux usagers leur mal en patience en leur assénant des contre-vérités dont l'essentiel consistait à considérer la rareté comme donnée une fois pour toutes' (Arrus 1985: 288).

4 The phrase is from the title of G. Bedoucha (1987).

5 Adapted from Cutler 2014.

6 Comité du Vieil Alger 2003. The Committee only published its writings all together as *Feuillets d'El-Djezaïr* in 1937. I do not necessarily support this claim, forgetting as it does the presence in Algeria of hydrologic works dating to at least Roman times while at the same time maintaining that technological innovation did not arise within Algeria itself. In this case, the suggestion is that Algiers' water supply could only modernize once people who had been living in Spain brought knowledge and technology with them. It is more probable that contemporary leaders in Algiers participated in networks of knowledge exchange before the arrival of expelled Andalusians. See Shaw (1995), note in particular the chapter 'Water and society in the ancient Maghrib: technology, property and development'.

7 Goubet 2001: 365–67. Let us not let this statistic go without mention, however, as in Algiers the water was much more plentiful in the winter rainy season, and thus yearly averages are bound to be skewed.

8 Again, we must take the sentiment with a grain of salt, as the mention of harsh punishment for minor infractions was a staple of Orientalist comment on despotic rulers of the 'East'.

9 See the dossier on colonial Gastu in the Archives d'Outre-Mer, Aix-en-Provence, France (AOM): FR AOM 93/1613.

REFERENCES

Anon. 1830. *Notice statistique et historique sur le royaume et la ville d'Alger. Résumé des meilleurs documents anciens et récents sur ce pays.* Paris: Thibaud-Landriot.

Arrus, R. 1985. *L'eau en Algérie de l'impérialisme au développement (1830–1962).* Alger: Office des Publications Universitaires.

Beaumont, P., Bonine, M., McLachlan, K. (eds) 1989. *Qanat, Kariz and Khattara.* Wisbech: Menas Press.

Bedoucha, G. 1987. 'L'eau, l'amie du puissant', une communauté oasienne du sud-tunisien. Paris: Ed. des archives contemporaines.

Berque, J. 1950. 'Un glossaire notarial arabo-chleuh du Deren'. *Revue Africaine*: 357.

Comité du Vieil Alger. 2003, 1937. *Feuillets d'El-Djezaïr*. Blida, Algeria: Editions du Tell.

Cutler, B. 2014. '"Water mania!": drought and the rhetoric of rule in nineteenth-century Algeria'. *Journal of North African Studies* 19(3): 317–37.

Despois, J. 1956. 'La culture en terrasses dans l'Afrique du Nord'. *Annales: Économies, Sociétés, Civilisations* 11(1): 42–50.

Goubet, J.-P. 2001. 'La ville, miroir et enjeu de la santé: Paris, Montréal et Alger au XIXe siècle'. *Histoire, économie et société* 20(3): 355–70.

Guerin, A. 2014. '"Not a drop for the settlers": reimagining popular protest and anti-colonial nationalism in the Moroccan Protectorate'. *Journal of North African Studies*.

Herbert, Lady. 1881. *L'Algérie contemporaine illustrée*. Paris: Victor Palme.

Hill, J. and Woodland, W. 2003. 'Contrasting water management techniques in Tunisia: Towards sustainable agricultural use'. *The Geographical Journal* 169(4): 342–57.

Lespès, R. 1930. *Alger: Étude de géographie et d'histoire urbaines*. Paris: Librairie Félix Alcan.

Lightfoot, D. 1996. 'Moroccan Khettara: Traditional Irrigation and Progressive Desiccation'. *Geoforum* 27(2): 261–73.

Perennes, J.J. 1990. 'Les politiques de l'eau au Maghreb: d'une hydraulique minière à une gestion sociale de la rareté'. *Revue de géographie de Lyon* 65(1): 11–20.

Saïdi, F.C. 2001. 'Alger: des inégalités dans l'accès à l'eau'. *Tiers-Monde* 42(166): 305–15.

Shaw, B.D. 1995. *Environment and Society in Roman North Africa*. Brookfield, VT: Variorum.

9 Colonial and Post-Colonial Irrigation Efforts in the Office du Niger, Inner Delta of the Niger, 1900-2000

Maurits W. Ertsen

> From the air one sees that civilization is order: a pattern imposed upon a shaggy landscape. To some extent it is simply straight lines. [...] we came suddenly to a patch of squares, as if we had an aerial view of the Red Queen's chess-board. The landscape had been laid out with a ruler, it was patterned with black and green rectangles, streaked by lines and dotted with neat toy compounds. [...] This was the Mwea-Teberre irrigation scheme, lying on the plains south-east of Mt. Kenya.
>
> Huxley 1960: 204[1]

INTRODUCTION

For Elspeth Huxley, who wrote the quote included above, it was clear what civilization should look like; after all, she was highly familiar with British African colonial policies as she had seen many material manifestations of them in the field. Her favourite place – or at least one high on the list – was the Mwea irrigation scheme in Kenya. This Red Queen's chess-board is located some 90 km north of Nairobi (Cambers & Moris 1973) and has been regarded as a most successful irrigation enterprise in more recent, post-colonial years. For the Food and Agricultural Organization (FAO) in the 1960s, the Mwea-system was a model to follow (Strong & Paton 1968).

In the 1980s, an FAO report still stated that during 'the first 10–15 years of its operation the Mwea scheme, which is an extreme example of a highly supervised project, has been very successful in terms of increased production and higher farmer incomes' (Sagardoy et al. 1986). The validity of the Mwea model, however, was considered limited for the 1980s, as 'experience elsewhere has tended to show [...] on projects where the mechanisms for centralized control are particularly powerful in the initial stages'. Systems along the Mwea model might 'fail to adapt to circumstances changing' (Sagardoy et al. 1986).

I have elsewhere shown that the Mwea system is a pivot in this process of continuity and change in the shifts from colonial to post-colonial irrigation policies and systems, including the perceptions of colonial models for African irrigation development (Ertsen 2008, 2006). The large-scale Gezira irrigation system in British Sudan was the source of inspiration for Mwea. Gezira has not only been Britain's most impressive colonial irrigation effort (at least by size), in the 1950s it was the metaphor for successful irrigation development in Africa, both regarding its cash crop (cotton) and its system management – before Mwea became the role-model (Chambers 1969, Ertsen 2016).

In this chapter, I will focus on the history of another huge irrigation scheme in Africa – even if the size of that irrigation system mainly existed in the fantasies of the planners and designers and has never reached the iconic status of Gezira after World War II or Mwea in the 1970s. The Office du Niger in modern Mali could not become the example to follow, as for a long time there was not a real success to show. At the very beginning of the new Millennium, however, the World Bank considered the Office du Niger an example of how to make 'a large irrigation scheme work' (Aw & Diemer 2005) (Figure 9.1).

Water from the Niger River is diverted into a system of canals from Markala (35 km downstream of Ségou) irrigating some 100,000 ha of flat alluvial plains. Nowadays, the irrigation system still functions. The main crop is rice – the scheme produces around 40 per cent of the country's rice production. Sugar cane is also grown on a commercial basis.

When the British plans for Gezira developed, the French were even more convinced than they already were (see below) that part of the former Inner Delta of the Niger should be a centre for cotton growth (Schreyger, see also De Wilde 1967b). France wanted to create its own Gezira on the Niger in the French Sudan, to grow cotton for the European market. The result of this colonial vision became the Office du Niger, a semi-autonomous government agency, although the name typically refers to the irrigation scheme in the inland Niger Delta that the agency administers. In this chapter, I will start with a brief summary of the history and early years of the Office du Niger, and then explain its development and contextualize its results with some statistics.

THE FRENCH IRRIGATION PLANS FOR WEST AFRICA

The French were actively preparing and executing colonial irrigation policies in Africa, especially in West Africa (Senegal, but mainly Mali) and northern Africa (Morocco, Tunisia and Algeria). Irrigation was a key element in a policy that aimed to transform exploitation of the colonies into productive imperialism, within which economic development would serve both the colonial powers and the colony itself.

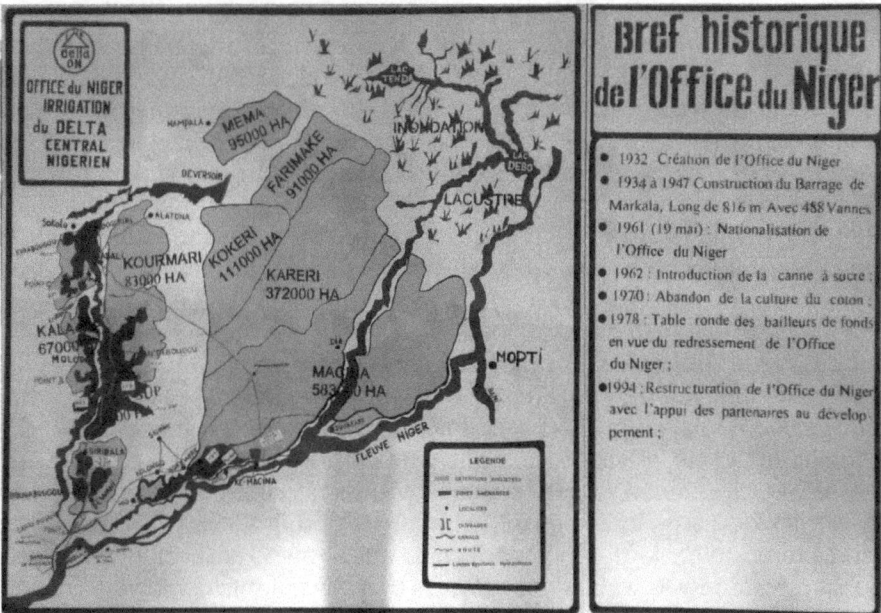

Figure 9.1.   The sign at Sansanding with the map of the Office du Niger (© Geertjo van Dijk).

The French drafted their first irrigation plans for their West African colonial territories in the early nineteenth century. The earliest policies were aimed at using the irrigation potential of the Senegal River valley, especially the delta – an area close to the sea and as such much better known than the far-away arid inland. An important stimulus for this was the abolishment of the slave trade. The French governor of Saint Louis received an order to study whether the people who could no longer be sold to American plantations could work in irrigated plantations producing export crops. This project idea would be very conflictual, because Arab groups claimed rights in the area and farmers refused to hand over their land despite promises by their rulers. In 1827, embankments protecting the irrigated areas in the low delta were destroyed by farmers. In 1831, this first attempt to develop an irrigation scheme was abolished.

It took some decades before new French plans to develop irrigation facilities in the area were drafted. In 1904, a series of smaller reservoirs in the Senegal River were proposed to deliver water to a system of canals parallel to the river. The canal system would submerge large areas when water levels in the river were high, as the Nile did in Egypt. Decades later, when in the middle of the 1930s Senegal's rice import grew to 60,000 tons per year, these plans were taken up again when the government appointed a commission with the task to stimulate irrigation facilities in the Senegal River valley and delta.

The new Mission d'Aménagement du Sénégal (MAS) remodelled the 1904 plans. The new irrigation approach for Senegal was to be based on the principle of "submersion controlee" (controlled flooding). The original version of this system seems to have been developed by French engineers in the Mekong valley, Vietnam – another major colonial area for the French – in the early 1930s. River water was diverted to embanked terraces to store water on the flood plains. In the Senegal design, additional pumping would provide the terraces with water in case the water levels in the Senegal River were low. In 1938, 100 hectares of terraces were realized at Demet and Diorbivol; at Guédé, a more impressive 1000 hectares were developed. The small area at Demet and Diorbivol was soon abandoned, but at Guédé the system continued to function up to the 1940s.

After World War II, the French colonial government – as many other European colonial powers – aimed for a decrease of the vulnerability of the colonies in terms of food availability. Increasing food production in the colonies became a new target. Increased production was on the one hand an answer to food scarcity issues in the colonies during the war. On the other hand, attention for local food production was a way to show both the larger global community as well as the colonies that the colonizing powers actually wanted to develop the colonies themselves, instead of just extracting resources for the colonizer. New areas on the principle of controlled flooding were prepared in the Senegal Delta. 50,000 ha were to be equipped with pumps and prepared for mechanized rice farming. MAS developed plans for an irrigation system of 50,000 ha in Richard-Toll (De Wit 1958); in the original plans, cotton was to become the main crop, but given the new food policy rice was finally selected. In 1949, Richard-Toll started a private European-managed system on a 15-year contract (De Wit 1958).[2]

THE FRENCH COLONIAL PLANS FOR THE NIGER DELTA

The results in Senegal convinced the French engineers that the flooding technology should not be limited to Senegal. The central Delta of the Niger River seemed perfect for large-scale flood irrigation. The magical number of 1,000,000 ha under irrigation was to be realized in the delta. The Niger had been the stage for colonial planning aspirations since 1883, when the first French troops arrived in the region (Spitz 1949). Soon, in 1899, in his novel *Fécondité*, Emile Zola expressed the hope that the Niger River, which conquered the invading desert and created a fertile valley, would become the French Nile (Van Beusekom 1989). The region was also referred to as the 'Mésopotamie Nigérienne' in an attempt to link it to the other ancient irrigation-based kingdom (Spitz 1949). It is clear that French expectations were high.

The Niger Delta is a flood plain of the Niger River and its principal tributary, the Bani. Together, these rivers flood an area of some 350 km length and 100 km width, with an annual average flooded surface of almost 20,000 km² (Van Beusekom 1989). These flood plains close to Diafarabé are flooded each year in October, when the rivers carry some 6,000 m³/s resulting in water levels up to five meters higher than in April (with a flow of only 50 m³/s) (Spitz 1949).

It was the other of the two Niger Deltas, created because of the rivers' changing course over time, that attracted the French. This old delta does not receive floodwater from the Niger, but with its enormous potential of cultivable land between two former river branches potentially available as main canals, the French looked at it as the natural habitat for large-scale, gravity-based irrigation – in a similar way the British saw the Gezira plain as being created for irrigation. The agricultural potential of the region was clear, millet and sorghum were widely cultivated by local farmers. Rice was cultivated as well. The farmers arranged their cultivation within the rhythm of a rising and falling Niger River, and the arrival of the rainy season. Even some irrigated farming was practised, especially on the riverbanks.

A high-potential area it might be, but there were some potential problems the French would like to solve. The area might be a former flood plain, but that did not mean that droughts did not occur. Pests of locusts were a threat as well. Overcoming these problems would create the opportunity to develop European managed plantations worked by African labourers, even though 'the prejudice and apathy of the Africans' (Van Beusekom 1989: 24) was to be overcome. Once the new delta was a reality, however, the increased number of potential consumers with the financial resources to purchase the goods manufactured in France itself would stimulate both the colonial and the French economy (Van Beusekom 2002, 1989).

The plans for irrigation and development in the Niger area were shaped by French colonial policies of 1920 as formulated by the Minister of Colonial Affairs. The French aimed for 'l'association' between colonizers and colonized and *la mise en valeur* ('development'), investment in public works instead of mere exploitation (Spitz 1949). The plans focused on realizing transport infrastructure, public health, and improving agriculture, for a large part through irrigation.

It was in 1919–20 that Emile Bélime, an engineer, travelled through the Niger valley to investigate irrigation possibilities (De Wilde 1967, Van Beusekom 2002, 1989). When working in French colonial Asia as an irrigation engineer, Bélime had visited British India where he 'could familiarize himself with the issue of irrigation in British India'. After that visit, he was appointed by the French colonial government to lead a Study Mission on cotton irrigation in the Senegal and Niger valleys, which took from 1919 to 1920 to report. The Mission concluded that the Senegal valley 'had little to offer' for cotton; however, 'the floods of the middle Niger were remarkably suited for the hydraulic needs of the cultivation of irrigated cotton' (Spitz 1949: 45, author's translation).

With the British plans for Gezira developing in the 1920s, the French were even more convinced that the Niger Delta had to become their cotton miracle. The focus before the First World War was on the apparent – or desired – potential of the Niger Delta to become as important as British Egypt and the future Gezira in growing cotton. Climatic conditions in the delta would favour cotton, assuming the missing link – regular water supply – would be found. Pilot cotton schemes at El-Oualadji, close to Tomboctou (1917–23) appeared to confirm these expectations (Schreyger 1984).

In 1920, a first general plan included 1,850,000 ha, with pasture, vegetables, rice, millet, wheat, and cotton. The irrigated area would be in the delta as well as in the areas to the north of it (Spitz 1949, Van Beusekom 2002, 1989); Schreyger mentions 1,300,000 ha, of which 435,000 ha were reserved for cotton). This plan drafted by Bélime consisted of several elements: a barrage at the Sotuba rapids close to Bamako, the Ségou and Nyamina canals with their distribution canals, a barrage at Sansanding with the Sansanding and Djenné canals associated with it, irrigation and drainage canals in the lake region and reclamation of Faguibine lake (Schreyger 1984: 26).

The program of the Bélime mission shaped the future projects exploiting the French Sudan (Schreyger 1984: 28–9). Cotton continued to play an important role in generating Niger irrigation plans, but large areas dedicated to cereals, vegetables, and pastures for cattle were never forgotten. As early as 1924, the French Governor-General in West Africa stated that the primary purpose for Niger irrigation development was ending famine by producing rice in French West Africa. The First World War had shown how vulnerable both France and the colonies were regarding food supplies. The Niger area should produce the food for itself, but certainly for Senegal as well, which had shifted its focus to growing peanuts – the cash crop feeding the French oil-seed industry.

Food grown in the Niger Delta would help to reach imperial autarky, for the population and for the French industry, as it would assure a constant supply of raw materials and expand the market for manufactured goods to increased numbers of potential customers for these goods in the colonial areas (Van Beusekom 2002, 1989). Such a policy of *colonisation indigène* ('indigenous colonization') was to generate an agricultural revolution in Africa.

NIGER VALLEY – A BREADBASKET?

Niger valley irrigation creating a grain basket for French West Africa was not inconsistent with the role this region had played before colonial occupation. Geographically and ecologically, the Niger Inland Delta was well suited for grain production and well located for trade. Rice, however, had not been the major cereal crop; millet and sorghum were the major crops. Rice growing in the Niger irrigation schemes was an outcome of a long-established colonial policy of favouring rice, with tax collection in rice stimulating increased demand for it (Van Beusekom 1989).

Although many had great expectations about the new irrigation scheme, some French actually strongly criticized the proposals of Bélime. The image of a region of incredible richness, with 'Eldorado [...] dawning once the irrigation systems are realized' (Schreyger 1984: 37; author's translation) was based on the assumption that the Niger Delta was similar to the Nile valley. Critics argued that Egypt, with its millennia of irrigation history and its about 100 times higher population than the French Sudan, could simply not be compared with the Niger. In addition, the Nile carried much more water, floods were bringing fertile sediments and was considerably closer to the European markets than the French Sudan.[3]

Despite these criticisms, French colonial policy went for irrigated cotton production in the Niger valley. In 1922, an experimental station for cotton production was opened in Niénébalé, 50 km south of Bamako (Spitz 1949); its maximum area was some 200 hectares.[4] In 1925, the Servive Temporaire des Irrigations du Niger (STIN) was established (Van Beusekom 2002, 1989). STIN began constructing the Sotuba barrage on the Niénébalé site with the aim to develop 3,000 ha for cotton, 3,000 for rice and 1,500 for pastures (Schreyger 1984, Spitz 1949). When the barrage was opened in 1929, it brought water to a main canal of 22 km with a maximum of 10 m^3/s. This first larger irrigation system in French West Africa, which included colonising more than 6,000 people, was considered a big success. Inspired by 'le champ d'expérience' for the Niger Delta (Spitz 1949: 53), an over-arching program for developing the Middle Niger was drafted (De Wilde 1967, Spitz 1949).

Before the economic crisis of the late 1920s and early 1930s, irrigation development in the French Sudan had been regarded as a private initiative. After the crisis, however, state-led development became the norm, as private companies would not want to invest (anymore, assuming they had ever wanted) in irrigation development (Van Beusekom 2002, 1989, Schreyger 1984, Spitz 1949). Bélime and other advocates proposed constructing a barrage at Markala, about 300 km downstream from Bamako, which would divert water to two former watercourses of the Niger River – part of the plans proposed in the early 1920s. In 1929, Bélime presented a fully elaborated plan for such infrastructure to the Minister of Colonial Affairs. The system would be good for irrigating 960,000 ha (510,000 for cotton and 450,000 for rice).

Bélime's 'Project d'Aménagement du Delta Central' consisted of Sansanding barrage at Markala, a 8 km navigation canal on the right bank, a 8 km conveyance canal on the left bank delivering water to the Macina canal (12 km) and the Sahel canal (24 km) and 69 km of embankments (Schreyger 1984: 49–50). The plan required 300,000 settlers to grow the crops and would cost some 300 million francs, plus an extra 40 million for the colonization program. On 16 March 1931, Paul Reynaud, Minister of Colonial Affairs, approved of the plans. On 5 January 1932, the Office du Niger was established as a financially autonomous governmental organization with the specific task to manage the irrigation scheme and its production. In a way, the Office can be regarded as the early version of the Sudan Gezira Board, even though the histories of the two are fairly different – for example, the Sudan Gezira Board was a governmental board taking over management of the Gezira Scheme from a private firm, whereas the Office du Niger started as a governmental agency from scratch (Ertsen 2016).

Actual construction activities at the Sansanding barrage site started in 1934. The conveyance canal, with a maximum capacity of about 500 m^3/s, fed the Sahel Canal (completed in 1935) and the Macina Canal (completed in 1937) (Van Beusekom 2002, 1989, Schreyger 1984, Spitz 1949). With Sandanding not completed yet, irrigation would only be possible when annual Niger floodwater levels would be high enough to reach the new canals. Therefore, a provisional barrage was built in 1941. In 1947, Sansanding barrage was completed (Figure 9.2).

As the irrigation scheme was planned and implemented with limited and sometimes incorrect technical information, however, agriculture was not too successful. Rainfall was actually too plentiful for long-staple cotton and topographic surveys lacked the required precision for adequate canal construction, land levelling, and field preparation.[5] These problems did not dampen the spirit, however. Completing Sandanding was a sign for new efforts to boost scheme development. The Commission for the Modernization and Equipment of Overseas Territories proposed a ten-year plan for the Office du Niger. Between 1947 and 1957 some 180,000 ha should be developed (105,000 for cotton and 75,000 for rice) (Schreyger 1984), enlarging the existing irrigated area (22,500 ha) eight-fold!

LOCAL ATTITUDES AND RESISTANCE

As many other African colonial irrigation schemes, the Office was designed as a settlement project with tenants. The farming community was expected (or forced) to cultivate exactly what and how colonial management prescribed. A main problem faced by the Office, however, was attracting settlers to command. The Office should draw settlers from the French Sudan and Upper Volta, but there were not that many people in these areas. In Upper Volta, most inhabitants preferred temporary migration rather than permanent settlement elsewhere. I would not be surprised if some of them went as a seasonal labourer to Gezira, where money could be made as a cotton picker.

In order to arrive at an acceptable number of settlers, the Office did apply 'a certain amount of compulsion to recruit settlers' (De Wilde 1967:

Figure 9.2. Impression and detail of Sansanding barrage (© Geertjo van Dijk).

252); others prefer the version that the Office 'forcibly recruited most settlers' (Van Beusekom 1989: 12). By 1945, about 20,000 persons had 'colonized' some 22,000 ha (in two different regions) of the central delta (De Wilde 1967). We read that until 1946, some 31,000 men, women and children had been compelled or forced to work in the Office. The French government prohibited forced recruitment in 1946, with as an immediate result that some 40 per cent of the settlers left the project. The Office itself started cultivating their abandoned lands, which throughout the years was always some hundreds of hectares.

In 1948, the Office developed land in a new sector, but instead of assigning parcels to farmers, the Office began to farm the land itself. Some 6,000 ha were developed for rice cultivation near Molodo by the 'Centre Rizicole Mécanisé' (CRM) (De Wilde 1967). But the story of sadness had not ended yet: the CRM experienced technical and economic difficulties and was abandoned in 1961. Its land was re-allocated to settlers. In 1957, the irrigated area in the Office du Niger counted 47,259 ha (Schreyger). 27 per cent of the farmers were rice producers.

It might have failed in terms of its high expectations, the Office had become an enterprise with comprehensive jurisdiction in commercial, administrative, and agricultural affairs. The Office was a contractor, a recruitment service for agricultural labour, an entrepreneur and a supervisor of settlers. The Office performed all these roles in an area scattered over several hundreds of square kilometres. Its headquarters was at Ségou, 40 km upstream from Markala. The Kolongo irrigated sector extended almost 150 km from Ségou toward the east; the Kourouma sector extended towards the north for more than 200 km. The main irrigation canals had a combined length of 280 km. In 1955, the Office had almost 7,000 employees. In 1960, at Mali independence, 54,000 ha had been developed (Van Beusekom 1989); that year Office administration was turned over to Mali. In 1964, the Office still had a large staff of about 4,700 employees, with Europeans in the higher positions (De Wilde 1967).

Colonial governments did have the legal power to expropriate land and they used this power when necessary. Forcing peasants to grow certain crops was a preferred option, but even in the African territories the 'degree of control exercised by the central irrigation authorities, and their manipulative powers with respect to agricultural improvement, were effectively very limited. This has to do with the technical aspects of the canal schemes and the broader administrative framework in which they operated' (Stone 1984; 8). Colonial rule was 'a juggling act' (Stone 1984: 8) or continuously in the making (Darwin 2013).

Local farmers might have had no influence on the design of the Office du Niger, they certainly did influence the success of the system by not showing up when obliged. Juggling between available labour from colonizers and colonized, available financial resources, and political goals shaped the story of the Office du Niger as much as in other areas. The French probably developed the most factory-like irrigation technology with its concrete, standardized canals in northern Africa (Ertsen 2007), but even there they did have considerable trouble before the local farmers adapted to French colonial efforts.

European colonial powers did everything they could to guide and/or command the Africans in their agricultural efforts. African colonial irrigation systems share the characteristics of an imposed production regime, a factory with Taylor-type control institutions (Ertsen 2006, Diemer 1990). Farmers were settlers, they were tenants, they did not own the land and as such were expected to obey orders from the higher management. The factory resemblance is also reflected in the mathematical layout of systems: canals were straight, plots were square, although this material dimension worked less well in the Office du Niger given the topographical circumstances.

With its colonial factory model, the Office du Niger was 'imbued with grand objectives aiming at transforming the rural social and material landscape' (Bolding 2004: 10). Civilization, state formation, modernization, and intensification of African agriculture are all terms applicable to the kick-start model to produce development. Colonial irrigated settlements were 'vessels of modernity labouring through a sea of superstition' (Bolding 2004: 10).

THE OFFICE DU NIGER AFTER WORLD WAR II

Irrigation development with the aim to modernize rural Africa continued and was further developed after independence. Actually, 'the wave of independence that swept the continent in the late 1950s and early 1960s, combined with a new found belief in the capacity of development administrations and an evolutionary model of radical change through agricultural modernization [...], led to a second heyday for settlement schemes, and irrigation in particular' (Bolding 2004: 11, see also Chambers 1969). Within the independent states, colonial irrigation programs and policies were continued. Irrigation development was a key activity in international aid programs aiming at increasing global food production.

After World War II, these huge development aid (or cooperation) programs created a new flow of technical activities, not for colonial officials but for the (mainly European and American) experts, engineers, and consultants that used to be active in the different colonial services on the globe. International development organizations within the United Nations – the Food and Agricultural Organization (FAO), the World Bank and the Asian Development Bank amongst others – recruited their engineering expertise from these former colonial recruits to develop new projects.

For Europe, colonial knowledge became an exportable commodity. The actual shape of the product (the specific know-how of irrigation

Table 9.1. Different infrastructure in Office du Niger.

Donor	Properties
French	Non-modular sliding gates at the intake of secondary canals; Neyrtec modules at cooperation the intake of tertiary canals; intake at tertiary level secured; tertiary canals over (Retail type) dimensioned with respect to peak water requirements; one field canal per 2 ha
Dutch	Neyrtec modules at the intake of secondary canals; semimodular overflow weirs at cooperation the intake of tertiary canals; intake at tertiary level not secured; one field canal (ARPON type) per 3 ha, additional field canals constructed by farmers
German	Non-modular sliding gates at the intake of secondary canals; Neyrtec modules at the cooperation intake of tertiary canals; intake at tertiary level secured; one field canal per 2 ha; (KfW type) a multiple of 7 field canals per tertiary canal in order to facilitate water distribution following a weekly rotation
World Bank	Neyrtec modules at the intake of secondary canals; Neyrtec modules at the intake of (World Bank type) tertiary canals; intake at tertiary level secured; one field canal per 2 ha
Not rehabilitated, original	Non-modular sliding gates at the intake of secondary canals; nonmodular sliding (not rehabilitated gates at the intake of tertiary canals; intake at tertiary level not secured; no field type) canals, unless constructed by farmers

design) differed for each country, as it had been developed in different regions (Ertsen 2010, 2008, 2007). In the Office du Niger, the variety of donor-related expertise is still visible in the irrigation infrastructure. Different types of infrastructure are found to bring water from the main canals to the clusters of fields at tertiary level (Figure 9.3, Table 9.1, Vandersypen 2007, Vandersypen et al. 2006).

The Retail-type was implemented by a French project, which is why so-called baffle modules are used. These baffles are the typical distribution structures from Northern Africa (Ertsen 2007). The hydraulic properties of a baffle allow for maintaining a constant incoming flow in the tertiary canal for a relatively wide range of upstream water-level variations. Locks should secure the baffle settings, with water guards being allowed to open or close them. In Office practice, the lock is often missing, or farmers have a copy of the key (Vandersypen 2007). In typical irrigation settings, baffles are accompanied by hydraulic gates that use a float on the required water level (Neyrtec gates) to regulate the upstream level for the baffles. In the Office du Niger, several of these devices are found, but we also encounter concrete structures obviously prepared for them, but without the actual floater and gate.

A Dutch-financed project introduced the so-called Arpon type, which are weirs that can be closed with a single sliding gate. The width of the weir relates to the irrigated area downstream. The weir is the preferred distribution type of Dutch irrigation engineers, based on their irrigation

Figure 9.3. Different division structures in the Office du Niger (© Geertjo van Dijk).

tradition that developed in the Netherlands East Indies (Ertsen 2010, 2007). Weirs are actually pretty suitable in areas without big potential drops in water levels (the technical term is 'head-loss') available, so in flat areas like the Office du Niger they could perform well. The downside is, however, that weir hydraulics mean that small variations in upstream water levels result in relatively large variations in flow rate. Apparently, the Arpon type has no longer been constructed after the early 1990s (Vandersypen 2005).

In smaller areas, other types can be found. In areas in the Office that have not been subject to rehabilitation (yet), the original water-distribution infrastructure consisting of sliding gates at the intakes of tertiary canals is still used. That same original design did not include field canals. Plots were irrigated from field to field. Nowadays, several units include field canals constructed by farmers. In all tertiary units, excess water is guided into tertiary and/or secondary drains. The flat area – again – produces specific flow patterns. Basically, Office drainage canals behave like small interconnected reservoirs, or communicating vessels (Vandersypen 2005). Some larger collector drains close to the Niger River discharge into the river, but the majority of larger drains flow out into natural depressions where the water infiltrates and evaporates. Quite often, farmers officially outside the Office du Niger re-use drainage water on their (formally illegal) plots.

DEVELOPMENT DISCOURSE AND THE OFFICE DU NIGER

The specific type of water-distribution structure might have been subject for debate between irrigation designers and possible financers, but the African colonial irrigation model, first with Gezira and later with Mwea on the first row, remained generally popular in the development discourse. Direct control as in colonial times, however, was impossible. Post-colonial production schemes had to build support for farmers (Ertsen 2008, 2006), translating colonial coercion and force into extension and training. Forced production turned to extension programs.

With farmers becoming independent actors responding to market incentives rather than colonial force, the need for strict management had to be shown with different arguments. Central control (support!) was defended with the statement that desired market incentives were still absent in African rural societies. African farmers might actually respond to prices, but they did not always behave as the desired homo economicus. 'Confronted by alternative opportunities Africans often display a keen ability to choose the least burdensome (in terms of labor) way of attaining the income they want, or, to put it in other words, the ones likely to give them the highest return for the amount of work they are willing to do' (De Wilde 1967a: 63).[6]

Thus, strong central management could remain an essential ingredient for many participants in the debate for tasks like maintenance, operation and purchase of machinery, managing the irrigation system, organizing supplies, organizing farmers, regulating marketing, etc. 'The effective discharge of such operations can be ensured only if the management enjoys a considerable measure of autonomy' (De Wilde 1967b: 241).

The Office du Niger may not have been an important scheme in the general development discourse,[7] but obviously still kept functioning and was important for Mali. The general line suggested in the 1986 FAO report quoted earlier, that the 'key to the success of a settlement project is that government officials must initially be responsible for and perform many of the tasks', may not have been inspired by the Office, but can certainly be applied to it. After all, 'only giving the farmers simple and straightforward tasks which they can carry out with their own means' was the starting strategy in African irrigation; only as 'time goes by, it may be possible to transfer more and more of the tasks and responsibility to the farmer' (Sagardoy et al. 1986). This is what happened in the Office du Niger as well.

In the original Office model, the scheme had a monopoly including land survey and development, civil works, land administration, recruitment and installation of settlers, motorized farm mechanization, agricultural extension, establishment and oversight of farmers' organizations, input supply, credit, marketing and processing of crops, sale and export of rice and cotton, transportation of agricultural inputs and produce by roads and canals, and management of guesthouses, seed farms, and a training centre (Aw & Diemer 2005). Vacant land within the Office area was declared state property.

The Office du Niger had become 'a state within the State' (Reste 1946: 31, quoted in Aw & Diemer 2005). Until 1947, the Office Board only included representatives of French institutions: ministerial departments, civil construction, and textile companies. From 1947 onwards, one farmers' representative was appointed on the board by the director general; in addition one delegate each from the European and the African employees'unions – nicely separated – were allowed on the board as well. The timing of these new representatives fits nicely with timing in Gezira concerning farmers' and labourers interests (Ertsen, 2016).

On 19 May 1961, after Mali's independence of 22 September 1960,[8] the Office du Niger was nationalized. In the short years after independence, the French remained co-owner of the Office, in order to transfer knowledge and management procedures to the independent state. Between 1960 and 1968, during the (socialist) regime of Modibo Keïta, the Office population decreased by 14 per cent – with normal growth at about 20 per cent for Mali (Aw & Diemer 2005). In the 1970s, with its many periods of drought, the Office attracted people again, even though there were still settlers leaving the area.

The original idea(l) of the French was to develop their own cotton miracle in the Inner Niger Delta, but cotton never prospered. Nevertheless, even after independence, cotton remained one of the main crops on paper. A new focus on food production – again coinciding with a larger wave in African agricultural policies, the same move was made in Sudan's Gezira system – brought the official decision confirming and conforming to realities on the ground: cotton was abandoned as Office-crop in 1971. The new focus was on rice production. Rice paddies had already come to occupy 80 per cent of the farmed area; the new focus on rice seemed a success, as between 1969 and 1978, paddy production rose from 46,000 to

101,000 tons, with yields increasing from 1,550 to 2,660 kg/ha (Aw & Diemer 2005).

At the same time, however, the Office farmers – who were tenants of the management – saw their debts to the Office increase from US$1.2 million to US$2.27 million[9] between 1972 and 1977 (Aw & Diemer 2005). The Office had a monopoly on the marketing of the rice yields; typically, rice had to be sold at prices set by the Mali government – usually under the market prices to be able to feed the urban population cheaply. 'In tacit protest, farmers defaulted on credit and service payments' (Aw & Diemer 2005). Obviously, this was not only a problem for the farmers, but also for the financial position of the Office and given the close relations between country and scheme, also for Mali as a whole. Again, this compares rather directly to the Gezira scheme within Sudan.

In the 1980s, such state-managed economic processes were no longer acceptable, at least not for countries that would like to receive international aid. The World Bank and other donors pressed for changes in scheme management, especially regarding the position of settlers. These should have land-tenure security and more freedom to produce what they liked and market how they saw fit. The Office could possibly support farmers' productive efforts, but should not direct them. More efficient water delivery through field and canal improvements was set on the agenda as well, which resulted in the many different distribution types discussed above (see the set of publications by Vandersypen and colleagues). Apparently, the Mali government liked the water delivery programs, but for policy reforms more was needed.

Aw and Diemer (2005) provide a detailed overview of the way institutional changes were realized in the Office du Niger. Small reform steps were realized hand in hand with programs for physical improvements. Short- and medium-term credit for all farmers was developed, village associations became involved, small threshers and small hulling machines were allowed. One donor 'negotiated farmer-friendly land-tenure rules that eliminated the agency's prerogative to force farmers to move to other villages or deny settlers access to land' (Aw & Diemer 2005). After the March Revolution (22–26 March 1991), when the one-party regime was toppled, further reforms were possible in the Office. Staff numbers were reduced, land-tenure security was made possible for farmers and sharp increases in yields, cropping intensity, and incomes were observed.

Figure 9.4 shows the results of all these changes and programs. Basically, we see sharp increases in rice production per hectare in the Office area, as well as a sharp increase in rice production in Mali.[10] Both the internal production in the Office and the irrigated rice area in the whole of Mali correlate to total rice production in Mali (Figure 9.4, lower graphs). A closer look at the graph and the numbers, however, suggests that the main possible cause for the increase in total Mali rice production is the increase in total rice area in Mali – which is largely outside the Office area.

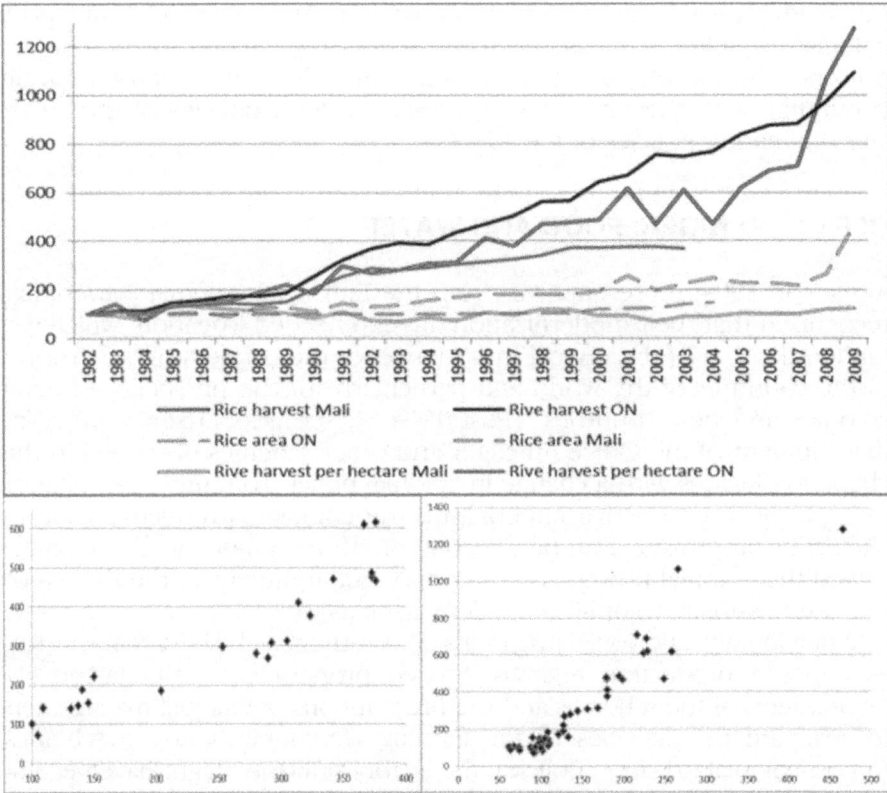

Figure 9.4. (Top) Rice growing trends in Mali and OduN; (bottom left) total rice production in Mali versus production per hectare in Office du Niger; (bottom right) total rice production in Mali versus total rice area in Mali (data for Mali from faostat.fao.org; Worldbank ADI data. Data for the Office du Niger from Aw and Diemer, 2005).

In the recent past, between 2005 and 2009, various investors obtained land leases in the Office du Niger area. In 2009, a staggering 870,000 ha of land had entered the allocation procedure. 500 national investors were granted leases for a total of 400,000 ha, with most areas between 1 and 50 ha. Three per cent of national investors requested leases for areas larger than 500 ha, but together these were for a total area of almost 300,000 ha. 470,000 ha was allocated to foreign investors, with project areas between 2,500 and 100,000 ha. There was a South African corporation aiming at developing sugar cane plantations, others were from government-owned corporations from countries as diverse as Libya, Saudi Arabia, China, and Burkina Faso. One project was even arranged through the United States' Millennium Challenge Account – MCA – project (Hertzog et al. 2012).

With this last stage of 'land-grabbing', the Office du Niger is the stage for one of the latest developments in African agricultural production – as land-grabbing is a phenomenon not isolated to Mali. Developments in the Office can readily be related to general debates in African colonial and post-colonial irrigation. We have seen the modernity mission of the colonial powers, the development area with their foreign experts basically continuing with their colonial expertise, the food policies of the 1970s, and the market policies of the 1980s.

OFFICE DU NIGER: FOOD AND WATER

Already in 1904, representatives from the European colonial powers had recognized that their modernization mission needed irrigation, which was 'undeniably [...] of the utmost importance for the richness of the majority of the countries of the world and particularly for the prosperity of many colonies and new countries' (Post 1904: 1, see also Ertsen 2010). The development of the Office du Niger and other schemes described in this chapter coincides with a change in colonial policy from mere exploitation to a policy of productive imperialism: the colonies' productive capacity should be improved. Naturally, economic development in the colonies served the colonial powers, but an important argument in the discussion was the possibility to uplift the colony itself as well.

When looking at colonial irrigations efforts, they can be best characterized as imposed production regimes. Shared properties are the factory-like organization of the schemes and the prescriptions of colonial management to structure the activities of the farming communities in the schemes. In post-colonial irrigation policies, the factory-similarity might have become less, but one could still discuss to what extent. As in other colonial schemes, the Office du Niger did not use the colonial discourse anymore; however, its roots and results are still relevant up to the present day.

Rice production in the Office has increased and will have contributed to the total availability of rice in the country, but the old colonial metaphor of a 'modernized island in a sea of superstition' still seems to hold. The Office might be less modern than has been presented throughout the project's history and the Mali sea was most likely not as superstitious as colonial narratives wanted, but nevertheless, the Office du Niger remains a special spot of producing high rice yields with privileged access to water on a relatively small area within the larger Mali rice cropping area.

NOTES

1   Compare the straight lines with the ideology of particularly the French irrigation engineers in northern Africa (see Ertsen 2006).
2   De Wit wrote his report in the first place because he came to Mali from Surinam on behalf of the 'Foundation for the development of mechanical rice cultivation' (De Wit 1958).

3 Spitz (1949) argues that it was impossible to develop an 'Egypte nigérienne', as the comparison with Gezira did not stand, because of climatic and river flow conditions.

4 The test station at El-Oualadji, about 800 kilometers north-west of the Office, has a different climate.

5 'The preliminary plans were made on the basis of topographic surveys on the scale of 1/20.000; and the final project was carried out on the basis of plans on the scale of 1/5.000 or 1/2.000. More detailed and precise data were really required. The difference in elevation, which is only a few meters between the extreme points of the Office, makes micro-relief particularly important. [...] Some drains have had to be converted into irrigation canals and vice versa. Some land has been too high for irrigation and other land too low for effective drainage' (De Wilde 1967: 263).

6 Don't we all? This rather stereotypical suggestion is replaced further in the book with a more business-based reasoning on labour compared to income (De Wilde 1967a: 223–224), and De Wilde even acknowledges explicitly that there are often seasonal labour bottlenecks in agriculture and that Africans simply work hard too (even though '[...] men may in many cases lag behind women in the effort they make at such times' (De Wilde 1967a: 224)).

7 Possibly because of its story of failure, it being in Mali and Mwea filling the gap.

8 Malinese independence is slightly more complicated, as there was the Mali Federation gaining independence from France on 20 June 1960. Senegal withdrew from the federation in August 1960, which allowed the Sudanese Republic to become the independent Republic of Mali on 22 September 1960.

9 Malian franc (MLF) 721 million to MLF 1,364 million (Aw and Diemer 2005).

10 There is also a sharp increase in development money, but what it is spent on is less clear – probably quite a bit in the Office, but it is not easy to find the exact relation between investment and rice areas/yields).

REFERENCES

Aw, D., Diemer, G. 2005. *Making a Large Irrigation Scheme Work: A Case Study from Mali. Directions in Development*. Washington, DC: World Bank.

Bolding, A. 2004. 'In hot water. A study on sociotechnical intervention models and practices of water use in smallholder agriculture, Nyanyadzi catchment, Zimbabwe'. PhD Thesis. Wageningen: Wageningen University.

Chambers, R. 1969. *Settlement Schemes in Tropical Africa. A Study of Organizations and Development*. London: Routledge and Kegan Paul.

Chambers, R., Moris, J (eds). 1973. *Mwea, an irrigated rice settlement in Kenya*. München: IFO-Institut für wirtschaftsforshung.

Darwin, J. 2013. *Unfinished Empire: The Global Expansion of Britain*. London: Penguin.

De Wilde, J.C. 1967a. *Experiences with Agricultural Development in Tropical Africa. Vol I. The Synthesis*. Baltimore: John Hopkins Press.

——— 1967b. *Experiences with Agricultural Development in Tropical Africa. Vol II. The Case Studies*. Baltimore: John Hopkins Press.

Diemer, G. 1990. 'Irrigatie in Afrika. Boeren en ingenieurs, techniek en cultuur'. PhD Thesis. Leiden: Leiden University

Ertsen, M.W. 2016. *Improvising planned development on the Gezira Plain, Sudan, 1900–1980*. New York: Palgrave MacMillan.

—— 2010. *Locales of happiness. Colonial irrigation in the Netherlands East Indies and its remains, 1830–1980*. Delft: VSSD Press.

—— 2008. 'Controlling the farmer. Irrigation encounters in Kano, Nigeria'. *The Journal for Transdisciplinary Research in Southern Africa* 4: 209–36.

—— 2007. 'The development of irrigation design schools or how history structures human action'. *Irrigation and Drainage* 56: 1–19.

—— 2006. 'Colonial irrigation: myths of emptiness'. *Landscape Research* 31: 147–68.

Hertzog, T., Adamczewski, A., Molle, F., Poussin, J.-C., Jamin, J.-Y. 2012. 'Ostrich-like strategies in sahelian sands? Land and water grabbing in the Office du Niger, Mali'. *Water Alternatives* 5(2): 304–21.

Huxley, E.J.G. 1960. *A New Earth: An Experiment in Colonialism*. London: Chatto & Windus.

Post, J.W. 1904. 'Rapport sur l'irrigation aux Indes orientales neérlandaises. Rapport préliminaire à la session de l'Institut Colonial International de Wiesbaden du 17 mai 1904', Brussels, [s.n.].

Sagardoy, J.A., Bottrall, A., Uittenbogaard, G.O. 1986. 'Organization, operation and maintenance of irrigation schemes', *Food and Agriculture Organization, Irrigation and Drainage Paper 40*, Rome.

Schreyger, E. 1984. *L'Office du Niger au Mali 1931 à 1982. La problématique d'une grande enterprise agricole dans la zone du Sahel*. Wiesbaden: Steiner.

Spitz, G. 1949. *Sansanding. Les irrigations du Niger*. Paris: Société d'Éditions Géographiques, Maritimes & Coloniales.

Strong, T.H., Paton, P. 1968. 'Economic appraisal and analysis of requirements for development of the Talata Mafara area', Sokoto Rima Drainage Basin, UNDP, FAO, Rome

Van Beusekom, M.M. 2002. *Negotiating Development, African farmers and colonial experts at the Office du Niger, 1920–1960*. Portsmouth: Heinemann.

—— 1989. 'Colonial rural development: French policy and African response at the Office du Niger, Soudan Français (Mali), 1920–1960'. PhD Thesis. Baltimore: John Hopkins University.

Vandersypen, K. 2007, *Improvement of collective water management in the Office du Niger irrigation scheme (Mali): Development of decision support tools*. Louvain: CIRAD.

Vandersypen, K., Raes, D. Jamin, J-Y. 2009. 'Simulating water management and supply effects at the Office du Niger collective canal irrigation scheme'. *Journal of Irrigation and Drainage Engineering* 135: 50–7.

Vandersypen, K., Keita A.C.T., Coulibaly, B., Raes, D., Jamin, J.-Y. 2007. 'Drainage problems in the rice schemes of the Office du Niger (Mali) in relation to water management'. *Journal of Agricultural Water Management* 89: 153–60.

Vandersypen, K., Keita, A.C.T., Kaloga, K., Coulibaly, Y., Raes, D., Jamin, J.-Y. 2006. 'Sustainability of farmers' organization of water management in the Office du Niger irrigation scheme in Mali'. *Journal of Irrigation and Drainage Engineering* 55: 51–60.

10 Water and Agricultures in the Niger Basin through the Twentieth Century

Andrew Ogilvie, Jean Charles Clanet,
Georges Serpantié and Jacques Lemoalle

INTRODUCTION

The Niger River basin gathers 100 million people over 2.1 million km^2, ten countries and six agroclimatic zones, ranging from Saharan in the north to Equatorial coastal around the Gulf of Guinea. Its scale and diversity provide a remarkable cross section of the efforts placed by African smallholders to produce the required 2,500 kcal/day/capita despite limited means and unreliable water resources, through traditional low input rainfed agriculture, livestock and localized fishing. Significantly, the trajectory of livelihood strategies over the twentieth century provides valuable insights into the ways the societies have adapted as a result of changes in water availability, rapid demographic growth (doubling of the African population over the past 30 years) and rising competition for land and natural resources. Historical accounts can help understand these complex hydro-socio-ecosystems and pave the way towards strategies on how to prepare for future climatic changes, rising pressure on land, water and resources, as a result of demographic pressure and external investments ('land grabs'), increased development of dams for energy and irrigation and complex transboundary water management.

This chapter seeks to describe in the first instance the change in climate over the past century, highlighting the severe drought which characterized the later part of the twentieth century. Its detrimental (though sometimes paradoxical) effects on surface and ground water resources throughout the region are described and the growing influence of anthropic changes such as surface water impoundments upstream and land use changes are discussed. The distribution of agricultural water uses and their transformation over the past century are then described. Livelihood strategies include agropastoral systems, nomadic pastoralism, and fisheries within aquatic ecosystems of river stretches and wetlands. The way that agricultural systems spread and diversified across the region, notably as a result of agrarian policies implemented by the range of institutional actors:

state, international aid and NGO programmes, are presented. Significantly, the way users adopted and abandoned techniques, adapted or modified their practices and strategies to cope with changes in water availability and to their natural environment are highlighted.

HYDROMETEOROLOGICAL CHANGES

When considering water and agriculture, the concepts of blue and green waters are now increasingly used. Blue waters refer to water available in streams, aquifers and lakes, while green waters designate the share of rainfall which is directly evapotranspired by vegetation. This distinction reflects a recent paradigm shift (Falkenmark 1995) whereby agricultural water management seeks for enhanced consideration of rainfed agriculture.

Green water characteristics

Bordered and influenced by the Atlantic Ocean to the south and west, and the Sahara desert to the north, the Niger Basin displays a sharp decline in rainfall as one travels north, related to the seasonal shift and intensity of the Intertropical Convergence Zone (ITCZ). Created from the convergence of trade winds from each hemisphere near the equator, the ITCZ moves into the northern hemisphere during its spring and the dry northerly trade winds such as the Harmattan in the Sahel are replaced by the south-westerly monsoon winds. This West African Monsoon generates advection of moisture from the Gulf of Guinea, reducing evaporation demand and bringing rainfall on the southern part of the basin from March onwards and around May in Sahelian regions. The ITCZ moves abruptly northwards in July–August from 5°N to 10°N latitude (Sultan & Janicot 2003) producing a peak in rainfall in the dry northern regions of the catchment during August. The ITCZ retreats south again during the months of September and October, after which rains cease and evaporation demand increases. In the far south of the catchment, the northerly position of the ITCZ in August produces a second, short dry season during one month. The intensity and trajectory of the phenomenon is relatively homogenous on an east-west axis (marginally drier in the east of the basin) and defines the contours of six climatic zones distributed along a latitudinal gradient (namely Saharan, Sahelian, Sudanian, Guinean, Subequatorial and Equatorial coastal; see Figure 10.2). From north to south, mean annual rainfall (over 1950–2000) ranges from under 300 mm in northern Mali and Niger to over 4,000 mm in parts of southern Nigeria and Cameroon. These stark spatial differences have essential implications on agricultural strategies (presented below), considering that 95 per cent of agricultural land in

Figure 10.1.   The West African monsoon (© T. Lebel, IRD).

the catchment is rainfed. Furthermore, the significant interannual variability observed on rainfall over the twentieth century led to lasting changes in the natural environment and agropastoral strategies. Especially in the Sahel regions where annual rainfall remains below 750 mm and water becomes the agricultural limiting factor, annual variations in rainfall can have devastating consequences on the yields and livelihoods of farmers.

Blue water characteristics

The Niger River presents the significant hydrological curiosity of flowing in a north-east orientation, directly towards the Sahara desert, before buckling back south towards Niger, Nigeria and the Atlantic Ocean. This particular morphology is understood to originate from two rivers merging together as a result of stream capture and the Sahara drying in 4000–1000 BC (Gupta 2008, Gasse 2000). Flowing as far north as Gao and Timbuctu, where rainfall is below 300 mm, the river constitutes a valuable resource for local smallholders who lack alternate water resources. Irrigation projects remain limited with only up to 5 per cent of cultivable land irrigated compared to 20 per cent on average worldwide. Blue waters are used punctually through direct river withdrawals and localized irrigation projects, notably small reservoirs in Burkina Faso and larger dams such as Selingué in Mali and Markala which diverts water

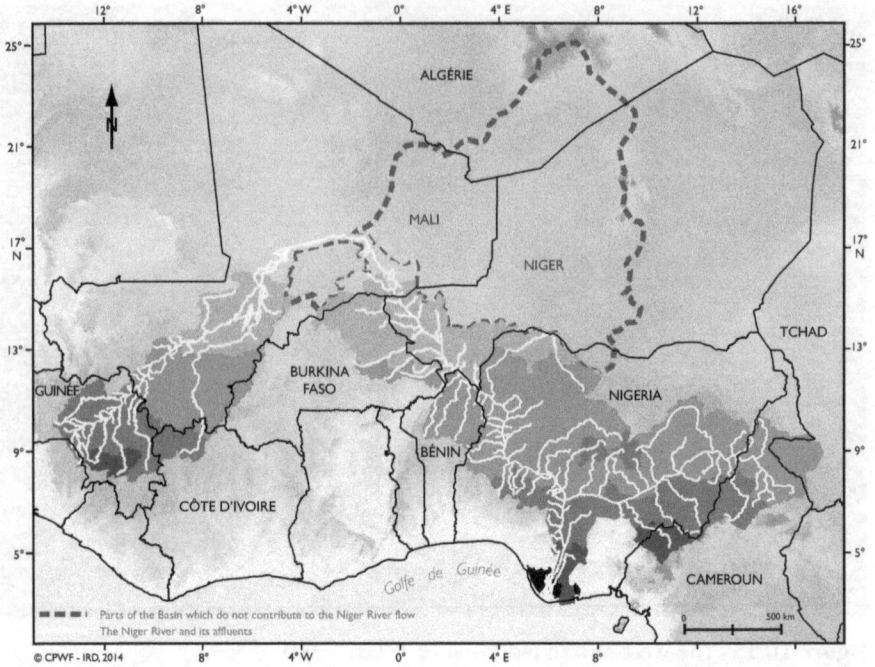

	Annual rainfall (mm)	Climatic zone	Plant associations	Agricultural activities
	< 300	Saharan	Desert steppe	No crops, occasional seasonal pasture
	301 - 750	Sahelian	Semi-desert grassland, shrubland and bushland	Occasional millet, seasonal pasture
	751 - 1 200	Sudanian	Grassland, woodland and savannah parkland	Millets, sorghum, green beans, groundnut, sesame
	1 201 - 1 800	Guinean	Semi-evergreen rainforest, or woodland, and secondary grassland	Sorghum, maize, groundnut, coco yam, peas
	1 801 - 2 500	Subequatorial	Lowland rainforest and swamp forest. Grassland above 1 000 m	Maize, cotton, sugar cane, rice, sweet potato, yam
	> 2 500	Equatorial coastal	Secondary forest, mangroves and grassland	Fruits, cassava, sugar cane, tea, plantain

Figure 10.2. The Niger River Basin and agroclimatic zones (modified from Clanet and Ogilvie, 2014).

to the 80,000 ha *Office du Niger*. Traditional irrigation, however, such as flood recession cropping, notably in floodplains (*bas fonds*) of the south of the basin and across the Inner Delta dominate in terms of surface areas and concern larger shares of the population. The river indeed fans out into a vast Inner Delta in Mali, spanning more than 40,000 km^2 (and a much greater area at its apogee around 6000 BC), where up to 1M herders and farmers depend on this annual flood. Changes in the flow regime, as a result of rainfall variations but also natural and anthropic changes within the catchment, can thus have stark detrimental effects on water availability and livelihoods of small scale farmers. Finally in selected areas, where shallow aquifers are accessible, such as in the Nigerian *fadamas* (lowlands in Hausa), agriculture also depends on groundwater.

Historical time series

Rainfall and hydrometric time series have been available in the Niger River basin since the end of the nineteenth century and provide insight into the modifications over time of rainfall and hydrological flows. Hydrometric gauging stations are scarce compared to other parts of the world and to climatic stations, but the importance of the large rivers of Africa, notably to the colonial powers, means that substantial records since the beginning of the twentieth century are available for the Niger, Senegal and Volta rivers. No flow measurements are readily available for the nineteenth century, and only occasional limnimetric gauges were installed (Sircoulon 1987) at the end of the 1800s for fluvial navigation purposes, as observed on the Senegal River at the end of 1890s on sills, and at Segou where gauge readings were noted during the flood in 1864–1865. By 1899, stations upstream and downstream of the Inner Delta had been installed and in 1907 the Koulikoro station was commissioned by the Compagnie Générale des Colonies. The station was maintained over the years and still operates today. During the 1950s, a coherent, organized monitoring network was developed partly with the support of French assistance and the Orstom (Office de la recherche scientifique et technique outre-mer). In the Inner Delta, considering its importance, several reports exist, however the measurement of the flood is inherently difficult and therefore relies on hydrological modelling or remote sensing (aerial photography and now satellite imagery). Punctual information also exists on lakes such as lakes Faguibine, Haribongo, Korarou in northern Mali, which provide qualitative information since 1908 for some years when lakes were full and exploitable (Haribongo and Korarou 1924–33, 1955–67, Faguibine 1894, 1930, 1955) or dry (1910, 1924, 1941, 1984, (Sircoulon 1987)).

Green water variability through the twentieth century

Serious prolonged drought periods occurred in 1910–16, 1940–49 and most emphatically, the whole region has been subject to a vast and significant decline in rainfall levels, since 1968. The drought after 1968 was the largest on records since 1896 in West Africa and the combination of its duration and spatial coverage extending across West Africa down to coastal areas make it one of the most significant climatic events worldwide in recent history (Hulme et al. 2001, Mahé & Paturel 2009). Mean rainfall over 1960–90 was around 20 per cent lower than previous decades in the century (Lebel & Ali 2009, Descroix et al. 2009) and this decline led to a shift in isohyets by 150–200 km south. The reduction reached its paroxysm during the 1980s, notably in 1983–4, leading to widespread famine notably in the Sahel regions. Rainfall levels recovered only slightly in the 1990s (more so in the west of the basin) remaining as low as during the 1970s (Mahé & Paturel 2009). Variability also increased, creating more dry spells during the cropping season. These variations are intimately coupled to changes in the monsoon dynamics (Eltahir & Gong 1996) and are believed to be driven by changes in sea surface temperature (SST) in the Atlantic, which follow natural low frequency (multi decadal, 65–80 years) oscillations, termed the Atlantic multi decadal oscillation. The precise dynamics and combination of factors driving decadal and multi decadal shifts as observed during the twentieth century remain somewhat uncertain (Hulme et al. 2001). Multiple large scale forcings and feedbacks are believed to exacerbate the effect of changing SST, notably changes in land cover and associated albedo. Conversely, significantly wet periods occurred over the years 1925–35 and 1950–65. These long lasting dry and wet periods were also observed over at least the past three millennia in paleohydrological data, and with often greater magnitude (Gasse 2000). More recently, a long term drought lasting 250 years (1450–1750) was detected during the Little Ice Age when temperatures were cooler than currently (Gasse et al. 1990).

In parallel, over the twentieth century, temperatures reportedly increased by 0.5°C over Africa, but this rise reached up to 1–1.5°C in parts of Mali and Niger (Hulme et al. 2001) notably since the late 1970s. The associated increase in potential evapotranspiration is estimated to be around 1–1.5 per cent (Mahé & Paturel 2009) and results in an increase in crop water requirements.

Blue waters: climatic and anthropic influences, the Sahelian paradox

The 20 per cent decrease in rainfall over West Africa since the late 1960s was reflected in the flow measurements of many large catchments, including the Niger and Senegal rivers where annual discharge reduced twice as much as rainfall (Descroix et al. 2009). The greater amplitude of the reduction in runoff compared to rainfall is remarkable and can be explained by the cumulative effect of reduced groundwater recharge over successive years (Olivry 2002). Reduced rainfall and recharge led to a progressive lowering of the water table, decreasing baseflow contribution and therefore annual flows. This phenomenon was confined to the southern part of the catchment (south of 750 mm) where baseflow is an important component of surface water flows (Mahé & Paturel 2009), notably around Koulikoro and on the River Bani where runoff over 1970–90 reduced by 55 per cent and 70 per cent respectively compared to 1950–70. In 1984, mean annual flow barely reached 640 m³/s at Koulikoro, compared to 1,400 m³/s over 1950–90 and up to 2,300 m³/s in 1925 (Olivry 2002, Descroix et al. 2009).

In drier Sahelian regions, a paradoxical phenomenon was observed on small and large catchments, whereby following the 20–30 per cent reduction in rainfall, runoff increased by 40–60 per cent (Descroix et al. 2009). This increase in the runoff coefficients (i.e. the amount of runoff generated from the amount of rainfall) observed after the 1970s was due to a change in the hydrodynamic properties of the soil, caused by natural and anthropic changes. Declining rainfall levels reduced the volume of water potentially converted into runoff (blue water) but were accompanied by predominant changes in land cover and land use. Droughts led to reduced biomass, while the increase in population densities resulted in deforestation (cutting or burning bushes, shrubs, trees, grasses, etc.), conversion to crop land but also shorter fallow periods, increased fuel wood collection, etc (Lebel et al. 2009). In the Nigerien Sahel but also in Burkina and northern Cameroon, cultivated land increased from 10 per cent to 80 per cent of available land (Cappelaere et al. 2009, Descroix et al. 2009). These changes, most visible around densely populated areas, led to an overall reduction in the water holding capacity and the soil compaction led to the creation of an impermeable crust. More influenced by the alterations in land cover than the changes in baseflow and rainfall, runoff increased in these Sahelian catchments.

At Niamey the combined effect of the reduced upstream flow from the Niger and the increased runoff from the numerous left bank Sahelian tributaries generated an overall reduction of 35 per cent over 1970–90 compared to 1950–70. Local flood dynamics also shifted, and in 1984 and seven occasions thereafter, the earlier flood produced by the Sahelian tributaries (100,000 km^2) even exceeded the amplitude of the flood produced from the upstream Niger river (300,000 km^2) catchment for the first time since records began in 1923 (Descroix et al. 2009). Such changes in runoff volume and in the timing and duration of floods have implications on the feasibility, planning and productivity of irrigation practices. The change in flows was so severe that modifications to hydrological design norms (for dams notably) were recommended (Mahé & Paturel 2009).

In the Niger Inner Delta, most of the inflow is provided by runoff constituted upstream in the Guinean Sudanian regions. Inundated surface areas, though hard to assess (Ogilvie et al. 2012, Zwarts et al. 2005), are estimated to have decreased to below 5,000 km^2 during drought years compared to up to 20,000 km^2 during wet years. The Inner Delta also comprises several lakes, including Lake Faguibine, which in 1983–4 remained dry as river levels remained below the threshold where these lakes become connected to the river (Sircoulon 1987).

Physical modifications to the Upper Niger River regime remain relatively modest compared to other large rivers and the nearby Volta and Senegal rivers, with notably few large scale dams upstream, except Selingué and Markala in Mali. Several dams were developed in the downstream Nigerian part of the river, but their influence on water availability remains limited, considering the significant rainfall and resources available. Dams foreseen on the upstream reaches of the Niger River, such as Fomi, will however have far greater consequences on the water availability downstream, notably on fisheries in the Inner Delta and on the *Office du Niger* irrigation scheme.

In the endoreic regions of the basin, where runoff does not flow out to sea, the increase in runoff also led to the increase in the number, size and depth of small ponds and to the raising of the water table (Favreau et al. 2009, Leduc et al. 2001), opening up and reinforcing new opportunities for farmers. In the exoreic regions of the Sahel, the greater proportion of rainfall converted to runoff reduces infiltration and lowers the water table (Descroix et al. 2009) notably in plutonic basement rocks of Western Niger and below endoreic Lake Chad. In Niger, this loss of groundwater resource led to the recent need for new surface storage solutions (Descroix et al. 2009).

Changes in temperature and increased potential evapotranspiration should have also reduced runoff, however in Sahelian tributaries the reduced water holding capacity and reduced biomass from land use changes may have conversely reduced actual evapotranspiration and increased runoff. The recent relative recovery of rains in the catchment led to a slight increase in runoff but as with rainfall levels these remain level with 1970s values. Conversely, the return of rainfall is also accompanied by a gradual recovery of vegetation, which may also reduce runoff coefficients in Sahelian regions.

AGRICULTURAL WATER MANAGEMENT TECHNIQUES AND PRACTICES

Agriculture relies on one hand on techniques and organizations that are designed to match water needs with available water resources. These are based on available knowledge and develop over the years, through inventions, borrowings, introductions and sometimes abandon. Practices (Blanc-Pamard & Milleville 1985) on the other hand represent the techniques adopted and employed by a particular society. New practices (or innovations) depend on local or introduced inventions as well as the effectiveness of techniques in meeting their targets considering available means, capacities and social standards. Innovations notably result in adjustments and repercussions on multiple levels. The evolution of agricultural water management techniques and practices from so called traditional early twentieth-century agriculture is the focus of this historical investigation and will show how in addition to continued adaptation of practices to a local context, these evolved during the later part of the twentieth century as a result of the extreme climatic conditions and the introduction of development aid. Analysis is based on detailed research work in Burkina Faso (Serpantié & Lamachère 1992, Serpantié & Milleville 1993, Douxchamps et al. 2014) as well as regional research and syntheses (Ogilvie et al. 2010, Venot & Krishnan 2011, Venot 2014).

The geographical diversity of practices

In the Niger River basin, the following range of practices, representative of those found at similar latitudes across West Africa, is observed from north to south (Figure 10.3). At one extremity, the oases of the south Sahara (annual rainfall under 300 mm) employ traditional irrigation techniques, using water pumped from wells by animals and social arrangements for farmers to each irrigate palm groves and market gardens alternately. Water pumps are nowadays sometimes mechanized and groundwater levels recharged from infiltration by dams situated along the *wadis*.

The semi-arid Sahel (annual rainfall between 300 mm and 750 mm) is divided between agropastoral land dependent on rainwater and areas irrigated from the Niger river. Farming traditionally focuses on rainfed millet (*Pennisetum glaucum*), the only tropical cereal able to withstand a short and intermittent rainy season. Faced with recurring droughts, the small lowlands (bas fonds) formerly reserved for grazing and gathering wild fonio (*Panicum leatum*), have gradually been turned over to cultivating sorghum. Ever since the colonial period, the *Office du Niger*, currently supported by private investment, has been planting Mali's *Delta mort* (an ancient branch of the Niger River) with irrigated cotton and rice crops under full control irrigation. Supporting two crop cycles, this area produces increasingly large harvests. Traditional agriculture in the Niger Inner Delta and other well-watered areas (the Nigerien dallols, the Nigerian *fadamas*

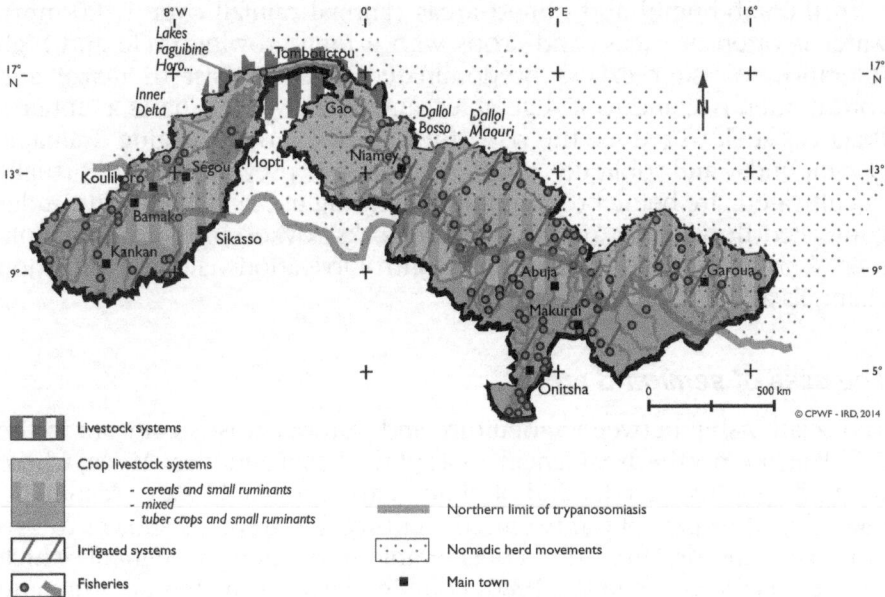

Figure 10.3. Agricultural systems (modified from Clanet and Ogilvie, 2014).

and the Sudano-Sahelian lowlands) developed several farming techniques on clayey soils, where intermittent floods turn water deficits into water excess. Traditional techniques include free flooding, a form of submerged rice farming carried out in floodplains, using local rice varieties with indefinite growth periods (floating rice) that are harvested by dugout canoes. The so-called recession flooding takes advantage of residual soil humidity and its capillary action as the flood recedes to grow rice and sorghum on the edges of rivers and lakes. Since the 1970s, dykes and canals have been built locally to improve water management, as in the Opérations Riz at Mopti and Ségou, to improve control of the submerged areas.

In the northern Sudanian areas (annual rainfall between 750 mm and 1,200 mm), rainfed sorghum combined with black-eyed pea cultivation dominate crop cycles, since sorghum is less sensitive to rainfall variations than maize. Maize yields indeed bear a positive relationship to rainfall, confirming that water remains a limiting factor. The seaboard climate of the Gulf of Guinea with its two rainy seasons supports similar crops though maize yields are more reliable. In the low-lying flood plains the traditional crop is African rice (*O. glaberrima*), combined with small market gardening (vegetable crops) during the dry season. Since the 1980s several projects (supported by NGOs and governments) developed small irrigated perimeters in these areas for rice growing and dry season market gardening. Likewise in Nigeria, the *fadamas* floodplains were intensely cultivated through the development of wells equipped with motorized pumps.

In the sub-humid and humid areas (annual rainfall over 1,200 mm), water is often in excess and crops with a long growing cycle and high tolerance to water excess predominate. These consist of maize and cotton, then rice and root vegetables, and in the wettest parts, plantains. Practices seek to reduce the risk of water excess by managing drainage, via mounds and oblique ridges. Lowlands, previously left virtually uncultivated, are being brought into increasing use as paddy-fields under controlled flooding by means of earthworks (dykes and small embankments) or more rarely, through full control irrigation via dams, irrigation channels and drainage works.

The case of semi-arid areas

The relationship between agriculture and water is most significant in the Sahelian and northern Sudanian areas of the basin and West Africa, where the large populations depend on limited and unreliable water resources. Over the latter part of the twentieth century this area was subject to two dominant trends: first the considerable reduction in rainfall which reached its paroxysm in the 1980s combined with rising temperature and increased climatic variability, and secondly the growing rural population resulting from high fertility rates, reduced mortality and reduced migration notably to urban centres. These affected the balance between land and water demand from the population and available resources. The increased pressure on resources led on one hand to a Malthusian dynamic of rural migration, land degradation and social tensions, and on the other hand to a creative positive pressure of innovation and intensification as suggested by the Boserup theory and facilitated by national policies and foreign aid. Both types of processes were observed simultaneously and alternately.

In these areas of low annual rainfall and with a short rainy season, traditional agriculture and community organization developed over generations to make the best of available water resources. The production system is based on manual rainfed diversified cereal agriculture, often combined with fruit and fodder trees and shrubs, extensive animal husbandry, and storage of grain and cattle. Storage reduces vulnerability to the interannual climate and biological hazards. Farming is undertaken in a family nuclear group or over several generations. It is designed to be self-sufficient, any surplus being stored or sold off. The farmer adapts his plots of land to the available water through adequate crop selection (timing of cycles, photoperiodicity, drought-resistance, rooting depth) and the cultivation system (sowing on first rains without tillage, sowing repeated in case of failure, low density, little fertilization, weeding and hoeing to reduce maximal evapotranspiration, fallowing). The water retained in the sub-soil during the dry season is also put to good use through catch crop cultivation (sorghum on clay soils), tree plantations and parklands (especially *Vitellaria p.*, *Parkia b.*, *Faidherbia a.* and other shrubs). Finally, reacting to the temporal and spatial variability of the climate combined with a diversity of livelihood strategies (gathering, extensive and intensive farming, semi-nomadic agropastoralism, tree harvesting) enable farmers to manage their water needs and exploit the diversity of available land (sloping ground, lowlands, rangelands). The flexible land rights system managed between descendants supports this mobility. Though traditional strategies are essentially based on adapting demand to the available water, some techniques also manage the water supply itself. They rely on favouring rainwater infiltration (*zaï*: funnel-shaped hoeing with manure to break up the soil, localized mulching, weeding and hoeing, building small stone lines and walls on steep slopes) and manual watering of small market gardens.

Climate variations and development projects represented great drivers of change

During the 1950s and 1960s when rainfall was abundant, water was not considered a priority. Development project leaning on bilateral cooperation focussed on cash crops and increasing yields and labour productivity through the introduction of draught power. Punctual projects introduced soil conservation works, notably GERES (Groupement

Européen de Restauration des Eaux et des Sols) [the European Group for Water and Soil Restoration] in Burkina Faso, but these were often poorly adopted by the local population, due to both the fragility of these earthworks and social reasons, namely, the lack of participation and appropriation by the local population as well as land rights issues.

Following the major droughts of the 1970s and the initial emergency measures, development programmes tended to favour improvements to rainfed subsistence farming and better access to water for livestock. Pastoral wells were drilled, populations were transferred to 'new lands' free of endemic diseases and projects to build small reservoirs were introduced. The proposed activities did not however always coincide with people's priorities or with the local organization of society. Those colonising the new lands rejected the preconceived strategies and the confinement to allocated plots of land, preferring to reproduce their original practices.

When the extreme droughts occurred in the 1980s, the failure of previous projects led to criticism over the techniques introduced and the modus operandi of initiatives which applied preconceived models and lacked partnerships. National governments became marginalized by foreign donors in their development operations and the NGOs proliferated as new substitutes for the state, acting as middlemen between donors and the local population. Traditional practices were seen as the required starting point and community participation was encouraged through local associations, with external financial and scientific support. The livelihoods approach was designed to address jointly the ecological, social and economic aspects through multi-disciplinary teams consisting of farming systems research, NGOs and the local people. Local techniques were recognized for their value, including zaï, stone lines and localised mulches. Intermediate techniques better suited to the means of local people and of civil society organizations emerged from these partnerships. These were designed to increase labour productivity, and relied on drought tillage or small-scale irrigation (rice and market gardens) with motorized pumps notably around the banks of small dams. Contour stone lines replaced the fragile soil ridges that had previously been used to reduce runoff and erosion. The dissemination of rediscovered traditional techniques, small mechanization and small dams was facilitated by numerous forms of financial support. At the same time, major conventional dams and great irrigation projects were launched to meet the needs of urban centres (water, electricity, fruit and vegetables, rice and even wheat).

Around 1990, bilateral projects gave way to international development programmes which resulted in a large scale dissemination of these intermediate technologies. Small reservoirs in Burkina Faso notably rose to more than 1,700 leading to an increase in market gardens (typically privately owned, less than 1 ha, watered by watering cans or motorized

Figure 10.4.   Watering onions in Burkina Faso (© J. Lemoalle, IRD).

pumps and supplying urban centres during the dry season with rice and vegetables) but also exacerbated conflicts with herders around watering points. International donors introduced decentralization and environmental policies, where sustainable resource management approaches were designed to optimize coordination between land users and optimize biophysical techniques. These led to proposals of new techniques and concepts (improved stone lines, improved *zaï*, mechanized half-moons, water use efficiency, conservation agriculture), but once again, these initiatives lost sight of development concerns, including participation, livelihoods, viability and the regulation of tensions.

Consequently, from the year 2000, there was a return to integrated land management, using the concept of *land husbandry* focussing at the scale of landscapes and communities. This involved integrating technical, commercial and institutional strategies, while addressing sensitive issues, such as land rights, the role of women, corruption and the hoarding of public investments by the elite, as well as conservation. The new arrangements for negotiating water allocation (IWRM, water user associations) put forward by projects represented a form of 'social engineering' and led to extremely diverse arrangements in practice. The involvement of multiple stakeholders led to the risk of fragmentation while the legal pluralism created by the overlap of traditional, state and project governance created confusion and conflicts (Ogilvie et al. 2010). In recent years, projects to increase water productivity ('more crop per drop') and reduce blue water losses have been instigated. Ponds for instance are used to provide supplementary irrigation to maize crops, drip irrigation is encouraged through social enterprises, while large dams for energy and irrigation are again being promoted (Barbier et al. 2009). More widely, the private sector is increasingly proposing access to technological innovations and financial facilities, tuning in to the considerable market and potential profits.

THE HISTORY OF WATERING LIVESTOCK

In the Niger Basin nomadic grazing systems are faced with two constraints: water and fodder, whose availability divides the region into two parts. In the south, where water and vegetation are abundant, small village herds of livestock have adapted to the often excessively damp climate and the predations of the tsetse fly. The animals are mainly dwarf guinea goats and shorthorn cows. In the Equatorial secondary forests and Guinean secondary grasslands (Figure 10.2), the small local herds have access to resources on the outskirts of villages without needing to be moved over long distances.

The second area covers the Sudanian savannahs (shrubland and bushland), Sahelian semi-desert grassland and the hydrologically inactive desert fringes. The rainy season maintains grasslands and creates temporary ponds, as far as the extreme north of the area. The herds graze the rangelands using the ponds as watering-holes, while in the dry season they are restricted to grazing land close to wells. Pastoralist herders must then structure the daily displacement of livestock to ensure that the benefits derived from exploiting a resource (water or grazing) are not counteracted by the physical efforts required by the herds to access it. Herders have therefore learnt to adapt to the seasonal and spatial variations of watering holes and pastures.

In order to focus on livestock-rearing in the Sahelo-Sudanian arid areas, this paper has not taken into account the small herds reared at the oases, nor those in the cotton-growing areas, where pigs, sheep and goats are bred and share the watering facilities designed to supply the villagers. In any case, wealthy landowners pay *Fulbe* herders to look after their stock. The introduction of pastoral hydraulic policies in the early twentieth century as a result of colonization disrupted the traditional pastoral systems. Veterinary services also opened up new land for herds by installing new facilities. However, since independence, signs of overgrazing and the difficulties due to emerging and changing social interactions have called pastoral hydraulic policies into question.

Traditional water use in pastoral areas, 1895–1945

At the start of colonization, in 1900, the agrarian society of the Niger Basin consisted of crop-livestock systems run by village communities who grew cereals or root-crops while managing small herds, and of nomadic pastoral systems migrating north in the rainy season away from the crop-growing regions (Figure 10.3) before returning to take advantage of the rains in the south over nine months of the year. Excluding the population living along the river banks, the year in the hinterland was divided into two distinct seasons. The wet rainy season when temporary ponds abound and it is unnecessary to dig wells as water stagnates in the smallest impluvia. In the dry season when ponds have dried up, only wells can access the water from the aquifers. The back-breaking pumping of water used to be performed by slaves, manually or with the help of pack animals, and could last for more than an hour to water a herd. All societies of the Niger Basin used to employ slaves (Bernus 1989, Hiskett 1973, Stilwell 2000) as domestic servants or to water their herds and flocks. By ending slavery, colonization *de facto* removed part of the breeders' workforce.

Household water was also drawn up from the well, another chore performed by women, who then transported water balanced on their head in clay jars or goatskin water bags, or with donkeys when wells were more than 1 km away. The traditional wells were narrow, sized so that the low-caste well-diggers employed by the dominant tribes could fit into them. Less than a metre wide in diameter, the wells were lined with logs roughly joined with plugs of rot-proof grasses. Depending on their topographical location, they could last for two to three years before collapsing and were rarely more than 40 m deep. Built using picks and shovels to remove the rubble, they could not be dug in unstable soils (pebbles, sandy ridges of the Koutouss in Niger) or on hard ground as in Tahoua, in the Azawad Region of northern Mali.

The traditional pastoral systems of the Niger Basin have exploited surface and ground water resources in five different ways. The pastoral Sahelo-Saharan Tuaregs use straw stacks and wells (Gallais 1975); to the south, the *Fulbe* follow the ITCZ to remain permanently within reach of ponds and new vegetation (Stenning 1994). Conversely, *Songhaï* and *Fulbe* livestock breeders are chased in the opposite direction by the floods of the Niger Inner Delta (Gallais 1975). Other herders execute small migrations to ponds and hollows between the dunes in the eastern part of the Basin or in the Soum (Barral 1974). Finally, the Tuareg camel-herders drive their herds long distances between the lowlands in the rainy season and the higher ground of the Aïr during the dry season (Bernus 1989), watering them at the *gueltas* (depressions fed by rainwater or karstic springs). Along the river reaches, ponds and lakes, very strict customary rights governed the seasonal use of water, as well as fishing

Figure 10.5. Fulani herders in the Inner Delta (© O. Barrière, IRD).

and access to fodder notably during the flood in the Niger Inner Delta (Gallais 1975). Tenacious, these customary rights continue to be upheld today throughout the Basin.

Colonial pastoral hydraulic policies, 1946–69

The colonial conquests only truly ended with the end of the the Kaocen Tuareg revolt in 1917, and in the south, when the British replaced the sultan of the Caliphate of Sokoto in 1903. Between the two world wars, colonial powers focussed on taking inventories and censuses, controlling trade and introducing cash crops (cotton and groundnuts) in the southern parts of the region. It was only after the 1945 armistice that the administration began to take an interest in the pastoral scene (Merlin 1951, Receveur 1960). This had previously been neglected in African agricultural policies and the various attempts to sedentarize nomadic breeders had failed.

Starting in 1948, pastoral hydraulic policies relied on the combination of various amenities (ponds, wells and mechanically pumped boreholes) to provide water throughout the dry months. Though it is relatively easy to calculate the watering capacity of wells and the required distance between successive wells, the livestock capacity that rangelands can support vary widely, leading to discrepancies in how policies were implemented and their success (Bernus 1991, Thébaud 2002). Furthermore, though an annual rainfall deficit may have only a moderate effect on the well's pumping rates, it can reduce by more than 50 per cent the number of cattle that can be fed from surrounding grasslands.

The waterworks built in Mali and Niger starting in the 1950s (Merlin 1951, Receveur 1960) notably consisted of deepened ponds with a minimum capacity of 6,000 m^3, 40 m deep wells located between 5 km and 10 km apart that could provide water for between 3,000 and 5,000 LU (livestock units)[1] and boreholes 20 km apart. All these were managed by herdsmen registered at the prefectures while the mechanized equipment was maintained by specialists. Boreholes drillings were opened between February and June according to a calendar fixed by the authorities, after consultation with local technical departments. The rules further provided that during this period a 20 km perimeter surrounding water works could not be exploited. A wider 40 km perimeter was reserved for grazing needs of residents only. The authorities were supposed to implement this legislation by decree, accounting for local conditions (Bernus 1991).

By independence, the programme had become a success and rangelands were covered with waterworks designed for watering cattle. In the northern Sahel (Tahoua, Agadez, etc.) the network was very dense and rangelands, had extended by two, and even three times their original size. Only a few limited sectors remained 'dry', namely the complex hydrogeological regions in north-eastern Sikasso (Mali) and Yaga and Seno (Burkina Faso). Unfortunately after independence, the budgetary constraints experienced by the budding states, their political instability and endemic corruption soon raised the issue of financing and maintaining these facilities. These issues were soon overshadowed by the drought starting in 1968 which 'benefitted' from unprecedented media coverage, displaying images of families dying of thirst and expanses covered with desiccated animal carcasses, while satellite images revealed circular patches of desertified land surrounding boreholes. These succeeded in moving public opinion and mobilizing donors.

Recent concerns over livestock hydraulic policies

The terrible droughts that hit the Sahel between 1969 and 2005 caused major environmental damage and decimated the herds. In 1973, Mali lost 50 per cent of its cattle and Niger 25 per cent (Maïga 1997). These drastic reductions in the size of herds ended the opulence experienced since 1950 when herd numbers grew constantly as a result of abundant rainfall and the pastoral waterworks programmes.

Worldwide concern over the repeated climatic disasters generated an influx of funds, mobilized international aid and led to various studies on climatic variations and the resilience of the pastoral and crop-livestock systems (developing notably the *Range Management* approach).

This attention shed light on the problems of financing waterworks, revealing financial mismanagement and implementation problems. Multiplied efforts to provide watering points for livestock were beneficial and were not the ecological disaster that they are sometimes claimed to be. However, the arrangements to protect the rangelands were poorly respected. Bernus (1991) found that certain boreholes supported herds three to four times greater than intended when nearby boreholes became no longer operational. On several wells, the influx of herds from other areas deprived them of their rangelands. When the situation became critical, several groups of herdsmen requested, and obtained, for their boreholes to be shut, preferring to pursue with their traditional manually dug wells. According to recent studies, sustainable governance and management of pastoral waterworks hinges on two aspects, namely allowing the local people to take ownership of the equipment provided and helping them correctly assess the availability of their resources.

FISHERIES AND AQUATIC ECOSYSTEMS

Inland fishing is widespread in Africa, since it supplies almost half the proteins found in local diets (Lemoalle & De Condappa 2009). Fish are caught using simple traditional techniques and with the exception of fishers in coastal waters, fishing is undertaken mostly by rural farmers drawn to this activity out of necessity or by dedicated fishing communities such as the *Bozo*. Together they make good use of the natural aquatic biotopes of the Niger Basin. Though fishing practices barely changed during the colonial era, the artificial biotopes created by the large dams built in the 1960s altered the way in which the aquatic ecosystems were used, attracting new fishers and promoting fish-farming.

Continued influence of traditional fishing

The Niger Basin is part of the ichthyological Nilo-Sudanese province extending from the Atlantic coast to the Indian Ocean and whose hydrographic basins have a large number of fish species in common (Lévêque & Paugy 1999). About 243 species of fish have been recorded in the Niger Basin. This diversity allows for extensive occupation of the various biotopes, and the richness of each area has always been protected by strict usage and custom. Tribal chiefs control land and water resources and regulate the seasonal fishing, which may only be practised under certain conditions and for limited periods.

In the upstream catchments, beside natural lakes and along rivers, fishing continues to be practised using the age-old techniques of lines, nets or very simple creels. On larger expanses of water, part-time fishermen, mostly rural farmers or herders for whom it's a supplementary or complementary activity, use small rafts or dugout canoes. Their catch barely exceeds a few dozen kilograms per year and serves for personal consumption, though in rare instances may be dried or even smoked in order to be sold.

In the river-floodplain aquatic ecosystem (Morand et al. 2012, Lemoalle & De Condappa 2009), created by the vast Niger Inner Delta, fishing is practised under very different conditions. There, the annual renewal of fish stocks depends largely on the extent of the annual flood and consequently on rainfall levels in the upstream catchment. The resulting quantity of fish caught is then heavily related to the amplitude of the flood which can vary by a factor of three over recent years. In these amphibious areas, fishing is still practised by ethnic groups traditionally recognized as professional fishers such as the *Bozo*, or ethnic sub-groups of the *Malinke* and *Songhaï,* for whom it's also a full time livelihood strategy. The Niger Inner Delta and its surroundings alone account for 62,000 full-time fishers out of the 100,000 fishermen officially identified within the basin (Lemoalle & De Condappa 2009). Although these fishers still use ancient fishing techniques and the handmade narrow boats that are characteristic of these inland waterways, their canoes are now motorized and their lines and nets are made of synthetic fibre. Drying, smoking and conditioning the fish catch for its transport remain largely traditional processes, although favoured species are now quickly transported to collection points where refrigerated trucks await them.

The colonial era and the large reservoirs

European colonization only indirectly affected fishing in the Niger Basin. Policies designed to regulate it mainly concerned fishing methods, rarely mentioning seasonal restrictions for juveniles. Laws were poorly implemented and the catch marketed through traditional outlets was rarely consumed by expatriates. During the first half of the twentieth century, most of the fishing resources remained under the control of local hierarchies.

The main changes to fishing sector were the result of several factors. These included the import of industrially manufactured hooks and the introduction of lines made at first from cotton then subsequently of synthetic fibres which in the 1950s started to replace lines made locally from plant fibres. These were followed by the import of outboard motors. Fish trading however was an ancient tradition in the Basin, especially in the Niger Inner Delta, unlike other parts of Africa. In the 1960s and 1970s, many African countries began building large reservoirs to develop irrigation and hydroelectricity. The upstream part of the Basin

Figure 10.6. Fishers transporting fish traps on the Niger River (© J. Lemoalle, IRD).

contains the sizeable Sélingué dam (1980; 410 km²) on the Sankarani River and the Markala dam further downstream, which diverts water to the large *Office du Niger* irrigation scheme. These impoundments produced a significant change in the river's hydrology and notably on the flood in the Niger Inner Delta, considering that a decreased inflow of 1m³/s is estimated to reduce fish catch the following year by nearly 28 tonnes. The numerous hydraulic structures built downstream, consisting of artificial lakes holding 28 billion m³ of water have also standardized the river's flow and that of its main tributaries.

These modern artificial lakes encouraged fishing practices since they can be more productive than other biotopes, even though the diversity of species caught may be lower. Traditional fishermen abandoned their sectors following droughts to settle along the lake shores, where roads also allowed them to export their catch. Other climatic refugees turned to fishing to supplement their income and gradually settled there as full time fishermen. All of these people (about 13,000 in total) use dugout canoes paddled by two to three people. Their equipment is very varied but dominated by traps, creels, nets and lines. Seine fishing carried out from the shore (beach seine) or in the water is thought to be responsible for destroying fish resources. In recent years, tilapia fish farming has developed in the Basin in small local or medium-sized enterprises, mainly in Nigeria. Fishing the restocked waters of small lakes and ponds has so far not proven to be economically efficient.

CONCLUSIONS

Agricultural livelihood strategies in rural areas of the Niger Basin have evolved following a long tradition of adapting to the often hostile environmental constraints. Precolonial empires, such as the *Songhaï* in today's Mali or the *Kanem-Bornu* in Chad, notably thrived over several centuries in the Sahelo-Saharan region of northern Africa, capable of adapting their techniques and practices to a variable and often scarce water supply. The second part of the twentieth century was marked by an extreme drought, remarkable in its amplitude, duration and geographical spread. Witnessed from Mauritania to Ethiopia, it was most devastating in the Sahel where rainfall deficits between 1969 to the early 2000s yielded large scale famines notably in 1972–4 and 1983–4. The prolonged drought had significant consequences on the dominant agropastoralism livelihood strategy, affecting rainfed agriculture and natural rangelands. The reduced rainfall also generated long term effects on river flows, groundwater levels, and the flooding of water bodies such as the Niger Inner Delta, further affecting millions of blue water users. Though the region experienced remarkable droughts and famine over previous centuries, the unprecedented worldwide concern relayed by modern media and the associated development aid it generated led to wider structural changes. Traditional, successful strategies to adapt practices to water supply were not affected and changes focussed largely on improved water access, through the construction of dams and small reservoirs, as well as waterworks and designated rangelands to support nomadic livestock. The introduction of imported solutions unsuited to local resources and customs combined with poor implementation of policies at the local level led however to mixed results. Recent independence, recurrent sociopolitical crises and inadequate governance further compounded these difficulties.

Recent policies and programmes have evolved as a result of further environmental changes (continued drought, population increase), exogenous ideas (environmental awareness, research in social sciences), and feedback from earlier projects, leading to more community participation, integration of traditional methods and gender consider-ations. Despite local successes, land and water productivity must continually improve and adapt to the growing population. Additional cognitive and financial investments accompanied by political and institutional changes remain vital to provide 2,500 kcal/day/capita sustainably and face further climatic changes.

NOTES

1 An LU (or tropical livestock unit) is a 'correspondence coefficient' concept, for example 1 camel LU =1.0; cattle 0.7; sheep or goats 0.1 (and subject to small variations).

REFERENCES

Barbier, B., Yacouba, H., Maiga, A.H., Mahé, G., Paturel, J.E. 2009. 'Le retour des grands investissements hydrauliques en Afrique de l'Ouest. Les perspectives et les enjeux'. *Géocarrefour* 84(1–2): 31–41.
Barral, H. 1974. 'Mobilité et cloisonnement chez les éleveurs du nord de la Haute-Volta: les zones dites d'endodromie pastorale'. *Cahiers ORSTOM. Série Sciences Humaines* 11(2): 127–35.
Bernus, E. 1989. 'L'eau du désert'. *Etudes Rurales* 115–6: 93–114.
———— 1991. 'Hydraulique pastorale et gestion des parcours'. In Grouzis, M., Le Floch, E., Bille, J.-C., Cornet, A (eds) *L'aridité, une contrainte au développement, caractérisation, réponses biologiques, stratégies des sociétés.* Paris: Orstom.
Blanc-Pamard, C., Milleville, P. 1985. 'Pratiques paysannes, perception du milieu et système agraire'. In *A travers champs agronomes et géographes.* Paris: Orstom, pp. 101–38.
Cappelaere, B., Descroix, L., Lebel, T., Quantin, G. 2009. 'The AMMA-CATCH experiment in the cultivated Sahelian area of south-west Niger – Investigating water cycle response to a fluctuating climate and changing environment'. *Journal of Hydrology* 375(1): 34–51.
Clanet, J.-C., Ogilvie, A. 2014. 'Water, agriculture and poverty in the Niger River Basin – Eau, agriculture et pauvreté dans le bassin du Niger', Challenge Program on Water and Food, Colombo and IRD, France, Agropolis International Editions.
Descroix, L., Mahé, G., Lebel, T., Favreau, G., Galle, S., Gautier, E., Olivry, J.-C., Albergel, J., Amogu, O., Cappelaere, B. 2009. 'Spatio-temporal variability of hydrological regimes around the boundaries between Sahelian and Sudanian areas of West Africa: A synthesis'. *Journal of Hydrology* 375(1): 90–102.
Douxchamps, S., Ayantunde, A., Barron, J. 2014. 'Taking stock of forty years of agricultural water management interventions in smallholder systems of Burkina Faso'. *Water resources and rural development* 3(1–13).
Eltahir, E.A., Gong, C. 1996. 'Dynamics of wet and dry years in West Africa'. *Journal of Climate* 9(5): 1030–42.
Falkenmark, M. 1995. 'Land-water linkages – A synopsis, in Land and Water integration and river basin management'. *FAO Land and Water Bulletin* 1: 15–16.
Favreau, G., Cappelaere, B., Massuel, S., Leblanc, M., Boucher M., Boulain, N., Leduc, C. 2009. 'Land clearing, climate variability, and water resources increase in semiarid southwest Niger: A review'. *Water Resources Research* 45(7).

Gasse, F. 2000. 'Hydrological changes in the African tropics since the Last Glacial Maximum'. *Quaternary Science Reviews* 19(1): 189–211.

Gupta, A. (ed.) 2008. *Large Rivers: Geomorphology and Management*. Chichester: John Wiley & Sons.

Hulme, M., Doherty, R., Ngara, T., New, M., Lister, D. 2001. 'African climate change: 1900–2100'. *Climate research* 17(2): 145–68.

Lebel, T., Ali, A. 2009. 'Recent trends in the Central and Western Sahel rainfall regime (1990–2007)'. *Journal of Hydrology* 375(1): 52–64.

Lebel, T., Cappelaere, B., Galle, S., Seguis, L. 2009. 'AMMA-CATCH studies in the Sahelian region of West-Africa: an overview'. *Journal of Hydrology* 375(1): 3–13.

Leduc, C., Favreau, G., Schroeter, P., 2001. 'Long-term rise in a Sahelian water-table: the Continental Terminal in south-west Niger'. *Journal of hydrology* 243(1): 43–54.

Lemoalle, J., Condappa, D. de, 2009. *Water Atlas of the Volta Basin – Atlas de l'eau dans le bassin de la Volta*, Colombo: Challenge Program on Water and Marseille: Food and Institut de Recherche pour le Développement and also online at http://hal.ird.fr/ird-00505116/fr/ (accessed 14 April 2015).

Lévêque, C., Paugy, D., 1999. *Les poissons des eaux continentales africaines: diversité, écologie et utilisation par l'homme.* Paris: Orstom.

Mahé, G., Paturel, J.E. 2009. '1896–2006 Sahelian annual rainfall variability and runoff increase of Sahelian Rivers'. *Comptes Rendus Geoscience* 341(7): 538–46.

Morand, P., Kodio, A., Andrew, N., Sinaba, F., Lemoalle, J., Béné, C. 2012. 'Vulnerability and adaptation of African rural populations to hydro-climate change: experience from fishing communities in the Inner Niger Delta (Mali)'. *Climatic Change* 115: 463–83.

Ogilvie, A., Mahé, G., Ward, J., Clanet, J.C. 2010. 'Water, agriculture and poverty in the Niger River basin'. *Water International* 35(5): 594–622.

Ogilvie, A., Belaud, G., Delenne, C., Bailly, J.S., Bader, J.C., Oleksiak, A., Ferry, L., Martin, D. 2015. 'Decadal monitoring of the Niger Inner Delta flood dynamics using MODIS optical data'. *Journal of Hydrology* 523(1): 368–83.

Olivry, J.C. 2002. *Synthèse des connaissances hydrologiques et potentiel en ressources en eau du fleuve Niger.* Washington D.C.: World Bank and Niamey: Niger Basin Authority.

Serpantié, G., Lamachère, J.M. 1992. 'Contour stone bunds for water harvesting on cultivated land in the North Yatenga region of Burkina Faso'. In Hurni H., Tato K. (eds), *Erosion, conservation and small scale farming*. Berne: Berne University, pp. 459–70.

Serpantié, G., Milleville, P. 1993. 'Les systèmes de culture à base Mil et leur adaptation aux conditions sahéliennes'. In Hamon, s. (ed.), *Le Mil en Afrique*. Paris: Orstom, pp. 255–66.

Sircoulon, J. 1987. 'Variation des débits des cours d'eau et des niveaux des lacs en Afrique de l'Ouest depuis le début du 20ème siècle'. *The influence of climate change and climatic variability on the hydrologic regime and water resources* 168: 13–25.

Sultan, B., Janicot, S. 2003. 'The West African monsoon dynamics. Part II: The "preonset" and "onset" of the summer monsoon'. *Journal of climate* 16(21): 3407–27.

Venot, J.P. 2014. 'Rethinking commons management in Subsaharan West Africa: public authority and participation in the agricultural water sector'. *Water international* 39(4): 1–15.

Venot, J.P., Krishnan, J. 2011. 'Discursive framing: debates over small reservoirs in the Rural South'. *Water Alternatives* 4(3): 316–24.

Zwarts, L., van Beukering, P., Kone, B., Wymenga, E. 2005. *The Niger, a Lifeline. Effective Water Management in the Upper Niger Basin*. Lelystad: RIZA.

Part III Agro-Water Variability and Adaptation

11 The Lake, Bananas and Ritual Power in Buganda

Andrew Reid

INTRODUCTION

An unusual manifestation of the relationship between water and food is to be found in the north-western corner of the Victoria Nyanza in eastern Africa (Figure 11.1). Here, the presence of a huge body of water, prevailing convectional currents running north-west across this lake and a low lying irregular coastline covered by rainforest created conditions in which humans were able to flourish, using the banana as a staple crop. Not only did 'banana cultures' emerge, in which the banana played a central role in the cultural construction of the society, but also the banana was in many respects associated with ritual power that held sway until the introduction of global religions in the mid to late nineteenth century. In exploring this dominance of bananas, it will be necessary to consider the landscape, or, more correctly, lakescape in which this emerged, through several different strands of evidence relating to the introduction and establishment of the banana and its later role in nineteenth-century polities. These strands come from disparate linguistic, historical, ethnographic, archaeological and geographical sources. Whilst none of the core elements – lake, banana or ritual – are exceptional on their own, in this part of Great Lakes Africa they came together over a long period of time and in a quite unique manner.

THE VICTORIA-NYANZA ECOSYSTEM

At 68,800 km^2 the Victoria Nyanza is the world's second largest lake by area, containing roughly 2,750 km^3 of water (Yin & Nicholson 1998). It was formed by tectonic activity, raising the Nile outlet at Jinja and resulting in the flooding of the vast low lying area behind the barrier. The resultant lake, with a catchment of 184,000 km^2, is fed by a number of relatively small rivers. The largest feeder to the lake, the Kagera, draining from Burundi, Rwanda, north-eastern Tanzania and south-western Uganda provides only 33 per cent of the flow into the lake. Moreover, this total basin drainage into the lake is hugely outweighed by the contribution rainfall makes, at 80 per

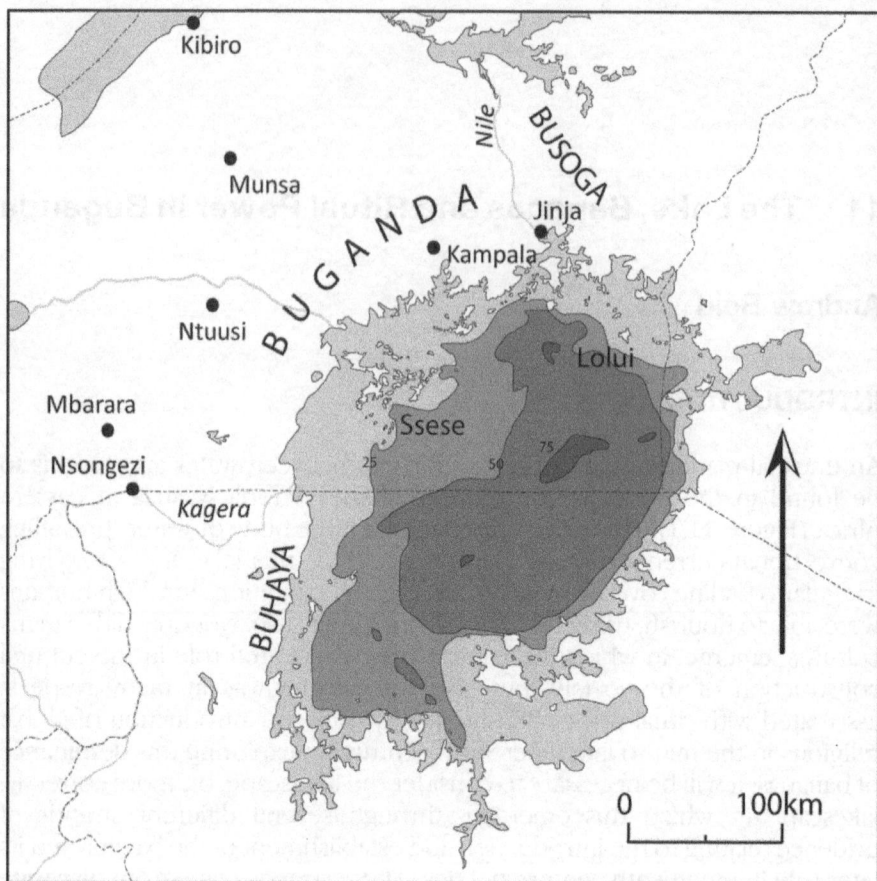

Figure 11.1.   The Victoria Nyanza in regional context (isobaths in metres) (© Andrew Reid).

cent of the lake's total water inflow. Situated on the equator, evaporation
from the lake is also severe.

Despite the vast extent of the lake it is only 84 m at its deepest point. This
means that the levels in the lake are very susceptible to fluctuations in water
supply. A one metre drop in the level of the lake was visible during
fieldwork seasons undertaken in the Ssese Islands in 2002 and 2003
and possibly related to the new HEP turbine at Jinja. More significant
fluctuations in the lake level have been noted in the more distant past.
Evidence for higher lake levels have long been recognized some way inland
(Temple 1964, Kendall 1969) and indeed there is an old lakeshore horizon
at Nsongezi, some 100 km or so up the Kagera from its entrance onto the
lake. In the last 18,000 years there have been at least three episodes of
higher-than-present levels, as evidenced by palaeoshores at 3 m, 12 m and
18 m above the current shoreline. This reflects both higher rainfall and

tectonic activity shifting the floor of the lake (Johnson et al. 1996, Stager et al. 2003). Conversely lower rainfall leads to rapid and widespread desiccation of the lake (Stager et al. 2002). There has been at least one, possibly two, complete desiccations in the last 18,000 years and a low water period as late as 1200–600 BP. Unfortunately, although there are important patterns in lake levels established for the last 1,500 years through correlation with the Rodah Nilometer in Egypt (Nicholson 1998), it is not yet possible to correlate these recent fluctuations with the human settlement of the lake. This reduction in water, combined with the very shallow nature of the lake, leads to the formation of a number of distinct and separate lakes and it is this ponding that has given rise to the exceptional diversity of fish species: during shrinking episodes fish populations were separated and evolved such that, on expansion and reestablishment of the single lake, distinct species had appeared. When Europeans encountered the lake in the mid to late nineteenth century there were more than 200 species recorded. Unfortunately, colonial fishing practices, particularly after the introduction of the Nile perch, and pollution have reduced this diversity by almost 70 per cent, with resultant well-documented damage to the lake due to the elimination of the niche roles performed by the lost species.

The large surface area of shallow water lying on the equator also has a marked impact on rainfall patterns through evaporation. Whereas annual rainfall in the south-eastern corner of the lake, fed only from prevailing weather systems coming up from the East African coast, may be as low as 800 mm, that in the north-western corner of the lake can be as much as 2,200 mm (Figure 11.2, Yin & Nicholson 1998). Furthermore, in this part of the lake there are annually as many as 255 thunderstorm days annually suggesting that the rainfall is relatively evenly distributed throughout the year. As a result, much of the shoreline of the north-west of the lake is, or has been, dominated by thick forest. The lowest lying ground, lacking impetus from small catchments and absence of elevation, are swamps, most often dominated by vast tracts of papyrus.

As will be seen below, this is not, however, simply a vast natural environment, but one which has demanded and drawn the attention of people over time and as a result it has been converted into a culturally constructed lakescape. This is readily reflected in the naming of the body of water for which there is no universal term. Kenya and Tanzania today, who between them control 55 per cent of the lake's waters, prefer the generic *kiswahili* term for lake, *Nyanza*, which is confusingly also applied to other major water bodies in the region. In Uganda, the colonial name – Lake Victoria – is preferred, not in celebration of the colonial past, but rather in rejection of attempts to rename many elements of Uganda by Idi Amin. Hence lacking a generally recognized name, the composite term Victoria Nyanza will be used here as a compromise. It is also worth recognizing that other names are used by the major languages to be found around the

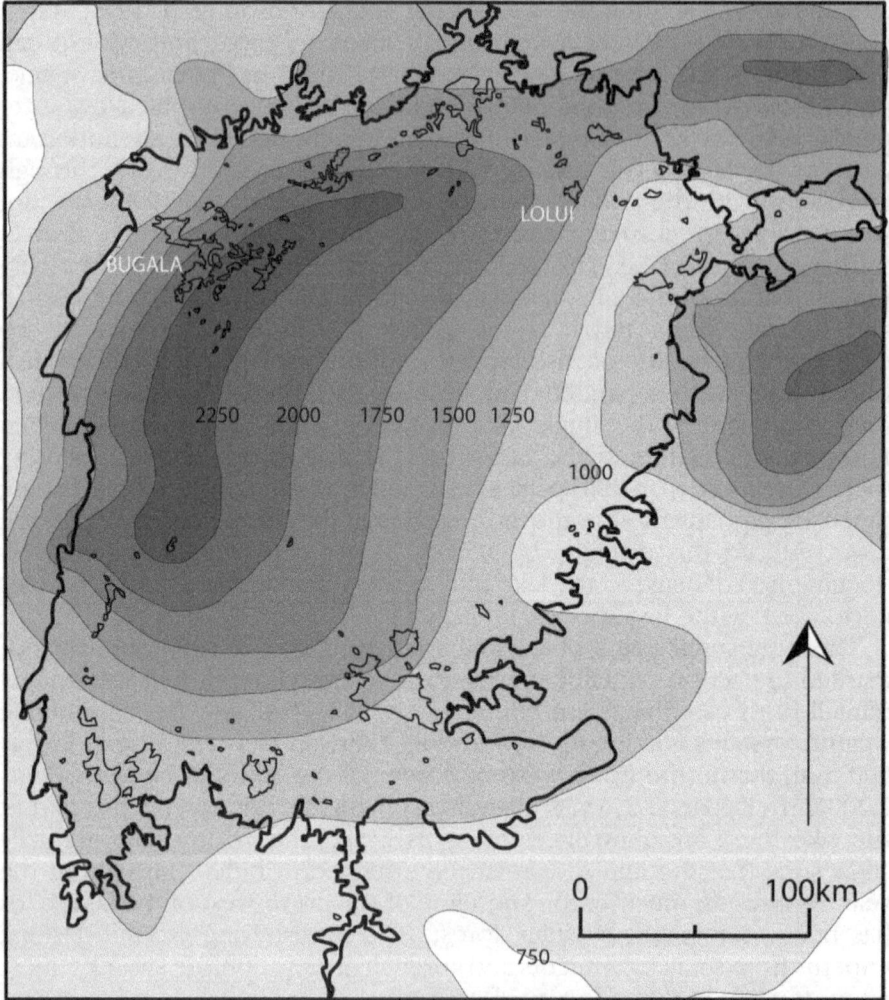

Figure 11.2. Rainfall map for the Victoria Nyanza (© Andrew Reid).

margins of the lake. Nam Lolwe is used by Luo-speakers in the north-eastern corner of the lake, who came to the lake relatively recently and which refers to a mythical giant of that name. In Buhaya, north-western Tanzania, which has a particularly straight featureless coastline, Lwelu, 'the great out there', is used to distinguish the contrast between the vast inland sea with its featureless horizon, and the green and encultured land.

The term used in Buganda, Nalubaale, which translates literally as 'the place or mother of lubaale' suggests a much more active engagement with the lake as will be seen in attempting to unpack the meaning of *lubaale*. The lakescape encountered in this part of the Victoria Nyanza is very different, being much more ragged, dominated by deep inlets and long peninsulas and also by frequent islands just off the mainland coast. These frequent inlets and islands mean that it is very difficult to gain a clear view of the lake and establish a sense of its true scale. There are also island chains extending further into the lake, most obviously the Ssese Islands, and also the Buvuma Islands further to the east. Movement on the mainland around the edge of the lake is difficult because of the frequent wide swamps and thick forest, not to mention the erratic nature of the coastline. In 1862, when Speke walked round the entire western side of the lake, from the southern end to the capital of Buganda, a distance of around 250 km, he only saw the lake on four occasions throughout his journey (Speke 1863). This restricted view provided Stanley with part of the excuse for one of his subsequent visits to the region, to circumnavigate the lake (Stanley 1878). As Stanley found out, movement across the lake, along the coast and from island to island would have been much easier than going by land. Hence, the lake would have been a crucial interface in this area for communication and interaction.

EARLY FARMING IN THE GREAT LAKES

Having introduced the first crucial component of this account, the lake, it is now necessary to consider the second, the banana, together with the broader context of cultivation. The earliest archaeological evidence for bananas in the Great Lakes region is no more than 200 years old, but comes from banana pseudostem impressions in iron slag, from sites in Kyagwe (Figure 11.3), demonstrating the breadth of uses that the plant, as well as the fruit, was used for (Reid & Young 2003, Iles 2013). There are possible remains of bananas in the form of banana phytoliths found in swamp cores at Munsa dating as far back as 4,000 BC (Lejju et al. 2006). As this predates evidence for habitation and cultivation in the area by more than 3,500 years it has to be considered problematic. Fortunately, that evidence is not crucial for this consideration of the banana and the lake. Linguistic reconstructions of history suggest that in early farming episodes in the Great Lakes region the banana was initially used as a marginal fruit, to be found on the perimeters of cultivation (Schoenbrun 1998: 79). This is quite distinct from the staple crop it became, not only dominating diets, but also providing a metaphor for the domestication and transformation of the entire cultural landscape. This process was achieved through the medium of the lake.

The early instances of cultivation, from the little evidence that exists, took place from around 2500 BP in the wet montane forests of the western Rift valley in Rwanda and Burundi, the wetter west and north margins of the Victoria Nyanza and the interconnecting river valleys such as the Kagera (Reid 1994/5). From limited palynological results and even more restricted macrobotanical remains as well as linguistic evidence, these early cultivators

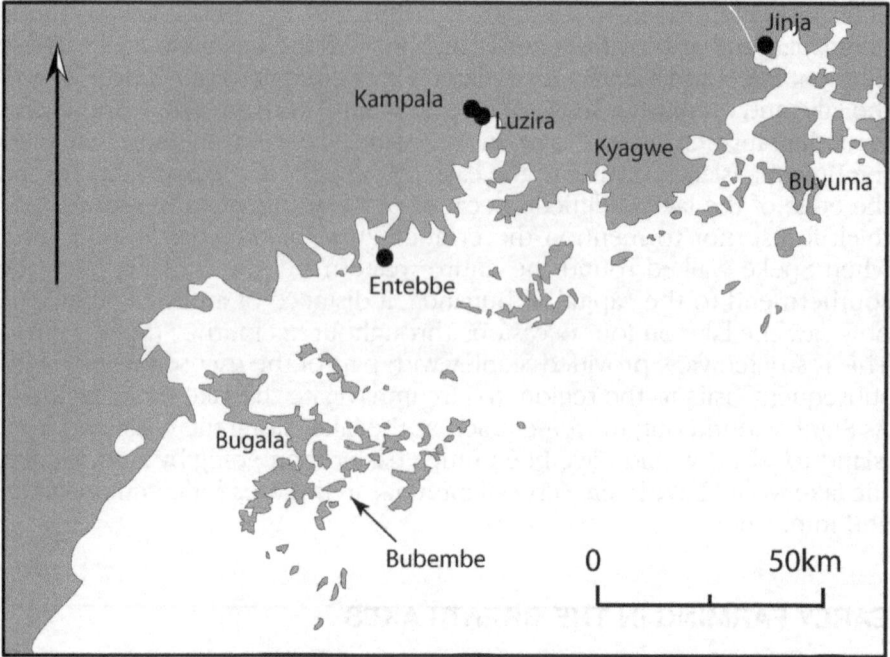

Figure 11.3. The north-western corner of the lake (© Andrew Reid).

were growing sorghum and finger millet and also yams in the wetter forests (Van Grunderbeek et al. 1983, Schoenbrun 1998, Giblin & Fuller 2011). The restriction to wetter locations for over 1,000 years suggests a reluctance to move into drier locations and/or a lack of need to move from conditions which were so reliable and favourable. These early efforts at farming are associated with communities using Urewe ceramics, a pottery tradition that involved considerable expenditure of effort and comparatively lengthy attempts at producing the finished item (Van Grunderbeek 1988, Ashley 2010). These ceramics also turn up in the very occasional burials that have been encountered and once under the base of an iron-smelting furnace, suggesting that there was a strong association between all ceramics and ritual purity. In recent archaeological surveys in Buganda, Urewe ceramics have been found around the coast but also throughout the interior, albeit sparsely distributed therein, indicating that early farmers had no particularly close association with the lake (Reid 2003). At the same time, Urewe ceramics have been found in the lake on Bugala Island and deeper into the lake on its north-eastern side on Lolui (Posnansky et al. 2005), making it clear that the people using these ceramics were not reluctant to move across open water.

The longevity of the cultural record for these early farming societies in particular locations suggests that they were relatively successful and stable (Reid 1994/5). The only real evidence for leadership comes from the analysis of language and the recognition at this time of shared words for leadership revolving around lineage, possibly fledgling clans and ritual authority (Schoenbrun 1998). The use of terms related to these institutions throughout this comparatively long time suggests that communities were broadly successful yet small in scale and limited in their ambition, leadership providing a moral guidance through the challenges that beset communities over time.

THE NEW LAKESCAPE

There are indications that these certainties that had lasted for so long may have been unravelling towards the end of the first-millennium AD. New terms and expressions were innovated in language. Environmental conditions may have been drier causing the breakdown of primary forest and the spread of grassland on the margins. Soils may have become poorer and flexibility of movement to new land increasingly restricted. It is debatable how much of an impact such conditions may have had on the inhabitants of the margins of the lake. The environment on the lake itself is still only evidenced by a single palynological core, from Pilkington Bay near the outlet at Jinja (Kendall 1969, Schoenbrun 1994, Stager et al. 2003), and given the extent of the lake and the catchment it is drawn from, it is of limited value, indicating merely broad patterns.

From around AD 800 there began to be some changes in material culture with ceramics being much less lavishly finished in their production and decoration being applied in simpler and less time consuming ways (Ashley 2010). These ceramics, of which there are several variants, clearly show continuity with the preceding Urewe tradition, yet are quite distinct. Within Buganda these ceramics are confined almost exclusively to the lake, both the coastal margins and the islands themselves. This indicates a cultural form existing for several hundred years that was lake-based, thriving on the opportunities provided by the relative ease of movement across the lake. The lake also provided means for communities to disperse and relocate, disrupting political authority and ties with specific landscapes. At Malanga-Lweru on Bugala Island four glass beads were recovered from a site dating to the eleventh century AD (Reid 2003). These demonstrate a participation in communication and exchange networks extending all the way east across the lake and down to the Indian Ocean coast and in the opposite direction into western Uganda at sites such as Ntuusi (Reid 1996) and later at Kibiro (Connah 1996) and Munsa (Robertshaw 1997).

The sites dotted around the lake shore are very limited in terms of the material culture present – overwhelmingly pottery – and also limited in the extent of sites, suggesting that habitation was short-term and dispersed and hence there is very little information on how these societies lived and organized themselves. There are grinding hollows frequently encountered which may be associated with grain processing, but far to the east on Lolui there are thousands of hollows which question if they were really simply for the purpose of grinding grain (Reid & Ashley 2014). Single rock surfaces on Lolui can be covered in more than 50 grinding hollows which overlap one another and are also sometimes placed in locations from which it would be impossible to effectively grind grain (Dismas Ongwen pers. comm. 2011).

Intriguingly, in association with these cultural contexts, linguistic evidence suggests that there is a radical change in political leadership taking place, with the longstanding concerns of social health and community at a village level being replaced by much more emphasis on personal embodiments of political leadership with the well-being of communities being determined by the health of their leaders (Schoenbrun 1998). It is also curious that from around this time several figures formed in clay have been recovered (Figure 11.4). The Luzira material features a head formed from an inverted pot, with all features embossed onto the surface of the vessel (Ashley & Reid 2008, Reid & Ashley 2008). It also includes at least two torsos of sitting figures which may be in positions of supplication, reverence or authority. A final element from Entebbe is a decorated column which appears to combine both male and female genitalia at its base (Posnansky & Chaplin 1968). There is a pressing need to find such figurative pieces in some form of stratified archaeological setting in order to attempt to understand their meaning. Unfortunately, both these existing collections came from contexts disturbed during building work, long before the present significance was appreciated. There may be scope in the future for further discoveries to be made. It can be argued that these demonstrably human figurative pieces reflect a new concern with corporeality and the embodiment of power within the more personalized associations of leadership that were emerging. Although both of these find locations are on the mainland they were encountered less than a kilometre from the lake and are therefore very much part of the radically different lakescape that flourished towards the end of the first millennium AD. Whilst ceramics were much less formal and much less polished, the extent of experimentation that is evident in different design elements is also suggestive of societies that are breaking away from established practices and are exploring potential new forms of expression. Movement on the lake was a significant part of that expressed freedom.

Another important innovation also fuels this change. Historical reconstructions based on linguistics reveal that there is at this time an 'explosion' of innovative new terms in the region which relate to cattle and to bananas (Schoenbrun 1993, 1998). The latter corpus has a two part history and includes the terminology needed to understand and communicate the intensive cultivation, cropping and preparation of bananas. This range of words relating to the broad management of bananas

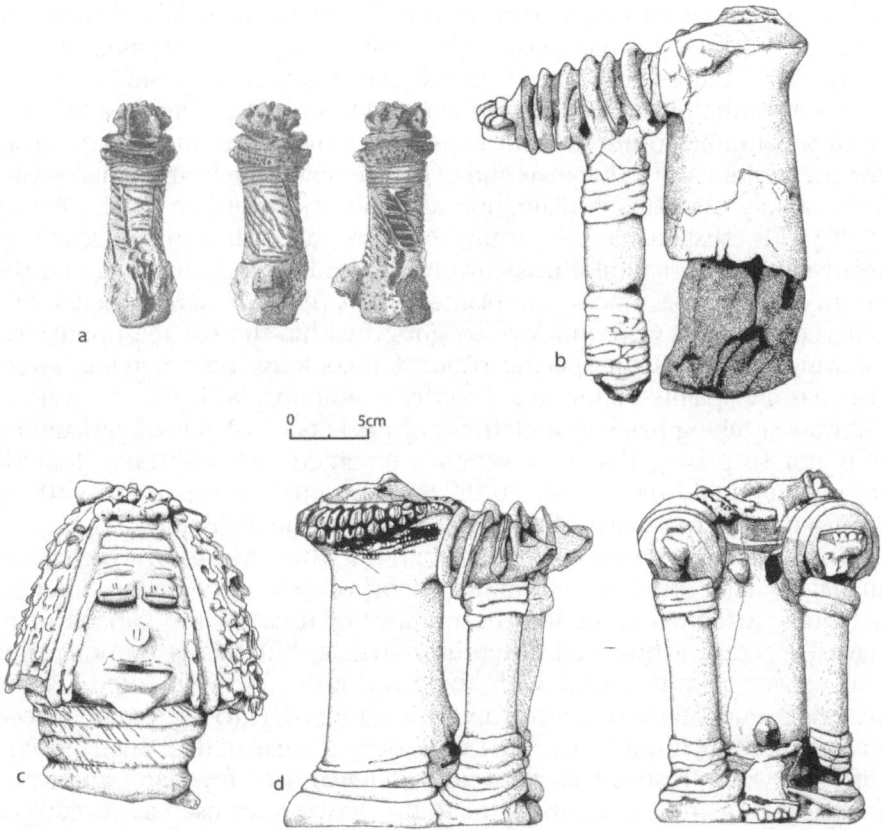

Figure 11.4. Figurative ceramics found by the lake: the Entebbe figure (a); and the Luzira figures (b–d) (© Andrew Reid).

indicates their initial introduction as a significant food crop, perhaps together with a range of cultivars which formed the core stock of the region and subsequently the development of banana plantation agriculture. The initial phase included much of the southern, western and northern shores of the lake from around AD 900. The second phase occurred several centuries later and was largely confined to Buganda and Busoga. The linguistic evidence indicates a very rapid spread of the initial new agricultural form around the wetter margins of the lake. Undoubtedly this spread would have relied on two crucial factors. First, the ability to uproot and transplant suckers growing out from the base of mature banana plants would have meant that new plantations would have been relatively easy to establish once the initial forest clearance had been undertaken. Secondly, these suckers would easily have been transported across the lake, involving very little manual porterage to establish new banana gardens. The islands themselves have varied fertility and a good proportion have extremely limited potential for crops. Throughout the Buganda mainland there are areas of good soils that encouraged cultivation. The Buganda area in particular, features year round rainfall and together with the good soils allowed plantations, if well looked after, to flourish needing no additional fertilizer or mulch other than the broken up plant stems, the peelings from the preparation of the bananas and of course manuring from human waste. Very rapidly this banana plantation agriculture established itself, offering high yields, sustainable year round produce, and, after initial clearance, relatively easy agricultural tasks involving weeding and mulching and the control of encroachment on plantation margins. It also appears that different qualities were quickly recognized within the banana population allowing the focus on specific plants for cooking, beer making, sweet bananas and plants better suited to dry conditions. With this intensity of cultivation taking place in a plethora of pockets or islands of agriculture, it is not surprising that new varieties emerged. Schoenbrun's Baganda informants could name over 60 different varieties of banana and 30–40 names for different parts of the plant (Schoenbrun 1998: 80).

The banana, its various plant parts and the whole means by which it was managed also became culturally incorporated into many different activities. Attempts to define the number of uses for the banana plant have exceeded a hundred different activities, but this is by no means exhaustive, as was found with the previously undocumented use of banana pseudostems to control and to act as barriers to the slag produced in iron smelting furnaces (Iles 2013). Besides the use of the banana fruit in different ways, fresh leaves were (and still are) used for plates and trays, for packaging and for steaming meat; dry leaves were used as curtains or to cover temporary shelters; fibre from the stem was used as an all-purpose string; sponges were made from the base of the stem; and stems were used to create a platform on which the body of a deceased person was lain. Perhaps more important than these myriad means through which bananas became an essential part of the life of the people who depended on them, is however the development of cultural associations relating to the plantation itself. The *lusuku*, the banana garden, was known for its coolness, referring to the repose, peace and shelter that it brought from the chaos of the outside world. For Baganda, the *lusuku* was the basic unit around which life was constructed. Settlements were dispersed with each home being built within its respective *lusuku*. People were born, grew up and were buried within their plantations. Houses were frequently rebuilt, shifting location within the garden, but the *lusuku* remained to sustain generation after generation. The *lusuku* became a point of return for those who had moved away, perhaps to create their own new gardens. The renowned coolness of the *lusuku* thus also relates to the ritual protection that ancestor spirits provided from the heat and danger of the outside world. These ties to particular places in the landscape were integral to the notion of clanship and lands to which larger groups of people were bound. This sense of possession was reaffirmed in ritual. During the funeral rites, the heir of a deceased person at the end of mourning would go into the *lusuku* and cut bananas for making the beer used in the final feast (Roscoe 1911: 122–3), thus confirming the heir's place amongst the clan, but also demonstrating the power of the *lusuku* in legitimising inheritance. In later years, as access to land became more difficult, the ancestors even became a useful ally:

> Chiefs had to be on the alert to prevent people from burying their dead in good gardens, because the gardens would thereby become freehold lands, and even the King did not like to turn out any family which had succeeded in burying three generations of its dead in the same place; he dreaded the anger of the ghosts (Roscoe 1911: 134).

Bananas, and the *lusuku* from whence they came, also became central to the construction of gender in Buganda. Typically, a young man seeking to establish himself would clear an area of forest and plant suckers. This land would then be given over to his wife or wives to cultivate. 'Girls were taught to cook and to cultivate as soon as they could hoe; to be a successful manager of the plantain grove and to be an expert cook were regarded as a woman's best accomplishments' (Roscoe 1911: 79). A woman's daily routine involved working in the *lusuku* until the 'second pipe of the day' (roughly 10 am), women being avid smokers using tobacco grown by themselves. This work in the *lusuku* involved pruning dead leaves and unwanted suckers, weeding and mulching. In this way the agricultural future was guaranteed for the lifetime of the man and woman. After the initial clearance of the land there was not much for the men to do and this abundance of time was profitably used in drinking banana beer, accompanying military campaigns and politicking.

Spread by the ease of movement across the lake and the availability of land close to the lake edge and also drawn by the relative productivity of the banana, widespread settlement of the islands and of the coastal mainland would have occurred. In the latter instance, the ragged nature of the coastline, interspersed with regular swamps will have encouraged isolated and disparate communities to develop on the gentle hilltops dotted around the lake. These will have gradually expanded to encompass the entire extent of suitable land in the particular locale, but broader political association will have been made difficult by the wide impenetrable swamps. By contrast, in the drier grasslands beyond the western shores of the lake there was a parallel development of cattle herding, with accompanying explosion of cattle-related vocabulary (Schoenbrun 1993) and archaeological evidence for the increase in animal size as well as evidence for large numbers of animals at and around Ntuusi (Reid 1996, 2013). On account of the more rapid accumulation possible with cattle and the more inclusive nature of the environment that allowed aggregation of communities, there appears to have been a much speedier move towards political centralization in these grasslands.

One of the great attractions of the banana is of course its high yields and comparative productivity (Tushemereirwe et al. 2001, Wrigley 1989). In terms of carbohydrate values, bananas are two or three times more productive than sorghum or finger millet and yields can be ten times that of yams. Finger millet in particular is a relatively labour intensive crop that has to be individually harvested. Yam cultivation encourages the pooling of water and thus the increase in likelihood of insect-borne diseases such as malaria. As bananas in Buganda are produced year round there is no need to store alternative food sources for the off season. On the other hand, bananas offer no protein and are deficient in vitamins, iron and calcium. Kwashiorkor and other protein deficiency illnesses will have been, and remain, a significant risk. The switch from yams and/or grain cultivation to bananas will have had a major impact on the nutritional status of human populations in the region. Those switching to bananas would have needed to augment their diet with other foods, principal amongst their needs being protein. In part this would have been addressed by the growing of beans and legumes (Wrigley 1989: 65). In the absence of significant numbers of livestock in these forested areas, fish, both fresh and dried, would have been very important and procurable as long as banana cultivation remained within easy access of the lake. Since many, but by no means all, of the islands had poor soils and limited potential for agriculture, the ability to trade fish for bananas would have been essential. As a result, 'there were many markets held along the shores of the lake, where the people from the islands brought fish and pots for sale or exchanged them for barkcloth and plantains' (Roscoe 1911: 456). As plantations spread further inland alternative sources of protein would have been needed.

Some Ganda traditions associate the initial development of the banana as a major crop with Kintu, others with the Ssese Islands. Kintu in Ganda tradition is the origin figure who is associated with creating most elements integral to the Ganda way of life. As bananas became so important, it is not surprising to find direct associations being drawn to Kintu. The attribution of the Ssese Islands may have deeper significance although not necessarily as the actual source of banana cultivation. The Ssese Islands may simply be a convenient means of associating bananas with the lake generally. This association with the Ssese Islands, accurate or not, may also be related to the emergence of ritual power in association with the islands.

RITUAL POWER AND THE LAKE

As noted earlier the Ganda name for the lake is Nalubaale which means the place of the gods. In this instance *lubaale* could possibly be interpreted as spirits or even ancestors. It is tempting to suggest that the use of the term implies a historical recognition of the early pioneers of the lakescape, who spread new ideas and the banana across its vast expanse. What is certain is that by the nineteenth century the Ssese Islands dominated the ritual world of Buganda. This ritual dominance had existed for some time and several of the shrines were recognized beyond the borders of Buganda. For instance the most important shrine to Mukasa was located on Bubembe (Figure 11.3). It was so powerful that the *kabaka* sent emissaries to the shrine rather than the priests operating from the capital. The powerful presence of Mugasha on the Ssese Islands is also recognized in Buhaya traditions (Schmidt 1978: 301). Indeed, the power of the Ganda *lubaale* was such that Schoenbrun believes Mukasa was originally a powerful, long-venerated spirit from western Uganda, who was brought to the Ssese Islands because of their emerging ritual power and converted into a Ganda spirit. Mukasa/Mugasha is generally associated with fertility and wealth and, in his lake setting, giving abundant rainfall, bananas and fish and ensuring safe passage on the lake. A frequently observed item at his shrines in Buganda was paddles. Schoenbrun (1998: 204–5) demonstrates how veneration of Mukasa drew on existing belief systems and the principal focus on fertility and social health. 'The creative power wielded in Mukasa's name by his medium and her priests, offered to commoners and royalty alike the possibility of overcoming the greatest obstacles to keeping the circle of life intact through children' (Schoenbrun 1998: 205). This was power wielded independently of the state, creating a potential support for, but also representing a significant challenge to, the power of the *kabaka*.

The ritual power of the Ssese Islands was a major feature of Buganda. Of the major Ganda shrines noted by Roscoe, 75 per cent were located on the Ssese Islands (Figure 11.5). This is quite remarkable, particularly when it is recognized that Buganda had only tenuous control over the islands. Bassese were recognized by many early commentators as a distinct population who were semi-autonomous from Buganda. The islands were not formally incorporated into the ssazas or counties of the mainland, although officials and Ganda clans did have established estates on the islands. The Ssese Islands also, together with Buvuma, provided the boatmen to operate Buganda's naval canoes which became increasingly important in the nineteenth century as Buganda sought to dominate trade across the lake (Reid 2002). These canoes did not really engage in naval battles, but rather transported land-based troops from the mainland from one place to another. Nevertheless, this represented a further uncertainty in the relationship between the mainland and the Ssese Islands. It is also worth noting that Buvuma and its neighbouring islands, although regularly contesting power with Buganda had no major shrines. Hence, ritual power was heavily dominated by the Ssese shrines, possibly in recognition of the

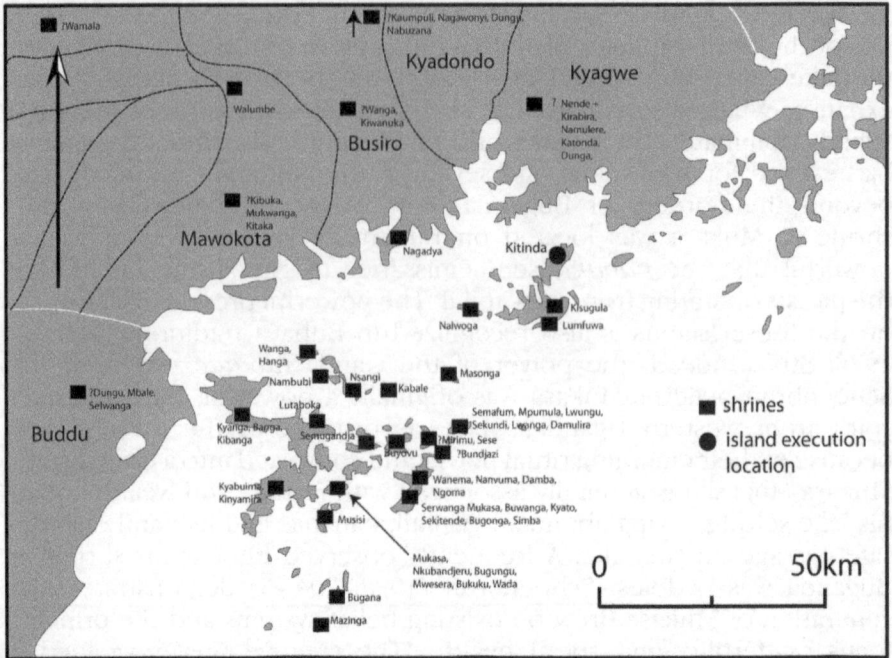

Figure 11.5. Major shrines of the *lubaale*, as noted by Roscoe (1911) and Kagwa (1969 [1934]) (© Andrew Reid).

historical role played by the islands in developing the lakescape and its potential. Intriguingly, although executions were said to have been ordered by gods or spirits, these actions were generally associated with Kibuka or Nnende, both of whom were associated with the mainland rather than the lake (Roscoe 1911). Of the 13 major execution sites mentioned by Roscoe only one – Kitanda – where victims were incapacitated and left on the shoreline to be taken by crocodiles – was situated in the islands. It is tempting to read great significance into this favouring of mainland locations for execution, but there may also have been a degree of pragmatism involved in preventing prisoners from having the opportunity to escape to the freedom of the islands.

The banana continued to play a full role, unseen and disregarded, in these negotiations of power and ritual practice. When re-building Mukasa's main shrine on Bubembe, outer stems were used to create guttering all the way down to the lake, to drain the blood from nine slaughtered cattle (Roscoe 1911: 292). Banana plants were used in healing by transferring disease from human to plant which was then taken away and disposed of (Roscoe 1911: 343). Spells and incantations were transferred in the same manner. Uncooked bananas were placed in a new canoe before launching to ensure its future bounty (Roscoe 1911: 390). Fines of bunches of bananas were issued to fishermen whose incompetence threatened the catch (Roscoe 1911: 396–7). Not surprisingly bananas were drawn into the symbolic and ritual realms of the state. When he died, the *kabaka*'s wives wailed their grief, painted themselves with ash and wore girdles of withered plantain leaves to symbolize the decay of the state (Roscoe 1911: 104).

Unfortunately, there is very little archaeological evidence for these transformations. From around AD 1000 a new decorative technique for pots appeared involving the use of roulettes. After around AD 1300 and the demise of Entebbe ware, the use of roulettes was ubiquitous through to the present day. The repetitive nature and speed of application of this decoration together with poorer firing suggests that there was not so much investment of time in the finished product. Quick and easy to apply, roulettes are also very difficult for archaeologists to subdivide and distinguish distinct periods of use (Haour et al. 2010). Furthermore, settlement in association with bananas, becomes dispersed and low density, offering little encouragement for more detailed archaeological enquiry. There are, however, indications of the change in meaning of ceramics. Pottery generally becomes much more utilitarian in nature, but specific items are imbued with ritual significance. Thus specialized production was responsible for making ritual vessels and also for manufacturing ceramics for royal use (Giblin & Kigongo 2012). Larger gatherings of people associated with clans, lineages, chiefs and even kings would have required much larger containers, particularly in association with beer drinking (Roscoe 1911: 440–1). Calabashes were frequently used for the latter, but ceramics are also likely to have been important.

THE LAKESCAPE DIMINISHED

Clearly Buganda in its formative years was a state dependent on the lake for communication, sustenance and ritual well-being. Without the lake Buganda could not have emerged. However, by the nineteenth century, Buganda had far outgrown its dependence on the lake. Not only had its territory expanded away from the lake and around the lake's margins but one of Buganda's most significant achievements was to link together the isolated banana islands on the hilltops of the mainland. This was accomplished by building and maintaining roads that crossed the wide swamps enabling the linking together of disparate territories and also the rapid movement of armies to control dissidence within Buganda and to raid neighbouring territories. The armies Buganda was able to field were a further consequence of the development of banana plantations. Established plantations freed up male labour in particular and allowed large numbers of men to join military expeditions. The control of land was integral to this process, with estates becoming associated with particular officers of the state. These estates were maintained in parallel with clan lands which probably maintained the older sources of power in the landscape (Kodesh 2010). A third kind of estate became increasingly important in the eighteenth and nineteenth centuries which were lands given by the *kabaka* as reward for services provided to him by individuals. This created an increasingly chaotic system of obligation between the *kabaka* and his people that threatened the stability of the state (Hanson 2003).

In the nineteenth century the state was coming under increased pressure from within. The *kabaka* was at odds with the clans who saw themselves as the rightful owners of the land, their power having been usurped by the position of the *kabaka*. Thus control of land and the bananas that were produced on that land was a crucial source of conflict. Claims regarding clan lands and the appropriation of lands by the state were to continue well into the twentieth century (Hanson 2003, Kodesh 2010). Furthermore, the *kabaka* in the nineteenth century was increasingly at odds with traditional shrines because of their largely incontestable power. Little wonder that first Mutesa and then his successors dabbled with Islam and Christianity to explore means by which to usurp the power of traditional religion. Civil war ensued in the 1880s and 1890s with factions dominated by Muslims, Catholics and Protestants. Traditional religion it seems was simply swamped by this political process and there was no political capital to be gained through standing by the old religions. Thus, the *kabaka* and indeed the Buganda state can actively be seen to have been dismantling the power of the lakescapes created 1,000 years earlier. The lake was still important in the nineteenth century, but as a means to access and control trade rather than as a foundational element of the state. Buganda was sending 2,000 slaves a year to the south of the lake and trade in iron and salt was being undertaken with communities living on the eastern shores (Kenny 1979, Reid 2002). Increasingly large forces of canoes, which may rightly be termed navies, were in operation on the lake. However, the means of warfare underscore the dominance of the land. Bassese and Bavuma powered the canoes, essentially transporting the mainland soldiers from one potential conflict on dry land to another (Reid 2002). Actual naval engagements seem to have been a rarity.

The lake also retained a tactical importance in the nineteenth century. In the 40 years before the declaration of the Uganda Protectorate, the capital was relocated ten times (Gutkind 1963). This reflected the political uncertainty of the time and the major internal conflict in the later years. The capital was laid out in formal terms with the royal enclosure dominating the landscape and compounds of functionaries, the various different categories of chiefs, and the many other components of the state laid out in front (Ray 1991, Hanson 2009). Behind the royal enclosure was a circle of functionaries, directly beholden, and therefore loyal to, the *kabaka*, protecting him. The royal enclosure had its back directly to the lake, for reasons of ritual exclusion but also to enable the *kabaka* to escape unhindered. Nearby 'several canoes were also kept in readiness, in case of emergency, for flight to the islands of the lake, where he could form his plans and restore order' (Roscoe 1911: 200–1). Hence, the freedom offered by the lake and the islands could also be harnessed by the *kabaka* as a means to avoid his enemies and regroup.

It is clear that at the formation of the Uganda Protectorate in 1894, the powers on the mainland had little love for the islands, their ritual power, or their people. Kagwa, the chief minister (*katikiro*) of Buganda, writing in 1918, contemptuously states that 'the people of Sese were not farmers and their women did not utilise the soil, which was very fertile. They gossiped too much of the time' (Kagwa 1969 [1934]: 158). Such views clearly influenced British perspectives: 'The natives of the Sese Islands have an ill fame among their fellow Baganda of the mainland as suspected cannibals' wrote Sir Harry Johnston, special commissioner of the Uganda Protectorate and avid ethnographer (Johnston 1902).

The final humiliation of the lake came shortly after the formation of the Uganda Protectorate. On three occasions, between 1902 and 1920, the islands were forcibly abandoned as a health measure to counteract sleeping sickness epidemics (Hoppe 1997). These heavy handed measures were presented as the only means by which to establish control over the disease and they were readily approved by the Buganda government. In this can be seen the last revenge of the mainland in isolating and abandoning the lake. Sleeping sickness clearances were also undertaken in Bunyoro and may also be seen as punishment for Bunyoro's failure to accept colonial authority. By contrast, at the southern end of the lake, in what was then Tanganyika, sleeping sickness control did not deem it necessary to force resettlement. The forced resettlement on the Uganda mainland finally broke the last elements of the lake's power. Bavuma communities were resettled in alien and uncomfortable conditions at least five miles from the coast in Kyagwe and their characteristic pottery (Jensen 1969) is evident on the ground today. Some islanders eventually returned, but on the Ssese Islands in particular they were joined by opportunists from other parts of Uganda, and the distinctive Ssese dialect has largely been lost.

POSTSCRIPT

Whilst the inhabitants of the islands, the Bassese and the Bavuma, clearly were disadvantaged by this process, losing their lakescapes and seeing the lake itself being marginalized and abused, the physical space has endured and indeed new lakescapes have been created. This resistance can be seen in twentieth-century contexts such as at Namusenyu in Kyagwe. Here a narrow overhang was occupied, hidden away 15m from the lake with evidence of fish bone interspersed with European cutlery and fabric. Exactly when the site was occupied in the twentieth century is actually immaterial as the people of the lake have been resisting authority from the colonial period into Independence'. As one informant from the area put it:

> We stayed in many new secret landings in the swamps, and moved only at night. These are still good places. We hid from the British doctors, then we hid from all the soldiers, now we hide from the revenue collectors (Asoman Wandoka quoted in Hoppe 1997).

This contestation of the lake has continued. At the beginning of the twenty-first century, Buvuma had a reputation for piracy. In Ssese, the main island, Bugala has been decimated by a huge government backed oil palm project. On the other islands, however, some isolated communities still value their freedom from compliance with the state and benefit from moving commodities unofficially across borders. It is rumoured that there are Congolese communities on the more remote islands, hundreds of kilometres from their homeland. On the other side of the lake, Uganda and Kenya are currently disputing ownership of the Migingo Islands, little more than rocks in the deepest waters of the lake, beyond Lolui. And there is also the development of a new lakescape building on some of the inequities drawn out in this account. The creation of refrigerated units at many places along the Uganda coast and in the islands have created new markets for fish, in particular the Nile Perch, now more limited to deeper waters from over-fishing. The comparatively huge returns for their catch have created cash-rich fishermen and have drawn in a wealth of opportunities for traders providing goods and services. Described rather prosaically as the 'Dubai of the lake' (Namugoji 2004), these services at places like Lolui begin with the more base standards of alcohol, gambling and prostitution. Basic though these may be, their florescence in these distant locations, far from the direct interference of government are testimony to the power of the new lakescapes being created.

Whilst these new lakescapes are flourishing, the bananas taken long ago to the mainland are failing. Most of Buganda has now been hit by the blight which has swept through world banana populations. As bananas do not reproduce, but are merely genetically cloned from one generation to the next, they are extremely susceptible to disease. Banana yields have been greatly reduced as a result of this and of declining soil fertility. Meanwhile, improved infrastructure and transportation has led to the development of new banana producers in drier areas previously associated with grain agriculture, such as the hills to the south of Mbarara. These plantations will inevitably also succumb to blight in time, begging the question of 'where next?' on account of the social status associated with eating bananas. There has also been very little desire to explore the link between bananas and malnutrition and equally to consider the beneficial qualities of neglected grains such as finger millet. If archaeology can actually encounter burials from the past it may be possible to consider the nutritional status of populations relying on the different crops.

A much larger shadow being cast over the landscape is population. Uganda's population, approximately 39.5 million in 2014 is projected to reach 103 million by 2050. It is forecast that by then Kampala and Jinja will have become one huge city. This will have a huge impact on the lake from pollution, deforestation and over-exploitation of fish stocks.

CONCLUSION

The Victoria Nyanza has had and continues to have a huge impact on the human communities who live on and around it creating a range of productive and fertile lakescapes. It has created vast opportunities: its yet abundant resources – fish stocks, coastal botany, the ease of movement across its waters and above all its year-round generation of rainfall – all benefit thriving human populations. It can also however bring disaster, its frequent storms endangering those on the lake, its rains spreading disease through virulent insect populations and its occasional high winds flattening banana plantations. The lake created the freedom for human populations to break away from established social organization, to seek out new lands and opportunities, easily accessed from the water, and to provide easy communication and exchange between these new communities, yet it also had the power to destroy.

It was into this world with its great potential that the banana made its entrance and rapidly provided the means for societies to transform their lakescapes still further. With its incredible productivity, the banana provided for larger communities and in being embraced by those societies very quickly became the basis of their cultural life. The coolness of the *lusuku* not only guarded and cared for its people, but it also became the centre of the human relationships played out in the lakescape, creating the building blocks for the ties of lineages and clans to the land and also the means through which kingship could wield and manipulate political relationships. In these latter contestations the ritual power and long-standing potential of the lake was challenged and ultimately overrun by the very communities of the mainland shores that the lake had first made feasible. Culturally marginalized and sidelined though the lake is today, representing disenfranchised border peripheries, it still retains its core qualities of independence, abundance and opportunity. It is these qualities that have lain at the heart of the many lakescapes that have been constructed within and around its waters.

296

A History of Water

REFERENCES

Ashley, C. 2010. 'Towards a socialized archaeology of Great Lakes ceramics'. *African Archaeological Review* 27: 135–63.
Ashley, C., Reid, A. 2008. 'A reconsideration of the figures from Luzira'. *Azania* 43: 95–123.
Connah, G. 1996. *Kibiro: The salt of Bunyoro, past and present*. London: British Institute in Eastern Africa.
Giblin, J.D., Kigongo, R. 2012. 'The social and symbolic context of the royal potters of Buganda'. *Azania* 47: 64–80.
Giblin, J.D., Fuller, D.Q. (2011). 'First and second millennium AD agriculture in Rwanda: archaeobotanical finds and radiocarbon dates from seven sites'. *Vegetation History and Archaeobotany* 20: 253–65.
Gutkind, P.C.W. (1963). *The Royal Capital of Buganda*. The Hague: Mouton.
Haour, A., Manning, K., Arazi, N., Gosselain, O., Guèye, N.S., Keita, D., Livingstone-Smith, A., MacDonald, K., Mayor, A., McIntosh, S., Vernet, R. 2010. *African pottery roulettes past and present*. Oxford: Oxbow.
Hanson, H. 2003. *Landed Obligation: The Practice of Power in Buganda*, Portsmouth: Heinemann.
——— 2009. 'Mapping conflict: heterarchy and accountability in the ancient capital of Buganda'. *Journal of African History* 50: 179–202.
Hoppe, K.A. 1997. 'Lords of the Fly: Colonial Visions and Revisions of African Sleeping-Sickness Environments on Ugandan Lake Victoria, 1906–61'. *Africa: Journal of the International African Institute* 67: 86–105.
Iles, L. 2013. 'The use of plants in iron production: insights from smelting remains from Buganda'. In Humphris, J., Rehren, T. (eds), *The World of Iron*. London: Archetype, pp. 56–65.
Jensen, J. 1969. 'Topperei und Topperwaren auf Buvuma (Uganda)'. *Baessler-Archiv* 17: 53–81.
Johnson, T.C., Scholz, C.A., Talbot, M.R., Kelts, K., Ngobi, G., Beuning, K., Ssemmanda, I., McGill, J.A. 1996. 'Late Pleistocene desiccation of Lake Victoria and rapid evolution of cichlid fishes'. *Science* 273: 1091–3.
Johnston, H.H. 1902. *The Uganda Protectorate*. London: Hutchinson.
Kagwa, A. 1969 [1934]. *The Customs of the Baganda*. Translated E.B. Kalibala, New York: AMS.
Kendall, R.L. 1969. 'An ecological history of the Lake Victoria basin'. *Ecological Monographs* 39: 121–76.
Kenny, M. 1979. 'Pre-colonial trade in eastern Lake Victoria'. *Azania* 14: 97–107.
Kodesh, N. 2010. *Beyond the Royal Gaze*. London: University of Virginia.
Lejju, B.J., Robertshaw, P.T., Taylor, D. 2006. 'Africa's earliest bananas?'. *Journal of Archaeological Science* 33: 102–13.
Namugoji, E. 2004. 'Dolwe Island: Lake Victoria's Dubai'. *The Sunday New Vision Magazine*, 3 October, pp. 6–12.
Nicholson, S.E. 1998. 'Historical fluctuations of Lake Victoria and other lakes in the Northern Rift Valley of East Africa'. In Lehman, J.T. (ed.), *Environmental Change and Response in East African Lakes*, London: Kluwer Academic, pp. 7–36.
Posnansky, M., Chaplin, J.H. 1968. 'Terracotta figurines from Entebbe, Uganda'. *Man* 3: 644–50.

Posnansky, M., Reid, A., Ashley, C. 2005. 'Archaeology on Lolui Island, Uganda 1964–5'. *Azania* 40: 73–100.
Ray, B.C. 1991. *Myth, Ritual and Kingship in Buganda*, Oxford: Oxford University Press.
Reid, A. 1994/5. 'Early settlement and social organisation in the Interlacustrine region'. *Azania* 29/30: 303–13.
—— 1996. 'Ntusi and the development of social complexity in southern Uganda'. In Pwiti G., Soper, R. (eds), *Aspects of African Archaeology: Papers from the 10th Congress of the Pan African Association for Prehistory and Related Studies*. Harare: University of Zimbabwe Press, pp. 621–7.
—— 2003. 'Recent research on the archaeology of Buganda'. In Mitchell, P.J., Haour, A., Hobart, J.H. (eds), *Researching Africa's Past: New Contributions from British Archaeologists*. Oxford: Oxbow, pp. 110–7.
—— 2013. 'The emergence of states in Great Lakes Africa'. In Mitchell, P., Lane, P.J. (eds), *Oxford handbook of African archaeology*. Oxford: Oxford University Press, pp. 883–95.
Reid, A., Ashley, C. (2008). 'A context for the Luzira Head'. *Antiquity* 82: 99–112.
—— 2014. 'Islands of agriculture on Victoria Nyanza'. In Stevens, C.J., Nixon, S., Murray, M.A., Fuller, D.Q. (eds), *Archaeology of African Plant Use*. Walnut Creek: West Coast, pp. 179–88.
Reid, A., Young, R. 2003. 'Iron smelting and bananas in Buganda'. In Mitchell, P.J., Haour, A., Hobart, J.H. (eds), *Researching Africa's Past: New Contributions from British Archaeologists*. Oxford: Oxbow, pp 118–23.
Reid, R.J. 2002. *Political Power in Pre-Colonial Buganda*. Oxford: James Currey.
Robertshaw, P.T. 1997. 'Munsa earthworks: a preliminary report on recent excavations'. *Azania* 32: 1–20.
Roscoe, J. 1911. *The Baganda. An Account of their Native Customs and Beliefs*. Cambridge: Cambridge University Press.
Schmidt, P. 1978. *Historical Archaeology: A Structural Approach in an African Culture*. Westport: Greenwood Press.
Schoenbrun D.L. 1993. 'Cattle herds and banana gardens: the historical geography of the western Great Lakes region, ca. AD 800–1500'. *African Archaeological Review* 11: 39–72.
—— 1994. 'The contours of vegetation change and human agency in eastern Africa's Great Lakes Region: ca. 2000 BC to ca. AD 1000'. *History in Africa* 21: 269–302.
—— 1998. *A Green Place, a Good Place: a Social History of the Great Lakes Region, Earliest Times to the 15th century*. London: Heinemann.
Speke, J.H. 1863. *Journal of the Discovery of the Source of the Nile*. Edinburgh: Blackwood and Sons.
Stager J.C., Mayewski, P.A., Meeker, L.D. 2002. 'Cooling cycles, Heinrich event 1, and the desiccation of Lake Victoria'. *Palaeogeography, Palaeoclimatology, Palaeoecology* 183: 169–78.
Stager, J.C., Cumming, B.F., Meeker, L.D. 2003. 'A 10,000-year high-resolution diatom record from Pilkington Bay, Lake Victoria, East Africa'. *Quaternary Research* 59: 172–81.
Stanley, H.M. 1878. *Through the Dark Continent, Vol. I*. London: Sampson Low.
Temple, P.H. 1964. 'Evidence of lake-level changes from the northern shoreline of Lake Victoria, Uganda'. In Steel, R.W., Mansell Prothero, R. (eds), *Geographers and the Tropics: Liverpool Essays*. London: Longmans, pp. 31–56.

Tushemereirwe, W.K., Karamura, D., Ssali, H., Bwamika, D., Kashaija, I., Nankinga, C., Bagamba, F., Kangire, A., Ssebuliba, R. 2001. 'Bananas (*Musa Spp*)'. In Mukiibi, J.K. (ed.), *Agriculture in Uganda: Vol. II, Crops*. Kampala: Fountain.

Van Grunderbeek, M.-C. 1988. 'Essai d'étude typologique de céramique Urewe de la région des collines au Burundi et Rwanda'. *Azania* 23: 11–55.

Van Grunderbeek, M.-C., Roche, E., Doutrelepont, H. 1983. 'Le Premier Age du fer au Rwanda et au Burundi: archéologie et environnement'. Brussels: IFAQ.

Wrigley, C.C. 1989. 'Bananas in Buganda'. *Azania* 24: 64–70.

Yin, X., Nicholson, S.E. 1998. 'The water balance of Lake Victoria'. *Hydrological Sciences Journal* 43: 789–811.

12 Dying Cows Due to Climate Change? Drought Can Never Finish the Maasai Cattle, Only the Human Mouth Can (Maasai saying)

Marcel Rutten

INTRODUCTION

In February 2010 reports appeared in the news of wildebeest, zebra and elands found outside the world famous Amboseli National Park in Kenya that were returned to their (unfenced) habitat. The 2008/09 drought had killed many of them and after the December 2009 rains the survivors had moved out looking for fresh grass. Lions and other predators followed their natural pray looking for a bite of game. In this search they also enjoyed the domestic stock of local Maasai pastoralists, or what was left of it as most herds had been diminished by up to 90 per cent. The media and certain groups of scientists attributed this disaster to climate change.

However, the death of livestock and wild animals was first and foremost caused by a reduction in the availability of biomass. This again is the result of a number of factors, such as below average rainfall in four subsequent seasons from late 2007 onwards. The short rains of 2009 came late, towards the end of December, but produced almost 40 per cent of the year's total average. The few weakened animals still alive might actually have died due to cold temperatures that came along with the rains. But is this a sign of climate change, often wrongly synonymously understood as desertification?

Maasai pastoralists, once called 'the lords of the plains', roaming around with their large herds on extensive pastures, have been admired as well as criticized by British colonial officers. The Maasai society has been labelled a beastly, bloody system, founded on raiding and immorality, disastrous to both the Maasai and their neighbours. The authorities had to halt 'the pernicious pastoral proclivities' and to encourage peasant agriculture as a first step towards 'civilized' forms of land tenure and usage. Yet towards the end of World War I it was estimated the Maasai were among the richest people in the world. The pastures available to the Maasai were now allocated to white settlers. Today, the Maasai welcome foreigners to their lands foremost as tourists interested in visiting one of the most wildlife rich places on earth.

Like these foreigners the droughts have also remained. Moreover, less land and water are available for a growing human population sharing more or less similar numbers of livestock. Mobility to overcome these shortages are crucial. Droughts are not new phenomena to Kenyan pastoralists. Periods of sustained lack of rain and subsequently depleted pastures occur repeatedly and will return again. Over the years, though, the options available to make it through to the next rainy season have changed, often for the worse. Rain is life for Maasai pastoralists occupying Kenya's southern rangelands and part of the 80 per cent of Kenya's territory receive less than 700 mm per year. Following a long period of drought, the rains rejuvenate the dried pastures allowing for fresh young sprouts to shoot, providing new biomass to graze. In addition, the rains will replenish the rivers, wells and pans (both natural and man-made). The rains, however, might come late or remain below average. If this occurs in two subsequent seasons, or worse, years, the Maasai are forced to apply and intensify their drought-coping strategies to safeguard their herds and flocks which supply their most important source of food, notably milk. Whether they succeed depends on the resilience of the pastoral system, and some luck. Wrong migratory decisions could mean losing all animals.

In this chapter, we will provide a detailed look into migration decisions made by Maasai households in trying to overcome the drought of 2008/2009. We will also weigh the reasons why livestock was lost. Is it climate change or are other issues at stake? But before doing so, let us take some steps back in history to better understand the Maasai livestock economy and the resources available.

MAASAI HISTORY OF LAND AND POPULATION CHARACTERISTICS

The Maasai have not always occupied the area they inhabit at the moment. Linguistic and oral evidence points to the Sudan–Uganda border area as the original birth-place of the Maasai (Mol 1980:4). The group moved southwards probably around AD 1400. During their migration they displaced and absorbed several peoples and reached Tanzania by the seventeenth century. In the nineteenth century the Maasai in today's Kenya ran into conflict among themselves. Livestock and human diseases further weakened the Maasai, who were split through the middle after the arrival of British and German colonizers in Kenya and Tanzania. In Kenya the Maasai lost their best grazing areas to white settlers and were restricted in one southern reserve from 1912/13 onwards (Sandford 1919: 3). Some parts were tsetse fly infested and lacking sufficient water and all year-round grazing. Settler restrictions on mobility of livestock, fearing transmission of livestock diseases and competition on cattle markets, were mostly felt during times of drought. Protests, including those towards the 1932 Carter Land Commission, did not help. The request to remove the artificial boundary between Kenya and Tanganyika Maasai was also pushed aside.

The Maasai repeatedly showed their interest in water development and inoculations to raise the carrying capacity and prevent livestock diseases, even providing funds themselves. However, the Administration tried to persuade the Maasai to accept some form of grazing control. By allowing a maximum number of livestock in the vicinity of a borehole it was attempted to gradually introduce a kind of ranch unit. The schemes introduced in the 1950s tried to enforce strict grazing control. On the other hand, the Administration stated that it realized that, owing to the vagaries of the rains, an even pattern for grazing control would never be achieved. Indeed, by 1959 every control had to be abandoned due to the drought and the Maasai trespassing to other districts.

Individual ranches were another new phenomenon of the 1950s. Influential Maasai (civil servants, politicians, and teachers) opted for individual ranches of 2,000 acres each. In addition, a pilot programme of land consolidation was begun by 1961; although opposed by Maasai elders this move was encouraged by the young and educated. It opened opportunities for non-Maasai who, since the war, had come to Kajiado (Kajiado, one of the Maasai-dominated counties in southern Kenya (see Figure 12.1)) in increasing numbers looking for land to be cultivated. Their eyes had fallen on the high-potential areas such as Ngong, Namanga and Loitokitok. At times they were removed though from Kajiado. Now being able to buy land took away this threat for the immigrants.

The Maasai also lost access to their dry season grazing areas near Nairobi and Mt Kilimanjaro/Amboseli as a result of wildlife management and tourism. Since the mid 1940s land was proclaimed for wildlife parks only which meant the loss of grazing area for Maasai cattle. Independence did not return the lost White Highlands neither were the Maasai compensated financially. Instead, the Maasai feared to lose more land to their educated elite and their agricultural neighbours eyeing land in Kajiado. To stop them Maasai elders accepted the idea of land adjudication.

During the late 1960s the World Bank and other donors introduced group ranches under the Kenya Livestock Development Project (KLDP). In short, the idea of a group ranch meant the setting aside of a piece of land communally owned by a group of people registered as the legal owners through membership of the particular ranch. Livestock movements would be restricted within specific boundaries and non-members would be forbidden to bring their animals to graze. Through the provision of loans for infrastructural development and steer-fattening an attempt was made to radically transform the nomadic subsistence-oriented production of the Maasai pastoralists into a sedentary, more commercial system.

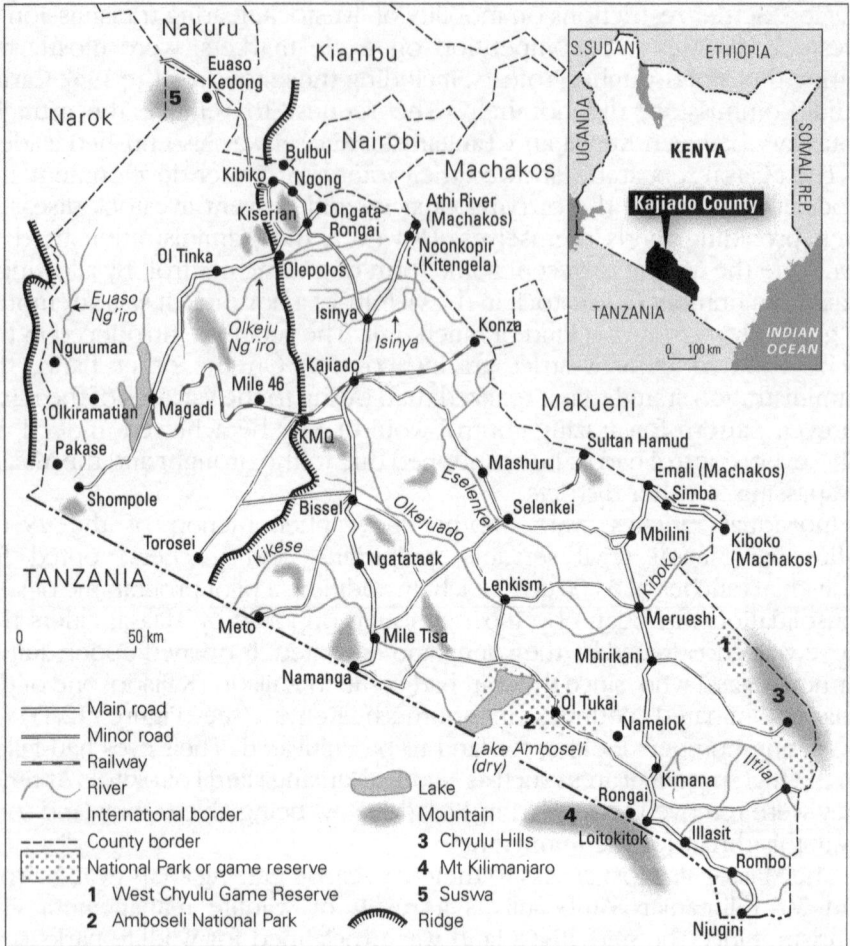

Figure 12.1. Kajiado County (© Marcel Rutten).

This market-oriented production was to bring about a destocking of the Maasai pastures while at the same time providing meat for the national and international market.

The Kajiado group ranches were effective in stopping the educated Maasai elite allocating huge chunks of former communal land to themselves, brought facilities such as boreholes, dams, troughs, tanks, pipelines and cattle dips, and stimulated the building of schools, shops and health centres. Besides these accomplishments the group ranch project had disappointing rates of investment and difficulties in loan repayment; continuing trespassing of group ranch boundaries; refusal to destock ranches; no real transformation to a meat, market-oriented livestock production and corruption among several group ranch committees.

Figure 12.2.   Floriculture obstructing grazing and drying shallow wells (© Marcel Rutten).

The final outcome of these problems was a growing wish among many Maasai for the subdivision of the group ranch into individually owned shares. By 1990 a total of 40 group ranches had made the decision to dissolve their ranches (Rutten 1992). The fencing of the shambas and ranches was expected to intensify and pose a threat to the mobility of Maasai livestock. Also the sale of land, which had now become a commodity, could take out grazing land when turned into other use such as cultivation. Initially, small plots were sold to a few cultivators and fences, an expensive investment, were limited. But from the late-1990s onwards large companies involved in horticulture and forestry plantations moved in. Now fences appeared all over. This reduced access to and control over land, rangeland degradation, and an unfavourable if not hostile political and economic environment are also among the main causes mentioned for repeated disasters next to weather challenges. Opinions as to the relative importance of each of these causes differ between several schools of thought within and between disciplines, now and in the past. Nevertheless, they all share the view that traditional nomadic pastoralism is losing ground (Figure 12.2).

Table 12.1 stresses Kajiado's relatively high population growth rate which is mainly the result of immigration of agricultural groups from neighbouring areas, although overall population density is still low. The percentage of non-Maasai in Kajiado has increased from 31.4 for 1969, via

37.2 in 1979 to 43.4 in 1989, respectively. Thereafter no ethnicity data is available but likely from 2000 onwards the majority of residents in Kajiado are non-Maasai, although these are mainly concentrated in the towns south of Nairobi. Non-Maasai can also be found near the scattered pockets of available good agricultural land, i.e. at higher altitudes or near swampy regions which allow some simple irrigation: Nguruman, Kimana and Loitokitok. Besides access to pastures, the availability of water is important to overcome a drought. In the following we will discuss the water situation in the area.

WATER: MAASAI CONCEPT, RIGHTS TO WATER AND DEPLETION FEARS

Water in the language of the Maasai is called *enkare* and forms a vital part of the Maasai (livestock) economy. For example, it is used to water, wash and spray the animals. Water is believed to be a natural gift from God and no other being can provide this gift. No person is left to die of thirst when other people have water. On the other hand, one can own a source when he or she has already worked on it (by providing labour, power or finances).

The Maasai dependence on rainwater in a semi-arid area forces them to live a nomadic way of life in search for water and pastures. Initially water was scoped from the (dry) riverbed. This method is known as *enturore oo'lchorroi*, which requires digging the river channels or any water source until the water table comes up.

The ownership and maintenance of the wells and the regulation of access to water are a complex matter. In general, every well is known as the well of a certain clan and is maintained and organized by the elderly owners. But working inside and any other contribution to the well makes everybody using it responsible and have certain user right claims. The major sanction underlying the Maasai system of water control is of course exclusion from water. Failure to supply labour at the well and failure to participate in the politics of water will soon lead to exclusion from the well.

The traditional pattern of user rights was interfered with due to the introduction of boreholes and dams. Government boreholes are managed by an employee who is to count the livestock taking water and receive the payments. The community is responsible to contribute a certain amount and to involve the local government for repairs. The boreholes are used by the whole community under communal ownership, irrespective of clan or ethnicity as long as fuel and maintenance is taken care off.

As a result of subdivision of group ranches land is becoming a commodity. In addition to an influx of small cultivators horticultural and ostrich farms, poultry ventures, schools and training institutes are buying land in the subdivided parts of Kajiado. They all need huge quantities of water for production or consumption purposes. As a result the number of

Table 12.1. Demography Kajiado County: 1989, 1999, 2009.

	1989	1999	2009
Population	258,659	406,054	687,312
Population density (in h/km²).	12	19	32
Pop growth (interdecadal)	1.74	1.57	1.69

Source: compiled from population census 1989, 1999, 2009. Population 1979=149,005; Kajiado areaal=21,756 km².

boreholes is on the increase (Figure 12.3). In addition some 1,800 shallow wells have been dug. These wells hardly ever dry up and are more cost-effective than boreholes.

DEPLETION OF WATER SOURCES IN MAASAILAND

Since the late 1970s Maasai frequently expressed their worries about diminishing flows in the rivers. Water availability dwindled due to an influx of agriculturalists destroying trees in water catchment areas and riverine zones. Almost all rivers in Kajiado were labelled permanent in a September 1938 water supplies survey report, only a few were said to have small flows in the lower reaches in the dry season (KNA/PC/NGO/1/5/3). The recharge capacity will have been lost in those rivers which were used for harvesting sand. Since colonial time great quantities of sand have, legally and illegally, been mostly taken by Indians for the construction industry in Nairobi (KDAR 1929; KNA/ HOR 1946). Equally, if not more, important was the water and charcoal demand by companies such as Magadi Soda. To improve water supplies hundreds of boreholes have been sunk and dams constructed as from 1927. A census in 1988 concluded that only about one-third were still operative (Sandford, 1919; Mwangi 1993).

More recent are reports of lowering ground water levels. Since 2005 shallow wells in the Athi-Kapiti plains are drying as a result of nearby boreholes providing water to floriculture and eucalyptus farming, with water levels suspected to be dropping by approximately 5 m/year (Rutten 2008). Another recent worrying phenomenon is the rapid reduction in the quality of water in the intensively cultivated areas because of effluent (illegally) released from the horticultural farms during the wet season when the rivers are flowing. The quality of groundwater is also affected in areas such as Ngong. Too many boreholes in too small an area have resulted in a situation whereby borehole water in the area in question is turning brackish. Other causes of hydrological droughts are sand harvesting from riverbeds (Figure 12.4).

A major water provision project (National Noolturesh Water Project) was started in late September 1987. Italian grants and soft loans for studies and

Figure 12.3.   Boreholes striking water at increasingly deeper levels (© Marcel Rutten).

construction were provided to develop the 262 km pipeline project bringing water from the Noolturesh springs located at the foot of Kilimanjaro to Machakos and Athi River Town and back to Kajiado Town. Technical consultants warned that the Noolturesh supply will, however, not be enough for the growing demands of the three towns.

Indeed, water flows from the source are increasingly reducing. The irrigation downstream has died and at the same time the swamps between Kilimanjaro and the Chyulu Hills have disappeared to the detriment of the local livestock keepers and wildlife. The main reason for the drying of the swamps is the high amount of water taken from the source (Masharen 1989:11).

THE MAASAI FOOD BASE AND STRATEGIES IN SECURING ACCESS TO FOOD

The main aim for every Maasai household is to secure food in sufficient amounts. Livestock keeping households use the animals as an intermediate to convert grasses and shrubs into milk, meat and blood. For a long time the Maasai were able to subsist entirely on the produce of their large herds, either directly (milk, meat, blood). or indirectly through (lucrative) barter for agricultural products. However, the number of stock units per capita

Figure 12.4.   Bare river bed after scoping of sand (© Marcel Rutten).

has dwindled over the years. More mouths have to be fed, while livestock numbers increase at a lower pace in an erratic manner.

The International Livestock Centre for Africa (ILCA) concluded that between 1948 and 1984 a consistent rise in the number of Maasai pastoralists in combination with a fluctuating cattle population has led to a steady decline in the cattle/people ratio. As a rule of thumb it is estimated that to subsist entirely on a diet of milk and meat, a minimum of about ten head of cattle/person is required. Thus, it can be concluded that since the early 1960s the Maasai of Kajiado, on average, no longer accord with this basic condition. After the 1984 drought cattle wealth dropped to a low of less than three cattle per person on average. In the early 1930s this stood at 15–20 cattle per person, while the 1988 Kajiado livestock census concluded that some 120,000 Maasai pastoralists shared approximately 633,000 head of cattle, a cattle/person ratio of 5.3 (MoLD 1988: iii). In early 2008 livestock numbers were estimated to be 457,863 cattle for some 300,000 Maasai pastoralist, a mean cattle ratio of 1.5 per person only (EWB February 2008). This drop is partly offset by the increasing importance of smallstock.

Milk is the Maasai staple food of choice, its availability is heavily dependent on rainfall and the number of cattle owned. Milk is consumed fresh as well as skimmed or mixed with tea. Milk is also stored and turned into sour milk. Fat is important, especially for young children.

The consumption of meat is low depending on the amount of (forced) slaughtering because of old or diseased animals or the occurrence of ceremonies. Blood, fresh or mixed with milk, is rarely drunk. The Maasai have decreasing numbers of livestock available per capita to feed themselves on a purely pastoral diet. Moreover, Maasai youngsters increasingly join schools where other types of food are served. Food habits are also changing due to assimilation of many outsiders with agricultural backgrounds into the Maasai households. All these factors seem to change the negative attitude to cultivation and its products. The change in food habits is exemplified by the gradual acceptation of chicken, rice and maize in the Maasai diet. Taboos on fish and pork are still strong, but also dwindling. The success to assure sufficient food depends on the well-being of the animals, and increasingly access to cultivation plots or wage labour.

Pastoralists are used to seasonal variations in food availability accompanied by a dwindling milk production in the dry season.[1] Besides a sufficiently large herd for direct subsistence needs, pastoralists need to maintain enough breeding animals for reproductive reasons, as an insurance against losses due to diseases and droughts, as well as for social obligations (bride wealth, assisting less well-off kinsmen, etc). This non-equilibrium environment necessitates the herd owners to engage in a whole range of stress-coping strategies against droughts, diseases, and predators as part of their overall livelihood strategy geared at accumulation of wealth, betterment, sustenance and, in exceptional circumstances of crisis, survival.

The strategies employed ultimately aim to secure the physical well-being of the household by defending the accumulated wealth or by restoring it. Thus both preventive and curative elements can be distinguished. Preventive elements are preservatives, applied (long) before the actual problem arises. Curative elements are applied as a 'cure action' when problems are there to be solved. Preventive strategies among Maasai pastoralists are mainly aimed at securing the underlying and basic factors affecting food security (i.e. herd management and resources management).

Emergency strategies in managing the Maasai food base are implemented during and following times of stress or severe crises. The emergency strategies include a change in food habits and herd and resource management that will come along with drastic reductions in food intake, and change of herding practices and locations. Support from social networks is also sought in case of need. The strategies mentioned above are characterized by an object and time variable. Some strategies are taken before, during and after a drought. In the following we will discuss some of these strategies in more detail following the 'time' variable: before and during a drought. But first let us discuss the concept of drought, both as used in science and by the Maasai.

DROUGHT: SCIENTIFIC DEFINITIONS AND THE MAASAI CONCEPT

Drought: a scientific definition

Drought is a condition of temporary aridity. Aridity (lack of moisture) is caused by five main processes: continentality, cold off-shore currents, topography, dynamic anticyclonic subsidence, and high pressure systems. About 30 per cent of the earth has a dry climate and all continents have dry ecosystems from hyperarid to subhumid (Mainguet 1990). The phenomenon of drought may be analysed in terms of meteorological, agricultural, hydrological, socio-economic, and organizational aspects. Scientific literature defines droughts accordingly. A broad classification is given by Vincent (1981). She distinguishes three types of drought: *permanent* drought of the driest climates; *seasonal* droughts in climates having well-defined wet and dry seasons; and *contingent* drought resulting from irregular and variable rainfall during a period when rainfall is anticipated. It is the latter we are concerned about here.

In general, drought is a protracted departure from normal water availability, a water deficit that exists for a long enough period to cause hardship. More specifically, whether or not drought occurs depends on the type of activity for which water is being used and on the expected or required availability of water (Farmer & Wigley 1985):

- *Meteorological drought* occurs when precipitation is significantly below expectations. Its definition involves only precipitation statistics and depends on the time of the year and on the location. Climatologists are still struggling to explain the phenomenon of African droughts. Some are able to find certain linkages. For example, El Niño, a periodic warm Gulf Stream in the Pacific Ocean, is strongly correlated with rainfall in eastern Africa and droughts in southern Africa.
- *Hydrological drought* occurs when water resources for industry, for human or animal consumption, or to support agriculture reach low levels. Hydrological drought is usually reflected in low levels of rivers or lakes, reservoirs, or ground water. These levels are determined not only by precipitation, but also by water usage and by evapotranspiration (direct and indirect drying) (Farmer & Wigley 1985).
- *Agricultural drought* occurs when the water supply necessary for agriculture becomes scarce. An agricultural drought depends on the amount of rainfall expected and the use to which water is put, hence a shortage of water is felt because of human activity, whereas a meteorological drought creates stress on plant life unconnected with people and their needs. Agricultural drought is a moisture deficit on a sufficient scale to cause disruption of the rural economy. In an extreme agricultural drought, crops fail, and animals and perhaps people die.

Soil types are of importance also. Deep soils offer good soil moisture-storage potential; shallow soils do not.

Maasai definition of drought

This brings us to the Maasai concept of drought. The variety in occurrence and intensity of a drought is reflected in the three categories of drought classification used by the Maasai. The first (*olameyu*) implicitly links failing rains to hunger. Rainfall is insufficient, pastures dry and animals become thin and produce less milk.

However, Herren (1991) states that *olameyu* starts from the impact side (i.e., lack of food, or *endaa*), rather than lack of rainfall. Furthermore, any dry spell is *olameyu* even if it does not (yet) affect production and nutrition. One might thus say that *olameyu* is more than 'drought', at least in the sense of meteorological drought; it is rather a crisis in the reproduction of both herd and family (Herren 1991).

Another term used by the Maasai is *emboot*. It describes a disastrous situation whereby due to a severe drought livestock as well as wildlife die in mass numbers because of a lack of grazing. An *emboot* is not a seasonal phenomenon covering just a few months but occurs over a longer period up to one year. Some of these disaster periods have been given their own names (see Box 12.1). It also mostly affects a larger area than in the case of *olameyu*. In addition to the severe and long dry periods sometimes a small caterpillar is mentioned as a characteristic of *emboot*. It is found in dry grasses only and causes the animal to die once they enter the animal's stomach.

Finally, *emperi* is thought to be the most serious type of drought of the three categories mentioned. During *emperi* the drought is very long lasting, over a year. Water holes dry up. The land turns to bare soil causing whirlwinds to blow dust into the air. People and livestock have to wander for long distances to fetch water and find pastures. In this form of drought they often fail and animal carcasses on the landscape indicate the severity of the disaster (see Figures 12.5 and 12.6). Human lives are also threatened and sometimes lost. This term of drought is dwindling and less known especially among young Maasai due to the fact that food relief nowadays prevents the loss of large numbers of human lives.

MAASAI METHODS FOR 'PREDICTING' DROUGHTS

Like the scientist trying to link El Niño with weather extremes in Africa, Maasai pastoralists also use the occurrence or disappearance of certain phenomena as signs for the coming of a drought period. It is knowledge based on many years of experience passed on to new generations. Some have become respected experts in predicting droughts. Their services are

Figure 12.5. Dead cattle; polythene bags that killed livestock after eating fresh grass and picture of cactus as an alternative source of food during droughts (© Marcel Rutten).

Figure 12.6. Dead sheep (© Marcel Rutten).

especially needed by those who have gone through formal education which is undermining the passing on of this indigenous knowledge among all members of society. There are several signs used by the Maasai to predict the coming or continuation of a drought spell.

- *Observation of the hemisphere and sky*: The movement of certain stars (the Pleiades (*inkokua*)) and Venus, the morning star (*kileken*) in relation to a star named *Olokirr ai* 'predicts' drought or rain. When the latter two do not cross on the right side rains are expected to fail.
- *Observation of (domestic) animals*: Intestines of small stock having air gaps and being thin towards the end are a bad sign. It is believed that drought will come when cattle urinate and pass dung while lying down, when they are not willing to leave the *boma* and need to be assisted to rise, or when they take food but do not fatten and/or crunch their teeth without eating food. Finally the look of the animals, i.e. standing hair and dullness is interpreted to mean drought will come.
- *Observation of trees and grasses*: The colour and shading of leaves of certain sacred trees will tell the experts whether rains will come or fail; the lack of flowering of the *oltepesi* trees (Acacia Seyal) is also used as an indicator. When grasses suddenly become easy to uproot a drought is near.

Box 12.1. Droughts in the Kajiado area.

1925–7: *Olameyu Loolonitok*
This very severe drought lasted about a full year and killed nearly all animals. People fed on skins and hides.
1929: Severe drought. Some 50,000 animals lost. Provision of famine relief.
1933–5: *Olameyu Looloyik*
Famine due to drought and locust invasion. A huge number of cattle got finished and many people died because of hunger.
1938–9: Serious drought.
1943–6: *Emboot Enkurma Nanyokie*
Famine after severe drought and hopper infestations. The government provided maize flour.
1953–6: Period of drought, famine relief provided.
1960–1: *Emboot Enkurma Sikito*
Extremely severe drought. Many animals died. The government gave (yellow) maize flour.
1963: Food By helicopter
Planes and helicopters were used to transport food relief.
1973–6: Severest drought since 1961. Some 60,000 people received famine relief. The government gave white maize flour. Modest number of cattle died.
1979–80: Rains absent for a short period. The drought did not take a long period and also the animals were not affected so much.
1984: Severe drought. It also killed animals but not like *Olameyu Loolonitok*.
1992–4: Rains were late. Animals did not die but people received food relief.
1999–2000: Severe drought with wildlife dying in large numbers as well.
2004–5: Period of drought.
2009: Severe drought.

Source: interview Ole Naikuni, Ole Sailenyi.

The importance of each of these signs during two major droughts (1984 and 1994) among Maasai pastoralists based in the Selengei region, north of Amboseli, is presented in Table 12.2. It shows that the stars and behaviour of animals are the key signs consulted.

Some Maasai regard drought as a natural catastrophe caused by certain evil actions that displeased *Enkai* (God/Rain). Praying and offering a sacrifice are methods used to please *Enkai*. These prayers are still practised, but mostly among traditional Maasai. The Maasai have many ways to ask God to give rain and long-lasting sources of water. One way is when the elders meet and agree to sacrifice a lamb to ask for rain. The lamb must be holy with the respected colour. It should always be black or brown and a male.

MAASAI WAYS OF PREPARING FOR DROUGHT

A Maasai proverb states that water cannot be drunk while it is flowing: you cannot do anything before it happens. In this section we summarize food

Table 12.2. Drought signs consulted by Selengei Maasai pastoralists in 1984 and 1994 (per cent).

Drought Signs Consulted before Drought	1984	1994
1. Movement of morning/evening stars (*kiliken/olokirrai*).	81.0	74.6
2. Movement of pleiades (*inkokua*).	78.4	72.0
3. Animals not willing to leave the *boma*	59.5	58.5
4. Animals urinating and passing dung while laying down	57.6	58.5
5. Animals crunching their teeth without eating food	57.6	53.4
6. Foggy skies and red sunset (*oloikulu*).	57.8	47.5
7. Smallstock intestines with air gaps/being thin at the end	50.0	46.6
8. Skincolour/hair condition of animals	38.8	36.4
9. Meat of slaughtered animals is watery	34.5	36.4
10. Moon tilting	25.9	30.5
11. Predictions by *oloibonok* (Laibon-ritual expert/prophets who 'throw' stones).	12.9	13.6
12. Predictions by *langeni* (clever elders, rainmakers, read intestines).	12.7	11.2
13. Not arriving of the *naokbibia* bird	9.5	13.6
14. Predictions by *laisi* (predict by observing stars).	8.5	11.2
15. Predictions by *ilaker* (study the stars-among Samburu and Maasai Illaarrosero clan).	7.8	9.3
16. Changed colour of grasses and leaves	3.4	7.6
17. Easily uprooting grasses	4.3	1.7
18. Short rains failing/wind blowing from west/cold weather in May/ termites finishing grass	3.4	1.7

Source: Rutten, 1999.

strategies as well as non-food related actions taken by Maasai households in preparation for a drought. Maasai implement certain preventive coping strategies to prepare for a period of severe drought. Two major tactics can be distinguished, i.e., management of the herd and management of (natural) resources. Considering livestock management the Maasai, like all pastoralists, aim to keep sufficiently large herds to accommodate potential losses from droughts and diseases. Likely, managing a variety of livestock species with specific ecological niches (grasses, shrubs) and production characteristics (i.e., different periods of milk recovery, disease susceptibility) should assist in overcoming periods of hardship. Herd mobility and allocating animals under various arrangements to friends and relatives also optimizes use of resources and reduces risks. Increasingly herders decide to destock and restock ahead of droughts and return of rains, avoiding a full collapse and seeking financial gains, respectively. Individual animals are also treated to prepare for droughts by fattening them and as a result making them more drought resistant, increasing output through crossbreeding or preventing diseases through applying acaracide or installing tsetse fly traps.

This last strategy could be considered also as a way of managing natural resources which is the other major approach available to prepare for

---

Box 12.2. Rain and grass prayers.

Rain prayer
Naai, interasaki yiook olkenkei omelok loo nkaik inono-Naai!
Enkai naaitalip iarat, tadamu iyiook minco iarat etoyio-Naai!
Naai, Enkai nashal inkilani tadamu iyiook minco iarat etoyiu-Naai!
O God, give us beautiful honey from your hands, amen!
God who makes streams to flow, do not let them run dry, amen!
O God, whose garments run with oil, remember us and make our streams to flow!

Grass prayer
Naai, intamelonoki iyiook inkujit metonyorra inkishu
Naai, intamelonoki iyiook enkare metonyorra inkishu o iltunganak.
O God, make sweet for us the grass so that the cattle may love it.
O God, make sweet for us the water so that cattle and people may love it.

Source: Mol 1978:127; Mol 1996: 62.

hardship. In this category is also the setting aside of certain dry season grazing areas in agreement with other households. At individual level households often also fence small spots (*olkeri*) used for calves and sick animals. Besides grass and browse, Maasai improve water availability through shallow wells, pans, boreholes. Diversifying the household economy through engagement in cultivation, wage employment or commercial business, and the opening up of bank accounts to store money for less fortunate days are on the increase. By contrast, food preservation through drying of meat or storing of cheese, are traditional practices on the decline.

Otherwise we should stress once more that the Maasai strongly believe in one (black) God (*Enkai narok*). He is omnipotent, the creator of the earth and the ultimate source of all human welfare and misfortune. Man's ability to 'plan for drought' is thus considered to be limited.

MAASAI COPING STRATEGIES DURING DROUGHT

During the time of a drought a range of extraordinary strategies are implemented. These are in the field of herd management, human food habits and assistance sought from others. The importance of each of these strategies varies in time and place.

For management of the herd more hands are required during a drought. Elderly people take over herding while younger Maasai search for distant pastures. Feeding of acacia tree (*oltepesi*) pods (*isagararam*) and leaves (*oltimigoni* (*Pappe Cappenis*), *oseki*) as fodder becomes important, so are special grasses (esp. *erikaru*) for young calves and rams. As for drinking, the watering regime for animals is changed from every day to every other day or less. Finally, more milk is left for the calves at the expense of human consumption.

Table 12.3 summarizes the most important ways Maasai pastoralists of Selengei sought feed for their livestock during the major droughts of 1984 and 1994. It can be concluded that Selengei pastoralists hardly collected fodder before the onset of a drought but do collect and provide extra feed during a drought. Especially *isagararam* pods, *erikaru* grasses, nests of birds and to a lesser extent leaves are looked for. These are indeed available in the Selengei area. Napier grass, hay and maize stalks by contrast are not.

Table 12.4 shows how Maasai pastoralists solve the problem of water provision during the dry and wet seasons in their home area. It stresses the seasonality of a number of sources (river, pans) as opposed to the groundwater resources stored at varying depths in wells and boreholes.

If the animals have been lacking food and water for too long the pastoralists will attempt, especially if the rains are near, to maximize the chances for survival of the weakened animals by lifting them during the night; if needed, also during daytime (Figure 12.7). The animals are made to stand for about 30 minutes and laid down facing the other side up. This is done to prevent dislocations of the bones. Their forelegs should not be stressed for too long. This rising of the animal is a natural phenomenon which healthy animals practice by nature every night. When the drought becomes that severe they are no longer able to stand up themselves, a human hand is needed. This is a gradual process whereby initially one person is able to assist the cow. In a later phase more people are needed, while in the final stage at least four people and a stick (placed under the belly of the cow) are required. If lifting is no longer possible the animal will be turned to the other side. The lifting is done twice or three times a night. Also at the start of the rainy season, when many animals are lost the practice of lifting continues. At the beginning of the rains, the water flushes away the little grass left. It also affects the cows if they keep on lying down for too long. The rain will affect them less if they stand.

Table 12.3. Extra forage collection by Selengei Maasai in 1984 and 1994.

	1984		1994	
	Before	During	Before	During
Isagararam pods	14.8	90.4	6.8	93.2
Erikaru grasses	15.7	89.6	5.1	91.5
Birds nests	7.8	80.9	4.3	76.9
Leaves (*oseki/oltimigoni*).	1.7	36.5	0.9	42.7
Maize stalks	4.3	8.7	0.9	2.6
Hay	0.9	7.8	0.0	3.4
Napier grass	0.9	1.7	0.0	0.9

Source: Rutten, 1999.

Table 12.4. Water sources used by Selengei Maasai, 1996.

Water Source	Always	Wet Season only	Dry Season Only	Main Problem		Other Problem 1		Other Problem 2	
1. Well (*ilcborroi*).	32.8	7.8	27.6	2	82.6	1	10.9	14,15	6.5
2. Shallow well (*ilumbwa*).	10.3	12.1	21.6	1	42.9	9	32.1	2,3,8	25.0
3. Borehole	48.7	8.5	2.6	3	78.1	10	15.6	2,11	6.3
4. Tractor pan (*silamkeni*).	1.7	59.8	0.9	2	*2*	14	*1*	16	*1*
5. Natural pan (*olbua*).	0.0	61.5	0.9	-		-		-	
6. *Iloogolala* pan	0.0	59.8	0.0	-		-		-	
7. Subsurface dam (*ituamati*).	0.0	60.7	0.0	-		-		-	
8. Shallow pool (*empakaai*).	10.3	69.0	0.0	14	*10*	13	*2*	2,11	*2*
9. Pool near river (*iltorot*).	0.0	81.9	0.0	14	*9*	13	*1*	2	*1*
10. Roofcatchment waterjar	0.0	0.0	0.0	-		-		-	
11. Watertank	0.0	0.0	0.0	-		-		-	
12. Pipeline watertap	38.8	4.3	8.6	18	54.0	19	32.0	7,17	14.0

Source: Rutten, 1999.

Note: Problem scores in italics are absolute, otherwise percentages. Code: 1. Sand sealed/digging needed. 2. Quantity is small, does not last long. 3. Solar energy fails during clouds. 7. Expensive. 8. Walls collapse. 9. Too deep. 10. Maintenance. 11. Congestion of livestock. 13. Far away. 14. Water dirty. 15. Wildlife disturbing livestock and destructing the source. 16. No nearby grazing near source. 17. Disconnection from tap. 18. Non reliable. 19. Pressure leakage.

Figure 12.7. Lifting of exhausted animals (© Marcel Rutten).

Another way for humans to cope during droughts with reduced availability of food is to change their food habits. Less food is eaten, especially by herdsmen, but also at the basic homestead less milk is available. Besides reduced amounts, the kind of food consumed is changed in that, for example, the less favoured parts of the animal such as tongue, pancreas and heart are eaten. Milk is mixed with herbs and (more) goats and sheep are milked. Blood (cooked or mixed with water) consumption increases. Food from other resources than livestock also becomes important such as gathering fruits and roots. Small antelopes, elands and other wildlife is hunted and animals and skins sales increase to buy food (e.g. maize meal and sugar mainly) from the shops.

If all of this fails, a Maasai household will seek outside assistance for money and animals from friends and relatives (see Box 12.3). Government and church relief programmes (either free hand-outs or Food for Work Programmes), if available, are also resorted to. This will help to reduce the need to sell livestock. A practice which is said to be on the decrease is praying for rains and sacrificing a ram of specific colours.

Campbell (1978) has shown that in the 1974–6 drought Maasai practising agro-pastoralism in the Loitokitok area were able to cope with the drought more successfully than pure pastoralists. The combination of livestock and crops did offset major difficulties and permitted many Maasai farmers to help less fortunate relatives. The Athi Kapiti plains region has also engaged in cultivation, but the drying of wells has made life less secure. Other regions like Meto are less affected and offer opportunities, but the driest areas will not allow successful cultivation or because of wildlife is a risky investment. Other sources such as wage labour, commercial livestock trading, remittances and the selling of land have become important in securing food availability. In addition, Maasai pastoralists try to intensify their activities (e.g. keeping improved breeds for higher milk and meat production). In the following paragraphs we will look at the specifics of the latest major drought of 2009 and the way mobility was applied by a number of Maasai households of Kajiado.

THE 2008–9 DROUGHT

Still recovering from the 2004–5 drought, Maasai pastoralists were confronted with poor rainfall in the November–December 2007 period. The food security could deteriorate to the highly food insecure category within the first quarter of 2008, should rains fail to establish and continue through January (USAID 2007). The Early Warning Bulletin for Kajiado reported in February 2008 that indeed the short rains expected in November and December came late, were erratic and poorly distributed and most of Kajiado received less than 50 per cent of normal rainfall which deteriorated key grazing fundamentals (pasture, browse and water). Worse, due to wildfire 20–30 per cent of dry season grazing pastures in several regions of Kajiado had been lost. Surface water sources were depleted. Watering distances had risen to 25 km in Kajiado's pastoral areas. More so, about 30 per cent of boreholes were not operational which caused people and livestock spending 8–12 hours waiting their turn

---

Box 12.3. *Ewali*: the Maasai social system.

Many pastoral groups are known to have a special social system that acts as a kind of insurance system that supports the poor. The Maasai system is called *ewali* (from *awal*: to respond). Also the phrases *Enkishorot* (gift) and *Iretunot* (assistance) are used. There are different forms of assistance and gifts (*aretu*). Special forms of assistance used after a family's herd was struck by a drought or diseases are:

1. *Enkiteng lepai:* one cow (often with a calf) is given to the receiver for an unspecified time. This kind of assistance is a form of borrowing livestock to provide milk and will be returned afterwards.
2. *Nkishu lepai:* a number of cows given for milking only. They can be recalled but this may take some ten years. The receiver cannot sell nor slaughter these animals. Also this assistance is a form of borrowing.
3. *Agel:* cows, sheep, goats, donkeys are given out without the condition to be returned.
4. *Awal:* this means 'to exchange'. Steers and smallstock could, for example, be changed for heifers. The idea is that the needy household gains in this process.

Source: Rutten, 1999.

at the watering points. Women left their homes as early as 3 am only to return at 4–6 pm in some areas. As a result, other household duties suffered at the expense of household incomes. Moreover, most of the planted crops succumbed to moisture stress when the rains ceased early, notably in the agro pastoral zones. Irrigated schemes performed better but in Magadi the water intake was destroyed by elephants (EWB February 2008). In nearby Torosei, baboons and humans were fighting over water. The pregnant women especially were attacked by the animals. The outbreak of Foot and Mouth Disease worsened pastoralists' prospects, after quarantine on livestock movement was imposed. Affected pastoralists were unable to trade livestock in exchange for cereals. By the end of 2008 more pastoralists started migrating into Tanzania, neighbouring cropping districts and into game reserves. Households joined labour in herding. Casual labour and selling of charcoal and sand were applied as coping methods. These opportunities to some extent helped to limit malnutrition in the area (Zwaagstra et al. 2010: 24).

Figure 12.8 shows the comparison made by Zwaagstra et al. (2010) of the monthly rainfall in Kajiado with the average monthly rainfall over previous years (2003–7) as provided by the Early Warning Bulletins. This long-term mean is based on too few years and it was concluded that data over at least 30 years should be analysed. In addition they point at the problem that the network of rain gauges is insufficient and shows a bias to urban and more humid regions.[2]

Long-term data from Isinya seem to indicate that the EWB data overstress the long-term mean rainfall in the November–December period (Figure 12.9). The figure also shows that both the 2008 long rains and short rains came too early and especially that the long rains were below the long-term average while December 2008 was completely dry. As a result, pastures do not regenerate sufficiently for more than one rainy season causing severe stress for livestock food intake, and subsequently human beings.

Therefore, even though we notice an upward trend in rainfall in the last decade for the Isinya region (see Figure 12.10 showing percentage deviation from the 605.6 mm mean), this does not necessarily translate into improved pastures.

Pastoralists complain of the fast-drying grasses as a major anomaly from the past. Part of the extra rainfall will be lost due to a rise in temperature. In addition, drying of rivers affects the riverine zone, creating an open canopy. Also, sprouting of the invader weed *Ipomea kituinesis* takes away good pastures (EWB January 2008). Cutting of trees for charcoal production, notably near Namanga and Mashuru, is contributing to the spread of the weed. The charcoal revenues, however, are of major importance to making ends meet during periods of hardship, as are remittances.[3] Wildfires and overgrazing as a result of concentration of livestock in some of the short rains regenerated pastures might also have added to the ILRI observation as pictured in Figure 12.11 that greenness in

Kajiado Monthly Rainfall

2008–2009 rainfall vs 2003–2007 mean

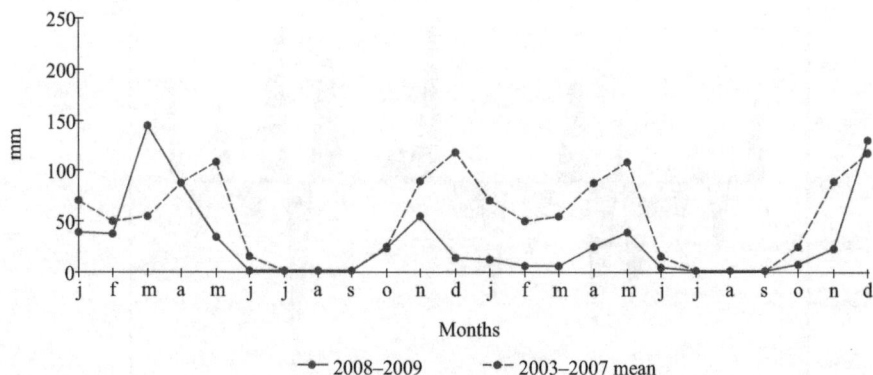

Figure 12.8. Average monthly rainfall in 2008 and 2009 (normal line) relative to the longer-term mean (dashed line) according to Early Warning Bulletins (modified from Zwaagstra et al., 2010).

Kajiado turned overnight from normal to emergency from December 2007 to January 2008 (Zwaagstra et al. 2010; EWB December 2007; EWB January 2008).

Rainfed crops failed, resulting in higher maize prices from late 2007 onwards at around Ksh 25 per kg to over Ksh 40, or double the long-term price (below Ksh 20/kg). A ban on maize imports from Tanzania did not help to bring the price down. It only did by March 2010 after the first

Isinya Monthly Rainfall

2008–2009 rainfall vs 1962–2010 mean

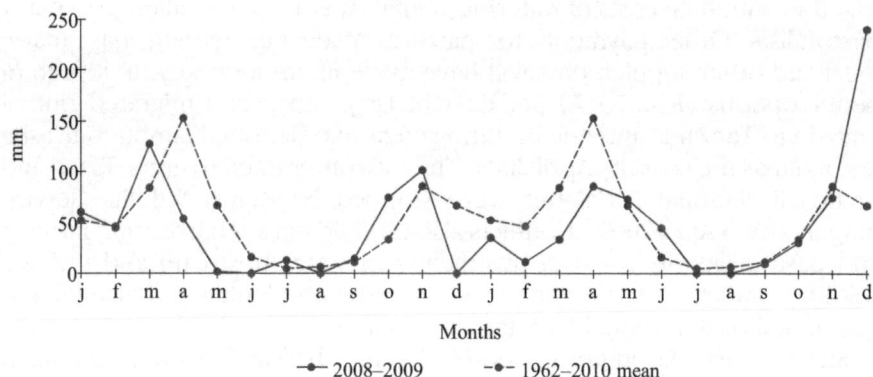

Figure 12.9. Monthly rainfall 2008–9 in comparison to the 1962–2010 average for Isinya station (© Marcel Rutten).

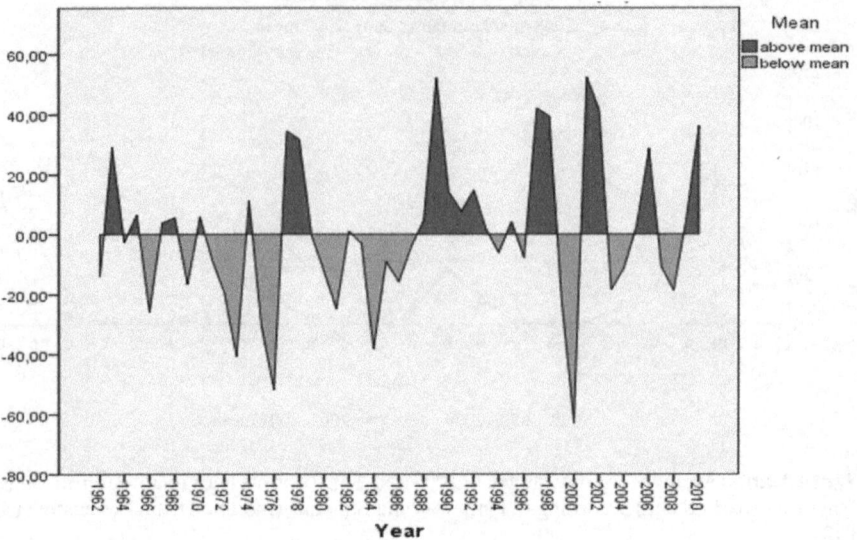

Figure 12.10. Annual rainfall during 1962–2010 deviation from the mean (percentage) for Isinya station (© Marcel Rutten).

successful crop was about to be harvested (EWB March 2010). By contrast, average cattle prices moved around the long-term average of about Ksh 10,000/head in the range of Ksh 8,000 to 13,000 during 2008 while in 2009 with a deepening drought it dropped from Ksh 8,000 to 4,000 at the onset of the short rains of December. Since than it has climbed very fast to Ksh 20,000 per head and has remained since. Prices of goats and sheep have almost tripled.

Overall the terms of trade were very poor during most of 2009, which together with high costs of watering animals was a major challenge to many pastoralists. Other payments for pasture, medicine, herding, hay, maize meal and other supplements will have made life more costly. In search of better options close to 70 per cent of large stock had migrated out of Kajiado to Tanzania and neighbouring areas like Nairobi, Kiambu, Naivasha and as far as the coast by April 2009. They also attempted to enter Tsavo and Amboseli National Parks but were stopped by Kenya Wildlife Service rangers (EWB April 2009). Conflicts also erupted between irrigation farmers and pastoralists due to over abstraction of water upstream and human/wildlife conflicts as farmers had to contend with elephants that migrated out of Amboseli National Park into their farms.

Strategic dry season grazing boreholes and shallow wells were the main water sources for livestock and domestic use. All temporary water sources (dams and pans) were dry. Use of borehole water for livestock was at a fee (Ksh 20–50 per head per month) by September 2009. The community

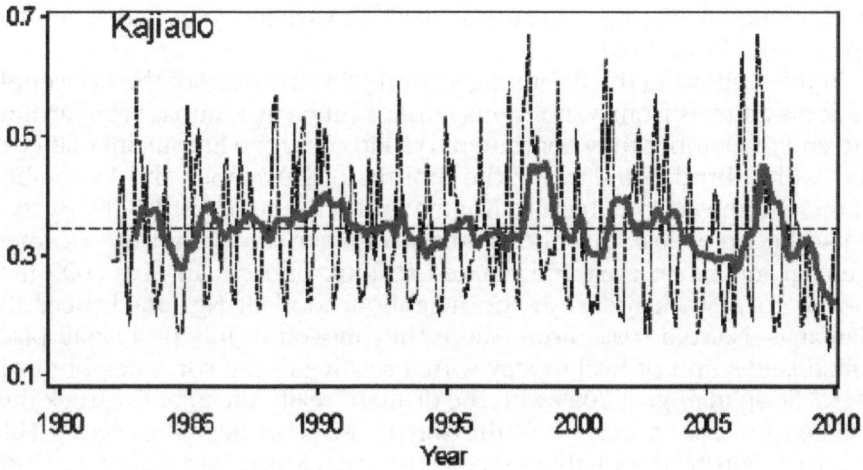

Figure 12.11.   (Mean) monthly Normalized Differential Vegetation Index (NDVI).[4]

water trucked by weak animals, calves and domestic use was at Ksh 300–500 for a 200 litres drum.

The government, international donors and local NGOs assisted in this exercise and through other ways. Of these, the subsidy of diesel for running boreholes and development of shallow wells was considered to be most effective. Provision of animal health was important, but often came too late (October 2009).[5] The commercial off-take programme was also not successful as the quota offered by the Kenya Meat Commission (KMC) was much too small and also came rather late (August 2009) (Zwaagstra et al. 2010). Moreover, pastoralists, especially if prices offered are too low, do prefer to try to keep the animals alive and make it to the next rains. The total costs involved over 1 million euros. Otherwise, it was up to the individual household to safeguard their livelihoods during the 2008–9 drought. To illustrate the above problems encountered during migration, let us turn to a narrative of a few Maasai households.

NARRATIVES OF MIGRATION BY METO/MAILUA MAASAI HOUSEHOLDS

Household A

For many years this family struggled to overcome poverty. Droughts, cattle diseases and losses due to the hilly area had decimated the herd repeatedly. Having engaged in cultivation since 1990 they were able only following the good rains of 1997 to harvest a bumper crop of maize that took them to the next wet season. The key water source was the Loretiti well in Tanzania, albeit owned by the Kenyan Maasai. Selling maize for goats and sheep had uplifted the household.

In July, following the poor long season rains the head of the household took his animals from Meto to his father's ranch at Kumpa. From around the end of October they continued to Nkito near the Empukani plains, to stay with a brother-in-law for the months of November and December. At the end they moved back to Kumpa until the end of April 2009. By then it had become clear that they could no longer stay within the (former) Meto group ranch area and needed to look outside. In May 2009 they moved to a stock friend in the neighbourhood of Ngatataek near the Namanga–Nairobi road. From there they moved in July to a small place called Lele south of Bisil to stay with a relative of the son's neighbour in Meto. Soon in August 2009 with the drought really affecting livestock they trekked for a long period all the way to Kiserian near the Ngong Hills south of Nairobi. It took them six days to reach Korna Baridi after a 217 km trip (see Figure 12.12). The cost to have the animals trekked all the way from Meto to Kiserian was between Ksh 300 to 500 per cow.

This is a high up area, relatively wet, but also windy and cold. The neighbour had a nephew near Kiserian but the key aim was to move up the forest in search of grass. They now moved as a group of four livestock owners, with some extra cows of others, making up a herd of 80 head of cattle. Thirty of these belonged to the head of the household. After a short while, the father decided to leave with two cows. The remaining cows stayed down in small earmarked areas during the day. Watchmen would prevent entrance to grazing areas of the forest areas during the day. However, during the night between 2 and 5 o'clock they would turn their heads. Being Maasai pastoralists themselves living at the foothill they did know the urgency and need to feed the animals. The Ngong Hills, though, came with three serious challenges:

1. In the forest there is not enough grass. Cows start eating from trees causing *imbenek* – a disease caused by animals feeding on leaves only.
2. East Coast Fever (ECF), in a different mode, was also around. The symptoms were less clear. Only a sign of coughing.
3. Pneumonia (cows start shaking).

No lifting of cows was needed as many cows died not so much from the drought, but of diseases. Calves died mainly because of diarrhoea. Another four got lost in the forest while grazing at night. Four rolled while descending and died within ten minutes because the stomach blocks the breathing system if the mouth faces downwards. Four old cows were sold at Ksh 4,000 each. In the end only six were left. The return to Meto started at the end of October, anticipating the short rains. In Meto, one more died. The others

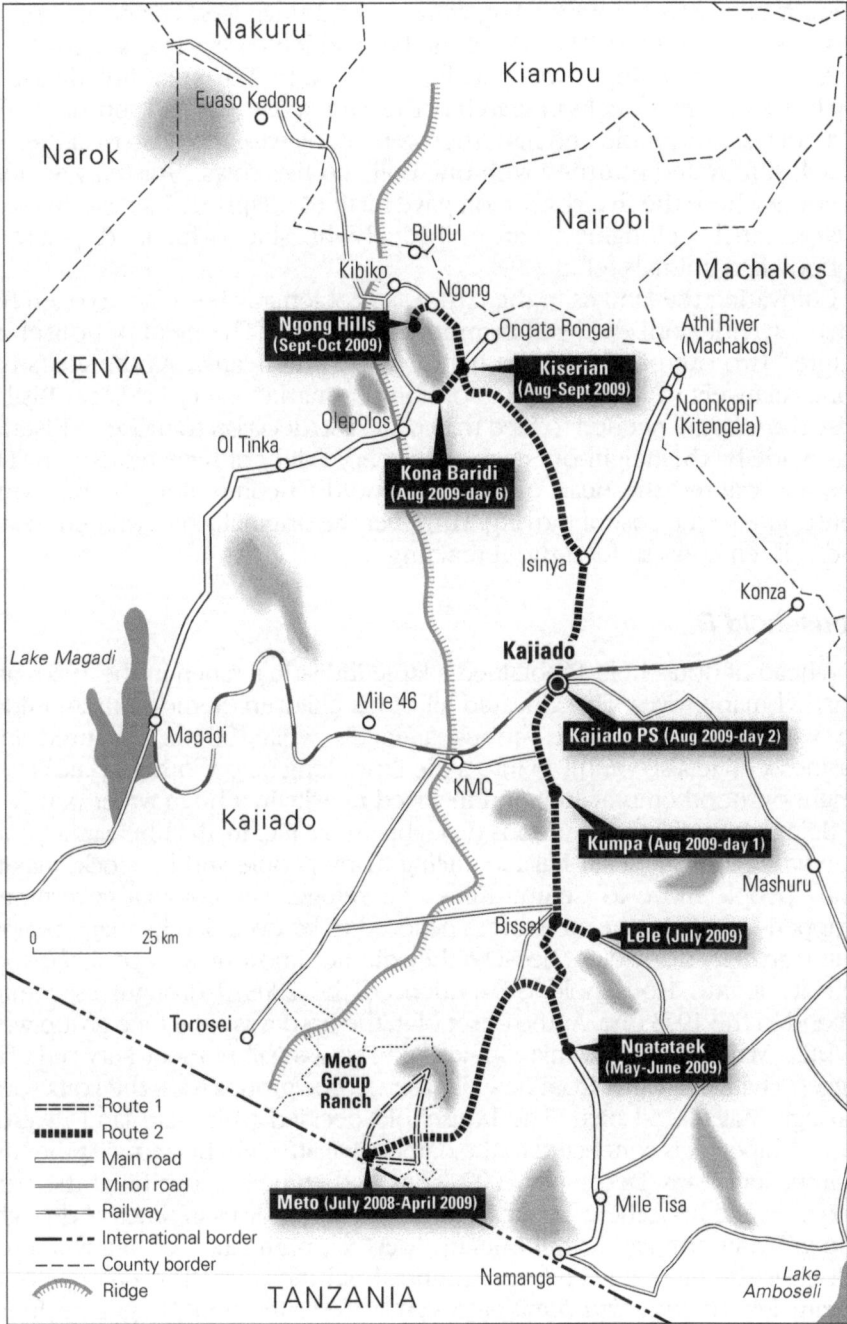

Figure 12.12. Migration track Household A during 2009 drought (© Marcel Rutten).

made it because of extra feeding with hay and unga feed (maize meal). Yet this was a costly affair. The two cows that had left earlier also survived.

After strengthening it was decided to move to Tanzania directly across the border for two weeks in search of the fresh grass that had sprouted. The head of the household and his father went along with ten head of cattle. Yet they lost four and returned with one bull and five cows. Grazing was done at home where the five cows soon gave birth to offspring. Use was made of a large ranch belonging to an individual who shared his land as almost nobody had animals left.

Cultivating the land as in the past was problematic because no oxen had been left, nor donkeys, seed or money for labour. The head of household planted two out of 12 acres by hand (maize and beans). As the goats had done relatively well, these were sold on the market every Friday at Bisil to raise the income needed to feed the cows. The decision to move to Kiserian was made by the neighbour as the household did not have herdsmen. The nephew wanted the head of the household to come along as he would contribute to the cost of moving. However, he originally had wanted to stay and sell ten cows to feed the remaining 20.

Household B

The head of household B obtained a large individual ranch in the 1960s and married many wives. He educated all of his children. Some of them ended up working abroad with various agencies while others ventured into businesses. He also ventured into agriculture long before other Maasai in his neighbourhood embraced it and invested heavily in a huge water pan.

Being old he has seen various droughts in his life. In 1943 he was a young man when a drought hit Kajiado, killing many people and livestock, making many people move to Central Kenya for refuge. The colonial government dropped food from the air in areas believed to be occupied by many people. Unfortunately, many died, because they did not know how to cook the food and ate it raw. Household B experienced the 2008–9 drought as coming second to the 1943 one. At the onset of it, the herd was split. One group went to Lake Manyara in Tanzania. Some died but a small number survived. The biggest challenge came from new diseases. Three groups took the Loita route through Magadi. All died. The household decided to irrigate land to grow food. A pipeline is connected to the dam to irrigate eight hectares. At the time of interview, early December 2009, rains had come but not filled the dam. However, the household expects agriculture to save lives in times of drought. The growing population will outstrip livestock numbers. Too many animals will clear the grass and the soils cannot hold moisture. The emergence of Oltiameleki (*ipomea kituinesis*) is a sign of desertification. No grass is found near the area where this evergreen, but unpalatable, plant grows.

The household fears that there is little chance of ever beating the droughts. But there is a need to act and plan before the drought strikes again; to engage more proactively in the search for solutions. For example, people have been cutting down trees for charcoal burning, an indicator of poverty. This calls for conservation also of the sand harvested so as to prevent overexploitation. Yet, despite the realization, there is little indication that the local people are ready to save the situation as this will affect their immediate livelihoods. The government and development agencies must put water development at the core of their concerns and focus on rainwater harvesting. Water is a panacea to sustainable land use. If no care is taken of the environment, it will take care of the country in a very harsh way. During 2008–9 household B used the income from the sale of goats and sheep to buy food and medicine for the cows and family members. The price of goats was higher than that of the cows and ranged from Ksh 2,500–4,500. Cows could sell for as low as Ksh 500 due to their severely emaciated condition that saw them at times unable to stand up on their own. The skin was worthless and thus the pastoralists did not even bother skinning the dead carcasses. The 2008–9 drought forced livestock to feed on a number of foodstuffs never experienced before. Livestock keepers searched for all sorts of foods to keep their animals alive. For example, cactus plants were cut, leaves burned to remove the thorns and offered in small pieces. Besides food the cactus also contained water. Moreover, without cactus, the milk production dropped by 50 per cent. In addition, other drought-resistant plants were collected such as sisal-like Ordupai. In the end the animals fed on anything they encountered, including plastic bags.

Household C

At the beginning of 2008 Household C had 25 cows, 15 steers and one bull. The animals were kept nearby the Mailua ranch north of Namanga, where the head of household acted as a head teacher at a primary school. But in mid October the animals were moved to the Oldonyio Orok mountain area. His animals stayed until 17 November, at no costs. The move had not been a success. Although there was free grass available there were diseases, notably ECF, as a result of buffalos spreading ticks, 21 cows died in spite of spraying. The remaining 20 were split in two herds, with ten kept at home. The others went to Euanata (near Mile Tisa). In Mile Tisa he had hoped to sell the cows to the KMC at Ksh 8,000, but the company never turned up. Four died because of drought and disease. The hides did not make a penny. Subsequently the remaining ones moved to Ngatataek to stay with a relative. Lifting of the animals was needed day and night. Only two cows and one bull survived, the others died because of the drought. The three remaining animals were provided with extra food. At home five died (plastic inside) and the survivors were also fed with molasses, unga, maize and hay. Cactus was bought from the Bartimaru area. The Maasai in that region only used cactus and Ordupai and did well (Box 12.4).

Box 12.4. How did others perform?

Household D: headed by a widow. She kept a few animals on her small fenced shamba. In September/October 2009 she added unga. All animals survived. After the drought she rented oxen out at Ksh 2,000/per acre for ploughing.

Household E: a rich household, having a large individual ranch. Hired six lorries to bring hay. Only lost calves sent to Tanzania.

Household F: a rich individual rancher. Lost some 320 head of cattle. Left with 80 (young ones).

Household G: moved straight to Simanyiro (the land of cows) and survived with 80 per cent.

Household H: moved all animals to Sonjo/Loita Tanzanian side. Tsetse fly killed all his cows. He ignored local advice not to move up.

Household I: moved to Lake Manyara in Tanzania with 400 heads of cattle and returned with 40 only. Payments were made to Tanzanian Maasai for grazing.

Source: Rutten, 1999.

CONCLUSION

This case study of Maasai pastoralists' encounters with droughts has illustrated that these famous cattlekeepers at times have to go to extremes in finding food and water for their animals. Mobility is crucial to be able to follow the rains producing grass, if need be up to hundreds of kilometres away from home. However, with land losing its communal tenure and becoming individually owned, paying for pasture, hay, maize stems and unga is on the rise. Collecting or paying for leaves, sagararam pods, climbers and birds nests are other options available. New foods such as cactus and sisal-like plants were added to this list in 2008–9. Maasai in other regions bought carnation flower stems, rose flower leaves, cabbages and waste food from hotels. Drought conditions killed many animals but other reasons also killed livestock: diseases, steep hills, polythene bags and dark forests all caused livestock to be lost.

The importance of borehole and shallow wells water is also clear as streams and pans did not last long into the dry season. Meto Maasai regretted that since 2000 they had lost access to their traditional wells, located in Tanzania. Subdivision of the ranch had triggered the development of new water sources such as public and individual pans as well as community boreholes. Cost sharing with NGOs, boreholes had been constructed and maintained by local committees. Outside the Meto area, though, access to water was limited and obtained by trekking long distances and paying high prices. Trucking of water, and subsidies on diesel to run the borehole engines turned out to be crucial.

Availability of food for humans, as a result, also dwindled. Many men moved with their animals and left women and children behind. Cultivation was not successful for several seasons in a row. School-feeding schemes, food rationing and food relief helped to overcome some of the problems.

Women started petty trade, such as selling firewood, charcoal or sand. Casual labour and remittances also helped to buy food and medicines for cattle. The crucial role of smallstock to earn food should also be mentioned, especially as the cost of living had risen as a result of low livestock and high food (maize, beans) prices.

In the end, the Maasai herds were rebuilt again. This was done partly through new offspring but also through the purchase of animals through livestock markets which were flooded in 2011 with cattle, going at very low prices. These cows had been trucked from northern Kenya which had been hit by a drought which had more or less bypassed the southern livestock ranges this time.

NOTES

1 Behnke (1994) has excellently outlined the management characteristics of these non-equilibrium environments as compared to the equilibrium grazing systems in more temperate zones. The latter are characterized by high levels of climatic stability resulting in constant levels of primary production.
2 EWB rainfall stations are not clear. However, EWB provides details mostly for Isenya, Namanga, Central, Mashuru, Upper Ngong and Loitokitok. This seems to be a bias to more humid zones of Kajiado.
3 During the 2008–9 drought charcoal became the overall most important source of income (35 per cent), ahead of livestock sales (20 per cent), petty trade (16 per cent), remittances (15 per cent) and sand harvesting (14 per cent). Only by April 2010 livestock trade had taken over again with 46 per cent, while charcoal and sand had gone down to 21 per cent and 8 per cent respectively. The remainder came from remittances (2 per cent) and casual labour (23 per cent) (EWB Dec 2009, April 2010).
4 The Normalized Differential Vegetation Index (NDVI) is a remote sensing based index (ranging from 0 to 1) reflecting vegetation greenness with NDVI < 0.20 – 0.25 for bare soil and dead vegetation and NDVI of 0.6 to 0.7 for green vegetation with closed canopy (Zwaagstra et al. 2010).
5 Drugs and animal feeds were sold by traders to the Ministry of Livestock Development at highly inflated prices, by as much as 1,000 per cent. The drugs would have been much cheaper if they had been sourced directly from the manufacturers. The entire practice exposed some traders in collusion with insiders in the Ministry of Livestock. There have even been accusations that some officers had formed their own companies which they used to sell goods and services to the Government (*The Standard* 18/11/2009).

REFERENCES

Behnke, R.H. 1994. *Natural Resource Management in Pastoral Africa*. London: Overseas Development Institute ODI, Institute for Environment and Development, Common Wealth Secretariat.

Campbell, D. 1978. 'Coping with drought in Kenya Maasailand: Pastoralists and farmers of Loitokitok area, Kajiado District'. *Working Paper no. 337*. Nairobi: Institute for Development Studies.

Carter, J. 1934. *The Kenya Land Commission*. London: HMSO.

Ecosystems Ltd. 1982. *Amboseli/Lower Rift Regional Study*, Final Report, report prepared for the Wildlife Planning Unit, Ministry of Tourism and Wildlife, Republic of Kenya.

EWB several years Early Warning Stages Bulletin Kajiado North, Central, Loitokitok Districts, Office of the Prime Minister. Ministry of State for the Development of Northern Kenya and Other Arid Lands, Arid Lands Resource Management Project ALRMP II.

Farmer, G., Wigley T. 1985. *Climatic Trends for Tropical Africa*. Report to the Overseas Development Administration, Research Project Number R3950.

Herren, U. 1991. '"Droughts have Different Tails". The Impact of and Response to Crises in Mukogodo Division, Laikipia District, Kenya'. In Stone, J.C (ed.), 'Pastoral economies in Africa and long-term responses to drought', *Proceedings of a Colloquium at the University of Aberdeen, April 1990*. Aberdeen: Aberdeen University, pp. 69–85.

Mainguet, M. 1990. *Desertification: Natural Background and Human Mismanagement*. Berlin: Springer Verlag.

Masharen, S. ole 1989. *The Green Revolution in Maasailand*, Arid and Semi-Arid Lands Programme Kajiado District.

MoLD (Ministry of Livestock Development Kajiado District) 1988. *Kajiado Livestock Census: Integrated Livestock Census and Infrastructure Survey Report*, Department of Livestock Production Kajiado District/Arid and Semi-Arid Lands Programme Kajiado.

Mol, F. 1978. *Maa: A Dictionary of the Maasai Language and Folklore: English–Maasai*. Nairobi: Marketing & Publishing.

——— 1980. *Maasai Mara*. Nairobi: privately published.

——— 1996. *Maasai Language & Culture Dictionary*. Limuru: Maasai Centre Lemek, Kolbe Press.

Mwangi, M.N. 1993. 'MPhil Research Report on Shallow Wells in Kajiado District, Kenya'. Unpublished thesis. Loughborough: Loughborough University.

Population Census several years Republic of Kenya 1989, 1999, 2009.

Rutten, M.M.E.M. 1992. 'Selling wealth to buy poverty. The process of individualization of landownership among the Maasai pastoralists of Kajiado district, Kenya, 1980–1990'. PhD Thesis. Nijmegen: University of Nijmegen.

——— 1999. 'Drought Planning and Rainwater Harvesting for Arid-Zone Pastoralists: The Turkana and Maasai (Kenya) and the Negev Bedouin (Israel): Social Constraints and Opportunities'. *NIRP project 92–1.3 Maasai Report* – African Studies Centre, March 1999.

——— 2008. 'Why De Soto's ideas might triumph everywhere but in Kenya: A review of land-tenure policies among Maasai pastoralists' In: Rutten, M., A. Leliveld & D. Foeken (eds), *Inside poverty and Development in Africa: Critical Reflections on Pro-poor Policies*, Leiden: Brill Academic Publishers, pp. 83–119.

Sandford, G.R. 1919. *An Administrative and Political History of the Masai Reserve*. London: Waterlow and Sons.

The Standard, Kenyan daily newspaper, several dates.

USAID. 2007. 'KENYA Food Security Update, December 2007', Kenya Food Security Network, Nairobi. Available at http://reliefweb.int/sites/reliefweb.int/files/resources/BE3008D910C45927C12573B8000EB313-Full_Report.pdf (accessed 16 April 2015).

Vincent, L. 1981. 'Drought risk and agricultural production in Maharashtra state'. *Transactions* 1(2).

Zwaagstra, L., Sharif, Z., Wambile, A., de Leeuw, J., Said, M.Y., Johnson, N., Njuki, J., Ericksen, P., Herrero, M. 2010. *An Assessment of the Response to the 2008–2009 Drought in Kenya. A Report to the European Union Delegation to the Republic of Kenya*. Nairobi: International Livestock Research Institute.

13 Rainfed Agriculture, Drought and Hunger in Tanzania

Terje Oestigaard

INTRODUCTION

The 2011 drought in East Africa was one of the worst in decades, affecting up to 13 million people in Djibouti, Ethiopia, Kenya, Somalia and parts of Tanzania. Many of the affected people were pastoralists, but this natural catastrophe also highlighted the role of the life-giving rains for farmers dependent upon rainfed agriculture for their subsistence, livelihood and welfare. Rainfed agriculture is by far the most dominant rural activity and livelihood in sub-Saharan Africa. Understanding the challenges of rainfed agriculture is therefore of utmost importance in a context of increased climate variability, population growth and higher food insecurities. Thus, this chapter will focus on the role of rainfed agriculture during the drought in East Africa in 2011 in a small Sukuma community along the southern shores of Lake Victoria in Tanzania. Although the direct physical and natural consequence of drought is the absence of water or the life-giving rains, the human and social implications are also consequences of a wider range of cultural factors which either strengthen or reduce the consequences of failing rains. Hence, I will focus first on the global and local discourses on water and food production as a background for understanding rainfed agriculture and drought; then on the agrarian question and overall challenges of rainfed agriculture in Tanzania, including the current land laws and national politics of nuclear villages; and finally an ethnographic in depth discussion of the 2011 drought among the Sukuma farmers in Usagara village from a water systems perspective.

GLOBAL AND LOCAL DISCOURSES ON WATER AND FOOD PRODUCTION

The Sukuma are the largest ethnic group in Tanzania, and are estimated to number more than 5 million people. Among the Sukuma:

disastrous droughts, epidemics and epizootics are not only hearsay and phantoms in the minds of the people but well-known facts of life. Life is precarious. Threatened by destruction through famine, sickness and death, life is always at risk. To stay alive is an achievement, something to work for incessantly through the whole array of technological, organizational and ideological means offered by culture and society: through cultivation and livestock-rearing, through cooperation with kin and neighbours and through the veneration of the ancestors (Brandström 1990: 168).

The Sukuma were traditionally agro-pastoralists and cattle were their main possession and form of storable wealth for procuring all of life's necessities. Although cattle still have importance in Sukuma society and cosmology, the role of farming has increased at the expense of cattle (Brandström 1985, Wijsen & Tanner 2002). In Mwanza region, smallholder agriculture employs about 85 per cent of the population. Usagara village is located about 25 km south of Mwanza, a 35 minute drive by public transport (Figures 13.1–13.2). In 2010, the population of the ward was 8,839. The number of cows was 2,057 and there were 42 oxen used for ploughing.[1] The annual precipitation in Mwanza is about 1100 mm/yr, but as seen from two precipitation data set, there are at times large annual variations, and the most important factor for farmers is the amount of rain arriving at which time and in which rainy season. Traditionally, water was regulated by intricate water laws among the Sukuma (Cory 1953, Drangert 2003). In today's world, rural farmers dependent upon rainfed agriculture are part of the global world, which puts emphasis on what to grow: food crops or cash crops?

Globally, about 12 per cent of the total land area is used for agricultural production and about 42 per cent of the world's population live on this land (Fraiture & Wichelens 2007). Only 19 per cent of cultivated land on the globe is irrigated, but this land produces 40 per cent of the world's food (Hanjra & Quereshi 2010: 365). Although agriculture accounts for about 70 per cent of the world's withdrawals of water, in many low-income countries the figure is about 90 per cent (WWAP 2012: 46). An estimated 60 per cent of Africa's population live in rural areas. In sub-Saharan Africa, more than 90 per cent of the farmland is rainfed (WWAP 2012: 177, 334). In sub-Saharan Africa, agriculture accounts for 35 per cent of GDP and employs about 70 per cent of the population. Agriculture is thus key for economic development and poverty reduction, and every 1 per cent increase in agricultural production reduces the numbers of the absolute poor by 0.6–1.2 per cent (Wani et al. 2009: 1–2). Due to population growth, however, the amount of cultivated land per person has globally declined from 0.4 ha in 1961 to 0.2 ha in 2005 (WWAP 2012: 46), directly related to hunger and food security.

Water and water availability is intrinsically linked to food security. Food security is defined as the point at which 'all people, at all times,

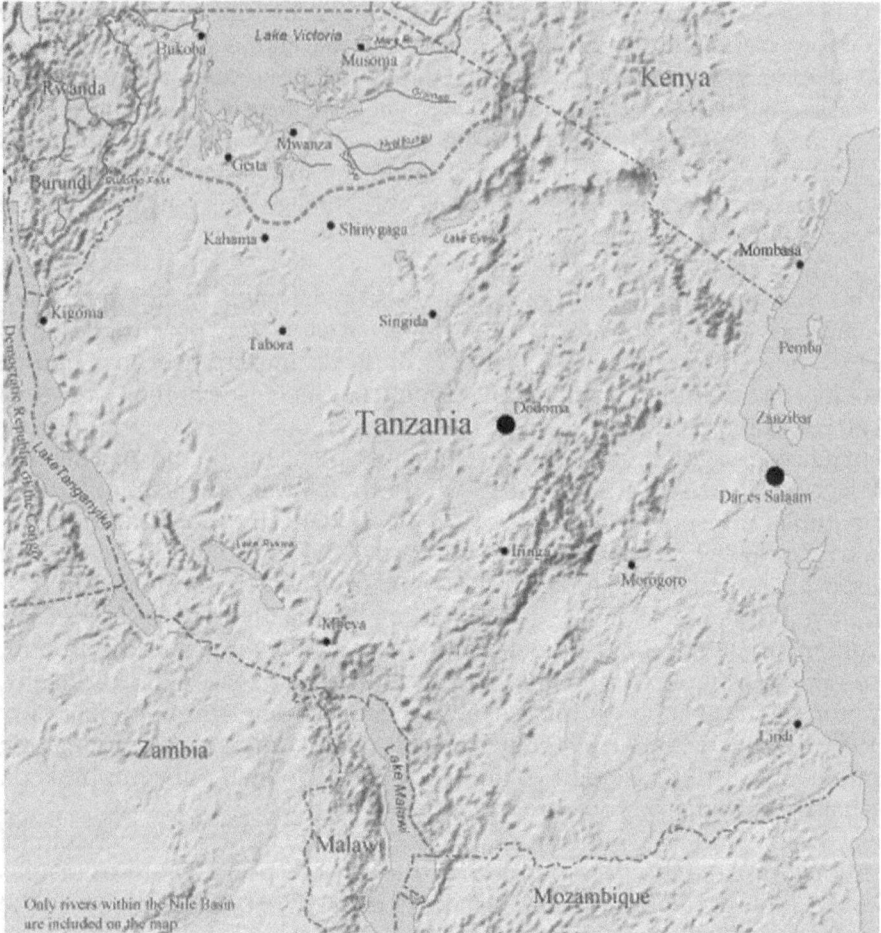

Figure 13.1. Tanzania (from the Nile Basin Research Programme, University of Bergen, Norway).

have physical and economic access to sufficient, safe and nutritious food to meet their dietary needs and food preferences for an active and healthy life' (FAO 1996), and as 'a necessary if not sufficient basis for poverty alleviation' (Cook et al. 2011: 2). According to the World Food Council:

> food security implies two things. First, ... that food is available, accessible, affordable – when and where needed – in sufficient quantity and quality. Second, it implies an assurance that this state of affairs can reasonably be expected to continue ... that it can be sustained. To put it simply, food security exists when adequate food is available to all people on a regular basis (World Food Council 1988: 2).

Figure 13.2. Map of Usagara and Mwanza in Tanzania (adapted by Terje Oestigaard from Google Maps).

In today's world there is a general consensus that even with the current land and water resources there will be enough food for even a world population of 9 billion. However, the current number of people living in hunger is increasing and today counts for more than 800 million people. In the more policy-oriented debates it is currently emphasized that one needs a holistic and global approach to the emerging water and food crises, including the whole value chain from the farm to the fork. Although these overall and global approaches are important, in practice they have little relevance for rural dwellers dependent upon the unpredictable rains. Many of the processes taking place during droughts and even more so during extreme droughts are already a reality for rainfed farmers in normal years, although to a lesser extent. In sub-Saharan Africa it was estimated in 2008 that 48 per cent of the population survived on less than US$1.25 a day. And as a paradox, in this region hunger is most widespread in rural areas among farmers and households producing food (Hårsmar 2010). In sub-Saharan Africa, being a smallholder farmer relying on rainfed agriculture means that you live through various degrees of poverty. In years of good rains, one may get a small surplus, but being

Figure 13.3.   Failed harvest in Usagara, early 2011 (© Terje Oestigaard).

dependent upon rainfed agriculture implies one thing for sure: droughts will come one year or another (Figure 13.3).

One may differ on or identify different types of droughts from an analytical perspective, although all of them are characterized by insufficient water or water at the wrong time for agricultural purposes. First, there is unpredictable drought, which occurs when total precipitation is comparable to normal years, but the harvest is exposed to growth stress as a result of unpredictable, erratic and uneven rainfall. Secondly, there is full-season drought, which occurs when overall precipitation patterns are much lower than in normal years and plants do not receive enough water. Thirdly, there is terminal drought, which occurs when initially there is enough water for cultivation, but later the soil is exposed to a water deficit. Fourthly, there is intermittent drought, which occurs when there is a short dry spell during the growing season and the harvest is exposed to drought only at one stage during growth (UNECA ACPC 2011: 19).

Variability in rainfall generates dry spells almost every season and hence shorter periods of water stress during the growing season. Dry spells are manageable and investments in water infrastructure can overcome these fluctuations, which may last from two to four weeks. Meteorological droughts, on the other hand, occurring on average once a decade in moist semi-arid regions and up to twice per decade in dry semi-arid regions, result in complete crop failure. When such droughts occur, they cannot be counteracted by agricultural water management, and other social coping strategies are necessary, such as food relief and grain banks (Wani et al. 2009: 8), if they exist. When the rain and the harvest fail, farmers have to buy their food for survival on the global market, where prices increase and fluctuate during periods of drought. This directs the question to food crops or cash crops.

For individual farmers, whether poor or relatively better off, it makes sense to generate as much income as possible by cultivating crops that can be sold on domestic or international markets and then to buy cereals at the same markets. From a poor farmer's perspective, however, there are no cheap cereals. In today's world, everyone needs money and volatile and increasing food prices make the need to generate more income ever more urgent. When the poor are barely able to make ends meet – and in most cases they cannot – there is only a marginal net difference between selling cash crops and buying life's necessities. This net difference is, however small, vital to farmers in sub-Saharan Africa living on an absolute subsistence minimum. Thus, even in arid regions farmers dependent on fluctuating rainfall patterns may choose to grow water-intensive crops well aware of the fact that if the rains fail, they are left with nothing. If they had chosen crops requiring less water, they could have some harvest even in years of bad rain and consequently some food. Subsistence farming means living on the absolute minimum – in other words, in extreme poverty and often starving. The possible net difference if rains and harvests are successful is a calculated reward and risk that comes at a high price. However, very few farmers take the risk of planting only cash crops; food crops are a security net (Oestigaard 2014). If what and when to grow is a risk mitigation strategy, *where* to farm is a political question institutionalized by law or land tenure regimes, and varying land systems may offer different possibilities as risk reducing or coping mechanisms in times of failing rains.

THE AGRARIAN QUESTION AND OVERALL CHALLENGE OF RAINFED AGRICULTURE

The classic agrarian question in Africa has been how capitalism can enable agricultural development that contributes to industrialization and decreased poverty. Thus, agrarian reforms have been seen from broadly three perspectives – the social, economic and political. The social perspective has a welfare focus and emphasizes possibilities to reallocate land to the poor. The economic perspective stresses the emergence of small commercial farmers who create employment with multiplier effects.

Finally, the political perspective argues for transforming the whole agrarian structure as a development strategy (Maghimbi et al. 2011: 19).

In 1895, the German colonisers declared that all land in Tanzania, occupied or not, would become Crown lands. In 1928, the British governor authorized land grants for periods up to 99 years. After independence, all land continued to belong to the state. The Land and Village Land Acts of 1999 state that all land on mainland Tanzania 'shall continue to be public land and remain vested in the President as trustee for and on behalf of all citizens of Tanzania,' meaning that the state is the ultimate owner of the land, and makes grants for the occupation and use of it (Cl. 4(1), op. cit. Maghimbi et al. 2011: 28).

President Nyerere introduced the concept of *Ujamaa* in the Arusha Declaration in 1967. This policy was populist and was based on what today is commonly seen as a utopian belief that a modern state could emerge from communal villages. The aim of *Ujamaa* was to increase agricultural productivity and at the same time reduce economic inequality. At the heart of the socialist vision of *Ujamaa* was a particular view of the world and its commodities. In this view, there was a limited amount of good and goods in the world, and one person's gain would be another's loss. Since Tanzania aimed to be 'self-reliant' in order to avoid dependence as a newly independent state, this implied a zero-sum national economy. Within Tanzania, wealth accumulation in any form would be at the expense of someone else, who would lose something (Sanders 2008: 115), thus running counter to the socialist vision.

Nyerere placed the nation at the centre of the framework for modernization and as a result repressed cultural, ethnic, social and religious diversity. By abolishing chiefdoms in 1963, the rural population's relationship to the state was changed. The chiefs had largely lost popular support in the colonial era, but with this radical break, in combination with villagization, the state-peasant relationship was fundamentally transformed (Havnevik 2010).

From 1967 to 1973, the number of people in Tanzania living in *Ujamaa* villages increased from about half a million to about 2 million, approximately 15 per cent of the rural population. In 1973 the government abandoned the policy of voluntarism and the president ordered that the whole population should be moved into villages. This took place through organized 'operations', basically a military metaphor for mobilization. 'Operation Villagization' forcibly removed about 7 million people into villages, or about 50 per cent of the rural population. Through this process, some 8,230 new villages were created (Mesaki 1993). However, as Bryceson remarks, villagization 'was not about socialist collectivity as much as nuclearizing scattered household settlements into villages where health dispensaries, schools and agricultural marketing services and productive infrastructure could be more efficiently provided' (Bryceson 2010: 75). Nevertheless, villagization also led to environmental degradation. This may have happened in any case, but the process of villagization accelerated it (Kikula 1997).

In practice, there was little link between the Arusha Declaration and the new *Ujamaa* policy of 1967 and the actual implementation of the policy, apart from the nucleated village unit. The farmers were still dependent upon fluctuating market prices for their crops rather than government stimuli, but more important; they still dependent upon the fluctuating rains. Although farmers were promised numerous benefits, the new policy was viewed with considerable scepticism (Havnevik 1993). As Kjekshus wrote in 1977:

> The implementational features have fundamentally undercut some of the preconditions of the program and harmed the outcome. Already the socialist features of the village plan have been compromised, and it is doubtful whether the move has been a school in grass-roots democracy that has strengthened the peasantry's sense of self-reliance and dignity. More fundamentally, it is doubtful whether the change of settlement pattern is consistent with the fundamental requirements for economic development in the Tanzanian ecology (Kjekshus 1977: 281).

The Sukuma used to live in widely dispersed households on the semi-arid cultivation steppe, and there were no villages in Usukuma until the forced villagization of 1974. This political decision to force the Sukuma into villages and to work on communal farms had severe repercussions. As a consequence, life for the Sukuma became harder in the postcolonial era and the environment and land was desiccated (Wijsen & Tanner 2000: 21–2).

The villagization process disregarded existing customary rights and land tenure systems, thereby completely undermining the security of customary landholders. It also created other problems that contributed to agricultural decline. Nuclear settlements and villages created an artificial land shortage in areas where land was in abundance. With farmers living together, all land in the vicinity was farmed, but beyond a certain point it became uneconomical for a farmer to walk and establish new farms. As a consequence, in many villages farmers faced land shortages while unoccupied farmland was available only some ten miles away (Maghimbi et al. 2011: 27–33).

Since 1989–90 people have been allowed to move back to their original lands, but most people have stayed in the villages. Nevertheless, although the intentions behind the vision were good, the programme failed to transform Tanzania into a modern self-sustaining state (Sanders 2008: 113). In Tanzania today, about 70 per cent of farming land is cultivated by hand hoe, 20 per cent by ox plough and only 10 per cent by tractor. Small-scale farms still predominate and are on average as small as 0.9 ha, with only a few farms of 3 ha or more (Maghimbi et al. 2011: 43–4). Thus, villages do not necessarily solve the problems, in fact they may create more problems when not access to land but water was and still is the most essential for rainfed agriculture. And irrigation did not follow the villagization process.

Water is an increasingly scarce resource, and not only at household level. Importantly and fundamentally, erratic rainfall patterns represent huge uncertainty and risk for rainfed agriculture. Water is the basis of all agricultural practices. The absence and presence of different types of water sources structure all societies, whether those sources are rain, rivers or lakes or a combination of water bodies at a certain place. Too much water at the wrong time of year, such as unpredictable and devastating floods, is as bad as too little water when it is really needed, the result being drought. Everything depends on the arrival of the right rains in the right amount at the right time (Figure 13.4), and their failure and the ensuing dire consequences underscore the fragility and vulnerability of all aspects of life (Oestigaard 2009). This total dependency on an unpredictable resource represents a huge uncertainty. Farmers aim to reduce and control this uncertainty in the best possible ways, but their means are in practice limited. Farmers rely on tradition, in this case the culturally accumulated knowledge derived from generations of experience involving the soil, crops and crop performance under varying and erratic rainfall patterns. Now, this centuries old knowledge is adapted to the modern world in which life and well-being largely hinge on growing the most valuable crops

Figure 13.4. The arrival of the rains in Usagara (© Terje Oestigaard).

that produce an income and secure a livelihood beyond subsistence (Oestigaard 2012).

RAINFED AGRICULTURE AS A WATER SYSTEM

Understanding the water-world and how different types of water create possibilities and limitations for livelihoods and agriculture is fundamental to understanding current agricultural practice and future development, and can be seen from a water systems perspective (see Tvedt and Oestigaard this volume, and for instance Tvedt 2010a, 2010b): (1) The physical water-scape at a given time, including rain and rivers etc., (2) human modifications and adaptations to the actual waterworlds; and (3) cultural concepts and ideas of water and water systems, including laws.

With rainfed agriculture as a particular water system, one may adapt and develop this approach (Figure 13.5). There is, of course, no direct and deterministic relationship between people's ideas about their world and how they cultivate. For instance one may grow millet using a hoe irrespective of whether one believes in Christianity, Islam or the ancestors.

Rain-fed agriculture in a water system perspective

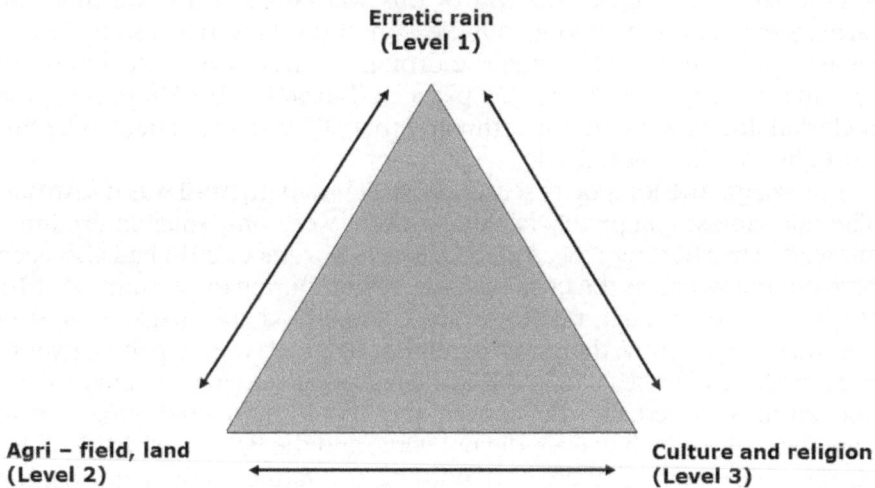

Erratic rain
(Level 1)

Agri – field, land
(Level 2)

Culture and religion
(Level 3)

Agriculture - from Latin *agricultūra, ager* is field, land + *cultūra* – culture

Figure 13.5. Rainfed agriculture in a water-systems perspective (© Terje Oestigaard).

Rain (layer 1) is, however, a constant and limiting factor that structures agricultural practice. Thus, by analytically using agriculture – from Latin *agricultūra*, from *ager* field, land + *cultūra* culture – one may distinguish the physical work and actual cultivation as belonging to layer 2 and the cultural elements as part of level 3. These levels overlap and are mutually dependent.

ERRATIC RAIN (LEVEL 1)

In Mwanza, as in many other places in Africa, there are two main rainy seasons: one in February–April and the other in October–November. The first is the 'long' rainy season and the latter the 'short' season. However, 'rain is scarce in this semi-arid region. When it does fall, it is often erratic and unevenly distributed. Too much or too little at the wrong time can and often does spell disaster for these agricultural people' (Sanders 2000: 473). Moreover, even when the rains come, they may fall unevenly: one village may receive sufficient rain whereas the clouds may pass over another village without rain falling. Even within a village, the actual precipitation may vary depending upon particular clouds and rain at a given time in the rainy season.

Although life-giving rain is of utmost importance, the two rainy seasons are not of equal weight, and it is during the long rainy season that most of the food surplus is secured. The long rainy season is thus fundamental to food security in a given year. When the short rainy season comes, additional food is produced. All of this will be eaten by the time the farmers expect the next long rainy season. If this fails, the results may be catastrophic like in 2011 when the drought in East Africa affected up to 13 million people, and also hit parts of Tanzania, the Mwanza region included. In Tanzania, it was estimated that 500,000 were affected by this drought (World Vision 2011).

In Usagara, the long rainy season from February to April was disastrous. The rain almost completely failed and there were only small intermittent showers. The short rainy season in October–November 2010 had also been bad, but not as bad as the ensuing long season. During the autumn of 2010, there had been enough rain to produce some food and small amounts of cotton. Consequently, there was barely enough food to tide people over to the next long rainy season. In 2011, the maize planted in late January died, and farmers hoped for the rain so they could start cultivating rice in particular. The rain, however, did not come and the fields were left virtually barren. Without rain, there was nothing the farmers could do. As one farmer put it, one cannot move the land and farmers have to live where their land is.

Still, the farmers work and have to be prepared if the rain comes. The preferred agrarian cycle is to prepare the land and plant in January and wait for the rain to come and harvest in March–April and then plant again and harvest in June. In a year of good rain, generous rain should fall at least three days a week from March to April. However, in 2011 the rains failed more or less completely, with only some small showers from mid March onward. When the rain fails, not only have farmers put in a lot of effort in vain, but the outcome is a failed harvest.

Dependence upon fluctuating rains is a gamble. The vulnerability of rainfed agriculture was once described by an Indian finance minister in his national budget as a 'gamble on the rains.' He went on to stress that 'variations in rainfall, or disruptions in water supply, can make the difference between adequate nutrition and hunger, health and sickness – and ultimately – life and death' (Human Development Report 2006: 174). Moreover, 'rainfed farming is not just risky; it involves long delays between investment and fluctuating returns. Sooner or later, steadily accruing loan interest outstrips a borrower's fluctuating capacity to pay' (Shipton 2010: 226). In 2011, the farmers faced the 'hungry season' with barren fields and hardly any food. When the rains came in Usagara village in late August in the short rainy season, they were met with great relief. Everybody had survived the difficult 'hungry season', but as one old woman remarked, this drought had been one of the three worst in her whole life. Famine is serious food deficit, and in the globalized world where one needs money, farmers face a difficult choice: should one grow food or cash crops?

Coping with the uncertainties of erratic rainfall in a time of climate change, when rainfall patterns fluctuate even more markedly, is an immense challenge. Farmers living on the absolute subsistence minimum are dependent on rain and on making the right decisions. How is it possible to be strategic when everything depends on the uncertain rains? Millet is a low water-intensive crop and in times of hardship may secure livelihoods. Rice and cotton, on the other hand, demand much water for cultivation, but if rain comes in abundance, they are the choices that bring the most money and prosperity. Farmers cannot afford to make wrong decisions, yet they do, because everything depends on the rain, which is impossible to predict.

The main solution is diversifying risks. Specialization may increase vulnerability, whereas diversification may reduce uncertainties. Some plots of land are used for maize, cassava, beans and vegetables and others for rice and cotton. Herein, however, there is also a paradox. If all land is used to grow millet, which is less water-intensive but generates little cash, there might be enough food if there was little rain. However, if the rains are good, farmers will lose opportunities because of their wrong choice of crop. Thus, when the rain fails, much arable land will be left uncultivated, because farmers will not want to put prepared rice fields under millet in case the rain does come. Hence, food deficits are also culturally made. Ultimately, this situation highlights the impossible choices farmers have to make when their lives and subsistence are totally dependent on unpredictable annual rains.

When the rains fail completely, all these strategies are in vain. The result is hunger and famine, which have huge impacts on society. People have to get money to buy food. Today, the global market on which they sell their agricultural products during times of good rain is also where they have to buy their necessities at fluctuating and increasing prices when there is drought. Money matters and changes lives.

After the long rainy season failed during the spring of 2011 in Usagara, during the following summer the farmers and their families experienced hunger and had a very hard and strenuous time. Those with the least food got some supplies of maize from the government, but otherwise social mechanisms in the village helped increase food security. Within families and among neighbours, people with a small surplus shared food with others. Those better off for various reasons also employed the poorest to do work in return for small wages, which enabled them to procure food. Moreover, Usagara village enjoyed a comparative advantage in being located close to Mwanza and other smaller cities. People from Usagara could thus migrate as day labourers to towns or search for agricultural work in other parts of the country. Thus, although the drought brought hardship and suffering to the Sukuma of Usagara, all of them survived.

This drought was hard, but the Sukuma have throughout history been used to droughts and have adapted to this way of living. Despite the difficulties, globalization and the market economy have brought some advantages in times of crises. The problem with drought, it was explained by one, was not really that harvests failed. The problem is the lack of money that enables farmers to buy food from elsewhere. By being day labourers, it was possible, even in the most recent crisis, to generate some income and thus buy food and survive. This was quite different from the crisis in the early 1980s when it was more difficult to buy food even if people had money.

The absence of rain influences all parts of society and life. During the drought the wells run dry. Villagers had to dig new wells at greater depths and further away. This implied that the women had to walk longer and spend more time collecting and carrying water. Girls are the first to be taken out of schools to carry water when there is a drought and more time is needed to secure water for the household's daily consumption. There was less drinking water for both humans and animals, and several farmers had to sell their cattle (Figure 13.6).

Although the short rainy season during autumn 2011 was good, rain can be a double-edged sword. If it rains continuously and intensively for a week, the crops will be swamped and die, since they also need sun. Rice can survive intense rains for longer periods, but maize and other staples are more sensitive. Good rains are those that ideally last

Figure 13.6. Cattle as commodity among the Sukuma (© Terje Oestigaard).

two–three days, followed by some three–four days of sunshine, and so on. Thus, as with rain so with sun: too much or too little at the wrong time is equally bad.

The short rainy season during the autumn of 2011 started early. The first rains came at the end of August and continued. Although the heavy rains necessary for rice cultivation did not come, the light rains provided enough water for maize, millet, finger millet, beans and other crops. Thus, after the catastrophic season in spring 2011 and the poor harvest of autumn 2010, the rains in autumn 2011 provided farmers with food for survival and even a small surplus for sale.

When the rain started, the farmers opted for different strategies. With the arrival of the first rains in late August, some farmers suspected they would not last. They therefore postponed maize planting, since they did not want to lose the planted seeds or waste time and money. Waiting to plant is a way to minimize risks, and by the end of October there were still farmers who had not started planting (the main period of rain during the short rainy season is October and November). Other farmers, however, calculated differently and took a chance in the belief that the early rains of late August would continue. Some started cultivating maize at the beginning of September, and because the rain kept coming they had the chance to have two harvests in this rainy season. Thus, given equal dependency on the seasonal rains, there are different ways to reduce risk and manage the uncertainties of rainfed agriculture. But in the end, these risk-managing strategies are a gamble when everything depends upon unpredictable and erratic rains.

AGRI-FIELD, LAND (LEVEL 2)

By the end of March 2011, when the rain had not arrived and the prospects for the forthcoming season looked miserable and people prepared for difficult times, farmers started to plant millet in various fields. The latest one can plant millet with a hope of a successful harvest is in April. However, they did not cultivate millet in the fields they had prepared for rice cultivation, since they believed the rain would eventually come. Still, even if millet is drought resistant, it nevertheless requires varying amounts of rain for a month. Unfortunately the rain failed in April as well, leaving farmers with almost no harvest or food until the onset of the next season's rains.

In the past, combinations of cash and food crops were grown: cotton together with chickpeas, millet, cassava and sweet potatoes, while in more swampy areas, which have now dried up, rice was cultivated. Millet was the dominant staple until the 1980s, but since then people have introduced crops they prefer to eat, and today millet is perceived as disgusting because of its murky brownish colour. From the 1980s and 1990s onwards, rice cultivation became more dominant.

Rice is preferred as a staple for several reasons. First, it is the dietary preference today. Secondly, if a surplus is produced or cash is needed, it can generate a good income, with prices per kilo ranging between Tsh. 1,600 and 1,800. Thirdly, rice can be stored for years after harvest and thus represents both food security as well as potential cash income. Maize, for instance, can only be stored some five or six months after harvest and must therefore be consumed within a relatively short time. Finally, even when there is insufficient rain for planting, as in spring 2011, the rice seeds may survive up to three years. Thus, the resistance of rice to spoilage is also a security mechanism.

Cotton is highly drought resistant, whereas maize and sorghum are more vulnerable to water shortage and drought. As such, cotton is preferable during dry periods. However, cultivation of cotton is water intensive and requires a successful rainy season, whereas millet needs little water. Where land is a limiting factor, farmers must calculate the proportion of land to be used for subsistence agriculture in relation to cash-crops. When they do not have enough land for cultivation, food-crops for personal consumption are more often given priority over crops that may generate limited cash in an uncertain market. Among the villagers it was generally agreed that the best way to improve agriculture is to introduce small-scale irrigation or rain-harvesting techniques, which would enable more secure food-crops, with some additional cash-crops.

Moreover, changing crops is not unproblematic. When smallholders change crops from year to year, the soil may lose some of its nutritional qualities. A major problem with maize is that the crop is attacked by striga – also known as 'witches' weed' – a highly destructive parasitic plant. Millet is also highly vulnerable to striga, and the government has instructed farmers who grow millet in their fields one year, to grow cotton or cassava the next in order to eradicate the parasitic plant. However, if the next season is also dry, farmers face an added problem, since they cannot grow millet again. Striga particularly affects maize, sorghum and millet, but not rice, which is located in wetter areas. Nevertheless, instead of using pesticides, if a field is infected by striga, farmers may cultivate cotton or cassava the next year, since striga does not survive alongside these crops. In short, crop rotation patterns also influence the crops chosen in a given year.

The Sukuma were the ethnic group cultivating most cotton and cash-crops in Tanzania. Mwanza region has been at the heart of cash-crop farming in Tanzania. In Usagara before 2011, the majority of farmers had shifted from cash-crops, and in particular cotton, to food-crops. Half a decade ago, about 50 per cent of the farmers were growing cotton, whereas in 2010 only 13 per cent had cotton as the major crop. The main reason for this decline in Usagara was the vulnerability caused by the free market system. If the cotton price is high, farmers switch from food to cotton production. When the prices of fertilizer and pesticides increase, the returns from selling cotton decrease: the growing costs may be higher than the net returns and farmers would not have enough income to pay for these inputs. The price for cotton was so low that food crops become the new cash crops, in particular rice. Moreover, individual farmers cultivating cotton on small farmlands cannot compete economically with large industrial plantations. Thus, increased expenses and decreased cotton prices caused farmers to return to food crops to survive.

However, during the short rainy season in the autumn of 2011, farmers started reverting to cash crops and cotton production, and many more were expected to produce cotton again during the main rainy season of 2012. The government offers preferential loans with good conditions to farmers who want to invest in cotton production, and farmers pay back their loans after the harvest and sale of the cotton. Overall, then, there has been a change from subsistence agriculture to cash crops and back again, and then possibly a return to more cash crops.

About 75 per cent of cotton is exported, mainly to China (approx. 50 per cent), but also to India, Bangladesh and Taiwan. Prices for cotton used to be Tsh. 500–600/kg, but rose rapidly to Tsh. 1,200 in 2011. There were two main reasons for this doubling of prices: the appreciation of the

US dollar and low global production. The farmers in Usagara were well aware of these changes, although not necessarily of the reasons for them. Some farmers suggested the Tanzanian government had increased prices in order to help them. The misunderstanding of why prices increase may lead to wrong choices about which crops to cultivate. In Usagara, cotton was largely abandoned because of the falling prices. With the rapid increase in those prices, many farmers were rethinking their options and considered growing cotton again. However, because the prices of cotton are governed by the international market and not the Tanzanian government, farmers may again fall prey to fluctuating prices if global production increases.

Nevertheless, although farmers were mainly relying on subsistence farming, as much as half the crop might be sold to generate income. Thus, if farmers were able to keep all the food they grew, there would perhaps be sufficient for the whole year. Normally, however, they do not hold back food for the next year, because they have to sell some of it to earn money. This highlights the impossible situation the farmers face: all decisions are a gamble on the rain to get the highest yield possible, failure to achieve which may have devastating consequences. But even with a successful harvest, farmers have to sell large parts of their food supply to generate income and cash. It is extremely expensive to be poor.

CULTURE AND RELIGION (LEVEL 3)

The previous descriptions of agricultural practices and choices are all, of course, within the realm of culture. Cultural choices and decisions, together with religious conceptions and understandings, shape all practices, but as shown, the cultural world is also dependent, restricted and enabled by the external, real world. And for subsistence farmers living off rainfed agriculture, rain is the first and most important parameter – without rain, no food. And as seen, the consequences of droughts are partly man-made given the Ujamaa policies and the villagization processes forcing people to live in villages. This created an artificial land-shortage and increased the pressure on land.

Importantly as well, food is not just food. There is hardly any other item endowed with so much cultural and religious significance, including taboos, as food. Suffice it to point out here that the cultural values of food often outweigh other rational considerations with respect to water consumption, agricultural production and caloric output. Thus, one key advantage of millet as a food crop is that it requires less water for cultivation than other crops and as such it is a surer food source during times of failing rain and drought. Generally, although millet has been the traditional staple, people do not like to eat it anymore and will not grow it, even though it requires less water for cultivation. This change in dietary preference has direct implications for what is grown. Thus, cultural taste and preferences have changed actual farming, and even during famines farmers rarely grow millet. And this is the case even though farmers were encouraged by the government to grow millet. Rather than planting millet on the fields prepared for rice cultivation, the fields were left uncultivated. Most farmers instead gambled that the rain would come and enable them to grow rice, which is preferred both as a cash crop for income and because it is perceived as tastier.

Regarding cotton, other social mechanisms are at work. Apart from the low selling price, there are further reasons cotton is a less attractive cash crop than others. In Usagara, cotton is sold to a local cooperative, but payments are frequently delayed. Often, farmers only get paid the first week and then the cooperative claims it needs to borrow money from banks. Subsequent payments may be delayed up to a month – a time when most farmers really need the money and cannot afford to wait. Other cash crops, such as tomatoes grown in gardens, are paid for immediately and rice in particular is seen as most lucrative as a cash crop. Milk is also sold, and keeping livestock, which are also a source of meat, is an additional source of income. However, even though cattle are a valuable resource generating extra income and providing security, particularly when the harvests fail, grazing land is limited and farmers cannot exploit the potential of livestock as much as they would wish.

During droughts and famines, government support and foreign aid are part of the strategy and planning for survival. Although the external help is limited and survival on it is hard, such help is nevertheless part of the coping strategy of the subsistence economy. There is a Sukuma saying 'We will pass', and the Sukuma have always suffered from severe famines, which occur every six or seven years. During famines, each person receives one big tin-can of maize from the government. This is not enough to survive on, but it gives the farmers some relief as they search for other job possibilities.

When drought and food shortage crises occur, the Sukuma usually turned to the ancestors for help. However, although the ancestors were propitiated in other Sukuma areas, this was not the case in Usagara. After they became Christians, propitiating the ancestors has been seen as akin to using witchcraft, and the church condemns both practices. Rainmaking rituals were also seen as sinful by the church and thus villagers did not conduct them.

The direct consequences of failing rain were that people suffered and starved and that poverty increased. This was interpreted in different ways. First, global climate change was seen by many as the cause of the scarcity or absence of rain, in particular among the young and those with a modern education. Indeed, through its access to information from television, radio and newspapers in varying degrees, rural Tanzania is part of the global world and the discourse about climate change is central in these rural areas too. Secondly, farmers have long been cutting trees in the forests, and consequently the land has become eroded and lost substantial water-retention properties. Deforestation is thus seen as an important human factor affecting the actual water environment. Thirdly, changes in weather and the absence of rain were also seen as stemming from declining traditions and the broken relationship with ancestors, so that it is no longer possible to communicate with and propitiate them, or make sacrifices (Figure 13.7).

With regard to the last theme, neglect of tradition and the non-propitiation of ancestors have given rise to numerous interpretations. Some traditional healers and diviners blame the Christians, in particular the Pentecostals, whom they perceive as evil because they have created more, and new forms of, witchcraft. Others saw the drought as a punishment by God because Sukuma values were no longer followed. In general, there was an element, or at least the perception, of generational conflict, since the youth were not honouring traditional values and respecting elders. Women, too, were criticized for dressing inappropriately and not wearing their traditional dress. The lack of rain in 2011 was thus also understood as a collective punishment by God for the moral decay of and misconduct and sin in society. As one informant explained, when people do not follow Sukuma tradition and values, God penalizes them with the absence of rain. Moreover, there has been an increase in witchcraft, and this is also believed to influence rainfall patterns in various ways. People use witchcraft to become rich, and one informant linked the drought directly to the killing of an albino in 2009. Nobody was arrested for the crime. According to this farmer, this misdeed led God to punish the society collectively by withholding rain.

In other villages with many traditional healers, however, rainmaking rituals were conducted, and if and when the rain comes, the rainmakers were acknowledged and greatly respected. When rainmaking rituals succeeded and the rain came, the life-giving waters were also seen as holy and a divine gift from God to all people. Although the practice of rainmaking had disappeared in Usagara, Christians did pray to God for rain, including the Pentecostals. People had long been praying for rain in the church, but it never came. Among Muslims, too, from a religious perspective rain is seen as a gift from Allah, but erratic and fluctuating rain was seen as a consequence of climate change and deforestation, which reduce the overall amount of water.

Although rainmaking as a ritual has disappeared, the whole rainmaking cosmology has not. There were still people in the village believed to have power to withhold the rain. The reason they might wish to harm other people and society was unclear, but somehow they wished to be respected and seen as important. In cases where rain is believed to be withheld by anti-rain witches, people may consult traditional healers who devise medicines to be sprayed on the fields where the farmers

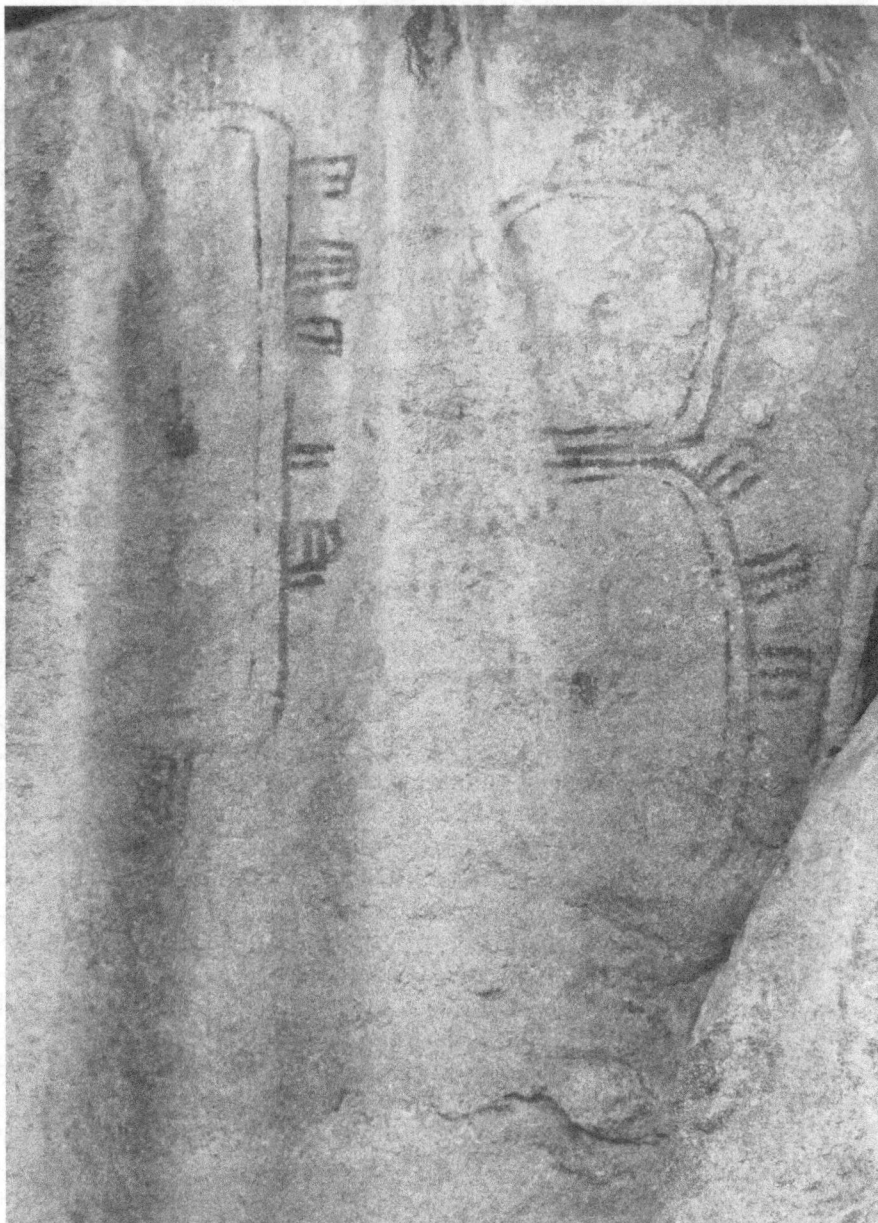

Figure 13.7. Rock-art in a rock shelter in Bukumbi village (see Figure 13.2) where rainmaking and ancestral rituals took place in the past (© Terje Oestigaard).

would like to receive rain. Thus, the absence or presence of rain was seen as an ongoing contest between those with the power to make it and those who could restrain it. If the rainmakers were the more powerful, they would punish the rain-witches, including by sending sandstorms to kill them. In 2010, in the neighbouring village of Nyahorongo, a man predicted before the short rainy season that there would be no rain that year. When the rain had still not arrived in November, the villagers grew very angry and beat the man to death. Withholding the rain was not seen as related to the world of ancestors as such, but as a result of witchcraft and malignant forces.

Thus, actual cultivation is not merely about planting, sowing and harvesting, but represents an intricate mixture of rainfall parameters and natural variables, including soil composition and qualities; customary and technological practices; as well as a wide and influential spectrum of cultural, ideological and religious concepts.

CONCLUSION

For smallholder agriculturalists in sub-Saharan Africa dependent upon the arrivals of the life-giving rains, climate variability will sooner or later result in droughts of different lengths and intensity. The risk-coping strategies will inevitably vary; from where and when to start preparing the fields to calculating the possible incomes of food or cash crops and how to balance the different crops requiring varying amounts of water for a successful harvest. But all this is of course a gamble since nobody can know when and in which amount the rains come. The right decisions have to be made, simply because the farmers cannot afford to make the wrong choices. Yet, sometimes the wrong decisions are made, and it highlights the impossible situation many farmers face and live by when their wealth and health – or suffering and possibly starvation, and in the worst case, death – is ultimately a matter of the arrival of the unpredictable rains. Political and institutional factors may reduce and increase the consequences of the farmers' choices given the possibilities they have. Diversifying risks is the optimal and the preferred choice, but when the rain fails completely all these strategies are in vain. And this is perhaps where the most depressing part of rural poverty and hunger lies; most of these farmers are strong, healthy and eager to work – and they do work hard. They prepare the fields and hope for a bountiful harvest, and when that is done they can only wait and look at the sky to see whether the clouds will come. During the drought in 2011, it was mainly the sun shining, and if there came some few clouds, most often they drifted away passing the village. When the rains failed that season, they knew very well how little food they had left while facing the extremely tough and long 'hungry season', only hoping that the rains the next rainy season would be better.

NOTES

1 Statistics from the ward office.

REFERENCES

Brandström, P. 1985. 'The Agro-pastoral dilemma: Underutilization or over-exploitation of land among the Sukuma of Tanzania', *Working Paper in African Studies 8*. Uppsala: University of Uppsala.

——— 1990. 'Seeds and soil: the quest for life and the domestication of fertility in Sukuma-Nyamwezi thought and social reality'. In Jacobson-Widding, A., van Beek, W (eds), *The Creative Communion: African Folk Models of fertility and the Regeneration of Life*. Uppsala: Almquist and Wiksell, pp. 167–86.

Bryceson, D.F. 2010. 'Agrarian Fundamentalism or Foresight? Revisiting Nyerere's Vision for Rural Tanzania'. In Havnevik, K., Isinika, A.C (eds), *Tanzania in Transition: From Nyerere to Mkapa*. Dar es Salaam: Mkuki na Nyota, pp. 71–98.

Cook et al. 2011. 'Water, food and poverty: global- and basin-scale analysis', *Water International* 26(1): 1–16.

Cory, H. 1953. *Sukuma Law and Custom*. Oxford: Oxford University Press.

Drangert, J.O. 1993. *Who Cares About Water? Household Water Development in Sukumaland, Tanzania*. Linköping: Linköping Studies in Arts and Science 85.

FAO 1996. *Report of the World Food Summit*. Rome: FAO.

Fraiture, C., Wichelens, D. 2007. 'Looking ahead to 2050: scenarios of alternative investment approaches'. I Molden, D (ed.), *Water for Food, Water for Life: A Comprehensive Assessment of Water Management in Agriculture*. London: International Water Management Institute, Colombo and Earthscan, pp. 91–145.

Hanjra, M.A., Quereshi, M.E. 2010. 'Global water crisis and future food security in an era of climate change', *Food Policy* 35: 365–77.

Havnevik, K. 1993. *Tanzania. The Limits to Development from Above*. Uppsala: The Nordic Africa Institute.

——— (2010). 'A Historical framework for analysing current Tanzanian transitions: the post-independence model, Nyerere's ideas and some interpretations'. In Havnevik, K., Isinika, A.C (eds), *Tanzania in Transition: From Nyerere to Mkapa*. Dar es Salaam: Mkuki na Nyota, pp. 19–56.

Human Development Report. 2006. *Beyond scarcity: Power, poverty and the global water crisis*. New York: UNDP.

Hårsmar, M. 2010. 'Why is agriculture so important to reducing poverty?' *NAI Policy Notes 4(2010)*. Uppsala: The Nordic Africa Institute.

Kikula, I.S. 1997. *Policy Implications on Environment. The case of villagization in Tanzania*. Uppsala: The Nordic Africa Institute.

Kjekshus, H. 1977. 'The Tanzanian villagization policy: implementational lessons and ecological dimensions'. *Canadian Journal of African Studies* 11(2): 269–82.

Maghimbi, S. et al. 2011. 'The agrarian question in Tanzania? A state of the art paper'. *Current African Issues 45*. Uppsala: The Nordic Africa Institute.

Mesaki, S. 1993. 'Witchcraft and witch-killings in Tanzania'. PhD Thesis. Minnesota: University of Minnesota.

Oestigaard, T. 2009. 'Water, culture and identity. comparing past and present traditions in the Nile Basin Region'. In Oestigaard, T (ed.), *Water, Culture and Identity. Comparing Past and Present Traditions in the Nile Basin Region.* Bergen: BRIC Press, pp. 11–22.

—— 2012. When everything depends on the rain. Drought, rain-fed agriculture and food security. *The Nordic Africa Institute Annual Report 2011: Africa's Changing Societies: Reform from Below*: 24-25. The Nordic Africa Institute. Uppsala.

—— 2014. *Religion at Work in Globalised Traditions. Rainmamking, witchcraft and Christianity in Tanzania.* Newcastle: Cambridge Scholars Press.

Sanders, T. 2000. 'Rains Gone bad, women gone mad: rethinking gender rituals of rebellion and patriarchy', *The Journal of the Royal Anthropological Institute* 6(3): 469–86.

—— 2008. 'Buses in Bongoland', *Anthropological Theory* 8(2): 107–32.

Shipton, P. 2010. *Credit between Cultures. Farmers, Financiers, and Misunderstanding in Africa.* New Haven: Yale University Press.

Tvedt, T. 2010a. 'Water systems, environmental history and the deconstruction of nature'. *Environment and History* 16: 143–66.

—— 2010b. 'Why England and not China and India? Water systems and the history of the industrial revolution', *Journal of Global History* 5: 29–50.

Wijsen, F., Tanner, R. 2000. *Seeking a Good Life. Religion and Society in Usukuma, Tanzania.* Nairobi: Paulines Publications Africa.

—— (2002). *'I am just a Sukuma'. Globalization and Identity Construction in Northwest Tanzania.* Amsterdam: Rodopi.

UNDP. 2012. *Assessing Progress in Africa toward the Millennium Development Goals. The MDG Report 2012.* New York: UNDP.

UNECA ACPC. 2011. *United Nations Economic Commission for Africa, African Climate Policy Center Agricultural Water Management in the Context of Climate Change in Africa.* Addis Ababa: Working Paper 9.

Wani, S.P. et al. 2009. 'Rainfed Agriculture – Past Trends and Future Prospects'. In Wani, S.P., Rockström, J., Oweis, T (eds), *Rainfed Agriculture: Unlocking the Potential.* Oxfordshire: Cabi, pp. 1–35.

World Food Council 1988. *Towards Sustainable Food Security: Critical Issues,* Report by the Secretariat, 14th Ministerial Session, Nicosia, Cyprus, 23–26 May.

World Vision 2011. *Horn of Africa. Response to Drought.* 90-Day Report, http://www.worldvision.org/resources.nsf/main/press-haiti/$file/hoa-90-day-report.pdf (accessed 2 July 2014).

WWAP. 2012. *The United Nations World Water Development Report 4: Managing Water under Uncertainty and Risk.* Paris: UNESCO.

14 Water-Food Security Systems in the Congo Rainforest

Raphael M. Tshimanga

INTRODUCTION

The geographical extent of the Congo rainforest encompasses an ecological region of rich biodiversity that sustains millions of people who rely on the ecosystem resources for their water supply, food security, shelter, livelihood and social welfare. Among the resources provided by this biotope, fresh water remains abundantly distributed. Studies conducted in the area have revealed a strong relationship between water resources and settlements (Vansina 1990, 2006, Kuper & Leynseele 1978, Bailey et al. 1991). Subsidence strategies, seasonal activities of the populations, pattern of territorial organization and social relations within communities and between communities seem much grounded on water resources. The relationship between water and food security in the Congo rainforest has also been a subject of many studies (Waehle 1986, Hart & Hart 1986, Bailey et al. 1991, Bele et al. 2014), which all clearly highlighted the significance of climate on regulating the systems of food production.

The activities undertaken to secure food in the Congo rainforest mainly include hunting, gathering, fishing and rainfed agriculture. It has been demonstrated that rainfall variability is a main factor that determines the seasonality, which in turn impacts on the trend of food availability (Waehle 1986, Hart & Hart 1986, Bailey et al. 1991). Because of this variability, the major challenge for most food security systems is temporal matching of supply with demand. While in typical agriculture the demand and supply can be controlled, in a traditional mode of subsistence (rainfed agriculture, hunting and gathering) people will adapt strategies as part of environmental management. Changing the pattern of activities to match the supply with demand, based on the pattern of seasonality, has been the main management strategy. Vulnerability associated with this system has been documented and it is clear that there are limiting factors that can contribute to unbalance supply with demand, thus increasing food insecurity in the Congo rainforest. These factors include extremes weather, land availability and local technology. In the context of environmental change, increased frequency and intensity of rainstorms, dry spells during wet seasons, increased dry spells during dry seasons, intense and sustained heat spells during dry seasons, high winds and wet spell in dry seasons are among major concerns. Understanding how the current systems of water and food security in the Congo rainforest will respond to predicted environmental changes is vital for the well-being of local people who fully depend on the forest services. The main objective of this study is to provide an analysis of past and current systems of water and food security in the Congo rainforest so as to allow a perspective view of their vulnerability in the context of environmental change.

THE CONGO RAINFOREST, WATER RESOURCES AND ECOSYSTEM SERVICES

The central African Congo rainforest is the second largest lowland tropical moist broadleaf forest in the world after the Amazon, extending over 180 million ha, between Cameroon (11.8 per cent), Central African Republic (3.4 per cent), Democratic Republic of Congo (54 per cent), Republic of Congo (12.4 per cent), Equatorial Guinea (1.3 per cent), and Gabon (17.7 per cent) (Bele et al. 2014).

The forest has about 20 per cent of the world tropical moist (Justice et al. 2001) and plays a major role in regulating the world's climate (Tshimanga 2012). It is, along with the western Pacific Ocean and the Amazon Basin, a main world rainfall centre that generates intense storms with a global reach (Eltahir et al. 2004). The presence of the tropical rainforest favours the moisture recycling capacity of the basin. An estimated 75–95 per cent of rainfall is reportedly recycled in the Congo basin (Cadet & Nnoli 1997) and evaporation from the Congo basin contributes about 17 per cent of West Africa's rainfall (Eltahir et al. 2004). Camberlain (1997) reported the existence of strong westerly winds that divert moisture from the Congo basin to Ethiopia and other parts of East Africa. This wind is a result of active monsoon conditions that enhance the west-east pressure gradient near the equator. Intense precipitations in Ethiopia are reportedly attributed to the moisture from the Congo basin (Shinoda 1986, Camberlain 1997). Studies have also reported patterns of moisture circulation from the Congo basin to as far as the United States American Great Lakes region (Avissar & Werth 2005, NLOM 2001).

The forest contains a large diversity of species, including some 400 mammal species, 1,000 bird species, 700 fish species and over 10,000 plant species (CBFP 2006, Bele et al. 2014). Water resources include tropical rainfall, which sustains perennial large rivers and vast aquifers. The forest is also home to millions of people, with more than 150 indigenous groups, who depend on the ecosystem for their shelter, water supply and food security. The majority of the population are characterized by low income, relying on subsistence agriculture for their livelihood. Rainfed agriculture is the main mode with slash and burn, forest clearing and shifting agriculture (de Wasseige et al. 2009). A United Nations' census for the period 2000–5 (UN 2007) shows a growth rate of 2.87 per cent per year for the population living in the region of the Congo Basin, with a potential to double in 25–30 years.

Anthropogenic activities, with sometimes remarkable consequences for land use change and natural variability of the climate systems, have induced major environmental changes which may have irreversible effects, at least at a certain time scale (Milly et al. 2008). Various studies have demonstrated that the major impacts on water resources of the Congo Basin would stem from land cover and land use changes (Hoare 2007, Ladel et al. 2008). Uncontrolled anthropogenic activities with potential impacts on water resources availability of the basin pose potential problems. In addition, there have been increasing reports of uncontrolled large scale deforestation and mining which are known to impact on the patterns of hydrological behaviour. Estimates for the deforestation with a focus on the evergreen forest zones of the basin for the period 1990 to 2000 show a net deforestation rate of 0.16 per cent per year (de Wasseige et al. 2009). A loss of about five per cent has been recorded in several catchments between these periods. These activities are sources of pressure on the basin's water availability and their cumulative impacts could result in change of the basin's hydrological patterns.

The focal area for this study is the north-eastern part of the Democratic Republic of Congo (Figure 14.1). The area of the north-eastern Congo, astride the equator, lies between extreme two inter tropical weather fronts. October is the wettest month with rainfall of over 2,000 mm, while December, January and February are the driest months. Bultot (1971), summarizing data collected from 1930 to 1959 at recording stations in eastern Congo, reports the mean annual precipitation between 1,700 and 1,800 mm in the Ituri Forest area, with an annual rainless period of no more than 40 days. The analysis of the rainfall in the region reveals a period of monthly rainfall intensity less than 100 mm that goes from mid December to February.

A HISTORICAL PERSPECTIVE OF WATER-FOOD SECURITY SYSTEMS

Throughout the lump studies on settlements of the north-eastern Congo (Vansina 1990, Kuper & Leynseele 1978, Bailey et al. 1991), three major theories can be spelled out. The first would consist of the early hunter-gather inhabitants of the rainforest, the second deals with the western Bantu expansion and the last will be summarized from other settlements that occurred in the region. It is important to notice here that water has

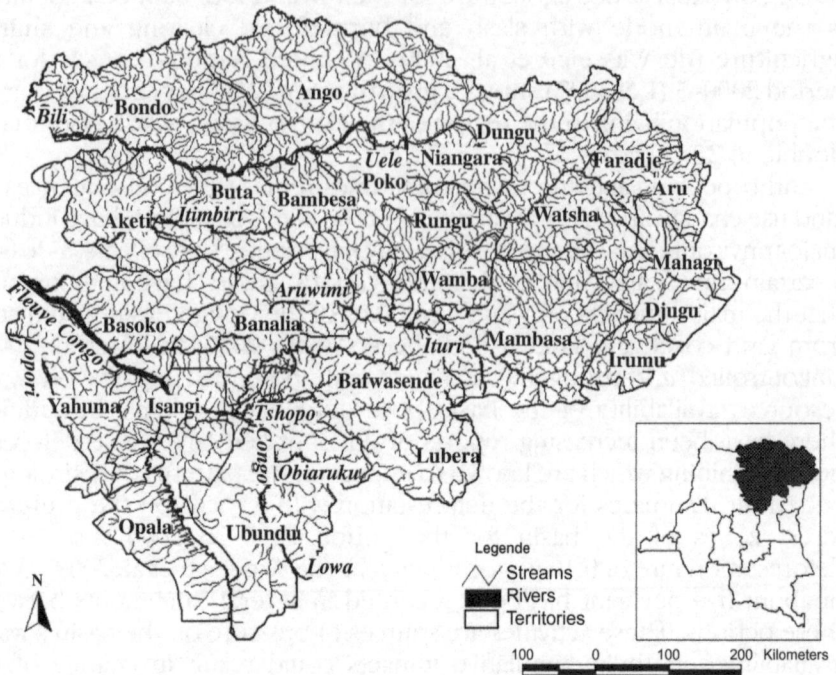

Figure 14.1. Map of the north-eastern Democratic Republic of Congo (© Raphael M. Tshimanga).

played a key role in these establishments as the evidences of settlements, expansions, and colonizations are in most cases related to water courses and water bodies.

Although the difficulty of establishing firm evidence of the ancient inhabitants of the rainforests has been recognized, some archaeological finds can be used to trace the link settlement-water of the early communities who would have occupied the rainforest some several thousands of years before our era. Vansina (1990) depicts the archaeological sites, most of which emanate from rivers, of the most ancient Stone Age hunters and gatherers. These rivers include, as for the north-eastern part of the Congo: Bomokandi River, Ituri River, Lindi and Aruwimi. Also Vansina (1990) observes that the activities of hunters and gatherers in the Congo rainforest began nearly 5,000 years ago. They became more sedentary, acquired ceramics, and began to supplement their hunting and gathering practices with new ventures in agriculture and trapping. Once perfected, it allowed them to spread far and wide as colonists in search of a mythical land of plenty, following rivers and elephant trails through the forest which was inhabited only by small numbers of scattered groups of hunters and gatherers and by fishermen.

Keim and Schildkrout (1990) observe that the presence of early hunter-gather inhabitants in the rainforest is indicated by several finds of polished tools, especially sturdy axes made of hematite, a very fine grained iron ore. Similar tools have been found between the Mbomu and Uele Rivers in the savannas, farther east to the rim of the mountains bordering the great lakes, and also south of the Uele and west of the Bomokandi valley in areas that were still heavily forested in 1880. The complex has been called the Uele Neolithic by archaeologists.

Although this theory has been challenged on ecological grounds as that over an extended period of the year, primary forest conditions did not allow for a food base sufficient and stable enough to support human foragers (Eggert 1992, Harts 1986). However, it is obvious from a number of scientific views that the area of rainforest was inhabited for many years. These early inhabitants had a subsistence mode of life related to hunting-gathering and fishing (Vansina 1990, Bailey et al. 1986, Preuss & Fiedler 1984).

While the theory of early hunter-gather inhabitants of the rainforest is being challenged and genuine evidence is scarce, glottochronological data and linguistic genealogy of the Western Bantu expansion along the Sanga-Oubangi-Congo swampy system has met a quasi satisfaction of researchers. Following this dispersion, the whole forest area of northern Congo, was then settled by speakers of the northern Congo languages. The original expansion of Western Bantu speakers in and around the rainforests resulted in the occupation of only small portions of the area that held the best potential for yam and palm growing and for fishing as a mode of subsistence.

Agriculture came to be developed in this area with the advent of Iron Age. The use of cutlasses axes for agriculture, and spears for hunting could now allow people to go beyond the sphere of food production. Ringing trees and burning them rather than cutting them down with an axe was the most efficient technique.

As if it was not sufficient, the advent of bananas introduced in equatorial Africa wrought major changes, as the advantages of its cultivation took over yams and oil palms previously adopted in the area. Unlike yams and oil palms, bananas are ideally adapted to evergreen rainforests. Unlike yams, the absence of a dry season does not hurt them. One needs to clear only about two-third of the trees on the field, rather than clear them completely as is necessary for yams. Compared with yams, the crop requires less care after planting and its preparation for food saves much time. Gradually yams were ousted as a staple over most of the area. Bananas became the ideal staple crop for agriculture in the rainforests, and allowed farmers to colonize its entire habitat everywhere. Farmers could now settle everywhere and populations increased faster, to the growing inconvenience of the hunter-gatherers. Then, now and for evermore, could start the socialization between farmers and hunter-gatherers. The farmers now produced food surpluses for exchange with hunter-gatherers and fisher folk in return for their products. The specialization of food production, so striking in this area, intensified. The growing advantage of the farmers over the hunter-gatherers gradually became such that autochthons were left with few choices. They could turn to agriculture and adopt the way of life of farmers, or they could intensify their relations with the villagers, accepting their bananas and their iron tools for the chase in return for meat (Vansina 1990). The evolution of this subsistence model has sustained the current mode of subsistence found in the area.

Along the way, the rainforest eco region continued to attract immigrants coming probably to take advantage of the resources offered, namely animal games, fish, water and agriculture productivity. This is the case of the southern central Sudanic speakers and Ubangian speakers. The first, identified as Mangbetu, came from higher and drier land at the confluence of Congo, Sudan and Uganda. They succeeded to expand far and wide from 3°N to the equator from 26°E to 28°E. Their area of settlement comprised the deep forest of the Lindi River valley, east of the Nepoko River, southwest of the Bomokandi River, south of Itimbiri and the banks of Aruwimi River. The southern central Sudanic groups were herders and growers of cereals, a way of life incompatible with life in the forest and they became so heavily indebted to Bantu speaking teachers that they borrowed almost their whole vocabulary relating to their new habitat from Bantu languages.

The Ubangian speakers were composed of the Mayoya and Bangba groups. They came from the west and brought a tradition of polished tools with them, among which the most remarkable were strong axes made of hematite with a very high content of iron. The Ubangianmayoyo and Bangba, the southern central Sudanic Mamvu, and the Bantu Buans all mixed during the second half-millennium of our era in the middle Bomokandi and Nepoko area, and by AD 1000 their different heritages had fused into a new common tradition. This combined much of the Buan forest heritage with features from the Mamvu savanna way of life. The mix of technologies is evident in hunting and farming. The practice of large-scale hunting associated with the burning of grasslands, typical of savanna, survived alongside hunting with nets, typical of the forest. The inventory of crops now included cereals and other savanna species, as well as forest crops. The association of crops, typical of forests, now went along with crop rotations from the savannas. The length of time a field would lie fallow was reduced to accommodate the Mamvu habit of more intensive agriculture, but, when no new fertilizing techniques were used, soils were overworked, forests turned into savanna lands, and the ecotone between rainforest lands and savannas was pushed southward.

Each of the main immigrant groups had brought its own social heritage. Together they came to shape a way of life based on hunting, gathering, fishing and rainfed agriculture for their subsistence. This way of life has subsisted for thousands of years along the upper Congo River and has significantly contributed to what is currently applied as mode of subsistence. Although the territory is actually shaped in accordance with the political and administrative structure, cultural mapping would still show that there is preservation of the social and cultural values inherited from the past.

Given the 1997 national census (PNSAR 1998), the output classification of ethnic groups for national statistics data lined up a broad spectrum of ethnic groups, which apparently goes beyond the sole criteria of linguistic genealogy as reported by Vansina (1990) for the early settlements. The census revealed a provincial distribution of ethnic groups composed of Azande, Mangbetu, Bati and Dongala groups in the Bas Uele district; the Azande and Mangbetu with their sub groups are also spread in the district of Haut Uele whereby they are associated to Lugbara and Pygmies. As for the pygmies, their main biotope is the Ituri forest situated in the Ituri district where they share the area with Balendu, Alure, Mabinti, Lombi and Lugbara ethnic groups. In the Tshopo district, the riverine people Lokele, Topoke and Wagenia share the area with the Itimbiri, Mangbetu, Mabinti and Bira ethnic groups.

A careful study of the current history of these ethnic groups would still establish a basic mode of subsistence based on rainfed agriculture, hunting, gathering, fishing and pastoralism. However, given a long term interaction between these categories of ethnic groups, it is sometimes a very difficult exercise to develop a standard classification of the mode of subsistence proper to an ethnic group as established in the early settlements. There have been cultural exchanges through intermarriages, internal migration, colonization, religion etc. Nevertheless, some people in the region are to date recognized for their specialized mode of subsistence. These are the forest people referred to as pygmies, who practice hunting, gathering and rainfed agriculture as their basic mode of subsistence and the riverine peoples referred to as Wagenia and Lokele, which have based their mode of subsistence on fishing.

Setting forth on the objectives of this study, what is important to notice here is that all these activities of human subsistence are grounded on water use. Agriculture in all its forms both irrigated and rainfed is a water user. The rainfed agriculture, as applied in the area, heavily relies on the seasonal pattern of rainfall and it is controlled by the intensity, duration, and seasonal variation of the rainfall. The techniques applied for its productivity: garden preparation, clearing, burning, planting, maturation and harvesting are adjusted according to the mean trend of rainfall parameters. As for the hunting and gathering mode of subsistence, it is demonstrated that environmental factors, which mainly depend on the seasonal pattern of rainfall, are very determinant for its productivity. Parslow (1999) observes from the definition of the relationship hunter-gather-settlement that it takes into consideration a symbolic relationship between the environment, subsistence activities, social organization and settlements. From this, it is therefore obvious that the changes of environmental parameters, in this case rainfall, within a component will ipso facto influence the changes in other components. As to the fishing mode of subsistence, water may not be required in the same form of rain but still, there is use of water for its productivity.

PAST TRENDS OF CLIMATE VARIABILITY AND SYSTEMS OF FOOD SECURITY

Climate is one of the most important limiting factors to societal well-being with respect to food security, human health, ecosystems maintenance, energy and most importantly to the mapping and characterization of the hydrological cycle. The earth's climate is driven by natural and anthropogenic factors such that changes to any of the forcing variables may cause fluctuations of the climate system resulting in a variable and changing system. Humans and the ecosystem depend on the availability of water for sustenance. The synergistic relationship between the climate and the hydrological cycle is such that if the climate becomes variable, the variability is also passed onto the water resources. Variability normally results in an unreliable water supply which tends to have significant impacts on the most vulnerable communities. Anthropogenic climate change also has the potential of imposing significant changes to water resources in the future (Sithabile 2013).

The IPCC (2001) defines climate variability as the short term fluctuations about the mean of the climatic variables and these fluctuations are largely caused by the natural climatic processes but they can be amplified by human activities. Climate change is defined as a statistically significant variation in either the mean state of the climate or its variability over an extended period of time.

Precipitation is the main determining factor of variability in the water balance and in observed flows over space and time. Because of the inherent association between the hydrological cycle and the climate system, hydrological variability is inevitably driven by climatic variability while at the same time variability in climate can be observed through changes in temperature and precipitation (Peel et al. 2002, 2004). Consequently, fluctuations in atmospheric circulation patterns which occur as fluxes of moisture and energy at the land surface have a large influence on the hydrological characteristics of a river system, the shifting of the ITCZ and the fluctuation in sea surface temperatures are linked to spatial and temporal variations in global runoff (Schulze 2005).

It has been demonstrated that seasonality in rainfall near the equator generally, and in the rainforest specifically, is less than in other areas of the world (Ricklefs 1973 in Bailey et al. 1989). However, Bailey et al. (1989), Eisenberg (1983) and Heatwole (1983) report that there are significant seasonal fluctuations in precipitation in rainforests, which, in contrary to what has often been written about tropical forests, contribute to mark seasonality in productivity at all trophic levels. This can be justified from the rainfall trend which shows variability in rainfall intensity of periods less than 100 mm per month. Eisenberg (1983) observes that most tropical rainforests have at least one period of the year when vegetable food resources suffer a decline. Variations in the onset and duration of the dry season can cause severe disruptions in the annual cycle of many biological systems, causing significant declines in food availability; faunal populations can fluctuate widely in response to unpredictable variation in food supply. Animals may migrate out of a stricken area. This variability of rainfall in the rainforests reduces their capacity to support hunting and gathering food for human foragers in tropical rainforests (Bailey & Peacock 1988, Hart & Hart 1986, Meggers 1973 in Bailey et al. 1989).

Figure 14.2 shows the trend in the mean monthly rainfall over the study area. The analysis of the rainfall in the region reveals a period of monthly rainfall intensity less than 100 mm that goes from mid December to February. Ecologists refer to this period as the dry season (Hart & Hart 1986). The rest of the year presents irregularities in rainfall intensity, which, however, does not fall under the ecologists' threshold of 100 mm. This is divided into early wet season (March to May), mid wet season (June to mid August), and late wet season (mid August to mid December).

Because of this variability, the major challenge for most food security systems is temporal matching of supply with demand. While in irrigated

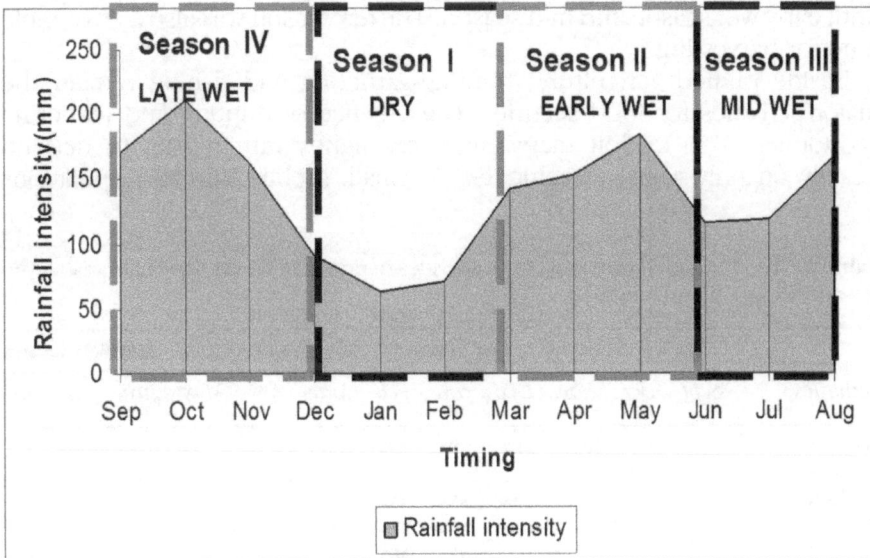

Figure 14.2. Rainfall trend and seasonality (© Raphael M. Tshimanga).

agriculture the demand and supply can be controlled, in this traditional mode of subsistence (rainfed agriculture, hunting and gathering) people will adapt strategies as part of environmental management.

This seasonality induces the variability that regulates many activities of food security, including rainfed agriculture, hunting, gathering and fishing. Rainfed agriculture in the Congo rainforest started with early settlers who came to the area in search for better living conditions. Agricultural expansion started with yams and oil palms, which were used by early farming societies for food security (Vansina 1990). The advent of the Iron Age greatly contributed to agricultural expansion through techniques such as clearing land, ringing trees, burning and poisoning arrowheads or spears. Much more agricultural expansion and settlements occurred with the advent of the banana, which allowed settlers to occupy vast areas of the forest ecosystem and adopt strategies to cope with seasonal fluctuation.

In tainfed agriculture, garden preparation, slashing, burning, planting and harvesting were used as the main techniques of food production. However, these activities were carried out so as to match productivity with seasonal fluctuation. Table 14.1 shows the temporal matching of the activities of production with seasonal variability in the rainfed agriculture mode.

In the shifting mode of agriculture, land availability and productivity (fertile land) were the main criteria used for selection. The search for typical land was undertaken during heavy rain, from September to December. Once this was found, garden preparation could take place, taking advantage of the dry season where slashing and burning was also carried out. Planting would then intervene in late dry season and continue until early wet season and mid season. The dry season was also a favourable time for harvesting.

Beside rainfed agriculture, hunting-gathering and fishing remain the major activities for food security. The essence of hunting and gathering economies is to exploit many resources lightly rather than to depend heavily on only a few (Waehle 1986). Small, mobile human populations

Table 14.1. Temporal matching of the activities of production with seasonal variability in the rainfed agriculture mode.

Activities	Late Wet season			Dry Season			Early Wet Season			Mid Wet Season		
	Sept	Oct	Nov	Dec	Jan	Feb	Mar	Apr	May	Jun	Jul	Aug
Garden preparation		xx	xxx	xxx	xxx	xxx	x					
Slashing				xxx	xxx	xxx						
Burning					xxx	xxx	xx					
Planting					xx	xxx	xxx	xxx		xxx	xxx	xxx
Harvesting		x		xxx	xxx	x					x	xxx

subsist on whatever resources are available within their territory. They adapt to conditions as they find them, using what is already there. They hunt game – whatever kinds are available, adapting their lifestyle to the conditions they face. In tropical rainforest, hunter-gatherers exploit many different plant resources for food, fibre, medicine, soap substitutes, etc. Again seasonal patterns of rainfall influence as much as it can this mode of subsistence as illustrated in Table 14.2.

The range of species is wide and throughout a year there are peaks of activity related to seasonal variations in the forest. Different roots and tubers are valuables. Various types of yams are among the more important wild foods. Various nuts are cherished, but they are only a snack and complementary food source. Mushrooms are especially abundant during the mid and late wet seasons. There is a peak of wild food in the last months of the year, and little ripe forest food from February through May. In the early part of the year, only yams and palm nuts (fruits of the oil palm) can be gathered from the forest. Waehle (1986) points out that when gathering, they (Bambuti) do not only go for wild plants resources, but eggs (from birds and tortoises), insects, molluscs, crabs and reptiles as well. Honey and termites are among the most prized of the forest resources. The availability of honey is extremely variable. Some honey is found in March/April, but the major season is in the second part of the year, after the blossoming of the threes: *Cynometraalexandrii* in February and March, and *Brachystegialaurentii* (from which comes the highly prized, clear honey). Because there are more Brachystegia than Cynometra in the forest and perhaps because each Brachystegia tree has more flowers or contains more nectar, the period of greatest abundance of honey is between July and September. The major season can last, in good years, from July throughout

Table 14.2. Temporal matching of the activities of production with seasonal variability in the hunting-gathering-fishing mode

Activities	Late Wet Season			Dry Season			Early Wet Season			Mid Wet Season		
	Sept	Oct	Nov	Dec	Jan	Feb	Mar	Apr	May	Jun	Jul	Aug
Hunting				XX	XXX	XXX	XXX	XXX	XX			
Snaring					XX	XXX	XXX	XXX	XXX	XXX	XXX	XXX
Fishing				XXX	XXX	XXX	XXX	XXX				
Honey	XXX	XX					XXX	XXX	XXX	XXX	XXX	XXX
Termite	X	XXX	XX				XXX	XX				
Yams							XXX	XXX	XXX	XXX	XXX	XXX
Palm nuts							XXX	XXX	XXX	XXX	XXX	XXX
Mushrooms	XXX	XXX	XXX							XXX	XXX	XXX
Ripe forest fruits	XX	XXX	XXX	XX			XX	XXX	XXX	XXX	XXX	XXX
Grubs	XXX	XXX	XXX	XXX	XXX	XXX	XXX	XXX	XXX	XXX	XXX	XXX
Caterpillars	XXX	XXX	XXX	XXX	XXX	XXX	XXX	XXX	XXX	XXX	XXX	XXX
Wild foods	XXX	XXX	XXX									

October. This is considered the honey season. If it is a particularly good season, it may last into November. While honey extraction can be labour intensive, it seldom necessitates labour input from more than four men, and two men are usually quite sufficient. Yet the honey is customarily shared more or less equally by all men present at the tree during the extraction, with only slightly more honey going to the man who originally sighted the hive (Bailey et al. 1991). During this season, there is a move of settlement to the honey camps in the forest. This is also a time when forest food like fruits and nuts are in plenty.

The honey season gradually blends into a heretic period of termite gathering in October and November. Grubs and caterpillars are available from time to time. Wild plants are important as a daily supplement. Several wild plants are highly prized and eagerly sought after. Men, women and children gather edible forest resources. Gathering is, however, mainly the women's task (Waehle 1986).

Hunting is traditionally accomplished with arrows, darts, or spears. These weapons are easily created from available plant resources and often tipped in poisons extracted from rainforest animals or plants. The addition of poisons assists the hunters by further disabling the animal targets, often by paralyzing the animals' muscles. It is after the honey season that rainfall makes hunting hard and the forest cold and wet to live in.

Fishing takes place, especially in the dry season and early wet season when the rivers are at their lowest, with a variety of techniques. These include hooks and line, small fish traps and large installations in the river. The latter will divert every large fish in the trap. Other techniques may include the use of darts, arrows, or spears to catch individual fish. Nets are sometimes used to catch more. Other times, special poisons are introduced into the waterways to stun or kill fish that can then be collected in larger quantities. Nothing but the fish appears to be affected. In this way, people roam around in the river catching unconscious fish by hand or with the help of a machete. Whenever large quantities of poison have been used, the river is more or less empty until the heavy rains arrive. It is possible to portion out poison in smaller quantities. In this way, fishing can take place several times along the same river.

Fishing is largely a collective endeavour including several households. Men manufacture simple fishing rods from saplings. They leave the camp in groups for a larger river nearby, but spread out along the bank and fish individually. Occasionally women and children will come along for such fishing trips. Some men borrow portable fish traps from a villager. They catch fish in smaller rivers (not necessarily restricted to the dry season). They do this in a 'share-cropper' basis. The owner of the trap has a right to a certain amount of catch.

Women have a way of fishing called *kusenga*, where it is explicitly stated that men cannot participate. *Kusenga* is held when the water level in the river is very low allowing women to stand in it and perform their catch.

Waehle (1986) observes that when there is more water, they build two dams of soil, sand and branches. The distance between the dams varies from 3 to 25 m. Equipped with large *mangongo* leaves, they work hard and fast, throwing water out of the area between the dams. When all the flow has been poured out, the fish would be caught by hands. The procedure is repeated several times during a single day. This very technique of fishing is called *Nzelensi* in the Cuvette Centrale of the DRC.

FUTURE CONTEXT OF WATER-FOOD SECURITY SYSTEMS IN THE CONGO RAINFOREST

According to IPCC (2001), the altitudinal distribution of dense humid forest types of the Congo basin and their projected climate risks include flooding, sea level rise, drought and temperature increase. In the context of environmental change, increased frequency and intensity of rain-storms, dry spells during wet season, increased dry spells during dry seasons, intense and sustained heat spells during dry seasons, high winds and wet spell in the dry season are among environmental concerns. It has already been observed that variations in the onset and duration of the dry and wet seasons can cause severe disruptions in the annual cycle of many biological systems, leading to significant declines in food availability. Perceived effects of the environmental changes on food security encompass cassava tuber rot, storage of seeds for subsequent planting being affected, falling of trees (a major danger for hunters), etiolating of plants such as rainfed rice, peanuts and corn, withdrawal of wild game during pockets of severe drought, blossoming of trees being affected (honey, caterpillars, etc.), drying of some water courses and swamps, fewer edible mushrooms, decrease in fish stocks, animal morbidity, increasing number of bush fires, failure of flowers, and rapid growth of some plants such as plantain, cocoyam and cassava (Bele et al. 2010).

Knowledge systems, practices and values related to water have evolved throughout history and have greatly contributed to shaping civilizations in the past and present. Settlements, food, politics, laws and religion have always been closely tied to societies and interactions with water. By tracing these developments through historical research, it is possible to understand experiences of the past and lessons to be learned, thus offering clues for sustainable development.

Water-food security systems are part of the cultural identities of people living in the Congo rainforest and sustainable solutions to related problems should be based on a deep understanding of the myriad interactions between people and water. Such interactions go beyond human uses of water such as drinking, washing, and fishing. It also goes beyond examining water-people relations in the framework of environmental services provided by water such as food, recreation and aesthetic values.

REFERENCES

Avissar, R., Werth, D. 2005. 'Global hydroclimatological teleconnections resulting from tropical deforestation'. *Journal Hydrometeorological* 6: 134–45.

Bailey, R., Peacock, N. 1988. 'Efe Pygmies of north-east Zaire: subsistence strategies in the Ituri forest'. In Garine, I. De, Harrison, G. (eds), *Coping with uncertainty in food supply*. London: Clarendon Press: pp. 88–117.

Bailey, R. 1991. *The behavioral Ecology of Efe Pygmy Men in the Ituri Forest, Zaire*. Ann Arbor: University of Michigan Museum Press.

Bailey, R.C., Head, G., Jenike, M., Owen, B., Rechtman, R., Zechenter, E. 1989. 'Hunting and Gathering in Tropical Rain Forest: Is It Possible?' *American Anthropologist* 91: 59–82.

Cadet, D.L., Nnoli, N.O. 1987. 'Water vapour transport over Africa and the Atlantic Ocean during summer 1979'. *Quarterly Journal of the Royal Meteorological Society* 113: 581–602.

Camberlain, P. 1997. 'Rainfall anomalies in the source region of the Nile and their connection with the Indian Summer Monsoon'. *Journal Climate* 10: 1380–92.

Eltahir, E., Loux, B., Yamana, T., Bomblies, A. 2004. 'A see-saw oscillation between the Amazon and Congo basins'. *Geophysical Research Letters* 31(23).

Hart, T., Hart, J. 1986. 'The ecological basis of hunter-gatherer subsistence in African rain forests: the Mbuti of Eastern Zaire'. *Human Ecology* 14: 29–55.

Hart, T.B., Hart, J.A, Murphy, P.G. 1989. 'Monodominant and species-rich forests of the humid tropics: causes for their co-occurrence'. *American Naturalist* 133: 613–33.

Hughes, R.H., Hughes, J.S. 1987. 'Zaire'. International Water Management Institute, available at www.iwmi.cgiar.org/wetlands/pdf/Africa/Region4.zaire.

IPCC, 2001. 'Climate Change 2001: Impacts, Adaptation, and Vulnerability'. In McCarthy, M.C., Canziani, O.F., Leary, N.A., Dokken, D.J., White, K.S. (eds), *Contribution of Working Group II to the Third Assessment Report of the Intergovernmental Panel on Climate Change*. Cambridge: Cambridge University Press.

Kuper, A., Van Leynseele, P. 1978. 'Social Anthropology and the "Bantu expansion"'. *Journal of the International African Institute* 48(4): 335–52.

Nlom, H.J. 2001. 'The Economic value of Congo Basin protected areas goods and services'. *Journal of Sustainable Development* 4(1).

Peel M.C., McMahon, T.A, Finlayson, B.L. 2004. 'Continental differences in the variability of annual runoff: update and reassessment'. *Journal of Hydrology* 295: 185–97.

PNSAR. 1998. *Monographie de la Province orientale*. Kinshasa: Gouvernement de la République Démocratique du Congo.

Shinoda, M. 1986. 'Rainfall distribution and monsoon circulation over tropical Africa in the 1979 northern summer: Their comparison between East and West Africa'. *Journal of Meteorological Society* 64: 547–61.

Vansina, J. 1990. *Paths in the Rainforests. Toward a History of Political Tradition in Equatorial Africa*. London: James Curvey.

——— 2006. 'Linguistic evidence for the introduction of Ironworking into Bantu-speaking Africa'. *History in Africa* 33: 321–61.

Waehle, E. 1986. *Elusive persistence: Efe (Mbuti Pygmy) Autonomy Strategies in the Ituri Forest, Zaire*. Oslo: University of Oslo.

15 Managing the Commons with Floods: The Role of Institutions and Power Relations for Water Governance and Food Resilience in African Floodplains

Tobias Haller

INTRODUCTION

Wetlands are key landscapes from the perspective of conservationists, such as IUCN and WWF. These areas represent not just water-protection issues but also complex humid ecosystems that play a crucial role for biodiversity conservation. Wetlands in dryland surroundings, in particular, provide the basis for survival to many, and often endemic, species during dry seasons (Howard 1992, Drijver & Marchant 1985, Roggeri 1995, Dugan 1990, Acreman & Hollis 1996, Thompson 1996). But wetlands are also dangerous areas because of the sicknesses they harbour and because of the damage caused by floods.

River floodplains play a central role among these wetlands as they bring constant fresh water and provide regular inundations, which are needed for the adjacent ecosystems to sustain their reproductive quality. These ecosystems, however, are not pure nature but have been used and transformed by humans for centuries. They are thus cultural landscapes and their conservation policies should incorporate this fact (Haller 2007b, Haller 2010 ed). From an anthropological perspective it is of central interest how local peoples have adapted to and also influenced flooding and dry–wet seasonal patterns in order to provide generally secure livelihoods.

Floodplain areas are under threat, however, because of great economic pressure on natural resources. On this fact much has been written from a natural science and conservationist perspective on African floodplains (Haller 2010a for an overview), often ignoring social anthropological research, which focuses on humans' adaptations and transformations that occurred in these areas before colonial times and external economic pressures. In order to bridge that gap, this chapter draws on findings and conclusions made in a comparative social anthropological research project, the African Floodplain Wetlands Project (AFWeP), which was conducted at the universities of Zurich and Bern between 2002 and 2008, including research in Mali, Cameroon, Tanzania, Zambia and Botswana (Haller 2007a, 2007b, Haller & Merten 2008, Haller 2010 ed, Haller 2013, see Figure 15.1).[1]

The focus of this work was on trying to explain why these floodplains are under pressure using a New Institutionalism approach for analysis: Starting from the view that these floodplains had been used and managed before colonial times by local communities, the project aimed at gathering more data on these institutions for the management of so-called common pool resources (CPRs) such as fisheries, wildlife, pasture, veld products, agricultural land and of course water.[2]

By institutions we mean laws, rules and regulations as well as norms and values related to resource use and property that are said to reduce transaction costs (information monitoring and sanctioning) (North 1990, Ostrom 1990, Ensminger 1992, Haller 2007a). We argue that locally crafted institutions played a key role in the management and the quality of the CPRs and the cultural landscape they belong to. While there are few direct institutions linked to water with, for example, irrigation systems, the research projects indicate that many institutions related to water are dependent and related to abundance and scarcity of water during an annual cycle.

The chapter outlines the reasons for the development of such complex management systems and the role they play for ecological, economic and social sustainability and resilience. In line with other scholars in social anthropology, we speak of resilience not just designating the ecological dimension, but the food-production-related dimension as well, and finally the social dimension. Thus we argue that floodplains, as wetlands in drylands, provide a key role for the survival of sedentary and mobile communities as these do not have to rely just on rain-fed humidity in the area, but can profit from water stemming from upstream areas. Therefore, river floodplains provide CPRs in times of humidity stress, when agricultural production is otherwise not possible and pastoral systems are under pressure. This quality makes floodplains attractive and thus often contested.

Our historical research in floodplain areas indicates that in pre-colonial times such contestations and conflicts have also led to solutions and institutional arrangements. These can best be analysed by using theories of common property, which have to be enlarged with issues of power and ideology in order to understand management structures that were in place before colonial times. We then focus on how the colonial and the post-colonial state misread and dismantled these systems through a process we call the shift from common to state property. This shift in property leads, as we will show, to transformations that can go in two directions: increasing open access or *de facto* privatization and a related process called fragmentation of resource management compared to an

Figure 15.1. African floodplains provide rich interrelated common pool resources (fisheries, pasture, wildlife and water for irrigation) (© Tobias Haller).

inclusive management of a cultural landscape ecosystem (Haller 2013, Haller et al. 2013). This latter fragmentation process leads to pressure on the resource and to the undermining of the resilience capacities of the cultural landscape ecosystems and of the people who live in these systems. This process then further triggers scarcities of CPRs, triggering more competition ('positive' feedback loops) in the social–ecological and as well as in the economic systems locally and regionally.

A central issue for water management then is the fact that this pressure is further aggravated by legal and institutional pluralism (plural state, civil society and local institutions) creating insecurities of tenure while also opening up to large-scale private investments in land and water facilities, commonly labelled as 'Land Grabbing or Large Scale Agricultural Investment'.[3] This is a second wave of transformations, which further undermines local use and creates many levels of conflict among and between so-called local, state and external private users. The chapter concludes by outlining possible ways out of this 'drama'. It shows that local initiatives and the support for collective action might be a way to mitigate the drama – however only if state management can be addressed properly.

FLOODPLAINS AS POOLS OF RESILIENCE

Resilience studies first focused on the ecological properties of a system to absorb perturbations before a system change is taking place. However, for resource management these have to be linked to social and property systems in order to make sense with regard to resource management (Berkes & Folke 2000). But what exactly makes floodplains such important ecological zones for resilience of wider regions, especially if we are looking at wetlands in drylands? In order to deal with this question, the ecological processes in floodplain areas have to be understood. These are summarized in several studies and publications. Here, we will only provide a short outline of the processes involved.

Following several scholars (Welcomme 1979, Ellenbroek 1987, Roggeri 1985, Howard 1992, Chabwela 1992, Stevenson & Frazier 1999, Thomas 1996, see summaries in Haller 2010a, 2010b, 2013), floodplains are peculiar because they show a seasonally changing pattern of inundation and dry constellations. There are roughly four phases: (A) during the rainy season the river stays in its bed, but is alimented by upstream water from higher and lower tributaries; (B) towards the end of the rainy season, as more water flows from upstream areas in addition to heavy local rains, the river leaves its bed and inundates the areas around it up to several 1,000 km^2; (C) after the peak of the inundation is reached, water recedes but often some is left in depressions; and (D) the water fully retreats to the main water bed followed by a dry season.

During floods, water moves into ponds and depressions and remains there as water pools for a longer period after the water retreats. In addition, main rivers may meander, creating oxbows and river arms often disconnected from the rest. This pattern has impacts on the flora and fauna. First of all, the water covers the vegetation and inhibits tree growth while grass vegetation can recover nicely. The vegetation is also alimented with nutrients from floods resulting in rich pastures after water retreat. The rich grass cover is then one of the major attractions for pastoralists, especially in the drier seasons. In addition, fish and reptiles move out of the main body of water during floods. Several fish species then move up the tributaries for spawning and, when they move back, some get trapped in the ponds. There they also breed and increase in size, while others return to the main river for the same process. Again other fish species breed in the main river or spawn on the river bed (certain bream species such as *tilapia or oreochromis pp.*).

Wildlife, such as the antelopes of the Kafue Flats floodplains in Zambia, might follow the pattern of floods and retreat as they move in herds, congregate on higher grounds and hide from the water. Afterwards, in the drier seasons, they look for drinking water in the ponds and are attracted to those. In addition, wild products are watered and mature during the drier periods. Last but not least, soil is fertilized in this way and renders land productive for agricultural production (see Haller 2013 for a summary).

But floods might also appear as threats, bringing with them the danger of inundations, which destroys fields. Nevertheless, floodplains are alimented with the water that creates a resource-rich landscape, serving as a seasonally important refuge in otherwise dryland areas. In extreme cases if the rains fail locally, the rivers might get water from other zones and still enable production within the cultural landscape. Having outlined these natural features we need to keep in mind that, following Berkes and Folke (2000) that we should look at the way resources used in this complex system have been institutionally ruled and managed by local groups and now increasingly by states and other actors.

WHAT IS NATURE, WHAT IS THE CULTURE OF THIS ENVIRONMENT?

While local people and people in the areas adjacent to the floodplains strongly depend on these resources and find means of adapting to them, the dominant view of governments, scientists and NGOs is to see floodplains as 'pure nature', a dominant and hegemonic ideology disregarding long-term human uses in the area. However, the way floodplains seasonally change discloses the multiple ways in which humans influence and transform the configuration of these ecosystems. Ellenbroek (1987) and others produced a model based on the ecosystem approach in which humans are the predators in the system. However, humans are not only predators for fish and wildlife without rules, but are crafters of biodiversity by regulated hunting and gathering strategies and the differentiated composition of wildlife and fisheries in species, size and age groups is reflected by this use. The other issue is domestication and its governance. Domesticated animals graze on the pastures and the composition of grasses is transformed by this use. Grasslands would look very different without pasture grazing by human-controlled herds. The differentiated number and quality of grasses depends on this type of grazing. Without grazing, or hunting and fishing, the floodplains would look very different than they do now, especially as cattle and wildlife do not often prefer the same types of grasses. Cattle dung then fertilizes the soil or is washed into the river system and fertilizes other areas of the floodplain (see Haller 2010 ed, 2013).

There is one puzzle regarding the question on how the water-interlinked fauna and flora use is related to the issue of sustainability. For hunting and gathering communities, many studies have confirmed what anthropologists such as Marshall Sahlins have argued earlier why their resource use can be labelled as being sustainable based on the solution to the problem of the imminence of diminishing returns they face (Sahlins 1972, for an overview, see Smith and Winterhalder 1990). These groups are mobile between different resource areas and remain only in an area as long as the catches are high compared to the effort made. When the effort involved gets too high, hunters and gatherers move to the next area where expected catches are higher than in the previously used area, allowing the remaining animal population to recover. However, this is not a conscious act of sustainability, but is based on what could be termed 'economic calculations' leading to a sustainable outcome. But to argue based on an economic principle alone would be misleading and would disregard local knowledge and emic views that the spiritual world rules in favour of the presence or absence of animals (for a debate on similar issues, see Berkes 1999, for a summary, see Haller 2007b).

Only in two cases of the AFWeP project were hunter-gatherers involved and they have been sedentary for a long time. Therefore, the issue of diminishing return does not apply directly but indirectly; it can be discussed regarding the movements of pastoralists, meaning that nomadic pastoralists – similarly as hunter-gatherers – move to richer pastures before the one used at the moment is degraded, as overuse would not make economic sense given better pastures elsewhere. What does apply more strongly however – and which is also part of a similar level of analysis of resource management by Berkes and others – is the focus on what impact religion has on institutions for the management of CPRs. This will be discussed elsewhere (see below). There are other features of the landscape related to water, which show how humans impact on their environment, and which leads us to call these cultural landscapes ecosystems (Haller et al. 2013). All groups studied in the mentioned project either have an active impact on the topography by creating depressions, ponds and channels or actively extend natural depressions, in which water can remain longer after the retreat of the main rivers. These depressions are of central importance for fisheries and wildlife and humans manipulate these cultural landscapes, and mark them as part of their landscape which is then related to a sense of ownership of the topography.[4] Regarding the issue of human-transformed environment, such as the composition of fauna and flora, this is a special feature of human transformations related to water management, which should have implications when it comes to ownership issues (Haller 2010, Fokou 2010, Beeler & Frei 2010, Saum 2010).

ENVIRONMENTAL PERCEPTIONS AND POLITICAL ORDER IN THE FLOODPLAINS

Seeing these changes and understanding that we are dealing with human-made land- and waterscape ecosystems is one issue. Another issue of interest in this context is the way these cultural landscape ecosystems are perceived by local people, who do not have the same ontological views as we have. These emic views are often related to instalment histories of local peoples and their political organization and linked to questions of identity. One question is how place of being, which includes land and altered water areas with their CPRs, can become important in newer debates in social anthropology and human geography. This debate focuses on the way ownership or stewardship is constructed via concepts of land and autochthony, especially when resources are getting scarce and more valuable.

The distinction between so-called first- and late-comers is highlighted by several authors writing on emic land views and so called customary resource tenure in Africa (Kuba & Lenz 2006, but also Benjaminson & Lund 2003 or Derman et al. 2007). This debate has gained contemporary relevance based on issues labelled as 'large-scale land acquisitions' or 'land-grabbing' processes, in which landowners shall be compensated (Toulmin 2008, de Schuter 2012, Peters 2013). Kuba and Lenz (2006) hint at a point that I would like to explore further and that is related to the 'first-comer–late-comer' debate: the perception of the environment does not just include local ecological knowledge related to water, but often also includes spiritual, especially traditional animistic religious views of floodplain landscapes. In this emic view water, land and the whole fauna and flora are

Figure 15.2. A Kotoko fisherman in the Waza Logone Floodplain using a channel for fishing. The Kotoko are the first-comers and controllers of the area and are centrally organized by a sultanate (© Tobias Haller).

not just interrelated living systems; they are under the control of spiritual beings. In the local worldview, they influence resource outcomes such as rainfall, floods, storms, wild animal attack, sickness etc., which all affect catch and yields (Haller 2007b). In social anthropology this has not been discussed as a way of 'primitive ideas' about the world but rather trying to give explanation that cannot be explained otherwise: this is why a certain environmental constellation happens or does not happen (for example, a concrete rainfall pattern), thus providing explanations and concrete options of action to be taken (ibid).

In all the cases studied in AFWeP, the people who call themselves first-comers often have a special relationship to these spiritual beings and their related places, which require rituals to be performed. These rituals can be preventive or reactive, but in most cases involve specific community subgroups, which have a special relationship with the spiritual world. In two cases studied, there are specific spirits of the water that have a certain kind of 'contract' with the living first-comers (among the Bozo and Somono fishing people in the Inner Niger Delta in Mali (Beeler & Frei 2010) and the Kotoko in Northern Cameroon) (Fokou 2010, Landolt 2010). In both groups, masters of the water by a specific ritual have to renew a 'contract' between the living and the spirits, who are seen as the real owners of the waterscapes.

Renewal of such 'contracts' is important as spirits have an influence on the abundance or seasonal scarcity of, for example, fish if certain regulations are not met (see below). In the other cases studied, the spiritual world is important for the relation with wild animals (Tanzania/ Rufiji: Meroka 2010, Botswana/Okavango: Saum 2010), land use and water flows (Tanzania/Pangani: Mbeyale 2010) or for rainmaking (Zambia/Kafue Flats, Haller & Merten 2010). In our comparative study, we can show that the local perception of landscapes with their resources includes this spiritual dimension as well as subsistence production, and coping strategies. These often include risk-reduction techniques and collective actions, which provide not the best but the most secure benefit (collective hunting and fishing, collective use of pasture, use of different species of domesticated animals, giving away some of the herd to poorer people in order to reclaim them back after loss of cattle as well as general diversification). All these strategies are not aimed at the best but at the most secure and needed benefit what can be called a minimax strategy. But ritual specialists provide also another important service in the emic view regarding the spiritual world that is insecure. These specialists help to reduce insecurity ritually by dealing with issues of prevention or dealing with insecure outcomes. This happens in situations in which people believe that control by minimax strategies is not possible anymore since outcomes of production are seen as being in the 'hands' of the spiritual world. Rituals are then an important way of communication and therefore insecurity reduction. It is often people from so-called first-comer groups who deal with this matter and therefore the identity and position of these people within the land and waterscapes are of central importance as they argue that they have the link to the spiritual world and are thus the legitimate stewards and coordinators.

Within these communities and groups, this distinction has implications for the political structure and order on an organizational level. This illustrates what Toulmin (2009) has summarized with the role of stewards of resource areas by first-comers relating to political control and management of land described above. The late-comers then have a right to use resources as members of the community of a territory (see section below). This strong relation with the spiritual world serves also as justification for the right for political control that can take different forms as our case studies show: It can be a hierarchical political system such as for the Kotoko in Cameroon, who retain central power of control and coordinate the use of late-comers and seasonal users. Other groups are organized with leaders who have the role of big men. These temporary leaders give access to resources to people who then become their followers. Through this process the big men increase their political power in relation to other powerful actors with whom they are in competition (see Ila in Zambia). Therefore, their power is not necessarily fixed and can thus remain highly contested. Others again are organized in age groups (Tanzania both cases, and Botswana). These different forms of political order hide the main structure: the first-comers as the ones coordinating the use of the cultural land- and waterscape ecosystems. They give access to others, based on rules and regulations related to the spiritual world. This is thus a central part of the institutional setting of common property tenure of resources in the cases studied.

COMMON POOL RESOURCES AND COMMON PROPERTY INSTITUTIONS

The commons debate includes a discourse of central importance, which should be newly looked at and related to issues of perception and political order. Up to now, the mainstream discussion has been contesting notions of tragedy of renewable resources held in common property. This led to not just denying Hardin's 'Tragedy of the Commons' notion – labelling commons as synonymous with open access – but at the same time underscores that common property meant shared property of a specific resource by clearly defined groups with clearly identifiable members whose individual rights and obligations in a tenure system are governed by institutions (see Haller 2007a for an overview, Feeney et al. 1990, Ostrom 1990, Ostrom et al. 2002, see also Ferguson 2015). The work of Nobel Prize winner Elinor Ostrom and colleagues showed that from a rational choice perspective, groups of humans can cooperate to craft institutions in order to solve problems of free-riding described as being inherent in the management of CPRs. In this context, and as mentioned before, institutions are defined as rules of the game (North 1992) including regulations, rules, laws etc. as well as norms and values that structure the do's and don'ts of resource use in a society (see Ostrom 1990, Ensminger 1992, for an overview, see Haller 2007a, 2007b, 2010a).

From a New Institutionalism perspective, institutions are therefore regulations and constraints that help to reduce transaction costs (information, monitoring and sanctioning) while also helping to coordinate and secure valued benefits. Based on this definition, researchers from around the world showed that, indeed, local groups are able to manage CPRs such as pastures, water for irrigation, as well as forestry and fisheries in common property institutions as long as membership is clear and monitoring and sanctioning rules are implemented. In the AFWeP, this could also be shown for pastures, wildlife and fisheries as well as for land, while for water related to irrigation only one example could be studied (see Pangani River system, Tanzania, Mbeyale 2010). This, however, obscures the role of water and therefore also of the other CPRs. Water bodies need to be managed because they provide vital resources for the other commons that are important to the people. Of all CPRs, water is the vital co-determining resource, crosscutting ties between these CPRs. Thus water from a local perspective is centrally included in the tenure institutions.

First of all, rules differ regarding flood or flowing or standing water and related resources. In traditional systems, areas under water are mostly but not always open access not just for water but also for fisheries.[5] However, these are still not fully open access, as there are cases of specific gender and fishing technique related issues so that even in times of high floods, certain areas have more fish than others and are thus under some type of regulation. For example, in the Kafue Flats in Zambia, some women mark first-comer spots with sticks, where much fish is expected. Normally, however, as soon as water retreats and remains in standing ponds, common to private rights apply in the flooding cycle. In this context, the notion of territory and membership is central. Nevertheless, these boundaries are permeable, with reciprocal rights and multiple resource use also between groups that need to be coordinated.

This is an interesting feature that can be seen in Zambia's Kafue Flats as well as in Cameroon's Waza Logone Floodplain. In accessing such seasonally changing tenure areas on the one hand, boundaries and membership are important as outlined. But on the other hand, reciprocal rights on demands and invitations provide help to buffer risks in the use of mobile resources such as fish in the water, or wildlife where there is unpredictability regarding the seasonal availability of the water-linked resources. Another issue is the multiple resource use aspect. By this I mean that waterways and waterscapes are close to pastures and pastoral use.

Such constellations demand coordination related to floods and movement of cattle to and from pastures as well as coordination of fishing activities. Cattle need water, which implies that standing water can be disturbed by cattle, a situation that has to be coordinated and mitigated as fishing cannot take place at the same time. Therefore, in all the case studies in our research where fisheries, water use and pasture use are involved, there are institutions, which are responsible for coordinating and regulating these multiple activities.

Another issue is flowing water, which is often linked to settlements and to the mobility of people as well as to boundaries of river sections. In this case, water as a mobile common pool resource increases the number of regulations for fisheries, drinking water and irrigation in many of the cases studied in the project.

It needs to be underlined that the outcome of these institutional settings for the management of CPRs was that local people managed water as well as water and land-related resources not in a vacuum but by commons institutions, in which complex cultural landscapes for secure livelihood gains were managed by leading people, whose coordinating function was religiously justified and where rituals provided coordinated access and governance because they provided timing and inbuilt sanctioning by higher religious actors if rituals were not performed or violated. One illustrative

Figure 15.3. In Waza Logone Floodplain, Cameroon, access to dry seasons pastures were regulated by institutions controlled by sedentary fishermen of the Kotoko first-comer group with religious ties to water spirits. The institutions comprised rules for coordinated access by fishermen, peasants and cattle herders (© Tobias Haller).

case are collective fishing events in ponds in the Kafue Flats where timing for fishing was announced by a ritual master, conducting rituals to ask ancestral spirits for permission before fishing would take place. Without such rituals, ancestral spirits would – so the local view – attack free riders in the form of crocodiles. As a matter of fact these dangerous animals can hide easily in ponds after migrating out of the main river during times of floods. Rituals here not only serve as a coordination tool but their violation – meaning free ridging in the sense of not waiting for the ritual to be performed before fishing – acts as well as a sanctioning device for the management of the fishery and is thus part of the fishery institution (Haller & Merten 2008, 2010). The issue at stake was not sustainability of fisheries or water use, but securing livelihoods. The outcome, however, proved to be sustainable.

But why are there now so many CPR systems under severe stress which previously seemed to have worked well? One answer lies in historical and institutional changes on which I will now focus.

Figure 15.4. A collective fishing event in the Kafue Flats floodplain, Zambia: access to common pool resources was often regulated by common property institutions including coordination of use by a ritual master. This also provided monitoring and sanctioning by the local view that free riders could be attacked by ancestral spirits sending crocodiles if fishing was done before the performance of the ritual (© Tobias Haller).

INSTITUTIONAL CHANGE IN WATER RIGHTS AND MANAGEMENT: THE COLONIAL PERIOD

Historical studies on common pool resource management in Africa often have to deal with the problem that the most important changes are excluded from the analysis. By including pre-colonial data our research team reveals that colonization is not just another change, but a central transformation process. However, this did not happen immediately, but rather as a blueprint for post-colonial state property tenure institutions. Important changes have happened before colonization, due to the European interest in a cheap African workforce and the consequent production of slavery for plantations in Latin America. Providing this workforce set local communities in tension. But if we focus on waterways and waterscapes, the transfer of rights from local systems to colonial state tenure systems was the transformation with most important impacts. Water and waterways are now no longer a local common property but a state property, especially in rivers, and all related resources then become state resources.

In the beginning of colonial times and in the cases discussed here, this process of institutional change becomes illustrative: from the start of the colonial process, centralized control was not easy to enforce and the British Empire's indirect rule strategy – also adopted partially in the more centralized French territories – seemed to be the most effective and cheap way of governance. But more and more, the British colonial administrations in Tanzania, Zambia and Botswana and the French colonial administrations in Mali and Cameroon fragmented the territories into two major spaces, one for white, large-scale farmers or for mining as well as reserves for timber and game and the other territories for local people, managed by local chiefs as customary areas. These local chiefs were often imposed by the colonial powers and their positions were created in order to provide cheap governance. Apart from the fact that this major divide then crosscut local territories, the same is true for larger areas and riverways, which were interrupted by this colonial divide.

This divide also caused the undermining of local institutions despite the belief that local problems should be dealt with locally. This fact can be nicely illustrated with the CPR water. Water was no longer a local resource, but had become a national resource, and its flow was to be captured first by the state and its actors. The same happened with water-related resources such as fisheries, pasture, wildlife and veldt products and forests that were also no longer legally local common property but, again, had turned into state resources. Therefore, state institutions began to be operational, undermining more and more local institutions, often as soon as CPRs increased in value for the state. A twofold process can be distinguished: not only are CPRs now taken away from local governance, that once based on local ownership and regulated by local institutions. But in addition, the interrelatedness of the resources and resource management, which local institutions provided, could no longer operate in the fragmented legal framework that the colonial and later post-colonial powers were setting up.

Despite many differences between colonial governance in Mali, Cameroon, Tanzania, Zambia and Botswana, our findings indicate that the state started to regulate these resources legally by only focusing on the resource itself and no longer based on the view of an ecosystem interaction. For example, water in Zambia was managed by a law that did not (and continues not to) recognize that fisheries, pasture, wildlife and veldt products are related to water. This is also true for all other CPRs in the other cases. But to worsen the problem, within a fragmented CPR such as water many contradictions in governance prevail. As water is a multiple-use resource, different sub-institutions fragment the use of water and inhibit coordination. Water has different meanings, values and governance rules for urban drinking water, for irrigation water (including large-scale land-acquisition issues) and water for energy production via dams are all dealt with in separate state institutions. Our case studies illustrate this fragmentation as legal and institutional pluralism and the consequences it entails for the governance of water generally and in interaction with the related CPRs.

These processes of legal pluralism are illustrated by cases from Cameroon and from Zambia. Fokou (Fokou & Haller 2008, Fokou 2010) and Landolt (2010) show that since the 1970s, the Cameroonian government perceived water as a means to implement a top-down food policy. It decided that water in the Waza Logone area in the north of the country should be used primarily for irrigation. In order to store water for a large rice production scheme, the large Maga dams were built in several sequences in order to expand the irrigated areas. While these technological changes were justified as a means to increase food security in the country, it soon became evident that not much of the rice was consumed locally, but instead was exported.

But more seriously, it impacted the flooding cycle as less water now reached dry areas as it did before the dams were built. This reduced pastures, fisheries and small land plots below the irrigated areas, thus undermining already weakened livelihoods. This modernization process thus redistributed water in a very unjust way, an issue that today would be addressed as a case of environmental injustice.

But while this contradiction of weakening local food security by implementing a large production scheme was ignored, the issue of justice was only brought up by conservationists. The World Conservation Union (IUCN) was concerned with the Waza Logone National Park, a protected area, in which wild animals were suffering from the lack of floods. As a consequence, they moved out of the park and got killed or damaged the fields of small-scale farmers. In order to mitigate these problems, IUCN launched a programme called 'The Return of the Water' with the aim to

Figure 15.5. Resource fragmentation in floodplains: large-scale dams and irrigation systems (such as in northern Cameroon) have reduced pasture and water commons of local people (© Tobias Haller).

push the government to release more water from the dams into the floodplain (see also Loth 2004).

However, local fishermen as well as farmers had already adapted to the lower level of water by, for example, building more and deeper channels for intensive fisheries, by increasingly owning cattle and lending these out to pastoralists and by increasing agricultural production as well as hunting activities that were now regarded as poaching. Therefore, the return of the water did not solve the problem since the institutional problems were not solved and in addition the adaptation strategies had also changed. The previous institutions were weakened and not robust enough. Coordinating measures between fishermen and pastoralists – part of the institutional design was to use the floodplain in a coordinated way between fishermen and pastoralists – no longer applied. In addition, local people lacked trust in the state and did not feel that relying on government agencies would enable them to get a fair share of the new resource. So, as more water came into the floodplain, everyone behaved as if the new resource was open access (like in Hardin's *Tragedy of the Commons*) and tried to profit as much as possible from water and water-related resources.

To make things worse, the state was extremely weak and state actors badly paid and not willing to 'make state'. Therefore, a situation developed called the 'paradox of the present-absence of the state' (Haller 2010a–c): immigrants and pastoralists argued that the land and its resources now belonged to the state and, if they adopted an identity as citizens, they were entitled to access the resources. But as the state was not present to enforce its laws and regulations the access became *de facto* a free access. Due to a lack of coordination, more conflicts arose between resource users, such as between farmers and pastoralists or between fishermen and pastoralists. This again was in the interests of badly paid state officials since they could get some extra income arbitrating in such conflicts. Fokou thus writes that officials were not really interested in mitigating these conflicts because conflict is 'like relish for the sauce', meaning that one could, as an official, get extra cattle and cash to pretend to arbitrate the conflicts (Fokou 2010). This case illustrates that by not paying attention to the transformed institutional settings and their transformations, mitigation measures can make situations even worse.

The second case illustrating my point comes from the Kafue Flats in Zambia where the situation is more complicated. Here, local institutions for all CPRs in the area had been undermined by the government declaring the resources as state property. This loss of control was worsened when the government declared that the area was very suitable for installing dams for hydropower. Two dams were necessary, one on the lower level (Kafue Gorge with the turbines) and a dam upstream called Iteshiteshi (as a storage dam to regulate the floods and to prevent that a very large area would be inundated). Behind this second dam the area was flooded and land and CPRs were lost there. But between the two dams CESCO, the electricity company run by the state, regulated the floods according to the requirements for hydropower production, turning upside down the seasonal cycle of the floods: in the dry season, CESCO released more water from the upper dam so that the turbines were working in the lower dam, causing inundation in times where these normally never happened. On the contrary, in the rainy season floods that should normally occur would be held back on the upper dam in order not to create overcapacities. This resulted in droughts when inundations were happening normally and had a major impact on other resource users and the resource cycle for fish, wildlife and pastures became disrupted (Chooye & Drijver 1995, Haller 2013). In addition, other claims were made on the water from nearby cities (such as Mazabuka and the capital Lusaka) and from large-scale irrigation schemes close to the city of Mazabuka.

These problems added to the already existing land conflicts as pastures and other CPRs had been taken out of local hands since colonial times by handing out farm land to white settlers and later on to large-scale land investors for sugar cane production. In addition, as the area suffered several food shortages and hunger crises local authorities argued that land and water for irrigation should be used in the future in order to enhance food security. Despite the fact that fishing, hunting and pastures would have been taken away from local people, chiefs did not check the national water laws in which hydropower production has priority over other uses of water.

These fragmentation processes led to further unexpected repercussions. Because of the lack of floods in the rainy season, the area became accessible for poachers and seasonal commercial fishermen from other areas of the country who were naturally excluded from the area due to inundations. Combined with low-level financial means to enforce state laws in the different government departments, CPRs such as fish and wildlife became an open-access resource due to the combination of lack of monitoring as well as the better physical accessibility to the resource areas. This again triggered several levels of conflicts between resource users regarding how land, water and related resources should be managed. Two examples from the Ila ethnic group in the Kafue Flats can serve as illustrations here. Due to economic changes fisheries became attractive and commercial fishermen from outside the area formed large seasonal fishing camps. While the Fishery Department was absent, due to the financial crisis, and the resource open access, outsiders countered the local critique of overfishing the resource with illegal methods that fish is a national resource and as citizens (discourse) they are free to use this resource, while the state was absent to control the fisheries (present-absence of the state).

Figure 15.6. The institutional change from common to state property can lead to open access and overuse of the CPRs as is the case in the Fisheries in the Kafue Flats (Zambia): Due to the lack local enforcement (because the fisheries are state property) *and* state enforcement (because of the state's economic crisis and lack of funds for monitoring) fisheries become an open access CPR overused by immigrant fishermen with small-meshed nets. But the state is not only absent, it is at the same time still present: Commercial fishermen claim fisheries to be a state resource to which they have access as citizens so they cannot be controlled locally (Paradox of the present-absence of the state) (© Tobias Haller).

This then had also further repercussions as collective fishing events were undermined by local young men, fishing out the ponds before collective fishing dates were announced. In addition, young men generally started to use fishing baskets reserved to women to fish for cash because it was a more productive fishing techniques than the spears which men used traditionally. All these developments led to serious conflicts between locals and outsiders, between the generations and on a gender level (Haller & Merten 2010). The second example illustrates elite capture of resources as some Ila chiefs were opting for smaller-scale irrigation projects to get funding, other leaders (often in opposition to the chiefs) were trying to stop such processes completely by arguing that the 'traditional way of life' (pasture, fisheries, wildlife) should be continued because they feared losing access to resources and thus their power.

To summarize the findings of the comparative study of AFWeP, the following points can be outlined for at least five out of six of the researched areas. The new way to manage water and related resources led to:

– local institutions for adapted access and resilience being undermined (seasonally adapted access to all water and land-related resources);
– new constellations of dams that release water or reduce flooding with consequences on several levels;
– new land rights after land reforms, often leading to a legal fragmentation of resources from each other, splitting up cultural landscape ecosystems;
– failed mitigation strategies through the formation of water boards and management plans, which in reality are unable to coordinate conflicting uses due to differences in interests and different bargaining power of actors and organizations within and between government departments;
– local struggles for scarce resources because of undermined local institutions and malfunctioning state institutions and increasing competition over the way floodplains are seen and perceived (while for the state it is just the water, for local people and other actors it is the water and the land, fisheries, wildlife, agro-production and protected areas (conservation and tourism) etc.).

An important issue in these cases studies is the changes in relative prices for resources, rendering a complex cultural land and waterscape ecosystem more valuable in the eyes of outside users and of the state. A New Institutionalism analysis in social anthropology (Ensminger 1992, Haller 2010a, 2010b) highlights external and multiple factors contributing to this increased value compared to other resources or alternative uses of resources. These factors include changing natural and political environments and markets, demography and technology. They contribute to an increase in the value of water, for example, or of another CPR (change in relative prices), which in turn increases the attractiveness of an area (see Figure 15.7).

Graph 1: Modeling change

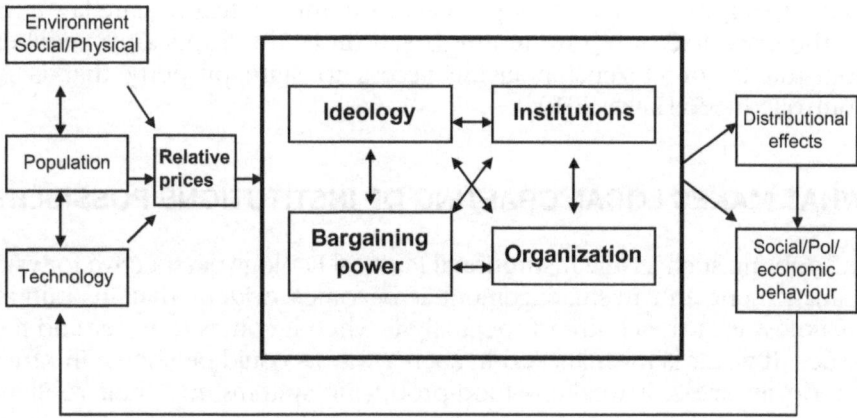

Figure 15.7. Modelling change (from Ensminger, 1992: 10).

In a further step, this process often leads to increased pressure on local resources itself and undermines resilient livelihoods as well as reduces bargaining power of local actors in the context of a weak state. Local actors lose their status as locals and just become citizens while the state is unable to protect their interests and due to their low bargaining power, the more numerous and more powerful external resource users and capital-oriented market interests are able to take over management of resources. These powerful actors (state officials, merchants, outside resource user groups as well as more powerful local actors) often use the ideology of modernity and state ownership of resources in addition to the notion of citizenship as a major discourse to legitimize their unlimited use of the resources. In fact they are able to persue what can be called institution shopping to select among the diverse institutional arrangements – such as different property rights – which suits them best (see also Toulmin 2009, Haller ed 2010, Haller 2013).

At the same time, governments' argumentation is based on a paradox notion: while wanting areas to be used in modern, sustainable ways based on citizenship, governments are unable or not willing to provide coordination, monitoring and sanctioning, which are key elements of transaction costs reduction that institutions should provide for sustainable and regulated use of resources. While the neoclassical economic perspective believes that institutions shall do just this, a New Institutionalism as well as a Political Economy and Ecology perspective argues that we have evidence that water management institutions are shaped and controlled by the actors with the most bargaining power. These actors also make their institutional choices according of the quality of the fragmented resources: if it is a mobile resource, open access prevails. If the resource is not liquid, results from the AFWeP indicates that private property and open access might be chosen simultaneously and legitimized ideologically by powerful groups of actors in the context of state law either by the discourse of private tenure based on land reforms after neoliberal adjustments or citizenship giving access to state property that is not controlled (see Haller 2010c).

WHAT MAKES LOCAL CRAFTING OF INSTITUTIONS POSSIBLE?

In adopting such a New Institutional Political Ecology perspective for water management and mismanagement it becomes evident that institutional weakness is at the centre of the analysis when it comes to water and food issues. If water is mismanaged in such a way as could be shown in African floodplain areas, it weakens food-producing systems and their resilience functions and as a consequence also undermines the capacity for sustainable food production. One important issue is the lack of coordination and the lack of trust between all actors in a market context, in which people cannot exist without financial means. Actors are forced to develop strategies to get access to cash, even in subsistence-oriented contexts. This is another driving force for immigrants, but for local resource users as well. Coordination becomes difficult and costly in such a context. Moreover, it becomes an even bigger challenge for sustainable use of resources when state institutions that have undermined local institutions fail to turn the promise of governance from theory into practice. In most of the cases in the AFWeP, local, NGOs and even low-level state actors tried to address these problems with several mitigation and adaptation strategies often linked to participatory approaches.

However, the problem is that in these cases, local power constellations and resource use patterns have changed. Scholars rightly argue that the ideal of the 'close-corporate community' (Cancian 1989) never existed, but what we can discover by studying local communities today is the fact that we have increasing levels of heterogeneity in interests between actors (according to class, patron–client relations, state power, age and gender). It is not just market adaptation, but a political process going on since colonial times and probably even before that, leading to further fragmentations and differentiations, rendering them more accentuated. The process by which colonial powers structured the indirect rule principle fragmented local societies further by introducing or including state created or transformed leaders, often called chiefs and local government official in the local communities, mostly for tax-enforcement purposes and cheap local control. These local actors have now gained much more power and are today in the position to further fragment local communities. At the same time, states lack money to make state as we could see in all the research projects with the exception of Botswana.

Through structural adjustment programmes by the IMF and World Bank, much of the management schemes for governments in Mali, Cameroon, Tanzania and Zambia were framed as decentralization and democratization schemes. These externalized costs of 'state-making' to the local level, usually without handing down political powers (Haller 2010, see also Poteete & Ribot 2011, for a similar process in Senegal, and Peters 2013 for general comments). Thus, despite the request for decentralization, recentralization occurred in many of these areas and often in a cheaper way than before. While the governments of the states in our cases were unable to pay for state employees, local villagers should now help to monitor a resource base to which they have no rights in the state's institutional setting. One exception in our case studies is Tanzania, which gives back rights to local villages. However, these face the problem that they have often been artificially created and were unable to craft their own institutions under the previous socialist government.

Village governance nevertheless might provide a basis for governance, as we will see. Many cases of so-called participatory initiatives have failed in all the areas studies (see also Blaikie 2006, Galvin & Haller 2008, Hara et al. 2010, Haller et al. 2008, Chabwela & Haller 2010), especially in the domain of conservation and protected areas. Cases from Cameroon, Tanzania and Zambia illustrate that the gains from the participatory management had been negative because local people had more costs than gains (loss of yields and access to CPRs due to wild animal attacks) while promised cash income from tourism mainly remains on a higher administrative level. The little that did trickle down is in the hands of local elites or is simply not enough to provide incentives, even if distributed on the household level. So instead of feeling themselves as 'stewards of the resource', such as wildlife and fisheries, people use discourses of not being owners anymore. Interviews in all the research areas reveal a discourse of 'being poachers now', poaching 'government animals' and to behave as Hardin predicted in his game theory axiom: get it before someone else gets it! These discourses indicate that many local actors lack a sense of ownership – be this private or collective – of the resource base, a fact that has been addressed by many scholars in African settings and elsewhere (Brockington et al. 2008, Poteete & Ribot 2010, Galvin and Haller 2008, Haller & Galvin 2011).

However, there are also winners. These winners are often large or middle-sized international nature conservation organizations, as well as governments and the tourism industry in Africa. For this reason, the political ecologist Piers Blaikie talks of 'participation as a Trojan horse' because the powerful and commercially strong actors have easy access to and control over CPRs previously managed in common property institutions (Blaikie 2006).

We need to address the resource puzzle for sustainable water and land management, particularly in asking the question of how to bring the local level back into the debate on institution crafting. This is a challenge, especially in light of local political differences and the fact that external and state actors have more bargaining power than local actors. There are of course no 'one-size-fits-all' solutions, but there are some rare 'social experiments' to be analysed scientifically. Our studies have shown that under certain conditions – and despite the fact that local communities were heterogeneous regarding bargaining power – local users were able and willing to reinstall institutions. These factors do not just echo Ostrom's design principles for robust institutions (Ostrom 1990), but deal more with the local perceptions of a constellation. Therefore, local bottom-up institution crafting was working when:

(a) local users recognized and felt the *strong need to reduce pressure* on the resource;
(b) people still had *knowledge on local "traditional" institutions* of resource management and basic features of these traditional institutions were still in place or known;
(c) *external actors* (often NGOs) *provided a platform* on which local interest groups and actors with different bargaining power (youth, poorer actors, women, economic interest groups) had a fair say and a sense *of fair play*;
(d) *local knowledge of resource dynamics* and cultural landscapes as well as creativity *was accepted* and served as *important precondition to draft new institutions* (mixture of old institutions and new views);
(e) people had a *sense of ownership* of the institution *in the drafting process* and could identify themselves with the institutions as their own institution despite power differences in the community;
(f) *recognition* of locally developed institutions *by the state* in order to help in situations of monitoring and sanctioning (see also Haller, Rist and Acciaioli 2015).

To illustrate this process, two examples will be shown, one from Zambia and one from Tanzania.

By-laws for fisheries in the Kafue Flats: The Kafue Flats fisheries had been regulated by local institutions of the Batwa fishermen and Ila agro-pastoralist before colonial times and a large variety of well-adapted rules related to flooding patterns were developed by these groups (See above and see Haller & Merten 2008, 2010, Haller 2013). Since colonial times and with state control and interest in the resource, local rules were undermined and proved to be ineffective. This happened when Zambia's copper crisis reduced state income and with this crisis unemployment and lack of state control due to lack of funding from the state created an open-access situation to the fisheries. Catches were going down and local people witnessed reduced access to fisheries as thousands of seasonal immigrants from the copper mine areas were attracted to the middle area of the flats, establishing seasonal fishing camps, as fisheries were still good in this section and prices were high. At the same time, local people were no longer in the position to control timing of fishing, fish bans and techniques as well as to enforce their regulations, while the state was physically absent. Therefore commercial fishermen were able to use illegal techniques and would at the same time reject local critique of overuse of the fisheries by using the discourse of citizenship to legitimize open-access use (present-absence of the state). In this constellation, the team of researchers, of which the author was a part, started a discussion process with local users and the local staff from the Department of Fisheries. It became evident that local staff were unable to monitor the area but at the same time that there were no incentives for local people to help in this. Nevertheless, there was the legal possibility to draft local by-laws to the national fishery laws in the context of decentralization. Because of the research, local people were sensitized on the issue. What was lacking was a neutral process since people were politically fragmented. Local chiefs did not represent whole communities due to political fights and the colonial past; their fathers or their lineage were seen as having been installed by the British and not representing the local people. People were articulating the problems with the open-access fisheries, but there was a need to develop a more or less neutrally perceived platform to discuss and negotiate new institutions.

The team of researchers and the local fishery department were then able to create a discussion platform, in which interest groups were identified for local collaboration based on financial aid from the World Fish Centre. In these discussions, people from interest groups and political factions first discussed elements of the new rules to be drafted among themselves. These first drafts were then pooled and discussed in a manner so that everyone could agree on the by-laws as a common product. The important issue was that all participants regardless of their political and economic position or gender felt they had been involved and had crafted 'their' own institutions. These then included elements of old fishery institutions for collective fishing and newly created regulations and laws in order to deal with new problems, such as, for example, law and order in the fishing camps, health and sanitation issues, or issues related to water (manage-ment as well). We called this process *constitutionality* by which we mean a sense of ownership in the institution building process that enables the embodiment of rules for which everyone was part in the crafting. Becoming a subject here in the Foucauldian sense does not refer to governmentality or environmentality but a previous process of being conscious of the rule making and later on embody the rule as one's own (Chabwela & Haller 2010, Haller 2013, Haller et al. 2015). While the management of the fisheries tended to become better after one or two years (commercial fishermen reduced fishing efforts or did not show up in the area, higher catches, return of fish species thought to be extinct), the problem was that the by-laws were not ratified. Thus they were not acknowledged on the national level and, as they were not backed by the government, this undermined the robust implementation process. The process itself however was very participatory.

Pasture regulations in Rufiji Floodplain Tanzania: An ultimately more successful situation is presented by Patrick Meroka in his work on the Rufiji floodplain (Meroka 2010). In one of the village cases studies, local people of the Warufiji ethnic group faced the problem of incoming Barabaig pastoralists to the Rufiji floodplain. This area had rich pastures that attracted the Barabaig, who were chased away from their northern territories due to large-scale agricultural investments. The Warufiji had never been much interested in pasture and animal husbandry but were hunters, gatherers, fishermen and peasants. The Warufiji group in this case was oriented towards agriculture and fisheries. However, this area is close to an urban centre and pressure on fisheries is high. Nevertheless, based on the new Village Land Act in 1999, villages used to have better rights than other groups in the AFWeP case studies. As the pasture was on village land it gave them the option to discuss and decide what they wanted to do with the pasture area, since they were not really using it. The interesting element of the process was that the two leading representatives of the two groups met and drafted regulations for the mutual benefit of pasture use by the Barabaig pastoralists – regulations, which had been discussed within the communities first. While the pastoralists again had access to grass in a rotation system, the local village had access to meat and milk, protein resources they were lacking due to the external pressure on the fisheries. Therefore, rules and regulations were set up and discussed in a participatory way, and local people of the Warufiji as well as the Barabaig also had a sense of ownership in the process, which seems to have been an important issue regarding sustainable use of the resource (Meroka 2010, Haller et al. 2013).

CONCLUSION

Water is the central element that enables growth and regrowth of land-related common pool resources. To manage these CPRs and the waterflows necessary in floodplain ecosystems, local people established institutions, which were not set up for conservation purposes as such, but to solve problems of cooperation and coordination. Such institutions are key to make local people more resilient in a complex water management system. They provide orientation *and* flexibility in this context. Local resilient common property institutions are crucial as they provide trust and robustness while allowing the flexibility needed for complex and risky resource systems.

But since these are being undermined by state laws and state, as well as by private property, serious situations of lack of coordination are created. In this way, local actors are stopped from being able to cope with bottlenecks in food production, which cannot be buffered through other systems. Therefore, undermining institutions' governing their access in constellations of risk, exposes local users to a high risk of losing food security. Flooding is not always the same, mobile resources (fish and wildlife) are not always available in all the areas every year, pasture quality can change, agricultural production can fluctuate, etc. As important as the design principle that Elinor Ostrom has developed, these need to be analysed with a focus on flexibility. While territorial boundaries and clear notions of membership in a community are key, flexible pathways of diversification and regulations for reciprocal access for survival in critical times are equally important. In addition, fairness and options to balance bargaining power of actors are central and the distribution rules of access and consumption patterns within the commons systems in many of the cases are evident.

However, for sustainable water and land as well as land-related resources management, we need to assess how local people dealt with the interconnectedness of all these resources based on local people's knowledge. This knowledge has to be adapted to environmental changes and also passed on to future generations and to non-locals. Resilience can be enhanced by including not only locals, but outsiders as well in a controlled way; floodplain areas can thereby again serve as a resilient food security for local and for outside users. But as these areas are open and of high economic interest, decentralization would also mean crafting new regulations based on local issues and participation in order to (a) strengthen local sense of ownership and governance of resources and its distribution and (b) incorporate outside users in a way where both sides can find grounds for collaboration that work well, based on mutual interests. If this is done, water-related resources could be sustainably managed.

There is another great challenge, however: floodplains with their land and water will be of greater importance for investments in the future, enhancing reliable food systems. This is the new threat – that because of changes of relative prices in land, large-scale land acquisitions (also labelled as 'land grabbing') will threaten people not just to lose land, but to lose their rights of and access to water (see Woodhouse 2012) and to all other related resources as well (Haller 2013). Strengthening local institutions and options for constitutionality would be key for adaptation strategies also under conditions of climate change, food insecurity and sustainable development issues. If this is not done – and if we continue to focus on one of these hegemonial discourses (climate change, MDGs and SDGs, biodiversity enhancement etc.), stemming from outside resource contexts – we will lose the most valuable resources we have in management: local knowledge and local institutions for the management of water and land-related resources.

NOTES

1 Three of the six research projects were part of the 'NCCR North–South: Mitigating Syndromes of Gobal Change' within the programme National Centres for Competence in Research (NCCR) financed by the Swiss National Science Foundation (SNSF) and the Swiss Development Corporation (SDC), lead by Prof. H. Hurni and Prof. U. Wiesmann, Departement of Geography, University of Bern, Switzerland between 2002–14.
2 Common Pool Resources are per definition resources that are subtractable (one unit of resource that is consumed is no longer available for the moment) and difficult to defend (for individual actors); meaning control and defence of a resource are easier to manage in groups than by individuals (defence costs are too high for individuals) (see Haller 2007b, Becker and Ostrom 1995).
3 See for example de Schuter 2012 as one of large wave of publications on the topic, see also Peters 2013.
4 See for example the fishing groups in Cameroon and in Botswana.
5 See most prominent in the Zambian Kafue Flats case, but also in other cases in the studies we refer to such as Botswana, Tanzania and Mali.

REFERENCES

Acreman, M.C., Hollis, G.E. 1996. *Water Management and Wetlands in Sub-Saharan Africa*. Gland: IUCN.
Beeler, S., Frei, K. 2010. 'Between water spirits and market forces: institutional changes in the Niger Inland Delta fisheries among the Bozo and Somono fishermen of Wandiaka and Daga-Womina (Mali)'. In Haller, T (ed.), *Disputing the Floodplains: Institutional Change and the Politics of Resource Management in African Wetlands*. Leiden: Brill, pp. 77–120.
Benjaminson, T.A., Lund, C. 2003. 'Formalisation and Informalisation of Land and Water Rights in Africa. An Introduction'. In Benjaminson, T.A., Lund, C. (eds), *Securing Land Rights in Africa*. London, Portland: Frank Cass, pp. 1–10.
——— (eds). 2003. *Securing Land Rights in Africa*. London, Portland: Frank Cass.
Berkes, F. 1999. *Sacred Ecology. Traditional Ecological Knowledge, and Resource Management*. Philadelphia: Taylor and Francis.
Berkes, F., Folke C. (eds). 2000. *Linking Social and Ecological Systems*. Cambridge: Cambridge University Press.
Blaikie, P. 2006. 'Is small really beautiful? Community-based natural resource management in Malawi and Botswana'. *World Development* 34(11): 1942–57.
Brockington, D., Duffy, R., Igoe, J. 2008. *Nature Unbound: The Past, Present and Future of Protected Areas*. London: Earthscan.
Cancian, F. 1989. 'Economic Behaviour in Peasant Communities'. In Plattner, S. (ed.), *Economic Anthropology*. Stanford: Stanford University Press.
Chabwela, H. 1992. 'The ecology and resource use of the Bangweulu Basin and the Kafue Flats'. In Jeffrey, R.C.V., Chabwela, H.N., Howard, G., Dugan P.J. (eds), *Managing the Wetlands of Kafue Flats and Bangweulu Basin*. Gland: Union Internationale pour la Conservation de la Nature (UICN), pp. 11–24.

Chabwela, H., Haller, T. 2010. 'Governance issues, potentials and failures of participatory collective action in the Kafue Flats, Zambia'. *International Journal of the Commons* 4(2): 621–42.

Chooye, P.M., Drijver, C.A. 1995. 'Changing views on the development of the Kafue flats in Zambia'. In Roggeri, H (ed.), *Tropical Freshwater Wetlands. A Guide to Currents Knowledge and Sustainable Management*. Dordrecht, Boston, London: Kluwer Academic Publishers, pp. 137–43.

Derman, B., Odgaard, R., Sjaastad, E. (eds). 2007. *Conflicts over Land and Water in Africa*. East Lansing: Michigan State University Press.

Drijver, C.A., Marchand, M. 1985. *Taming the Floods: Environmental Aspects of Floodplain Development in Africa*. Leiden: Leiden University Press.

Ellenbroek, G.A. 1987. 'Ecology and productivity of an African wetland system. The Kafue Flats, Zambia'. *Geobotany* 9.

Ensminger, J. 1992. *Making a Market. The Institutional Transformation of an African Society*. Cambridge: Cambridge University Press.

Fokou, G. 2010. 'Tax payments, democracy and rent seeking administrators: common-pool resource management, power relations and conflicts among the Kotoko, Musgum, Fulbe and Arab Choa in the Waza Logone Floodplain (Cameroon)'. In Haller, T. (ed.), *Disputing the Floodplains: Institutional Change and the Politics of Resource Management in African Floodplains*. Leiden: Brill, pp. 121–69.

Feeny et al. 1990. 'The tragedy of the commons: twenty-two years later'. *Human Ecology* 18(1): 1–19.

Haller, T. (ed.). 2010. *Disputing the Floodplains: Institutional Change and the Politics of Resource Management in African Wetlands*. Leiden: Brill.

—— 2007a. 'Understanding institutions and their links to resource management from the perspective of new institutionalism'. *NCCR North-South Dialogue* 2.

—— 2007b. 'Is there a culture of sustainability? What social and cultural anthropology has to offer 15 years after Rio'. In Burger, P., Kaufmann-Hayoz, R. (eds), *15 Jahre nach Rio – Der Nachhaltigkeitsdiskurs in den Geistes- und Sozialwissenschaften: Perspektiven – Leistungen – Defizite*. Bern: Schweizerische Akademie der Geistes und Sozialwissenschaften, pp. 329–56.

—— 2010a. 'Common-pool resources, legal pluralism and governance from a new institutionalist perspective: lessons from the African Floodplain Wetlands Research Project (AFWeP)'. In Eguavoen, I., Laube, W. (eds), *Negotiating Local Governance. Natural Resources Management at the Interface of Communities and the State*. Munich: LIT Verlag.

—— 2010b. 'Institutional change, power and conflicts in the management of common pool resources in African floodplain ecosystems'. In Haller, T. (ed.), *Disputing the Floodplains: Institutional Change and the Politics of Resource Management in African Floodplains*. Leiden: Brill, pp. 1–76.

—— 2010c. 'Between open access, privatisation and collective action: a comparative analysis of institutional change governing use of common-pool resources in African floodplains'. In Haller, T. (ed.), *Disputing the Floodplains: Institutional Change and the Politics of Resource Management in African Wetlands*. Leiden: Brill, pp. 413–44.

—— 2013. *The Contested Flood Plain Institutional Change of the Commons in the Kafue Flats, Zambia*. Lenham: Lexington (Rowman & Littlefield).

Haller, T., Acciaioli, G., Rist, S. 2015. Constitutionality: Conditions for Crafting Local Ownership of Institution-Building Processes. *Society and Natural Resources*. http://www.tandfonline.com/doi/full/10.1080/08941920.2015.1041661#.VZUhKnpvdd0, 1–20.

Haller, T., Fokou, G., Mbeyale, G., Meroka, P. 2013. 'How fit turns into misfit and back: institutional transformations of pastoral commons in African floodplains'. *Ecology and Society* 18(1): 34.

Haller, T., Galvin, M. 2008. 'Introduction: the problem of participatory conservation'. In Haller, T., Galvin, M. (eds). *People, Protected Areas and Global Change: Participatory Conservation in Latin America, Africa, Asia and Europe*. Bern: University of Bern, pp. 13–34.

Haller, T., Merten, S. 2008. '"We are Zambians – don't tell us how to fish!" Institutional change, power relations and conflicts in the Kafue Flats fisheries in Zambia'. *Human Ecology* 36(5): 699–715.

———— 2010. '"We had cattle and did not fish and hunt anyhow!" Institutional change and contested commons in the Kafue Flats Floodplain (Zambia)'. In Haller, T. (ed.), *Disputing the Floodplains: Institutional Change and the Politics of Resource Management in African Floodplains*. Leiden: Brill, pp. 301–60.

Haller, T., Galvin, M., Meroka, P., Alca, J., Alvarez, A. 2008. 'Who gains from community conservation? Intended and unintended costs and benefits of participative approaches in Peru and Tanzania.' *Journal of Environment and Development* 17(2): 118–44.

Howard, G.W. 1992. 'Floodplains: utilisation and the need for management'. In *Wetlands Conservation Conference for Southern Africa. Proceedings of the Southern African Development Coordination Conference, Botswana*. Gland: Union Internationale pour la Conservation de la Nature (UICN), pp. 15–26.

Kuba, R., Lenz. K. (eds). 2006. *Land and the Politics of Belonging*. Leiden: Brill.

Lenz, K. 2006. 'Land and the politics of belonging. An introduction'. In Kuba, R., Lenz. K. (eds), *Land and the Politics of Belonging*. Leiden: Brill, pp. 1–34.

———— 2007. 'First-comers and late comers: indigenous theories of land ownership in the West African Savanna'. In Kuba, R., Lenz. K. (eds), *Land and the Politics of Belonging*. Leiden: Brill, pp. 34–57.

Loth, P. (ed.). 2004. *The Return of the Water. Restoring the Waza Logone Floodplain in Cameroon*. Gland: Union Internationale pour la Conservation de la Nature (UICN).

Meroka, P, Haller, T. 2008. 'Government wildlife, unfulfilled promises and business: lessons from participatory conservation in the Selous Game Reserve, Tanzania'. In Galvin, M., Haller, T. (eds), *People, Protected Areas and Global Change: Participatory Conservation in Latin America, Africa, Asia and Europe*. Bern: University of Bern, pp. 177–219.

Mhlanga, L., Nyikahadozi, K., Haller, T. (eds). 2014. *Fragmentation of Natural Resources Management: Experiences from Lake Kariba*. Berlin: LIT Verlag.

North, D. 1990. *Institutions, Institutional Change and Economic Performance*. Cambridge: Cambridge University Press.

Ostrom, E. 1990. *Governing the Commons. The Evolution of Institutions for Collective Action*. Cambridge: Cambridge University Press.

Peters, P. 2013. 'Conflicts over land and threats to customary tenure in Africa'. *African Affairs* 2013(1): 1–20.

Poteete, A.R., Ribot, J.C. 2011. 'Repertoires of domination. Decentralization as process in Botswana and Senegal'. *World Development* 39(3): 439–49.
Roggeri, H. (ed.). 1995. *Tropical Freshwater Wetlands. A Guide to Currents Knowledge and Sustainable Management.* Dordrecht, Boston, London: Kluwer Academic Publishers, pp. 137–43.
Stevenson, N., Frazier, S. 1999. 'Review of wetland inventory information in Africa'. In Finlayson, C.M., Spiers, A.G. (eds), 'Global Review of Wetland Resources and Priorities for Wetland Inventory'. Supervising Scientist Report 144, Canberra.
Thomas, D.H.L. 1996. 'Fisheries tenure in an African floodplain village and the implications for management'. *Human Ecology* 24(3): 287–313.
Thompson, J.R. 1996. 'Africans floodplains: a hydrological overview'. In Acreman, M.C., Hollis, G.E. (eds), *Water Management and Wetlands in Sub-Saharan Africa.* Gland: Union Internationale pour la Conservation de la Nature (UICN), pp. 5–20.
Toulmin, C. 2009. 'Securing land and property rights in sub-Saharan Africa: the role of local institutions'. *Land Use Policy* 26: 10–19.
Saum. R. 2010. 'Promise and reality of community based natural resource management in Botswana: common-pool resource use and institutional change in Ikoga, Okovango Delta (Pandhandle)'. In Haller, T. (ed.), *Disputing the Floodplains: Institutional Change and the Politics of Resource Management in African Floodplains.* Leiden: Brill, pp. 361–412.
Welcomme, R.L. 1979. *Fisheries Ecology of Floodplain Rivers.* London: Longman.
Woodhouse, P. 2012. 'New investment, old challenges. Land deals and the water constraint in African agriculture'. *Journal of Peasant Studies* 39 (3–4): 777–94.

16 Water, Labour and Politics: Land-Use Dynamics along the Niger River in Mali

Tor A. Benjaminsen

INTRODUCTION

According to a Tuareg proverb, 'Water is life and milk is food' (*aman iman, ax isudar*). This saying reflects the main tenets of pastoral life in the West African Sahel. Without available water, pastoral production is impossible – and among pastoralists milk is the preferred food, although their main staples are normally grains such as millet, sorghum and rice.

Water obviously shapes all life in the Sahel, not only that of pastoralists. West Africa can be divided in climatic zones following the rainfall pattern, with increasingly wetter conditions from the desert in the north to the coast in the south-west. The Sahel can be defined as the savanna zone south of the Sahara with 100–600 mm of long-term annual rainfall (Le Houérou 1989). The 100-mm isohyete corresponds approximately to the border area between concentrated and dispersed vegetation. In the desert, the vegetation is concentrated in the topographical depressions (oueds, wadis), while in the Sahel, it is more dispersed with a grass cover of annual species and scattered trees and bushes. However, the depressions still contain key resources during the dry season (for example, *fonio* (*Panicum laetum*) and *bourgu* (*Echinochloa stagnina*)).

The further north one gets in West Africa, the shorter and more unpredictable is the rainy season. In the Sahel, the rain falls between June and September in 10–30 showers. The high variations in rainfall in time and space, and the adaptation of the annual grass species to the climate have created the two main features of the Sahelian ecosystem; instability and resilience (Benjaminsen 1997). While instability is a result of rainfall variations and unpredictability, resilience implies, in the Sahelian case, that the grass cover in particular has a high capacity to recuperate after drought or heavy grazing. These two characteristics have led people to develop opportunistic and flexible coping strategies. Pastoral production is often highlighted in this context (for example, Sandford 1983), but also farming (of rainfed millet or rice along the rivers), gathering, fishing, labour migrations and the use of food/development aid when accessible form part of the opportunistic survival strategies of Sahelian populations.[1] When resources are unstable, unpredictable, transitory, but also resilient, it pays to have a strategy based on flexibility and the ability to move and to use the resources as long as they are available. In the long term, it also pays to spread and minimize risks. Following this line of thinking, investment in land productivity is not a good idea, because it is too risky, unless the environment can be controlled. In practical terms, such control would normally imply irrigation and would usually have to take place by a river. However, irrigation with full control of the water level in the fields is generally still beyond the scope of most Sahelian populations, because of lack of capital, poor access to markets and in many cases also lack of sufficient available labour.

Even though a duality between farmers and pastoralists exists, most people in the Sahel follow an agro-pastoral adaptation, despite being classified as 'farmers' or 'pastoralists'. In this region, many herders also cultivate, and many cultivators also herd. Therefore, integration and overlapping between the two productions is the rule, rather than distinction and separation (Raynaut, 1997).

But while livelihood sources may be combined, identities such as 'farmers' or 'pastoralists' linger on. And while some people may move back and forth between these identities, others are fairly locked up in one or the other. Agro-pastoral relations have a long history of both cooperation and conflict in the Sahel. With this caveat, I still use the terms 'farmers' and 'pastoralists' for practical purposes.

In this chapter, I will take a closer look at some of these agro-pastoral land-use dynamics along the Niger River in Mali with a focus on how water fluctuations, access to labour and the broader political context influence these dynamics. I frame these cases within a political ecology approach (for example, Robbins 2012), which tends to focus on issues of power in land and environmental governance, or differently put on the interplay between political agency and the agency of nature.

The cases I discuss are taken from fieldwork carried out at various periods since the late 1980s in the inland delta of the Niger River and the area around the bend of the river in northern Mali. First, I present land-use dynamics in these two areas in their geographical and historical context. Secondly, I discuss how rainfall and climate variability shape the availability of natural resources including what research tells us about 'desertification' and the prospects of the impact of climate change. Thereafter, I move on to discuss the role of labour for agro-pastoral land-use dynamics, before I finally discuss how these dynamics may play out in the form of conflicts formed both by water availability and the political context.

THE INLAND DELTA AND THE NIGER BEND: HISTORICAL AND GEOGRAPHICAL BACKGROUND TO LAND-USE DYNAMICS

The inland Niger delta in Mali is the largest wetland area in West Africa (Figure 16.1). This is a vast floodplain that in good years covers up to 20,000 km^2. The flooding in the delta depends on the annual rainfall in the upper catchment area of the Niger River in Guinea and southern Mali. For centuries, the area has provided rich resources for rice cultivation, fishing, and pastoralism.

The pastoral system in the delta is based on livestock staying in the dry season grazing area during December–June combined with the use of dryland pastures in the rainy season. The length of pastoral migrations among the Fulani may range from 40 to 500 km (Turner 1999). From the beginning of the rainy season in June/July, many delta pastoralists move north-east or north-west before they return to the delta sometime in the dry season.

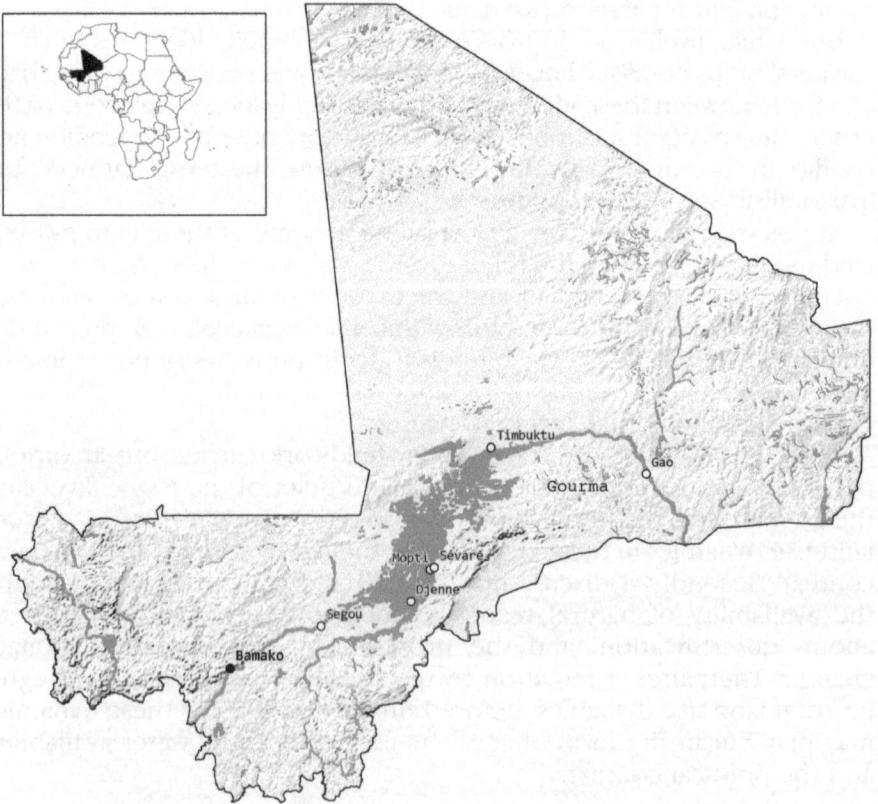

Figure 16.1. Mali and the Niger River with the inland delta (from Benjaminsen et al., 2012).

The main fodder resource in the delta is *burgu*, which grows in deeper water than paddy rice. Paddy fields have during the last decades expanded enormously at the expense of *burgu*. According to Kouyaté (2006), about one quarter of the *burgu* areas in the delta have been converted to rice fields since the 1950s. This can partly be explained by decreased levels of flooding in the Niger river (Figure 16.2) especially as a result of the droughts of the 1970s and 80s as well as the construction of the hydro-power dam downstream at Sélingué that was completed in 1982 (Turner 1992, Cotula & Cissé 2006).

Since the Sahel drought of the mid 1980s, flood levels have, however, increased somewhat again. According to official censuses, the population of the Mopti region, which covers most of the delta, increased from 910,713 in 1964 to 1,478,505 in 1998 (Cotula & Cissé 2006). This means that with the growth rates of the last decades, the population in the delta would be slightly above 2 million in 2015. The region consists of a mix of sedentary farmers (Bambara, Songhay, Malinké, Fulani), pastoralists (Fulani, Tuareg, Maure), and fishers (Bozo). These communities have a long tradition of cooperation and coexistence, though inter-communal raids and clashes have also happened from time to time.

The current land tenure system in the delta is based on the principles introduced by invading Fulani warriors in the fourteenth century, the so-called Ardobé (Ba & Daget 1962). The Ardobé constituted the military and political leadership in the area for more than 400 years from about 1400. They 'provided floodplain land to subordinates for farming, which led eventually to the bounding of their spheres of influence into *leyde*' (Turner 1999: 110). Furthermore, the Ardobé introduced local chiefs (*Jowros*[2] to manage the *leyde* (sing. *leydy*). The *Jowros* who were noble Fulani (*rimbé*) pastoralists were 'owners of grass', and, hence responsible

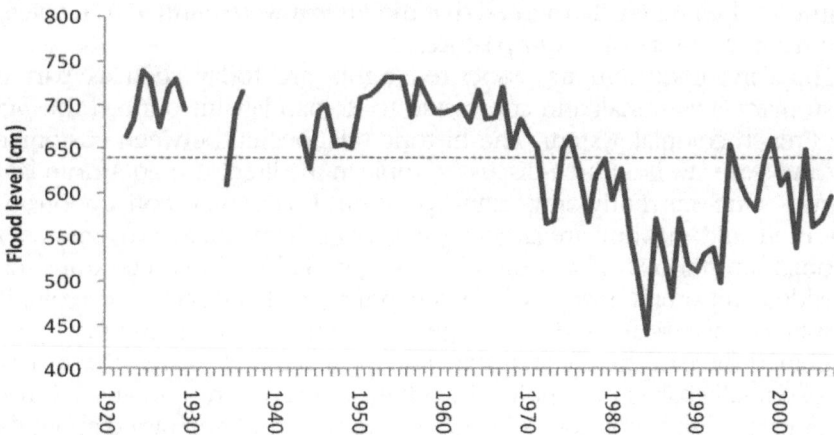

Figure 16.2. Niger River flow variability in Mopti, 1922–2006 (© Direction Nationale de l'Hydraulique, Bamako, Mali).

for the management of pastures in these territorial units. In 1818, Islamic clergymen mobilized a jihad and conquered the delta region through the leadership of Cheikou Amadou. This resulted in the establishment of an Islamic theocratic state, the Dina, based in Hamdallahi south of Mopti. 'Cheikou Amadou adopted the territorial model of the Ardobé to force Delta households to settle into *leyde* composed of the pre-existing *leyde* of the Ardobé and new *leyde* carved out of the remaining floodplain' (Turner 1999: 110). The Dina also codified and formalized many of the resource management principles and rights introduced by the Ardobé. For instance through this codification, the *Jowros* were formally granted the authority to manage the *leyde*, rights to *burgu* fields were defined, and a list of livestock entry routes in the delta was established in order to keep farmers' fields at a distance (Gallais 1967). While the *Jowros* managed pastures, the allocation of agricultural land was devolved to a *bessema*; a chief of the low caste *Rimaybé* farmers. All users of *burgu* pastures paid a fee to the *Jowro* and a clear ranking order of herds was established deciding the order in which they would enter the *burgu* areas. In addition, a village chief (*jomsaré*) would be responsible for handling administrative and political issues.

In 1893, the French defeated the Toucouleur who in 1862 had conquered the Dina, and in 1895 the colony French Sudan was established. The colonial government was centred on the *cercle* as its main administrative unit. The *cercle* was governed by a French *commandant* whose directives passed to the local population through a *chef de canton*, who was picked out among the local chiefs in the area.

The principles of spatial organization and resource management inherited from the Dina were sustained by the French administration (Vedeld 1997, Barrière & Barrière 2002). Some *Jowros* became chefs de canton, while the *Jowros* in general maintained their role as managers of pastures. The French also agreed that the *Jowros* were entitled to receive a rent from the users of *burgu* pastures.

The Dina code and its associated rights are today seen as part of 'customary law' in Mali and contrasted to Roman law introduced through the French colonial system. The historic relationship between customary law and state law is a much-discussed topic in the literature on African land tenure. One currently important position holds that conceptions of 'tradition' and 'custom' are largely results of constructions or inventions by colonial authorities (for example, Ranger 1983). This literature has provided important insights into the practice of indirect rule, colonial cooptation of African chiefs, and the exploitation of the peasantry by the combined force of the colonial state and local elites. However, these views are essentially based on empirical work in southern Africa. In studies from other parts of the continent, questions have recently been raised about the validity of these conclusions. Informed by his own historical studies from Tanzania, Spear (2003: 3) argues that 'the case for colonial invention has often overstated colonial power and ability to manipulate African institutions to establish hegemony'. It therefore makes little sense to talk about 'invention', because '(t)radition was both more flexible and less subject to outside control than scholars have thought' (p. 26). While Spear focused on chieftaincy and ethnicity, Lentz (2006) has extended the argument to the issue of land tenure based on studies in West Africa. She contends that not all statements about past authority in land matters are results of recent 'inventions'.

While a revised and milder form of the Dina code was endorsed by the colonial government, it would be an exaggeration to state that the current remnants of the Dina are colonial inventions. Even though the Dina is seen in Mali as a great product of African pre-colonial state formation, applying its principles today is also a contested issue. It is perceived as biased towards the Fulani as well as being reminiscent of a past when a few pastoral leaders controlled land, resources and people. This depiction of the Dina and the *leyde* system as an indigenously developed pastoral system is also ironic and 'somewhat misleading since its agriculturally rooted territorial form was imposed upon and openly resisted by Fulani pastoralists' (Turner 1999: 110).

Further north, in the Niger bend, the dominating land-uses along the river are still rice cultivation and *burgu* pastures. The main ethnic groups in this area are the pastoral Tuaregs and the agricultural Songhay. In addition there are the Bozo who fish in the Niger and the lakes, and the pastoral Fulani in the southern part of this geographical region called Gourma (Figure 16.1).

The Gourma has been hit by drought several times during the twentieth century, most importantly in 1903, 1913–14, 1930–2, 1944–8, 1972–5 and 1982–7. Nomads who survived the great droughts during the 1970s and 1980s say that based on what their parents told them, the drought that hit the area in 1913–14 must have been the worst during the whole century. While nomads in good years stay within a relatively limited area, they have to move more during bad years to find water and pastures.

In 1960, Mali gained its independence, and the country's first president, Modibo Keita, was inspired by socialist ideas of industrialization and agricultural modernization. Pastoralism was looked upon as an obstacle to development in general. An aim of the Keita government was to convert pastoralists into 'productive' citizens by taking up farming (Benjaminsen & Berge 2004). The socialist government also saw the *Jowros* as well as Tuareg chiefs as feudal landlords and generally tried to undermine their authority.

After a *coup d'état* in 1968 and the establishment of the military government of Moussa Traoré, the position of these traditional chiefs was gradually rehabilitated and towards the end of Traoré's reign, before he was toppled in another *coup d'état* in March 1991, the chiefs had again become powerful local actors through alliances with officials of the single party.

After the introduction of democracy from 1991, a Constitution committed to decentralization was accepted by referendum in 1992, and in 1993 and 1995 laws on decentralization were adopted by the National Assembly. However, how to deal with land tenure conflicts remains one of the key challenges of the decentralization reform.

'DESERTIFICATION', RAINFALL AND CLIMATE CHANGE

Due to a strong dependence on highly variable rainfall in the Sahel, an analysis of land-use dynamics will easily need to reflect on the current knowledge of 'desertification' and climate change. Due to strong livelihood dependence on renewable resources in the Sahel, changes in the availability of these resources will necessarily impact on land-use. Claims about desertification in the Sahel are in fact as old as European presence in the region. Already in the early 1900s, there were debates about whether desertification in French-occupied West Africa was a man-made process or caused by desiccation (Benjaminsen & Berge 2004). With time, however, the view that it was created by local overuse of natural resources prevailed and even during the droughts of the 1970s and 1980s this view dominated in research, policy and media presentations.

From the late 1980s, claims of widespread degradation and desertification in the Sahel have been undermined by a number of studies. For instance, scientists at the National Aeronautics and Space Administration (NASA) in the USA have studied satellite images of the southern limits of the Sahara and concluded that the edge of the desert moves back and forth as a direct result of annual rainfall (Tucker et al. 1991, Tucker & Nicholson 1999).

Furthermore, a number of studies published from the late 1980s led researchers to increasingly question the idea of desertification in the Sahel. Some of this research was reviewed in the *New Scientist* in an article entitled 'The myth of the marching desert' (Forse 1989). This research led to what was termed a paradigm shift in drylands research (Warren & Khogali 1992, Behnke & Scoones 1993, Benjaminsen 1997). It recognizes the resilience and variability of drylands and stresses the need for flexibility in coping with a highly unstable environment. These ideas have led to the questioning of ecological theory based on notions of equilibrium, carrying capacity, succession and climax as applied on tropical drylands. Instead, non-equilibrial ecological theory states that the vegetation in drylands varies with the annual rainfall and that external factors such as climate, rather than livestock numbers, tend to determine the vegetation composition and cover (Ellis & Swift, 1988; Behnke, Scoones & Kerven 1993). Moreover, unavailability of forage in bad years may depress livestock populations to the point where the impact of grazing on vegetation is minimal (Sullivan & Rohde 2002). Therefore, in areas of fluctuating climates, rainfall rather than density-dependent factors related to herbivore numbers may ultimately be the most significant variable determining herbivore populations. Wet season pastures such as in the West African Sahel, with its short rainy season, domination of annual grass species, and high resilience, would be a good example of a non-equilibrial system (Hiernaux 1993, Turner 1993). The herders' use of pastures is adapted to the seasonal changes in these drylands. During the rainy season, when the grass grows, herders often move, and therefore exercise little pressure on the vegetation (Hiernaux 1993).

Since it is largely rainfall that drives the Sahelian ecosystem, global warming might in the long run lead to desertification – if it reduces rainfall. However, as demonstrated by Buontempo et al. (2010), there is currently considerable uncertainty about current rainfall trends and projections in the Sahel. Not only are there uncertainties about future scenarios, but there are also some disagreements about how to read available climate data. For instance, Hulme (2001) and Chappell and Agnew (2004) disagree on how to interpret rainfall data from the Sahel for the period 1930–90. While Hulme (2001) holds the position that there was a 20–30 per cent decline, Chappell and Agnew (2004) question this claim arguing that this decline was largely produced by historical changes in the climate station network.

Climate researchers in general stress that there is a great deal of uncertainty as to how global warming will affect the climate in the Sahel. This is also underlined by the IPCC in its fourth (Boko et al. 2007: 444) and fifth reports (Niang & Ruppel 2014). The former points out that the various models do not concur concerning future climate scenarios for the Sahel. While some models support the theory that this region will become drier, other models suggest that it might rain more in the future in the Sahel (for example, Haarsma et al. 2005, Odekunle et al. 2008).

Buontempo et al. (2010) also highlights the problem of current generation climate models not being able to capture processes driving Sahelian climate in the twentieth century, in particular as concerns precipitation. In addition, given the disagreement between models, Buontempo et al. advise against basing assessments of future climate change in the Sahel on the results from any single model in isolation. Until the processes responsible for the projected changes can be understood and constrained, the long-term future climate change impact on Sahel rainfall will remain uncertain. Reviewing recent climate modelling for the Sahel, Biasutti (2013), however, finds that most models conclude that the rainy season will be 'more feeble at its start' and 'more abundant at its core'. Hence, the overall trend seems to be towards wetter conditions, but with rainfall more concentrated in time and with higher average temperatures. Of 20 models only four are outlier models coming to other conclusions. The *Fifth IPCC Report* assesses, however, that there is low to medium confidence in these projected changes of heavier rainfall (Niang & Ruppel 2014).

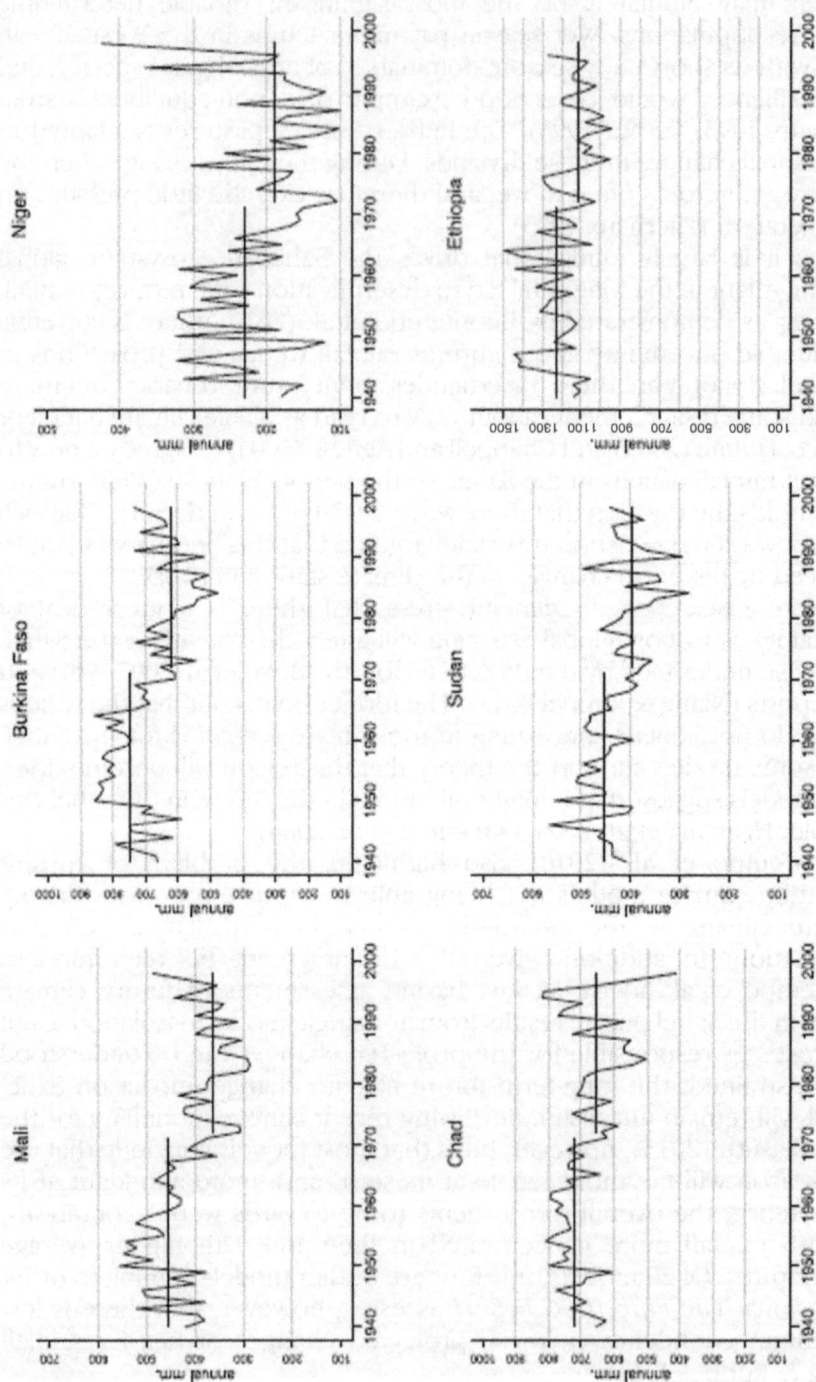

Figure 16.3. Long-term rainfall variability in the Sahel (© Mike Hulme, University of East Anglia).

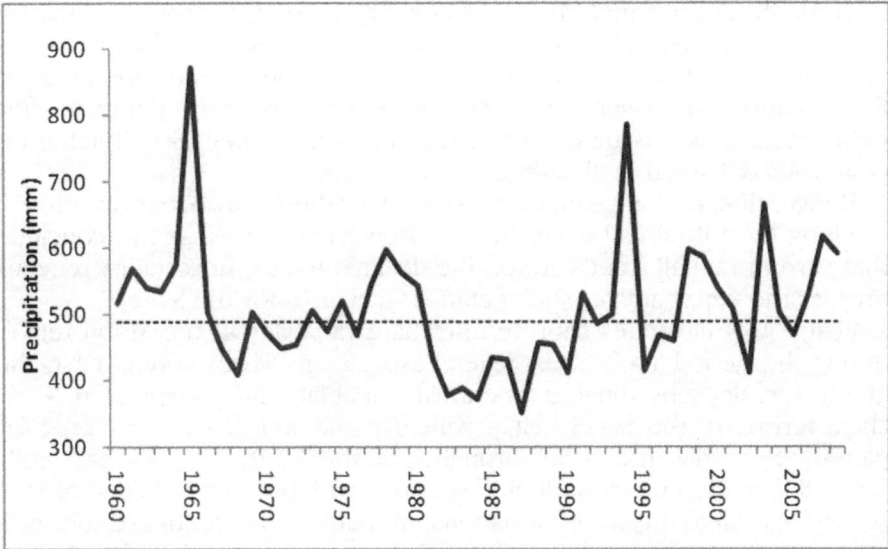

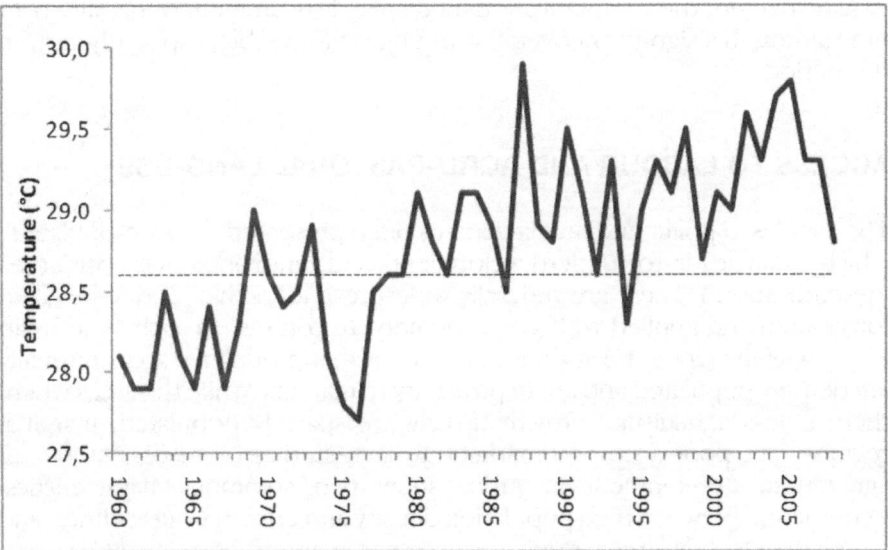

Figure 16.4. Climatological trends in Mopti, 1960–2008. The figure shows total annual precipitation and mean annual temperature for the Mopti region with dotted lines representing the mean values for the 1961–90 period (from Benjaminsen et al., 2012).

Throughout the Sahel, there has now been a partial recovery of rainfall over the last 20 years (Figure 16.3). Research on the Sahel is thus no longer discussing desertification, but the fact that the Sahel has become greener. For instance, in November 2005, the *Journal of Arid Environments* published a special issue on 'The Greening of the Sahel' (See Hutchinson et al. 2005 & Olsson et al. 2005).

Hence, climate change may lead to drier conditions and desertification in the long term if rainfall declines. It is, however, problematic to conclude that current rainfall trends are on the decline. In fact, uncertainty remains thus far the key characteristic of climate scenarios for the Sahel.

Figure 16.4 illustrates short-term climate variability in the Mopti region during the period 1960–2008. Several aspects are worth noting. First, the graph reveals substantial inter-annual variability in precipitation – so characteristic of the Sahel belt – with dry and wet years in an irregular pattern. Secondly, the Sahel droughts of the 1970s and 80s are easily detectable in Mopti meteorological statistics, with the early 1980s displaying particularly large negative deviations in rainfall. These trends do not, however, provide any evidence of global climate change having played a role in driving the droughts of the 1970s and 80s (Christensen et al. 2007).

As expected, the temperature data display less dramatic variability over time although a significant overall warming is discernible, especially early in the period.

ACCESS TO LABOUR AND AGRO-PASTORAL LAND-USE

The Sahel is in policy and media debates often presented as 'overpopulated', which again leads to 'underdevelopment' and 'environmental crisis' (i.e., 'desertification'). These are generally widespread ideas that tend to be taken for granted and applied with great certainty. In contrast to such Malthusian ideas, Wiggins (2005) for instance points out that productive areas normally are densely populated and are in proximity to markets, while the areas where there is less agricultural growth usually are sparsely populated, marginal regions with poor access to markets. In fact, studies of success stories of agricultural development in Africa show two common characteristics: a correlation between high population density and growth in agriculture; and a growth, which is largely driven by access to markets (Wiggins 2005).

The close relationship between population density and growth in agriculture that has been observed from case studies in Africa supports the theory of agricultural development introduced by Boserup (1965) who did not view population growth as a problem, but as a resource. According to Boserup, a certain population density is a prerequisite for technological innovation and productivity improvements in agriculture.

In line with the experience from some success stories in African agriculture and the Boserup model, we may argue that a certain population density is a necessary condition for increasing agricultural production in Africa and other countries in the south, where first and foremost it is the labour of the people that drives development – in the absence of investment capital. As a second factor, there is the necessity of access to markets to sell surplus production.

Following up on this line of thinking, I will in this section argue that the Niger bend might just as well be seen as underpopulated for agricultural development. This is an arid region on the border of the Sahara where the river represents a rich agricultural resource in an otherwise barren landscape and where herding of livestock dominates. In the rainy season, the waters of the Niger spread out over large plains, where rice has been cultivated for thousands of years (Ki-Zerbo 1978: 56). This is a sparsely populated area and the methods of rice cultivation are still relatively simple. Because of lack of labour and capital, the farmers can only build simple levees and floodgates to attempt to check the river water. The system is therefore fragile and it is totally dependent on the river water flowing smoothly and gradually without drowning the floating rice. Often, the water flows quickly onto the rice fields or the levees burst, so that the water disappears from the rice fields and the rice plants dry out. It is perhaps only in one of three years that this traditional system provides any substantial crops. Several aid organizations, such as Norwegian Church Aid (NCA), have attempted to improve the system by investing in more robust levees and floodgates for better control of the floodwater (Figure 16.5). These investments have to an extent reduced, but not eliminated, the risk.

I will argue that low population density has been, and still is, one of the biggest obstacles to agricultural development in the region. This means that large areas of fertile land along the river are left fallow or not very intensively cultivated, due to lack of capital and labour. The labour shortage is particularly evident where Tuareg pastoralists control the land. These are places along the river where the Tuaregs have been encouraged to settle and cultivate rice. As long as the NCA paid people through a 'food for work' programme during the 1980s, labour was available to build levees and floodgates and to stay and cultivate rice. But as soon as this support was discontinued in the 1990s, pastoralists had the choice between resuming their nomadic life and staying in one place to farm the land. The latter entailed investing their labour in a very risky and vulnerable agricultural production. Usually, they chose to resume their nomadic life (Pedersen & Benjaminsen 2008).

A concrete example of this is the hamlet called Banguel, which lies a few miles west of Gourma Rharous – the administrative centre of the *cercle*. The Tuaregs based in Banguel are from the group of Igouadaren Ouest. This group consists of some families with high status warriors (*imoushar*), some craftsmen (*inhaden*) and a large majority of former slaves, called *bella* (or *eklan*). The *bella* make up the workforce and are still loyal to the *imoushar*, who make all the decisions. The group controls two large rice

Figure 16.5.   Floodgate built by Norwegian Church Aid (© Håkon Lislerud, Norwegian Church Aid).

fields – one of 97 ha and another of 248 ha. When I visited Banguel in 1997, one of these two rice fields lay fallow because of lack of labour to maintain the levees. However, *imoushar* in Banguel are powerful and control a large number of *bella*. If rice cultivation was a priority for them they could have ensured enough labour for maintenance. But this would have meant sacrificing a relatively secure livestock production – based on herding – for a far more uncertain agricultural production. Even though the technology and therefore the cultivation systems have been improved through the intervention of the NCA, it is still too risky for people to take the gamble. There is simply not enough labour available in the area to keep livestock as well as cultivate rice. Therefore, the *imoushar* prioritized letting the *bella* herd animals in the vast grazing regions south of the Niger River. Animal husbandry is, after all, so profitable that the most prosperous pastoralists in northern Mali possess four-by-four vehicles and satellite telephones.

However, pastoralists who in this way prioritize livestock herding and leave land along the river fallow risk over time losing control over that land. Secure land tenure requires maintaining land-use. If land is left fallow for some years, there is the risk of encroachment from neighbours, in this case sedentary Songhay farmers. Such encroachments may also lead to land-use conflicts. As we shall see in the next section, however, whether conflicts arise depends to a large extent on the politics of land-use and land tenure.

Figure 16.6.   Bella preparing a rice field (© Carsten Sørensen).

THE POLITICS OF LAND-USE CONFLICTS

A number of reports hold that in recent years the delta has been marked by numerous land-use conflicts (Ba 1996, 2008, Barrière & Barrière 2002, Benjaminsen and Ba 2009, Benjaminsen et al. 2012, Cotula and Cissé 2006, Moorhead 1991, Turner 1992). Given the area's sensitivity to rainfall fluctuations and its dependence on the seasonal flooding of the Niger River, one might surmise that the high frequency of land disputes in part reflects a decline in the water level in the river. A possible, and often repeated scenario in Mali, is that declining water levels lead to less land available for rice cultivation and *burgu* pastures, which in turn leads to more land-use competition, especially between farming and pastoralism.

   In order to give some insight into the details of ongoing land-use dynamics, the history of the land-use conflict in the village of Saremala may serve as an example (This case is reported in Benjaminsen and Ba (2009)). This village is located in Kounary leydy in the heart of the delta. In the village lives the local *Jowro* and a small number of *Rimbé*, while the vast majority of the villagers are *Rimaybé*. The Office Riz Mopti (ORM) – the state organization of promotion of rice cultivation – is active in this region. Since the state formally owns all the land in Mali, the ORM can confiscate land at will. In particular, much of the *burgu* pastures controlled by the *Jowro* have been confiscated and turned into paddy fields. These fields have been divided up into equal-sized parcels of land and leased out to people who have applied to the ORM for land. In addition, there has been widespread random cultivation of *burgu* pastures by local *Rimaybé* farmers. The massive loss of *burgu* pastures, which constitute the power base and main source of income for the local *Jowro*, is leading to a gradual transfer of local power from the *Jowro* to the *Bessema*. A positive aspect of this is that the previously underprivileged *Rimaybé* now gain more power and a higher standard of living. A negative aspect is that important pastures used in the dry season are disappearing and being replaced by paddy fields. These are pasturelands that the entire pastoral system in the delta region depends on.

In the wake of the transition to democracy in 1991, the state's presence in rural areas was reduced for a period. This was in general a time of great uncertainty about the future direction of the political and administrative system in Mali. State bodies were reorganized, and plans were laid for a new, major decentralization reform. Many local actors took advantage of the power vacuum that arose in the early phase of the decentralization process, by taking possession of land in various ways. This also happened in Saremala. Farmers extended their fields into pasturelands, while the *Jowro* tried to regain control over the lost *burgu* areas that had been converted into cultivated farmland. Farmers usually give a small symbolic share of the harvest (usually about 5 kg) in recognition of the person who owns the land according to customary law. However, the political changes and the state's temporary withdrawal since 1991 had whetted the *Jowro's* appetite, and he decided to try to take control of the cultivated land and introduce a clearer tenant farming system with a larger share of the yield for the owner of the land. This strategy failed because of strong resistance from the *Rimaybé*.

Frustrated at his loss of power and the loss of the *burgu* pastures, the *Jowro* decided to take the *Rimaybé* to court in 1994. On 25 August 1994, the local court in Mopti ruled in favour of the *Jowro*, establishing that he had customary rights over all the land in Saremala, not just the pastureland. However, the *Rimaybé* appealed the case to the Appeal Court in Mopti, which ruled on 31 May 1995 that while the *Jowro* had customary rights, the *Rimaybé* had usage rights to the same land. While both parties interpreted this ambiguous ruling in their own favour, the *Jowro* appealed the case to the Supreme Court. At the same time, the *Jowro* of Saremala, started acting as if his ownership rights had been finally confirmed by the legal system, banning *Rimaybé* farmers from cultivating the land. At the harvest in December 1995, he announced a general ban on harvesting rice, stating that all harvested rice would be confiscated by force. The *Rimaybé* (farmers) then paid 18 armed guards to protect them while they harvested their crops. Despite the guards, there was an armed confrontation between the *Rimbé* (herders) and the *Rimaybé* on 23 January 1996 resulting in two dead and 16 injured farmers and herders. The village chief, a *Rimaybé*, claims that the *Jowro* had bribed the guards to look another way and not intervene when the *Rimbé* tried to force the *Rimaybé* to stop harvesting the rice.

On 18 February 1997, the Supreme Court declared that the Appeal Court's decision was invalid and sent the case back to that court, which upheld its earlier decision on 2 July 1997. Indeed, it even increased the ambiguity of its former ruling by ascribing the *Jowro* all three aspects of ownership under French law: *usus*, *fructus* and *abusus*. *Usus* is the right to use, *fructus* is the right to enjoy the fruits of the land (harvest, rental income, etc) and *abusus* is the right to get rid of the property by giving it away or selling it. This judgment actually gave the *Jowro* more extensive rights than any reasonable interpretation of customary law might ascribe to him. It is also self-contradicting in another way, because according to Malian law, only the state has the *abusus* rights to land without deeds. To top it all off, The Appeal Court's decision means in practice that the *Jowro* and the *Rimaybé* farmers were granted basically identical rights (*usus* and *fructus*), which both parties interpreted as a victory. However, we can probably conclude that the *Rimaybé* have more to celebrate because the court granted them rights they would otherwise have had little chance of establishing under the customary system.

It seems that the legal wrangling back and forth and the court's ambiguous rulings are the result of both parties having paid bribes to the judges and their entourage. Bringing a conflict before a court of law is often a final desperate attempt on the part of one of the parties. Prior to this, the parties have usually already spent vast sums of money on trying to influence the administration.

This case study reviewing the political and economic context of a herder-farmer conflict in Mali shows that the drought in the 1980s was one of several factors that contributed to the loss of *burgu* pastures and marginalization of pastoralists – and thus indirectly to more conflicts. The drought contributed temporarily to the shrinking of available *burgu* pastures for pastoralists. With less pastoral space available, herders and livestock will more easily trespass and damage agricultural crops, and conflicts might emerge.

However, when comparing data from 820 court cases in the Mopti Court of Appeal with rainfall data, Benjaminsen et al (2012) found little support for claims that weather patterns drive social conflicts among pastoral and agrarian communities in central and northern Mali.

So, while a currently popular idea in policy circles holds that resource scarcity drives land-use conflicts and that the Sahelian zone is in particular a good example of this linkage[3] Benjaminsen et al. (2012) found that the state's policy, which led to marginalization of pastoralists – and in turn to increased scarcity of pastoral land – plays a far larger part in explaining the increase in the number of conflicts in the inland delta region of Mali. In addition, corruption by a rent-seeking bureaucracy is also perpetuating land-use conflicts in the delta. In order to get support from state technicians and administrators communities and individuals need to pay (Benjaminsen & Ba 2009, Benjaminsen et al. 2012). As mentioned, this also goes for having the court to rule in one's favour. However, by receiving payments from both parties, the courts' decisions become ambiguous. This again contributes to perpetuating conflicts. It is often stated in interviews with peasants in rural Mali that they have become milk cows for the bureaucracy.

Hence, this research in the delta published by Benjaminsen and Ba (2009) and Benjaminsen et al. (2012) concludes that land-use conflicts in the area are primarily caused by political and economic factors rather than climate variability. These factors include first agricultural encroachment on productive key resources for pastoralism and on livestock corridors, obstructing the necessary mobility of herders and animals. This trend is primarily caused by agricultural policies and laws promoting farming at the expense of pastoralism. Secondly, decentralization from the early 1990s caused a political vacuum that led rural actors to follow opportunistic strategies to claim ownership of land and natural resources. Thirdly, rent-seeking among government officials has undermined rural people's trust in government institutions and the willingness and interest of officials to solve conflicts. This lack of trust may have contributed to some actors taking action on their own, including using violence to lay claim to resources. Climate variability may only play a secondary or tertiary role in increasing or decreasing the conflict level.

CONCLUSION

In this chapter I have investigated the role of water through flooding, access to labour and the general political context in understanding land-use dynamics along the Niger river in Mali. The Sahelian droughts of the 1970s and 80s led rice farming to move down the riverbed and encroach on the dry season *burgu* pastures. In this sense, drought may play a role in confronting farmers and pastoralists and increase inter-communal tensions and, quite possibly, escalate latent conflicts to the use of violence. The fact that pastoralists are absent from high potential land along the river during large parts of the year, whether this land is *burgu* pastures or rice fields, tends to lead to encroachment on their land, which again may cause further conflicts. In addition, while pastoralists are encouraged by governments and development aid to combine livestock keeping with farming, they often do not have enough labour available for successful combination of these two activities. This lack of labour leads to very productive land being left fallow.

Finally, several structural conditions shape a large number of land-use conflicts across the Sahel. First, there is agricultural encroachment on productive key resources for pastoralism and on livestock corridors obstructing the necessary mobility of herders and animals. This has led to massive loss of dry season pastures that are essential for the survival of the pastoral system. This trend is primarily caused by agricultural policies and laws promoting farming at the expense of pastoralism. Secondly, decentralization from the early 1990s caused a political vacuum that led rural actors to follow opportunistic strategies to claim ownership of land and natural resources. Thirdly, rent seeking among government officials has undermined rural people's trust in government institutions and the willingness and interest of officials to solve conflicts. This lack of trust may have contributed to some actors taking action on their own, including using violence to lay claim to resources.

NOTES

1   In fact, the more recent success of salafist groups in recruiting young men in northern Mali for an armed rebellion against the state and foreign 'crusaders' may also be interpreted as resulting from a basic opportunistic attitude – of course in addition to the fact that poor and unemployed young men may be easy victims to wealthy Jihadist groups that have made millions of euros on hostage taking and drug trafficking.
2   We here use *jowro* (sing.) and *jowros* (pl.), which correspond to oral use in French and English. The alternative would be the Fulfulde terms *jowro* (sing.) and *jowro'en* (pl.)).
3   As an example, Al Gore and the IPCC received the Nobel Peace Prize in 2007 based on the idea that there is a close linkage between climate change and conflicts. In the Nobel Committee's justification for the award, the Sahel was highlighted as a case where this relationship has been repeatedly played out. The idea is that climate change leads to desertification and scarcity of resources, which again lead to more conflicts. This idea continues to hold a strong position in policies and the media, despite poor support in empirical research (see Benjaminsen et al. 2012).

REFERENCES

Ba, B. 1996. 'Le conflit meurtrier entre Sossoobe et Salsalbe (cercle de Tenenkou, Mali)'. In Le Roy, E., Karsenty, A., Bertrand, A (eds), *La sécurisation foncière en Afrique. Pour une gestion viable des ressources renouvelables*. Paris: Karthala, pp. 280–6.
——— 2008. *Pouvoirs, ressources et développement dans le delta central du Niger*. Paris: L'Harmattan.
Ba, A.H., Daget, J. 1962. *L'empire peul du Macina (1818–1853)*. Abidjan: Nouvelles Editions Africaines.
Barrière, O., Barrière, C. 2002. *Un droit à inventer. Foncier et environnement dans le delta intérieur du Niger*. Paris: Institut de Recherche pour le Développement (IRD).
Behnke, R., Scoones, I., 1993. 'Rethinking range ecology: implications for rangeland management in Africa'. In Behnke, R., Scoones, I., Kerven, C. (eds), *Range Ecology at Disequilibrium. New Models of Natural Variability and Pastoral Adaptation in African Savannas*. London: Overseas Development Institute and International Institute for Environment and Development.
Behnke, R., Scoones, I., Kerven, C. (eds) 1993. *Range Ecology at Disequilibrium. New Models of Natural Variability and Pastoral Adaptation in African*

Savannas. London: Overseas Development Institute and International Institute for Environment and Development.

Benjaminsen, T.A., 1997. 'Natural resource management, paradigm shifts and the decentralization reform in Mali'. *Human Ecology* 25(1): 121–43.

Benjaminsen, T.A., Berge, G. 2004. 'Myths of Timbuktu: from African Eldorado to desertification'. *International Journal of Political Economy* 34(1): 31–59.

Benjaminsen, T.A., Ba B. 2009. 'Farmer-herder conflicts, pastoral marginalisation and corruption: a case study from the inland Niger delta of Mali'. *The Geographical Journal* 174(1): 71–81.

Benjaminsen, T.A., Alinon, K., Buhaug, H., Buseth, J.T. 2012. 'Does climate change drive land-use conflicts in the Sahel?' *Journal of Peace Research* 49(1): 97–111.

Biasutti, M. 2013. 'Forced Sahel rainfall trends in the CMIP5 archive'. *Journal of Geophysical Research: Atmospheres* 118: 1613–23.

Boko, M., Niang, I., Nyong, A., Vogel, et al. 2007. 'Africa. Climate Change 2007: Impacts, Adaptation and Vulnerability'. In Parry, M.L., Canziani, O.F., Palutikof, J.P., van der Linden, P.J., Hanson, C.E. (eds), *Contribution of Working Group II to the Fourth Assessment Report of the Intergovernmental Panel on Climate Change.* Cambridge: Cambridge University Press, pp. 433–67.

Boserup, E. 1965 (1993). *The Conditions of Agricultural Growth.* London: Earthscan.

Buontempo, C., Booth, B., Moufouma-Okia, W. 2010. 'The climate of the Sahel'. In Heinrigs, P., Trémolières, M. (eds), *Global Security Risks and West Africa. Development Challenges.* Paris: OECD, pp. 58–71.

Chappell, A., Agnew, C.T. 2004. 'Modelling climate change in West African Sahel rainfall (1931–90) as an artefact of changing station locations'. *International Journal of Climatology* 24: 547–54.

Christensen, J.H. et al. (2007) 'Regional Climate Projections'. In Solomon, S. et al. (eds) *Climate Change 2007: The Physical Science Basis. Contribution of Working Group I to the Fourth Assessment Report of the Intergovernmental Panel on Climate Change.* Cambridge: Cambridge University Press, pp. 847–940.

Cotula, L., Cissé, S. 2006. 'Changes in "customary" resource tenure systems in the inner Niger delta, Mali'. *Journal of Legal Pluralism and Unofficial Law* 52: 129.

Ellis, J.E., Swift, D.M. 1988. 'Stability of African pastoral ecosystems: alternate paradigms and implications for development'. *Journal of Range Management* 41(6): 450–9.

Forse, Bill, 1989. 'The myth of the marching desert'. *New Scientist* 4 February: 31–2.

Gallais, J. 1967. 'Le delta intérieur du Niger: étude de géographie régionale'. IFAN thesis, 78, two volumes. Dakar: Institut Français d'Afrique Noire.

Haarsma, R.J., Selten, F., Weber, N., Kliphuis, M. 2005. 'Sahel rainfall variability and responses to greenhouse warming'. *Geophysical Research Letters* 32.

Hiernaux, P., 1993. *The crisis of Sahelian pastoralism: Ecological or economic?* Addis Ababa: International Livestock Centre for Africa.

Hulme, M., 2001. 'Climatic perspectives on Sahelian desiccation: 1973–1998'. *Global Environmental Change* 11: 19–29.

Hutchinson, C.F., Herrmann, S.M., Maukonen, T., Weber J. 2005. 'Introduction: The 'Greening' of the Sahel'. *Journal of Arid Environments* 63: 535–7.

Ki-Zerbo, J. (1978) *Histoire de l'Afrique noire.* Paris: Hatier.

Kouyaté, S. 2006. 'Etude des enjeux nationaux de protection du basin du fleuve Niger'. Report to Groupe de Coordination des Zones Arides (GCOZA). Bamako: GCOZA.

Le Houérou, H.N. 1989. *The Grazing Land Ecosystems of the African Sahel*. Berlin: Springer.

Lentz, C. 2006. 'Land rights and the politics of belonging in Africa: an introduction'. In Kuba, R., Lentz, C. (eds), *Land and the Politics of Belonging in West Afr*ica. Leiden: Brill, pp. 1–34.

Moorhead, R.M. 1991. 'Structural chaos, community and state management of common property in Mali'. PhD Thesis, Brighton: University of Sussex.

Niang, I., Ruppel, O.C. 2014. 'Africa', IPCC's Fifth Assessment Report.

Odekunle, T.O., Andrew, O., Aremu, S.O. 2008. 'Towards a wetter Sudano-Sahelian ecological zone in twenty-first century Nigeria'. Weather 63(3): 66–70.

Olsson, L., Eklundh, L., Ardö J. 2005. 'A recent greening of the Sahel – trends, patterns and potential causes'. *Journal of Arid Environments* 63: 556–66.

Pedersen, J. Benjaminsen, T.A. 2008. 'One leg or two? Food security and pastoralism in the northern Sahel'. *Human Ecology* 36(1): 43–57.

Ranger, T.O. 1983. 'The invention of tradition in colonial Africa'. In Hobsbawm, E., Ranger T.O. (eds), *The Invention of Tradition*. Cambridge: Cambridge University Press, pp. 211–62.

Raynaut, C. 1997. *Societies and Nature in the Sahel*. London: Routledge.

Robbins, P. 2012. *Political Ecology*. Oxford: Blackwell.

Sandford, S. 1983. *Management of Pastoral Development in the Third World*. New York: John Wiley and Sons.

Spear, T. 2003. 'Neo-traditionalism and the limits of invention in British colonial Africa'. *Journal of African History* 44: 3–27.

Sullivan, S., Rohde, R.F., 2002. 'On non-equilibrium in arid and semi-arid grazing systems'. *Journal of Biogeography* 29, 1–26.

Tucker, C., Dregne, H.E., Newcomb, W.W. 1991. 'Expansion and contraction of the Sahara desert from 1980 to 1990'. *Science* 253: 299–301.

Tucker, C., Nicholson, S. 1999. 'Variations in the size of the Sahara Desert from 1980 to 1997'. *Ambio* 28(7): 587–91.

Turner, M. 1992. 'Living on the edge: Fulbe herding practices and the relationship between economy and ecology in the inland Niger Delta of Mali'. PhD Thesis. Berkeley: University of California.

——— 1993. 'Overstocking the range: a critical analysis of the environmental science of Sahelian pastoralism'. *Economic Geography* 69(4): 402–21.

——— 1999. 'The role of social networks, indefinite boundaries and political bargaining in maintaining the ecological and economic resilience of the transhumance systems of Sudano-Sahelian West Africa'. In Niamir-Fuller, M. (ed.), *Managing Mobility in African Rangelands. The Legitimization of Transhumance*. London: Intermediate Technology Publications.

Vedeld, T. 1997. 'Village politics. Heterogeneity, leadership, and collective action among Fulani of Mali'. PhD Thesis, Ås: Norwegian University of Agriculture.

Warren, A., Khogali, M. 1992. *Assessment of Desertification and Drought in the Sudano-Sahelian Region 1985–1991*. New York: UNSO.

Wiggins, S. 2005. 'Success stories from African agriculture: what are the key elements of success?' *IDS Bulletin* 36(2): 17–22.

17 Fishermen, Herders and Rice-Farmers of the Inner Niger Delta Facing the Huge Challenge of Adapting to Weakened Floods: A Social-Ecological System at Risk

Pierre Morand, Famory Sinaba and Awa-Niang Fall

INTRODUCTION

In Africa, rivers represent a significant portion of the available water resources. This fact has not been ignored by past civilizations (such as Ancient Egypt), nor by colonial and African modern governments of the twentieth century, who built large infrastructures to control rivers' water, in order to increase agricultural production and – more recently – to create new hydropower resources.

Despite these facts, most African rivers have long remained characterized by rather low level equipment and a so-called 'under-exploitation' of their hydropower and agricultural-water potential, at least when we hear the views expressed by development experts. That is why these experts now advocate policies to revive the investments in equipment and watershed management (UA-NEPAD 2009). Moreover, such an orientation recently joined the global agenda of adaptation to climate change.

Among the basins that have long been regarded as having a lack of equipment and development, we must include the middle and upper parts of the Niger River Basin, from Guinea to the Mali and Niger countries. But things have recently changed since this basin is now under strong political pressure for the installation of new infrastructures (development plan leading by (ABN 2007)). This will lead in the future to increasing environmental changes with respect to which the local/rural communities of the basin must adapt. The present work aims to understanding the challenge of adaptation that arises for these communities in a particular region of the basin: the Inner Niger Delta.

THE INNER NIGER DELTA: MAIN PHYSICAL FEATURES

The Niger River runs through a large part of West Africa, over a length of 4,200 km. The first half of its course crosses the plateaus and plains of Guinea and Mali. At this stage it is paradoxically oriented toward the northeast, moving away from the sea to come to brush against Sahara desert. The course of the river then turns East and then South, before discharging into the Gulf of Guinea. At the end of the first part of its course, when leaving the savanna zone and approaching the desert, the river flows through a plain with a very weak slope, causing a slower flow rate and separation into several branches, creating a vast wetland called 'Inner Niger Delta' (IND).

Given the seasonality of tropical rainfall that takes place on the upstream part of the basin, extending from (May) June to September, the Inner Niger Delta zone receives lot of water as of September and it is then flooded over large areas for a period of several weeks, sometimes up to December – or even January in a distant past. Thus the 'Inner Niger Delta' constitutes the largest wetland of West Africa, with a flooded area that could reach more than 22,300 km^2 in the case of rainy years (Mahé et al. 2011) or even more in the distant past.

Although the seasonal timing of peak flood is very constant, flooding extension is highly variable and unpredictable from year to year, since the inter-annual variability of rainfalls and runoffs in the basin of Niger are of great amplitude – that is a well-known characteristic of climate in this part of Africa. The latter half of last century offers a good example of this general feature: after 20 years of continuous high rainfall (1950–69) followed by more than 20 years of decreasing/low rainfall (the so-called 'Sahelian Great Drought' of 1970–93), the last 20 years (1994–2014) have been characterized by a succession of medium and low rainfall years.

In particular, from 1973 to 1993, average amount of rainfalls from May to September exhibited deficits of 180 to 200 mm in most of the West African hinterlands (Le Barbé et al. 2002), representing drops ranging from 20 per cent (in the Sudanian climatic zone) to 60 per cent (in the Sahelian zone). When such conditions prevail, it inevitably leads to a decline in the annual flooding in the Inner Niger Delta, regardless of the possible effect of other factors. For instance, in 1984, the rainfalls and discharge were so weak that the river has not overflowed above the alluvial bank, so that the plains have hardly been flooded.

THE INNER NIGER DELTA: AN OLD-AGE HUMAN-NATURAL SYSTEM

The Inner Niger Delta is regarded by archaeologists as one of the original places of African rice cultivation 3,500 years BP, an activity which was based on the African rice (*Oryza glaberrima*) species (Portères 1950). However, it is not known what became of these proto rice-farmers, neither who their descendants are today. Notice that the Asian rice *O. sativa*, which was brought in the sixteenth century on the West African coasts by Portuguese merchants, has today partially replaced *O. glaberrima*, especially under intensive conditions of cultivation.

Among the ethnic groups inhabiting the Delta until today, the one which has been traced as having arrived the earliest was originally focused on hunting the aquatic mammals and harpoon-fishing of large fish. Today we call this group the '*bozo*' community, a name which has become synonymous with 'fishermen of the Inner Niger Delta'. Another group called '*nono*', supposed to be brothers of the previous one, arrived at nearly the same time and was engaged in rice farming. Over the unfolding of history and as a result of the pressure or gains of the neighbouring empires and kingdoms (Mali Empire, Gao Empire, Kingdom of Segou), other new groups have arrived and settled in the large floodplain of the Delta – or on its riparian zones (Kassibo 1994). Each of these groups has found its place by engaging in specific livelihood strategies (Fay 1995). The '*somono*' group primarly fished on the river, the *sonraï-sorko* group had developed flood recession agriculture and fishing in the northern part of the Delta, the *bambara* group had undertaken the farming of rain-fed millet on the margins of the Delta, the *fulani* group was guiding seasonal migration of herds of zebus, while their 'captives' the *rimaïbe* were practicing rice crops inside the floodplain.

From the eighth century to the eighteenth century, the Delta area and its cities of Djenne, Mopti and Timbuktu were steps along the great trans-Saharan route of merchants going from Black Africa to the Arab and Mediterranean world, who used donkey caravans to travel through the Sudanian savanna, canoes to navigate the Niger River and finally camel caravans to cross the Sahara.

The nineteenth century saw a period of intense political structuring called '*dina*' or '*Fulani* Empire of the *Macina*' (Daget & Ba 1955). Under the leadership of Islamic religious power, harvesting practices led by the different groups and communities of the Delta were at that time precisely established, with many rules and trade-off to manage sharing issues, either on aspects of time (for example, the sequence of migration of herds) as on the spatial dimension (access rights and land tenure). Such consolidation of the activities rules has resulted in an encompassing regional production system that combined milk, fish and rice, the first two – the protein products – being mastered by the leading ethnic groups, first the *fulani* herders (the masters of land) and second the *bozo* fishermen (the masters of water), while the cultivation of rice remained an activity open to different groups (*nono, rimaïbe, bozo*, etc.) under the consent of the master of land. Such a sophisticated system was considered a model of 'ethnic-occupational specialization' (Gallais 1967).

It must be admitted that such a system was not egalitarian since some groups dominated others. But we must acknowledge that it allowed several human groups and communities to operate sustainably the multiple resources of a single ecosystem and finally to coexist in relative harmony, while allowing them to maintain with each other some intensive exchanges (including through barter) thanks to their complementary products. Thus, this system really deserves to be considered as an integrated 'social ecological system'.

Moreover, this social-ecological system also had the advantage of rusticity to the extent that all productive activities are extensive ones (neither fertilizer nor man-made feeding of livestock).

The outcome of all these assets allowed the Inner Niger Delta to remain, long after the collapse of the trans-Saharan commercial route, a more densely populated area than most of the surrounding areas in the Sahel region. Thus the average human density within the Niger Delta is about 30 people per km^2 – based on 1.2 million inhabitants for 40,000 km^2 in a broad definition of the Delta – which should be compared with figures of 10 to 20 inhabitants per km^2 in the same latitude band ($14°$ to $16°30$ N Lat) within the surrounding area to the west and to the east.

It also allowed the Delta to benefit from the development of trade opportunities during the twentieth century and so to become an exporter of large amounts of food products to other regions and countries of West Africa. Thus, from the late 1920s to the 1960s, a significant part of the meat and fish eaten in Ivory Coast and 'Haute-Volta' (the former name of Burkina Faso) came from herds and fisheries of the Delta. At that time, the pathway from Mopti to the 'Haute-Volta' was called the 'route of (smoked) fish' (Weigel & Stomal 1994).

However, such economic importance subsequently declined. It is such a negative trend over the last 40 years that we have to explain it.

HYDRAULIC DEVELOPMENT POLICIES APPLIED TO THE UPPER AND MIDDLE NIGER THROUGHOUT THE TWENTIETH CENTURY UNTIL TODAY

The Inner Niger Delta itself is a region where flooding occurs naturally without human intervention and without control. Only the *sonraï* farmers' communities of the Northern part of the Delta were historically accustomed to dig little canals in order to facilitate the filling of small agricultural areas. Thus, broadly speaking, the Delta region was the subject of very few irrigation schemes in the course of history, with the exception of the very recent past since the 1970s.

But when considering things on a wider scale, we find that the rivers of the Sudano-Sahelian zone have been the subject of hydraulic development since nearly a century ago, and some of them have impacted the Delta environment. Indeed, the French colonial authorities in West Africa began during the 1920s to plan infrastructure on the Senegal River and on the Niger River for the purposes of agricultural production. However, the 'Sarraut plan' was not fully brought into execution. On the Niger basin, the major projects were carried out by the engineer Belime and they consisted of the building and commissioning of the sills-dams of Sotuba (1925) and Sansanding/Markala (1947), both located upstream of the Delta. The latter led to the creation of the large irrigated area (about 800 km^2) called the 'Office of Niger' (ON), located along an ancient water path that comes off from the current valley to the north.

During the period 1950–70, including the shift to the Independence of Mali (1960), the hydraulic development policy stalled. At this time,

1: Sotuba (1925)
2: Sansanding-Markala 1947)
3: Selingue (1981)
4: Talo - sill (2007)
5: Kandadji (2010)
6: Djenne - sill (2015)
7: Fomi (planned)
8: Taoussa (planned)

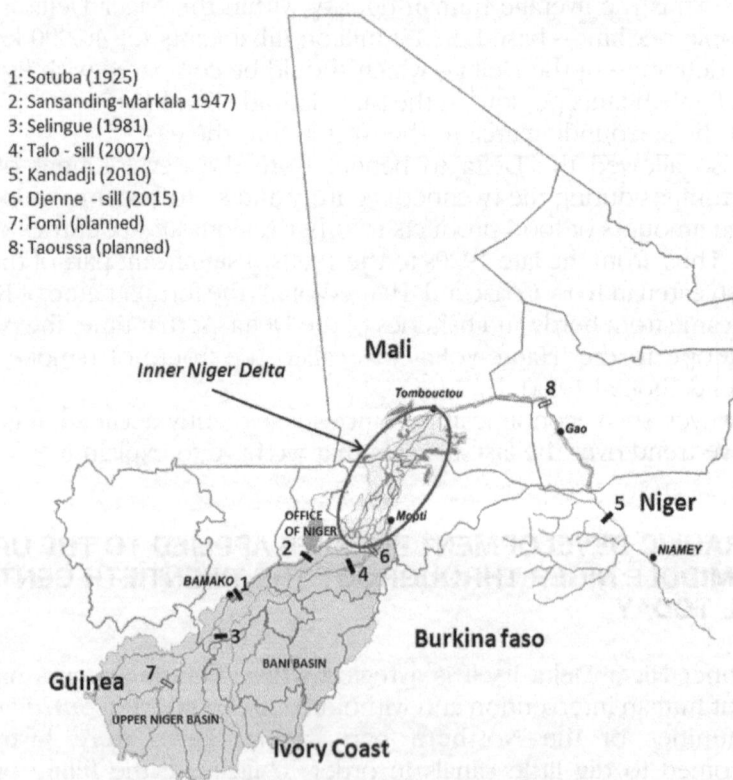

Figure 17.1. Map of the regions of Mali showing the borders and the neighbourhood countries, the Upper Niger basin (mainly extended in Guinea), the basin of Bani (tributary of the Niger river), the Inner Niger Delta floodplain, the irrigated zone called 'Office of Niger' as well as the existing and planned dams and sills (modified from J. Marie in Marie et al., 2007).

concerns were focused mainly on increasing the yields and maintenance of the existing infrastructure. However, the period from 1970 to the present day has seen the revival of the hydraulic development projects, by including henceforth the target of power production (Selingue dam in 1981, with a volume of about 2.5 billion m^3). This trend has accelerated since 2005, because of the questioning of the findings of the World Commission of Dams (2000) and the global crisis of food markets in 2008. Thus, two medium-size sills were recently built (Talo in 2007, then Djenne – planned to be commissioned in 2015) associated with the creation of irrigated areas. In addition, a more than doubling of the irrigated area of the 'Office of Niger' is planned and is already in progress to supply new irrigated areas for the foreign companies which invest to produce sugar cane and fuel plants for the worldwide market (Hertzog et al. 2012). Thus, the 'Office of Niger' zone will soon reach 2,000 km^2 of irrigated areas. To supply water to such a large area, especially during the low flow months (January to May), and also to generate more power for cities, a new large dam was scheduled by ABN on the upper Niger basin, located on Guinean territory: the 'Fomi' dam. It will have a retained volume of about twice the Selingue dam.

THE EFFECTS OF WATER INFRASTRUCTURES ON THE NIGER RIVER REGIME AND THE FLOOD CONDITIONS IN IND

Infrastructures built before 1980 (Sotuba and Markala) were dams-sills designed to create water bodies raised a few meters to allow an easy water catchment for feeding a gravity flow towards the irrigation area. The effects of such structures on downstream flows were negligible in flood season but quite sensitive in the season of low water. Thus, the water catch at Markala to supply the 'Office of Niger' (ON) accounted for 150 m^3/s in flood season (i.e., less than 3 per cent of the river flow at that time) and 70 m^3 in the low water season (more than 50 per cent of the river flow at that time).

With the commissioning of the Selingue Dam, which is equipped with a big reservoir that must be completely filled using the flows of the rainy season, a new phenomenon has emerged: the weakening of the annual flood wave downstream. Thus, the amount of water needed to fill the Selingue reservoir is about 400 m^3/s during six weeks at the end of the rainy season. When added to the water catch exerted at Markala for O.N., the whole represents a weakening of about 10 per cent over the flood peak flow for a typical average year. Such effects of water infrastructures on the strength of the flood have long been minimized and denied by engineers, who are accustomed to consider any regulation effect on downstream flows as a good thing.

In addition, the worst period of the Sahelian Great Drought was contemporary with the first ten years of the Selingue dam operation, so that the perceived impact of this dam was crushed under the impact (much stronger) of the natural deficits of precipitation and flows. More generally speaking, we must recognize that the up and down variations of the strength of annual flood peak due to natural climate variability (SD of about 5,000 km^2) fit into a range of variation which is wider than the variations due to the effects of the existing infrastructures (Selingue *plus* Markala).

Nevertheless, the fact is that dams and water catches have a real weakening effect on the flood of the Niger River. Since 2005 some environmental expert teams (Zwarts et al. 2005, Marie et al. 2007, Liersch et al. 2013) have started to consider this issue, especially focusing on the change in inundation conditions within the Inner Niger Delta. They also attempted to take into account the impact of the foreseen dams (for example, Fomi).

According to the published figures of Zwarts et al (2005) and Marie et al (2007), the Markala and Selingue dams – which were commissioned in 1947 and 1981 – decreased the flood peak in the Delta by 20–25 cm. As the new dam to be constructed at Fomi in Guinea (upper Niger Basin) will hold about two times the volume of the Selingue reservoir (although there are still some debates about the precise size of Fomi reservoir), we thus foresee an additional drop of 30 cm to 45 cm during the peak flood season.

According to hydrological studies, a 1 cm fall in water height during the flood peak translates into a loss of inundated area of 65 km^2 (Zwarts et al. 2005) or 95 km^2 (Marie et al. 2007). Assuming the various upper hypotheses regarding flood peak drops, on one side, and the conversion factor into loss of inundated area, on the other side, we can estimate a current drop of inundated area ranging from 1,300 to 2,400 km^2 and expected to reach between 3,250 and 6,500 km^2 by the construction of the Fomi dam.

In addition to these large dams, several smaller dams and man-made sills are yet established (for example, Talo) or planned in the upper Niger basin (Ferry et al. 2012). Individually these small dams have limited impacts, but exacerbate those of the larger dams. In addition, the doubling of irrigated areas in the 'Office du Niger' irrigated perimeter is expected to lead to an increase in the quantities of water abstracted by the Markala dam throughout the annual cycle.

SENSITIVITY AND DIFFICULTIES OF ADAPTATION EXPERIMENTED BY TRADITIONAL PRODUCTION SYSTEMS FACING LOW FLOODS

Traditional production systems of the Delta, whether agricultural, pastoral or fisheries are the result of a long evolutionary process that made them suitable to cope with large variations of the hydro-climatic conditions, both at the level of the year (seasonal cycle) and at the inter-annual scale.

However, climate conditions and anthropogenic pressures that exist today exert increasing constraints on these production systems, leading to a risk of exceeding their capacity to adapt. To assess this risk, it is first necessary to describe how these traditional systems work and how they have responded to known environmental changes that occurred in the recent past.

Agriculture and farmers

First, we consider traditional agriculture, which involves about 100,000 households in the Delta. Some of them are full-time farmers, while others are engaged in agriculture only as a secondary activity (e.g., in addition to the fishing activity).

In most parts of the Delta, the rice is the dominant culture and is mainly operated through extensive methods, using the species *Oryza glaberrima* for which the growth of the stem of the plant naturally adjusts to the rising water: it is the 'floating rice'. This system does not involve any irrigation infrastructure. Farmers choose a location for their parcels according to where they believe the water levels will be adequate during the next high-water season. They may also move their parcels from one year to the next. They plough during the dry season (February to May), then scatter the seeds in late June/early July. The rainfall triggers the growth process, and then the floods take over. This rice growing spanned from 100,000 ha (Zwarts et al. 2005) to 160,000 ha (Marie et al. 2007) in the IND. The mean yields are very low (less than 1 t/ha) and unpredictable: rainfall may be insufficient or late, and flooding may be low or even (sometimes) in excess.

In addition to this ancient type of cultivation, another floating rice cultivation method was developed from the early 1970s: it is the scheme for culture in controlled submersion which was set up on approximately 34,000 ha in the vicinity of Mopti (mid part of the Delta). This method is slightly more sophisticated since it involves building small walls and inlets at precise levels to govern the submersion. Thus the farmers can scatter later in the year (in late July) and therefore make sure the seedlings have received enough rain and have started their growth before the flooding arrives – since authorities controlling the perimetre can delay the opening of the inlet if and as required.

The total yearly production of rice obtained by these two latter extensive cultivation methods varies between 20,000 tonnes and 130,000 tonnes according to hydro-climatic conditions (ORM data in Zwarts et al., 2005). Most of the harvest is locally consumed, either by farming households or by other Delta communities, but it may also lead to small exports outside the Delta that can reach 10,000 tonnes in good years (Kuper & Maiga 2002). According to Marie et al (2007), the total usable area for floating rice could vary by X4 factors depending on inundation extension. That is why there is a correlation between the total production of floating rice and the strength of the flood in the Delta, even though other factors may affect yields (Zwarts et al. 2005). However, it is not easy to provide an accurate estimate of the loss of production in floating-rice due to the existing upstream large infrastructures, as it greatly depends on many circumstances such as rainfall distribution and strategy of farmers, especially related to the choice of location of the parcels sown – which themselves depend on the crop degree of success (or not) at given places during the previous year. Nevertheless, one can state that the degree of success for a campaign of rice cropping collapses when flooding is very low (Liersh et al. 2013).

Traditional rice farmers can hardly anticipate the intensity of the next rainy season and of the coming flood. But if the years of low rainfall/low floods keep coming, as in the 80s, farmers' strategy is to move their crop fields toward the lowest and best flooded places of the plain. By doing this, they encroach on wet grasslands, namely the *bourgou* pastures, that can deprive the herds of passage ways and food when they are back in the Delta in December. Such a situation leads to an increasing number of conflicts between farmers and herders, a phenomenon clearly observed during the Great Drought period (Benjaminsen & Bâ 2009).

Fishing activity and the fishermen

The fishing activity is likely the second one in the Delta regarding the number of households involved, that is around 30,000, among whom a minority (of about 20 per cent) do not have any farming activity (Marie et al. 2007, Morand et al. 2012).

Fishing activity takes place mostly during the receding and the low-water season until the beginning of the next flood, i.e. between late November and early July. At the opposite end, the fishing activity and catches are low when fish are dispersed in a large mass of water during the high-water season (August–early November), but it is at this time that the fish reproduce and grow. Thus the fishing activity is to harvest during the receding and the low-water season the most part of the biomass of fish that has been generated during the last flood. Moreover it is noted that over 75 per cent of fish caught are juveniles in their first year. Whatever the intensity of fishing activity and depletion of the fish biomass in the late dry season, there is never any sign of trouble in fish biomass renewal, because it succeeds in regenerating each year from the very small breeding stock remaining in the water at the end of the fishing campaign/onset of the next flooding.

By this way, the resource is fully exploited but not endangered. Such a dynamic also implies that there is no hope to get a greater harvest by increasing the fishing effort (Kodio et al. 2002). Keeping in mind the way in which this fishery works, one deduces that total catches during the fishing campaign from November of year$_t$ to July of year$_{t+1}$ are simply controlled by the success of the last reproductive period, which itself depends on the strength of the flood that occurred between August and October of the year (Morand et al. 2012).

According to the estimates (Laë 1994), total annual catches range from 37,000 tonnes to 87,000 tonnes in the 1969–92 period and they are strongly correlated with inundated surfaces of the previous flood season. Moreover, the flood area-catches relationship is nearly linear, showing a slope of increase of 4–5 tonnes per km^2 inundated. Assuming the decrease of the flooded area due to the effects of the existing infrastructures is about 1,300 km^2 to 2,400 km^2 (see above), we can then consider that the current loss of the annual catches is 5,200 tonnes to 12,000 tonnes, i.e. more than 12 per cent of the unaltered volume. Thus, the worsening effect is already significant. Surely it will be even stronger when the Fomi dam and full extension of the 'Office of Niger' take effect.

The review of facts observed in recent decades, especially during the Sahelian Great Drought, shows that fishermen do not remain unresponsive in the face of years of weak floods and low fishing production. Among their responses, the most widespread is the increase in fishing pressure through the increasing use of predatory and unmanageable fishing gears (such as cast nets and overlying nets) and the reduction of the net mesh size – that leads to a predominance of one to two fingers mesh size (Laë et al. 2004). Despite this increase in fishing pressure, the total harvesting of fish remains necessarily low when floods are weak and the only concrete result is a decrease in the average catch per unit of effort, which negatively affects the economic performance of the activity. Moreover, all these practices strengthen competition and conflicts between fishermen, especially between sedentary fishermen and fishermen who make seasonal migrations within the Delta.

Another response consists in definitive emigration of part of the young Bozo fishers toward far-flung fishing areas, usually dam reservoirs located either in Mali or in other African countries such as Ivory Coast (Jul-Larsen & Kassibo 2001). But these departures often lead fishermen to experiment with harsh living conditions and new risks in their host areas.

Livestock farming and the herders

The Inner Niger Delta is the Mali's leading livestock-farming region. In the 1980s, a census suggested that there are some 1.2 million head of cattle inside and around the IND. It spans only three per cent of the country's land but counts about 20 per cent of Mali's cattle. This is because its flooded pastures are exceptionally fertile and stay 'green' for seven to eight months. The best pastures, the *bourgou*, produce from 20–30 tonnes of dry matter per ha per year and they provide green feed from December to July (by comparison, Sahel rain-fed pastures rarely produce over two tonnes of dry matter and are only green for four months a year).

The livestock farming activity relies on migrations schemes, by following a strict calendar. At the beginning of the rainfall season, in July, herders of the IND leave the floodplain for their long seasonal migration to the Sahelian pastures (Farimaké, Léré and Nampala zones on North-West, Gourma on the East, Seno on the South). They return in November or December, depending on how fast the water recedes. They lead their herds across the Delta through a system of pathways and shelters, which they use according to traditions and rules long established (often since the Dina period). To some extent, herders depend on the annual flooding, since the peak-water level and the inundation surface dictate, in a part, how much fodder they will find for their herds.

According to Marie et al (2007), a very weak flood (peak of 5.10 m at the Mopti gauge) may cut IND annual fodder production by about 40 per cent compared to the fodder production resulting from a good flood (6.60 m). Notice that this is less than the corresponding decrease in the inundated area, which is about 75 per cent. So the loss in fodder production is considerably less significant than the drop in flooded areas. This is because some vegetal formations, namely those of the pastures that are not inside the floodplain, are not 'aquatic species' and thus can exhibit normal growth thanks to rainfall. So the variability of the Niger River flood does not affect the whole fodder production in a close and single way: the changes in the flooded area only account for some part of the environmental change affecting total fodder production. Nevertheless, when considering all the environmental hydro-climatic terms, we can assume that the Delta can support about 1.5 million cattle during a series of good years (high flood and high rainfall on the Sahel region), but that it can only sustain 900,000 cattle during a period of bad environmental conditions. Notice also that, just like the fish biomass, the pastures of the Delta seem to be resilient: after a period of bad years, they can recover and thrive again if good environmental conditions come back (Hiernaux & Diarra 1986).

When periods of bad hydro-climatic conditions occur (for instance during the Sahelian Great Drought), the pastoral system of the Delta does not fit easily because the cultural tradition of *fulani* herders precludes the early slaughtering of livestock. This is why substantial livestock mortalities were then observed, which led to a strong impoverishment of the herding communities. However, during the 1980s, some of the livestock were moved to regions located further south, which helped to save part of it. Such a move has also led to an increase of livestock in the cotton farming areas of southern Mali (Koutiala and Sikasso zones), leading to beneficial effects on land fertility. In spite of this fairly positive side-effect, one must recognize that the pastoral system of the Delta and surrounding Sahel is highly affected by any prolonged decline in the strength of floods resulting in a reduction of *bourgou* areas, which makes it vulnerable in the same way as the two above cases of traditional production systems.

SOME INNOVATIONS AND NEW OPPORTUNITIES BUT MUCH DIFFICULTIES TO GRAB

According to a modern agronomic vision, the adaptation of the Delta to a lower amount of water received in flood should pass through a partial renunciation to traditional extensive production systems dependent on the natural inundation (fishing, pastoralism and floating rice) in favour of more intensive production systems mainly based on water control.

Such a development orientation was not entirely absent from the Delta in the last decades, since new opportunities and innovations in production modes have indeed emerged in this direction. Some of it occurred spontaneously, others were supported by NGOs projects and donors. By describing these new opportunities and by measuring the degree of success they reached as well as their actual dissemination within communities, one attempts here to assess their possible contribution in tracing a sustainable future for the communities of the Delta.

One of these innovations, which is spontaneous and began to be employed in the late 1980s (at the end of the Great Drought Period) by certain farmer and fishermen communities, is to dig or deepen trenches in the alluvial slope to promote the inundation of the floodplains (Courel & Chamard 1999). This practice locally accelerates the flooding of rice plots, which can enhance their yield. When the floodwater recedes, the trenches can be blocked by fish traps that provide an additional catch for fishermen. However, since this practice is generally not accompanied by the construction of gates to retain water, it also accelerates the drainage of the floodplains in the falling stage, which is undesirable in the long term both for rice and fish productivity. At the scale of the Delta as a whole, the many trenches dug in the alluvial slopes do not seem to have improved the relationship between the peak water level at Mopti and the inundated area.

More recently, as from the late 1990s, new auxiliary productive activities have emerged in the livelihood strategies of households of all communities. One can mention here poultry, small-scale livestock raising and market gardening along the river's edge. But these activities are just intended to get some little monetary income supplements. For instance, they are often managed by women to cope with the needs and care of children's health. Anyway, these activities remain at a small scale level and they do not deeply change the pattern of use of village territories.

Another innovation, which constitutes a more radical departure from the traditional lifestyle of certain village communities, is intensive irrigated rice farming during the off-season, from March to July, in paddies watered by motor-driven pumps that draw water from the river (Ducrot et al. 2002). This new form of rice farming is costly in terms of inputs but makes rice production less subject to hydro-climatic contingencies and offers much better yields. However, it requires considerable labour for the puddling of the paddies and transplanting of seedlings. Such a constraint may discourage some communities, such as fishers, who prefer to devote intensively to fishing activity in this season, especially when it requires mobility (seasonal migration).

Moreover, off-season rice farming is possible only where a large water source exists in the dry season, i.e. only in areas along the river itself. Thus the irrigated perimeters may represent a strong grip on these riverside areas where the land needs are high, especially for the passage of the herds that go to drink. That is the reason why this form of intensive agriculture could aggravate the tension over land occupancy in these areas, which have long seen conflicts over land use rights between livestock herders and farmers.

For all these reasons – costs, problems of integration in the activity schedule and risks of land conflicts – the communities of the Delta less widely adopted this new form of rice farming than development workers had hoped. According to Fossi et al (2014), over a sample of 210 small rice irrigated areas established since the 1990s throughout the Delta, 50 per cent have been abandoned. When examining a more specific and well monitored area (Figure 17.2), the data show that the importance of this intensive form of rice crop remains low in comparison to the traditional floating-rice crop system, both among fisher communities and farmers (Mainguy et al. 2015).

Another agricultural innovation was the creation of tree plantations, usually Eucalyptus, in plots along the river. Young trees benefiting from irrigation grow fast and provide after a few years timber and charcoal, two highly profitable products because there is high demand in the internal market, especially in large cities. However, these plantations require upfront investments, which are usually out of reach for the Delta households and communities and are therefore paid by external (urban) investors. Nevertheless, the processing and trade of charcoal creates jobs and income for some rural people of the Delta, including the poorest.

Some other innovations consist of changes in practices. For example, fishermen of the Delta newly use semi-emergent vegetation branches devices to attract fish, to fix them and then to capture them more easily. But only families and communities who hold customary rights on water can implement such practices of fishing intensification. In addition, when this practice is new in a given region, it is known to reinforce ownership behaviours on aquatic areas with a risk of increased social disparities (Welcomme 2002).

In addition, an activity of transport and guiding the Western tourists on the river was developed in the decade 2000–10, but it then collapsed with the increasing of insecurity problems.

The above review of technological innovations and new opportunities that have emerged since the late 1980s in the Inner Niger Delta show that they have not deeply changed the activity profile of households and communities. In some cases they have been engaged by a fairly large

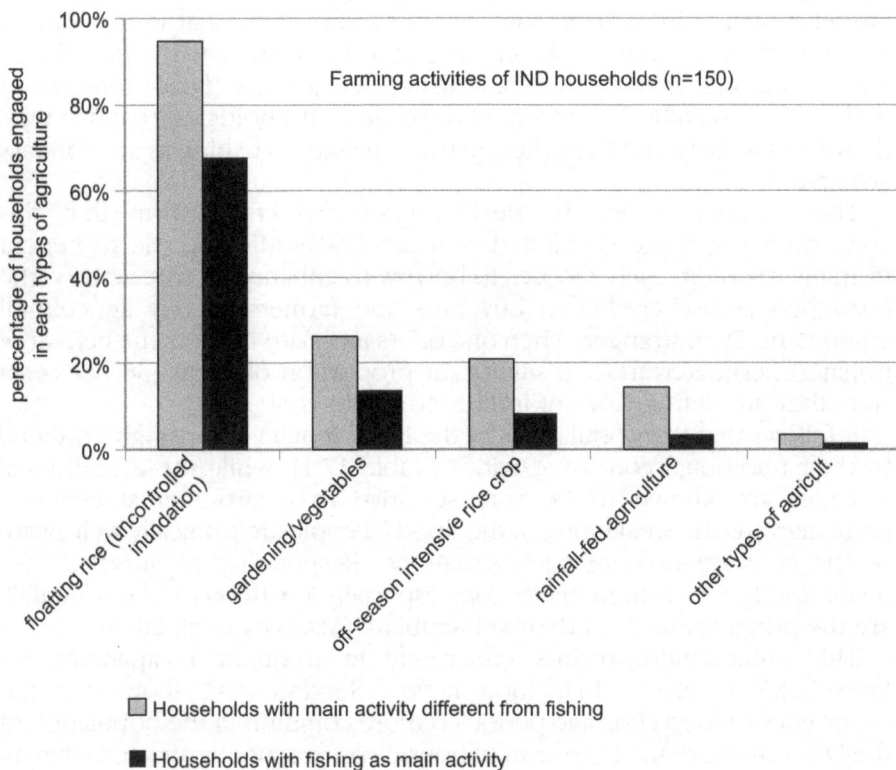

Figure 17.2. Farming activities engaged by households in the Batamani area (45 km North Mopti), based on data collected through VUPOL 2011 survey (modified from Mainguy et al., 2015).

number of households but remained confined to a little role in the family economy – as the ancillary activities of small livestock and gardening. In other cases, innovations have been adopted by a limited number of households because their implementation has proven to be costly or raised new problems, such as increasing social tensions and land conflicts, which communities do not wish to face.

FROM THE OBSERVATION OF A DEADLOCK SITUATION TO A 'RIGIDITY TRAP' AS EXPLANATORY HYPOTHESIS

As a result of the multiple impediments to grab new opportunities and to build significantly on them, a large majority of households of the Delta retain until today a livelihood strategy based on traditional activities tied with their ethnic group, in line with the ethnic-occupational specialization model (Gallais 1967). Data from recent surveys show that until today 98 per cent of households reporting fishing as the main activity claim for an identity belonging to traditionally fishermen groups (*bozo*, *somono* or *sorko*). Reciprocally, 75 per cent of the *bozo* households say that fishing is their main activity and the other quarter engages in fishing as a secondary activity.

The experiences led by development projects confirm that this conservative dynamic is still active: when NGOs offer people to benefit from micro-credit, each chooses to borrow to enhance its core activity: the *bozo* people use credits to buy nets and farmers to buy agricultural equipment. Even stranger, when one offers to *fulani* herders the benefit of irrigated perimeter areas, a significant proportion of them (38 per cent) state they are definitively not interested (Fossi et al. 2014).

It follows that the populations of the Delta remain surprisingly confined to their traditional core of activities (Table 17.1), while these traditional activities are known to be very sensitive to environmental changes, particularly to the weakening of the floods. People are ironically well aware of the consequences of such situations. Responding to surveys, they recognize that environmental factors, especially low floods and low rainfall, are the primary source of their vulnerability (Mainguy et al. 2015).

This vulnerability results concretely in frequent incapacities for households to meet their food needs. Surveys data show that the occurrence of food shortage periods is more common in the population of the Delta (more than 70 per cent of households frequently affected) than in the rural population of Mali inhabiting the regions of cash crops such as the cotton zone (Koutiala and Sikasso). This is a paradox when one points out that the Delta was a large exporter of food products 50 years ago.

Rather than blaming the Delta communities for their low apparent ability to deeply change their livelihood strategies, we have to try to understand the roots of such a deadlock situation.

If we consider *fulani* and *bozo*, i.e. the two groups that were leaders in the three goods traditional system (milk, fish and rice), one must recognize that we are dealing with two communities which are strongly anchored in their practices and professional cultures. Such a characteristic can be

Table 17.1. The current activities of ethnic groups in the Delta: very few changes from the ethnic-occupational specialization model described by Gallais, 1967.

	Livestock Rearing (herder system)	Fishing	Traditional Floating Rice System	New Intensive Rice Crop system based on Irrigated Perimeters
Peul (Fulani)	***			
Bozo, Somono		***	**	*
Nono, Marka, Rimaïbe		*	***	**

connected with a lifestyle that is partly nomadic. Indeed, in these two ethnic groups, many young people and young households spend part of the year in camps, engaging trips planned from one season to the other, either to accompany the herds or to track aggregations of fish resource. This lifestyle is so special and constraining that it requires early initiation of young people, especially young boys, which goes along with a strong avoidance behaviour of school attendance. For these two ethnic groups, there are thus very low school enrolment rates. For example, the school enrolment net rate at primary school remains below 25.0 per cent among fishermen communities in the Mopti region (UEMOA 2013) compared to 50.3 per cent in the whole of rural Mali and 57.4 per cent at national scale (UNICEF 2014). In addition, fishers' populations are often characterized by a negative anomaly of the rate of boys enrolment versus the girls rate (UEMOA 2013).

All these features reinforce the compliance with the rule of ethnic endogamy which remains largely in effect in the Delta according to Fay (1995) as well as the persistence of cultural barriers that prevent young *bozo* and *fulani* to engage in activities different to their parents. It is then difficult for them to obtain employment in urban commercial activities or in NGO projects, all involving a certain level of literacy – see also Nielsen and Greenberg (2010) which analyse this phenomenon for *fulani* in Burkina Faso.

These features explain why the adaptation of the Delta to long periods of drought and low floods, such as the period of Sahelian Great Drought, has been hampered and inadequate, leading to the marked impoverishment of its inhabitants (Bauman et al. 1994).

Although the climate situation is not as bad today, the region has not regained its role as a supplier of protein that fully satisfies the needs of fish and meat throughout Mali. The country is now forced to make increasing imports of frozen marine fish. Certainly, this downgrading of the Delta is not solely due to lower floods but also to the increase in population and food needs, both within the Delta region and in the whole country. But the weakening of floods combined with the limited adaptability of Delta production systems seems to condemn its position as a major food-producing region, compared with other regions of Mali where agricultural systems are changing rapidly and intensify, welcoming new productions focused on the domestic market (onion) or export (cotton, cashews, karite).

Beyond such a pessimistic conclusion, a more historical and more holistic vision should help us to understand why such a social-ecological system is proving so hard to change. One hypothesis may be found in its high degree of sophistication in the crossed definition of roles and activities of the various social groups. This may result from a stacking process of social and technological trade-offs that have occurred in the past among the groups and between practices. Such a historical process may have been historically driven by the search for social peace under strong environmental constraints. One has to remember here the biophysical complexity and the strong seasonality of the Delta ecosystem. Such a finding brings out the hypothesis of a 'rigidity trap' that tends to stay in place.

Such rigidity has been described in other cases in the world, it is for example a characteristic of the caste system in India. Gadgil and Malhotra (1983) documented 'that castes more directly dependent on natural resources had so organized their mode of subsistence as to avoid excessive overlap with other castes in their demands of various resources'. They also state that such organization 'favored cultural evolution ... leading to the sustainable use of natural resources'. Although Delta ethnic groups are not exactly regarded as castes, there are great similarities to the case described in India. Then one realizes how it would be unwise for policymakers to impose a plan of transformation of lifestyles and modes of production in the Delta in order to allow its adaptation to lower floods in the near future.

CONCLUSION

Therefore, it seems that the future of the Inner Niger Delta can only be ensured by a careful combination of adaptation policies and conservation policies. Adaptation policies should be very gradual, envisaged at village territory scale and in a participatory way for the emergence of new compromises between communities. As for conservationist policies, they should aim to save the motor of the ecosystem productivity, i.e. the annual flooding of the Niger River. It must be kept in the largest extent possible given the overall development needs of the basin.

This should guide policy makers and donors at the launch of new hydraulic development projects in the Niger River Basin, reminding them that they are responsible for safeguarding a system that is one of the last of its type in the world to run.

REFERENCES

ABN (Autorité du bassin du Niger). 2007. 'Plan d'Action de Développement Durable du bassin du Niger'. Report synthesis.

Baumann, E., Fay, C., Kassibo, B. 1994. 'Systèmes de production et d'activité: trois études régionales'. In Quensière, J (ed.), La Pêche dans le Delta Central du Niger. Paris: IER-ORSTOM-Karthala, pp. 305–405.

Benjaminsen, T., Bâ, B. 2009. 'Farmer-herder conflicts, pastoral marginalisation and corruption: a case study from the inland Niger delta of Mali'. The Geographical Journal 175(1): 71–81.

Daget, J., Ba, A.H. 1955. L'Empire Peul du Macina. Bamako: IFAN.

Fay, C. 1995. 'Car nous ne faisons qu'un: Identités, équivalences et homologies au Maasina (Mali)'. Cahier des Sciences Humaines 31(2): 427–56.

Fossi, S., Bakouan, N.D., Traoré, A., Barbier, B. 2014. 'Variabilité de la crue du fleuve et options agricoles dans le delta intérieur du Niger: riziculture ou bourgouculture ?' *Sciences Eaux and Territoires. Special issue* 15: 1–6.

Ferry, L., Muther, N., Coulibaly, N., Martin, D., et al. 2012. *Le fleuve Niger: de la forêt tropicale guinéenne au désert saharien*. Montpellier, Bamako: IRD et Unesco. Available at http://unesdoc.unesco.org/images/0021/002155/215565f.pdf

Gadgil, M., Malhotra, K.C. 1983. 'Adaptive significance of the Indian caste system: an ecological perspective'. *Annals of Human Biology* 10: 465–78.

Gallais, J. 1967. *Le Delta intérieur du Niger, étude de géographie régionale*. Paris: Larose.

Hertzog, T., Adamczewski, A., Molle, F., Poussin, J.C., Jamin, J.-Y. 2012. 'Ostrich-like strategies in sahelian sands? Land and water grabbing in the Office du Niger, Mali'. *Water Alternatives* 5(2): 304–21.

Hiernaux, P., Diarra, L. 1986. *Bilan de cinq années de recherches (1979–1984) sur la production végétale des parcours des plaines d'inondation du fleuve Niger au Mali central*. Bamako: CIPEA.

Jul Larsen, E., Kassibo, B. 2001. 'Fishing at home and abroad: access to waters in Niger's Central Delta and the effects of work migration', in Benjaminsen, T., Lund, C. (eds), *Politics, Property and Production in the West African Sahel. Understanding Natural Resources Management*. Stockholm: Elanders Gotab Publications, pp. 208–33.

Kassibo, B. 1994. 'Histoire du peuplement humain'. In Quensière, J. (ed.), *La pêche dans le delta central du Niger: approche pluridisciplinaire d'un système de production halieutique*. Paris: IER-ORSTOM-Karthala, pp. 81–98.

Kodio, A., Morand, P., Dienepo, K., Laë, R. 2002. 'La dynamique de la pêcherie du delta intérieur du Niger revisitée à la lumière des données récentes. Implications en termes de gestion'. In Orange, D. et al. (eds), *Gestion Intégrée des Ressources Naturelles en Zone Inondable Tropicale*. Paris: IRD, pp. 431–53.

Kuper, M., Maïga, H. 2002. 'Commercialisation du riz traditionnel dans le Delta intérieur du Niger (Mali)'. In Orange, D. et al. (eds), *Gestion Intégrée des Ressources Naturelles en Zone Inondable Tropicale*. Paris: IRD, pp. 639–60.

Laë, R. 1994. 'Effect of drought, dams and fishing pressure on the fisheries of the Central Delta on the Niger River'. *International Journal of Ecology and Environmental Sciences* 20: 119–28.

Laë, R., Williams, S., Malam Massou, A., Morand, P., Mikolasek, O. 2004. 'Review of the present state of the environment, fish stocks and fisheries of the River Niger (West Africa)'. In Welcomme, R., Petr, T. (eds), *Proceedings of the Second International Symposium on the Management of Large River for Fisheries 1*. Bangkok: RAP Publication 2004/16, pp. 199–227.

Le Barbé, L., Lebel, T., Tapsoba, D. 2002. 'Rainfall variability in West Africa during the years 1950–90'. *Journal of Climate* 15: 187–202.

Liersch, S., Cools, J., Kone, B., Koch, H., et al. 2013. 'Vulnerability of rice production in the Inner Niger Delta to water resources management under climate variability and change'. *Environmental Science and Policy* 34: 18–33.

Mahé, G., Orange, D., Mariko, A., Bricquet, J., 2011. 'Estimation of the flooded area of the Inner Delta of the River Niger in Mali by hydrological balance and satellite data'. *Hydro-climatology: variability and Change* July 2011(IAHS Publ. 344).

Mainguy, C., Ballo, B., Bidou, J.E., Cissé, I., et al. 2015. 'Vulnérabilités et politiques publiques en milieu rural au Mali: les exemples du Bassin cotonnier et du Delta intérieur du Niger'. In Brunet-Jailly, J., Charmes, J., Konaté, D. (eds), *Le Mali contemporain*. Paris: Editions IRD et de Tombouctou, pp. 198–229.

Marie, J., Morand, P., N'djim, H., 2007. *Avenir du fleuve Niger – The Niger River's Future*. Paris: IRD Editions.

Morand, P., Kodio, A., Andrew, N., Sinaba, F., Lemoalle, J., Béné, C. 2012. 'Vulnerability and adaptation of African rural populations to hydro-climate change: experience from fishing communities in the Inner Niger Delta (Mali)'. *Climatic Change* 115: 463–83.

Nielsen, J.Ø., Reenberg, A. 2010. 'Cultural barriers to climate change adaptation: A case study from Northern Burkina Faso'. *Global Environmental Change* 20 (1): 142–52.

Portères R. 1950. 'Vieilles agricultures de l'Afrique intertropicale. Centres d'origine et de diversification variétale primaire et berceaux d'agriculture antérieurs au XVIe siècle'. *L'Agronomie tropicale* 5(9/10): 489–507.

Quensière, J. (ed.) 1994. *La Pêche dans le Delta Central du Niger*. Paris: IER-ORSTOM-Karthala.

UA/NEPAD 2009. 'Plan d'action pour l'Afrique de l'UA/NEPAD 2010–2015 (Union africaine/Nouveau partenariat pour le développement de l'Afrique)'. Available at www.nepad.org

UEMOA 2013. 'Rapport national sur les enquêtes cadres "Pêche artisanale continentale"'. Accessed on http://atlas.statpeche-uemoa.org/

UNICEF (accessed on 12/2014): http://www.unicef.org/french/infobycountry/mali_statistics.html

Weigel, J.Y., Stomal, B. 1994. 'Consommation, transformation et commercialisation du poisson'. In Quensière, J. (ed.), *La pêche dans le Delta Central du Niger*. Paris: IER-ORSTOM-Karthala, pp. 165–89.

Welcomme R.L. 2002. 'An evaluation of tropical brush and vegetation park fisheries'. *Fisheries Management and Ecology* 9: 175–88.

WCD 2000. *Dams and development. A new framework for decision making. The report of the World Commission on Dams*. London: Earthscan.

Zwarts, L., van Beukering, P., Kone, B., Wymenga, E. (eds) 2005. *The Niger, a lifeline. Effective water management in the Upper Niger Basin*. Amsterdam: RIZA, Lelystad.

Part IV   Contemporary Water and Food Regimes

18 'Where there is Water, there is Fish'. Small-Scale Inland Fisheries in Africa: Dynamics and Importance

Jeppe Kolding, Paul A.M. van Zwieten and Ketlhatlogile Mosepele

INTRODUCTION

The importance of fish, and in particular small fish, for sustainable and healthy livelihoods in Africa, as well as their strong relationship with climate driven water dynamics, is generally undervalued and little understood: most small fish species are consumed locally and go unrecorded in catch statistics. Fish are vital providers of animal protein and indispensable micronutrients in many African societies, but modern food production and policies are almost unilaterally associated with terrestrial agriculture and livestock production. The majority of fish species are carnivores and by primarily targeting large adult fish, such as Nile perch in Lake Victoria, humans feed about two trophic levels higher in water than on land, which in terms of energy is a very inefficient utilization of available food. Small fish are lower in the food web and much more abundant and productive but their capture is inhibited by outdated selectivity regulations. Small fish are ubiquitous in all aquatic environments from large lakes to seasonal ponds and their productivity is highly correlated with rainfall patterns and water level fluctuations. Catching small fish, which are sun-dried and consumed whole, is the most high-yielding, eco-friendly and nourishing way of utilizing the natural food that aquatic ecosystems provide.

The role and importance of fish in securing food and nutrition for humans, particularly in developing countries, has frequently been overlooked. Fisheries and aquaculture are often arbitrarily separated from other parts of the food and agricultural systems in governance, food security studies and policy making (HLPE 2014). Furthermore, nourishment is no longer only a question of calorie availability and access: food security should be broadened to include alimentary aspects as well. There is now robust evidence that a lack of essential micronutrients such as zinc and vitamin A affect hundreds of millions of malnourished people around the world (IPCC 2014). Nutritional value is particularly important in sub-Saharan Africa where approximately 28 per cent of all deaths are attributed to malnutrition (Benson 2008). Fish is especially rich in essential Omega-3, long-chain polyunsaturated fatty acids and micronutrients, including bioavailable calcium, iron and zinc (HLPE 2014, Longley et al. 2014), which all play a critical role in cerebral development, immune defence systems and general health. Fish is an important source of animal protein in human consumption (Delgado et al. 2003), and features prominently in the diet of many people in large parts of Africa. But even when per capita fish consumption is low, small quantities of fish can have a significant positive nutritional impact by providing essential amino acids, fats and micronutrients that are scarce in vegetable-based diets (FAO 2012). However, the importance of fish as regards sustenance and household economy is largely neglected (Béné et al. 2006).

For the most part, fisheries policies are approached from the connotations of fish stocks being overfished, exploited unsustainably and in dire need of management (Allan et al. 2005). But in fact we have only limited knowledge about the actual status of inland African fisheries and the important interplay between climate, water dynamics, fisheries and food production. Fish provides the main source of animal protein for some 200 million people on the African continent (Heck et al. 2007). Fisheries also provide a direct source of livelihoods to over 10 million Africans, while five to ten times more engage in fisheries as a secondary activity for food security in rural areas. The official statistics of inland landings in Africa are around 2.5 million tonnes per year; of which Lake Victoria alone produces 1 million tonnes. But the actual total catches are likely to be significantly higher (perhaps closer to 20 million tonnes), assuming that the total area of freshwater resources (lakes, rivers, reservoirs, floodplains and swamps) on the continent is approx. 1.3 million km^2 (Lehner & Döll 2004, de Graaf et al. 2012) and the average annual production of fish is around 150 kg/ha (Marshall & Maes 1994, Kolding & van Zwieten 2006). As a rough rule of thumb it is highly probable that the official records are underestimated by an order of magnitude. Inland fisheries and aquaculture contribute at least 40 per cent to the world's production of fish at the global level (Dugan et al. 2007, Lynch et al. 2016) produced from a tiny proportion (approximately 0.2 per cent) of the global aquatic surface area (Kolding & van Zwieten 2006).

WATER, FOOD AND FISH

In this chapter we will outline and document the importance of fish and fisheries for sustainable and healthy livelihoods in Africa, as well as their close relationship with climate driven water dynamics. Actually, the environment plays a much more important role for fish production than the largely futile attempts of regulating fishing activity. On the contrary, current dominant management approaches, based on the ideology of controlling the fishing pattern, mainly have adverse effects on both the ecosystem resilience and the potential production (Kolding & van Zwieten 2011, 2014). This may have serious consequences for the growing focus on food security and ecosystem resilience in face of climate change. One of the key recommendations of the latest IPCC Assessment Report is that key adaptations for fisheries and aquaculture shall include policy and management to maintain ecosystems in a state that is resilient to change (IPCC 2014). The overwhelming ecosystem dependent factor for fisheries is water, and how natural processes around water resources are distributed, managed and conserved. After all, in Africa, like in Asia (van Zalinge et al. 1998), the saying goes 'where there is water, there is fish'.

THE IMPORTANCE OF SMALL FISH

One reason for the high level of production in African inland fisheries is the focus on small fish. The most productive fisheries are all aimed at small fish species weighing from one to a few grams. The 'Kapenta' (*Limnothrissa miodon* and *Stolothrissa tanganicae*) fisheries in lakes Tanganyika, Kariba, Cahora Bassa and Kivu are the most important in all these lakes. Likewise the 'Chisense' (a mixture of *Potamothrissa acutirostris*, *Microthrissa stappersii* and *Poecilothrissa moeruensis*) fishery in Lakes Mweru, Bangweulu and Mweru-Wa-Ntipa; the 'Usipa' (*Engraulicypris sardella*) and 'Kambuzi' (many small demersal haplochromine species) fisheries in Lake Malawi, Chilwa and Malombe; the 'Dagaa', 'Omena' or 'Mukene' (*Rastrineobola argentea*) fishery in Lake Victoria, the 'Mazeze' (mainly small sized cyprinids) in the Okavango Delta; and the similar 'Kapesa' in the Bangweulu swamps, are all high-yielding and extremely important for local consumption. Most of these species go unreported into the local markets and are not captured in official catch statistics. Processing and conservation is straightforward as they are simply sun-dried in a few days, in contrast to larger fish which need gutting, salting or smoking for preservation. The simple preservation technology and easy transportation makes them available at most local markets at low cost where they are sold in small portions by weight, often fetching the same price as large fish (Brummett 2000). Heaps of small fish (Figures 18.1–6) are ubiquitous on local markets far from the original source: Lake Victoria Dagaa is found all over the riparian countries (Hoffman 2010); Lake Mweru Chisense and Lake Kariba Kapenta are found in all large cities in Zambia and southern Democratic Republic of Congo (Overå 2003, IOC 2012). As small fish are sun-dried whole, with heads, bones and organs intact, they are a concentrated source of multiple essential nutrients, in contrast to large fish which are usually not eaten whole and therefore do not contribute as much to micronutrient intake (Longley et al. 2014).

Figure 18.1. Kapesa (mixture of small fish) being sun-dried in Bangweulu swamps, Zambia (© Carl Huchzermeyer).

Figure 18.2. Dagaa (*Rastrineobola argentea*) being sun-dried at Lake Victoria, Tanzania (© Modesta Medard).

Figure 18.3. Dagaa being sun-dried and packed at Lake Victoria, Tanzania (© Modesta Medard).

Figure 18.4. Packed sun-dried dagaa distributed to local markets (© Modesta Medard).

Figure 18.5. Sun-dried dagaa at a local market, Tanzania (© Modesta Medard).

Figure 18.6. Traditional Kenyan dish of sun-dried omena with maize porridge (Ugali) (photo reproduced with permission from Msupa, Nairobi http://www.msupa.com/).

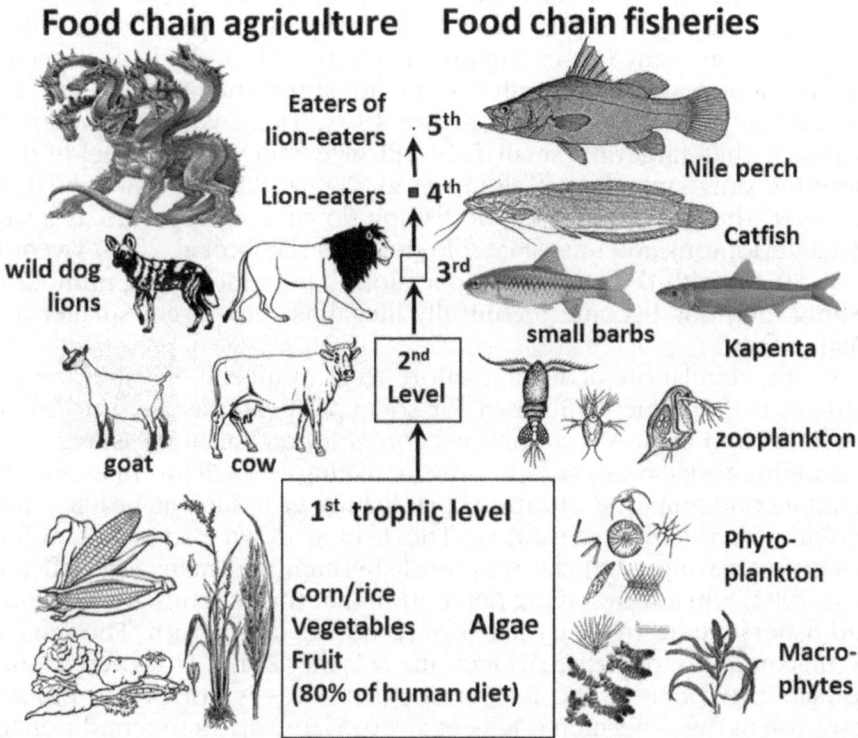

Figure 18.7. A comparison of the terrestrial agricultural and the aquatic fisheries food chains. The average human trophic level (TL) is 2.21 (Bonhommeau et al. 2013), meaning we are about 80 per cent terrestrial vegetarians (only slightly higher than aquatic zooplankton). In contrast, we are feeding about two TLs higher in most fisheries targeting large fish, resulting in around 99 per cent of the corresponding energy being lost in transfer inefficiency (modified from Duarte et al. 2009).

THE CONTROVERSIAL 'FISHING DOWN PROCESS'

In addition to the targeted fisheries on the small, mostly pelagic, high productive fish species listed above, a general feature of African small-scale fisheries is the so-called fishing down process (Welcomme 1999), which also results in catching small fish. This process is based on the serial reduction in the sizes of individual fish and fish species as fishing pressure increases, by a corresponding successive reduction in mesh sizes, and diversification of fishing gear and methods. The process is induced by the inevitable decline in individual catch rates as the number of people fishing increases with general population increase (Jul-Larsen et al. 2003, Kolding et al. 2014a). The individual decline, however, is accompanied by a corresponding rise in the total catch from the combined fishery as smaller, faster growing, more productive species and sizes replace larger, slower growing, less productive ones. In addition, as many fish-eating predators are among the larger species, the reduction of these will boost the abundance of species and sizes lower in the food chain (Figure 18.7). There is therefore strong evidence that targeting small fish will give much higher yields than targeting only large fish (Kolding et al. 2003a, 2003b). Unfortunately, however, the general process of 'fishing down' is interpreted as a sign of a deteriorating and unsustainable situation (Pauly et al. 2008, Tweddle et al. 2015), with the added complication that an increasing number of fishing methods become technically illegal as they target smaller and smaller fish.

As the regulation of fishing effort (e.g. numbers of fishermen) is difficult to implement in Africa for socio-political reasons, the fishery regulations in most African nations consist of technical measures, such as minimum legal mesh sizes to prevent fishing of small juveniles and the general condemnation of unselective fishing gear such as beach seines (Kolding & van Zwieten 2011). The fishing down process therefore causes increasing conflicts between fishermen and managers (Misund et al. 2002) and a snowballing perception that the fisheries are 'doomed' and fishers are destroying their own resources in line with 'The tragedy of the commons' doctrine (Welcomme & Lymer 2012). However, as shall be elaborated below, the fishing down process is not only a rational response of the fishermen (Plank et al. 2015), but also a precondition for maximizing food production while maintaining the health and structure of the fished ecosystem (Kolding & van Zwieten 2014). Thus, in spite of rules and regulations, the overall result of these ongoing processes is that African inland fisheries are increasingly providing large amounts of small fish (sizes and species), which from a nutritional point of view is highly beneficial (Kawarazuka & Béné 2011, Beveridge et al. 2013, Longley et al. 2014). It is also highly advantageous from an ecological point of view as catching small fish in proportion to their productivity conserves the aquatic ecosystem structure (Law et al. 2012, 2014, Kolding & van Zwieten, 2014), as well as maintains the terrestrial ecosystem by reducing the cutting of firewood necessary for smoking and preserving large fish. There are even indications that access to affordable fish protein is contributing to the conservation of endangered mammal species hunted for bush meat (Wilkie et al. 2005, Junker et al. 2015). Still, the major governance focus on inland fisheries at the moment is not on their essential contribution to food security, but on their alleged self-inflicted destruction of the resources from overfishing and illegal fishing methods. Most management effort at present seems oriented at constraining fishing, particularly on small juvenile fish, instead of studying and understanding the dynamics of local fishing patterns, and quantifying their importance for nutrition and impact on the ecosystem.

THE TERRESTRIAL AND AQUATIC FOOD CHAIN

Humans have been both farmers and fishers for millennia, but modern food production is almost unilaterally associated with terrestrial agriculture. Most people are predominantly vegetarians; the average trophic level is 2.21 (Bonhommeau et al. 2013), which means that around 80 per cent of our food is from trophic level 1 (plants), and the rest is mainly meat-based from our domesticated herbivore livestock animals (trophic level 2). It is therefore easily forgotten that the life history of fish is quite different from the farmed creatures we use in animal husbandry all over the world. All the livestock we farm for food are relatively large herbivores that can feed directly on primary vascular vegetation, or so-called higher plants, which dominate the terrestrial ecosystem. The agricultural food chain is therefore short, consisting chiefly of two trophic levels (Figure 18.7), and nobody would dream of farming terrestrial carnivores for human food production as we lose about 90 per cent of the available energy each time we move one trophic level up the chain. The aquatic food chain, however, is fundamentally different. Most importantly, the majority of primary producers (plants) are microscopic suspended algae, and only small amounts of larger vascular macrophytes inhabit the fringes of the aquatic ecosystems. In general, organisms have to be small to consume minuscule algae, and the major herbivores (cows and goats) of the water are therefore tiny filter-feeding zooplankton, though there are some important categories of fish that can feed on detritus and algae (for example, tilapia and carp species).

However, the majority of all fish species are primarily carnivorous but, in contrast to terrestrial predators, they have very high fecundities and minuscule progenies that all start their life at the bottom of the food chain. In short, fish breed like plants with lots of seeds but feed like lions. Consequently, aquatic communities are different from terrestrial in that nearly all fish, even the largest predators, start their life as small prey for larger fish (Kolding et al. 2014b). This means that most fish during their ontogeny often traverse across several trophic levels before reaching adulthood. By primarily targeting large adult apex predators, such as Nile perch in Lake Victoria, humans feed about two trophic levels higher in water than on land, which in terms of energy is a very inefficient utilization of available food. Nearly half of the global primary production is aquatic, and yet only about 2 per cent of the global human food is derived from fisheries and aquaculture (FAO 2006). In theory this discrepancy should envisage a huge potential for increase, but looking at the difference between the aquatic and terrestrial food chain (Figure 18.7), it is only possible if we target lower trophic levels, which means 'fishing down' and catching small sizes, as they do in Africa.

THE DOOMSDAY PERCEPTIONS, THREATS AND LACK OF RECOGNITION

Notwithstanding that the reported African inland capture fisheries are still steadily rising at about 3.7 per cent per year, there is a general pessimistic view of the future of inland fisheries following the numerous threats to aquatic ecosystems posed my man's activities (Welcomme & Lymer 2009, Lynch et al. 2016). This 'inland fisheries are doomed' view is supported by many individual studies and reports from all continents including Africa (Allan et al. 2005, Friend et al. 2009). Catches are alleged to be falling, species disappearing and many of the symptoms of chronic overfishing at the level of individual species or whole communities are being reported (Allan et al. 2005, Welcomme et al. 2010). The problem with many of these statements, however, is that they are based on highly ambiguous indicators (Kolding et al. 2014a). The omnipresent reports of 'declining catches' are a typical example. The statement almost never differentiates between individual catches (which will always decline as more and more fishers share the same resource), and the total catches (which mostly increase and rarely decline). When asking a fisher if his catches have declined, he will almost invariably agree as this is in accordance with his own personal observations: fishermen live in a world of eternally decreasing individual catches as overall effort increases. They are rarely aware of the fact that the combined catches over the whole system at the same time have increased. Thus reported inland catches in Africa are still increasing linearly (FAO 2012), and there are no examples of decreasing total catches from any system outside the North Atlantic region that can be attributed unequivocally to fishing. When total catches in small scale fisheries are falling, it is always in association with decreased amounts of water or deteriorating water quality and aquatic habitats (droughts, floodplain conversion, river regulation, abstractions and dams).

Another example of ambiguous indicators is the typical observed decrease in the size of fish caught (as described above), which in general is interpreted as a sign of overfishing (Kolding et al. 2014a). However, a decrease in size, as in the common 'fishing down process', may be a sign of fishing, but not necessarily a sign of overfishing. Actually, it can just as well be interpreted as a healthy sign of a redistribution of fishing effort over larger parts of the fish community and thereby keeping the structure intact (Essington et al. 2006, Kolding & van Zwieten 2014, Kolding et al. 2015). Still, the general perception favours a sense of hopelessness that leads to the neglect of the sector as a whole and renders policy makers to focus on other, more optimistic, sectors for growth and development (Welcomme & Lymer 2009).

As a result, the important and beneficial contribution of wild caught inland fish to food security has been largely ignored and undervalued (Mills et al. 2011), priorities for food studies have been switched to other sectors, and aquaculture is being promoted as the solution to sustaining production in the face of the perceived inevitable decline and eventual disappearance of freshwater fish stocks (Welcomme & Lymer 2012). This view is prominent in many African countries and, together with the general unawareness of the importance of fish for human nutrition (HLPE 2014), has led to a lack of means assigned to inland fisheries, a lack of informed approaches to managing many aspects of the resources, as well as an apparent failure to include the sector in national and regional development policies (Welcomme & Lymer 2012, FAO 2012, HLPE 2014).

Adding to this pessimistic view on small-scale fisheries in Africa, as elsewhere, is the large number of perceived and real anthropogenic threats to inland aquatic systems. Foremost among these are (Welcomme & Lymer 2012):

(a) *Poor fishery management*: Including uncontrolled and excessive fishing, increasing use of illegal gear (catching small fish!), and the introduction of invasive exotic species.
(b) *Water abstractions*: There is a growing trend in Africa for rivers to be regulated and flows to be diverted for irrigation either directly or from reservoirs (Dudgeon et al. 2006, Richter et al. 2010).
(c) *Land drainage*: There is an increasing trend to drain wetlands and separate floodplains from the river channel. This results in a loss of habitat and breeding areas of many fish species and will negatively affect productivity (Kolding & van Zwieten 2012).
(d) *Dam construction*: With the rising demands of energy there is a surge in proposals for, and construction of, large dams. The impacts of such dams on the fish fauna downstream are rarely assessed although there is often a compensatory effect with the creation of new fisheries in the reservoirs upstream (Kolding & van Zwieten 2006, Richter et al. 2010).
(e) *Pollution/eutrophication*: Pollution has important local effects in rivers and in lakes. In lakes eutrophication is an increasing threat from the growing levels of human population around their shores, a lack of proper waste water treatment system or ancient agricultural practices involving seasonal burning of vegetation on fallow land (Tamatamah et al. 2005). Lake Victoria, the largest lake in Africa, is a typical example of this development (Kolding et al. 2008).
(f) *Climatic variability/change*: Climatic variation has always been a severe problem especially in the drought-prone belts of the Sahel and southern African region. There is a clear correlation between fish productivity and rainfall in all African water bodies (Jul-Larsen et al. 2003, Kolding & Zwieten 2012).

'FISH COME WITH THE RAIN'

For local fishermen, their biggest concern is about water. 'Where there is water there is fish' or 'Fish come with the rain' are often heard statements when asking African fishermen about the drivers of fish production. A good example are the many isolated endorheic water bodies in Africa, such as Lake Ngami (Botswana), Lake Chilwa (Malawi), Lake Mweru Wa' Ntipa (Zambia) or Lake Liambezi (Namibia), which periodically dry out completely, becoming muddy swamps or even dust pans. Upon refilling, however, during years of good rains, the fishery immediately bounces back and resumes very high productivity within a very short time – usually less than a year. At the moment, both Lake Ngami, which was dry from 1982 to 2002, and Lake Liambezi, which was dry between 1986 and 2009, are highly productive and characterized by outstanding fish yields. Currently, both fisheries are exporting large quantities of fish to the Zambian and Democratic Republic of Congo markets. These lakes are typical examples of how rain and water are the main factors controlling fish production in inland fisheries (Jul-Larsen et al. 2003). Africa is well-known as the continent containing some of the largest lakes in the world (Victoria, Tanganyika, Malawi and Turkana), but in relative terms the high number of small, often ephemeral, water bodies, dams, swamps, marshes and floodplains are much more important in terms of food production (Anderson 1989, Marshall & Maes 1994). Due to the high surface-to-volume ratio, small shallow water bodies, are much more productive and have a much higher standing fish biomass per unit area than large lakes (Marshall & Maes 1994).

However, following the prevailing climatic conditions, large parts of Africa have fluctuating and often unpredictable rainfall patterns (Nicholson 1996), which together with a high evaporation rate make many small water bodies liable to dry out intermittently, or at best result in pronounced water level fluctuations. Such unstable environments which constantly undergo successive resetting, are only inhabitable for small, short-lived and fast growing species with high generation turnover. Consequently, the fisheries in small water bodies are characterized by small fish and high seasonal and inter-annual variability. However, alternating wet and dry conditions seems only to boost productivity as it recycles nutrients through oxidation and leaching (Kolding & van Zwieten 2006, van Zwieten et al. 2011). Another example is the floodplain fisheries, which are considered among the most productive in the tropics. Here the strong positive relationship between hydrology and fish production has long been recognized (Welcomme 1979, Junk et al. 1989, Welcomme et al. 2006). More recently, the general dependency of larger lakes and reservoirs to the ambient hydrological regime has been documented (Kolding & van Zwieten 2012).

In general, for all water bodies, there is evidence that fish productivity increases exponentially with the amplitude of annual water fluctuations relative to the mean depth. Thus, water alone is not enough for high

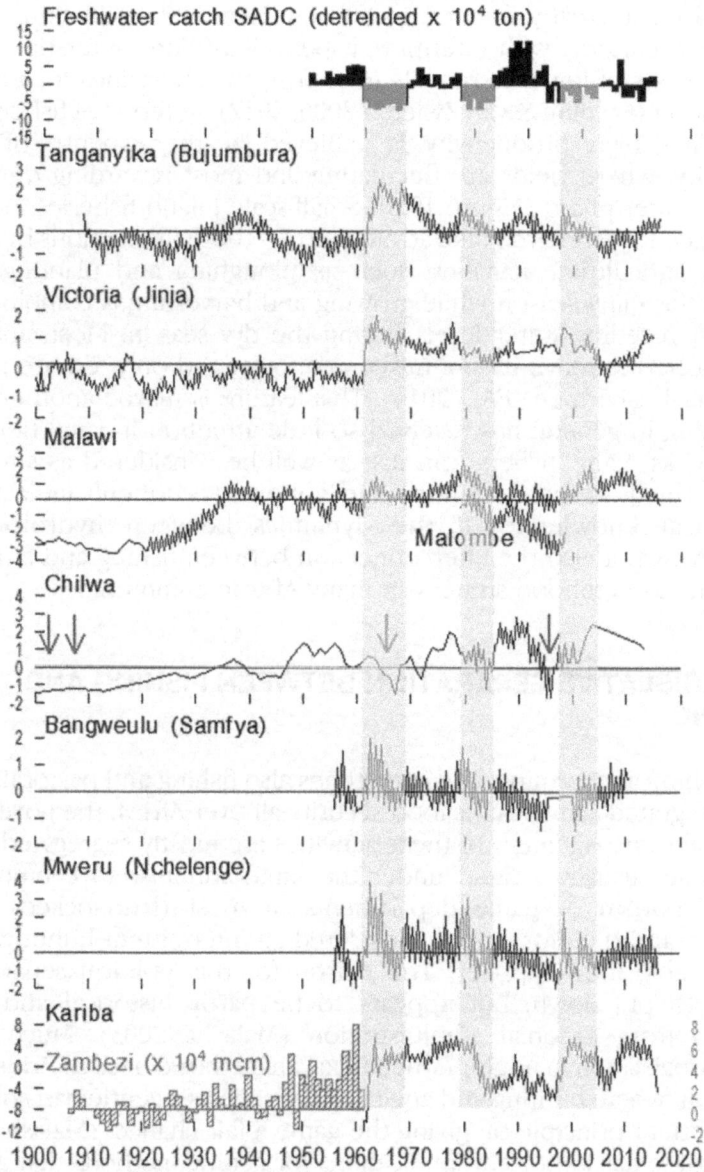

Figure 18.8. Relative water levels of Lakes, Tanganyika, Victoria, Malawi, Malombe, Chilwa, Bangweulu, Mweru and Kariba expressed as deviations from the long-term mean of annual mean levels over the period for which data were available. Light gray bars are the deviations of the 89 year mean annual total inflow of the Zambezi at Victoria falls (mcm = million cubic meter). Arrows indicate the years Lake Chilwa was reported to be dry. Malombe is hydrologically considered a satellite of Malawi: when this lake has low stands Malombe completely dries up. The top panel shows the variability around the trend of the total fish catches of the SADC region with grey bars comparing periods of low catches relative to the trend (modified from Jul-Larsen et al., 2003).

productivity; a greatly enhancing effect is seasonal oscillations (driven by the rain pattern) with intermittent periods of flood interspersed with fallow periods of low water levels to expose the shorelines for terrestrial rejuvenation (Kolding & van Zwieten 2006, 2012). In terms of fisheries, this means that high productivity is achieved at the expense of strong seasonality, where yields are fluctuating and most rewarding during the receding water phase (Figure 18.8). Small-scale inland fisheries, however, are adapted to – and even take advantage of – these fluctuations by shifting between agricultural activities, such as ploughing and planting at the onset of the rainy season while growing and harvesting is combined with fishing at receding water levels during the dry season. Most small-scale African fishermen thus have a mixed economy, and only few are full-time commercial fishermen (FAO 2014). This feature is maybe another reason why fishing in general has received so little attention in rural household economy, as many 'fishers' can just as well be considered as small-scale farmers. The 'Fish comes with the rain' aphorism is not only indicating the deep local knowledge of the dynamics between hydrology and productivity, but also the interconnection between fishing and farming as an integrated livelihood strategy in many African communities.

THE LEGISLATIVE SEPARATION BETWEEN FISHING AND FARMING

While fishing and farming, and sometimes also fishing and pastoralism, are often integrated activities for food security all over Africa, the governance, policies and management of these activities are mostly segregated. Inland fisheries are usually located under the same administrative umbrella as wildlife, tourism or game departments in most (land-locked) African countries, and therefore more considered an intermittent hunting activity than a stable food supplier. The reason for this political separation is difficult to pin down, but appears to be partly historical and mainly inherited from Colonial administration (Malasha 2003). Much of the fisheries legislation in Anglophone Africa can be traced back to British game legislation, where hunting and angling were seen as a gentleman sport with the important principle of 'giving the game a fair chance' (Malasha 2003). This attitude has important implications for fishing methods that are seen as 'herding', 'indiscriminate' and 'unselective' and considered particularly unethical when immature individuals are targeted. In addition, during the last decade of the Colonial period, a new and ground-breaking fisheries theory was developed in the UK, which rapidly became the doctrine of modern rational fisheries management (Kolding & van Zwieten 2011). The theory (Beverton & Holt 1957) stipulated minimum size limits on exploited species in order to maximize yields, and the principle was soon exported to the colonies (Beverton 1959), resulting in widespread mesh-size regulations and the condemnation of catching small and immature fish (Kolding & van Zwieten 2011). Traditionally, however, African fishers have always targeted all sizes of fish as there is no selective preference for large sizes as in Europe (Tsikliras & Polymeros 2014).

Together with the above described 'fishing down process' as a result of increased effort, the overall outcome is an increase in conflict and distrust between managers and fishers (Misund et al. 2002) and a growing perception in the wider society that the traditional fishing pattern is destructive: fishers are seen as irresponsible law-breaking buccaneers instead of good citizens providing much needed and essential nutritious food to supplement the starch-based staples from the fields. If African fishers had followed the current regulations, and only fished selectively on the legal large sizes, there would inevitably be a decrease in catch rates and in the average size of fish caught. It is therefore a great paradox in fisheries governance that the predictable result of fishing within legal requirements (a decrease in size and abundance) is simultaneously used as a diagnostic of unsustainability and depletion (Kolding et al. 2014a). Ironically, there is now increasing evidence that the traditional African balanced fishing pattern focusing predominantly on small fish is much more ecologically sustainable and provides more food than predicted by the conventional fisheries theory (Kolding & van Zwieten 2011, 2014, Law et al. 2012, 2013, 2014). Together, this calls for a revaluation of the current legislation and a need for a paradigm shift in management (Mosepele 2014, Kolding et al. 2015). However, the political division between fishing as a hunting activity in the wild and farming as a domestic food supplier may not only prevent such changes, but also help to explain the negative perceptions and recurrent management problems that African inland fisheries suffer from.

GOVERNANCE, LIVELIHOODS AND THE ECOSYSTEM APPROACH TO FISHERIES

Freshwater availability is predicted to be a serious constraint to achieve future food requirements (Postel 1998, Molden 2007). As a consequence we are likely to see an upcoming intensified competition for water, and most likely an increased regulation and extraction of natural inland water bodies. Most of the research on increased food production under limited water supply is focused on agriculture, with very little regard to the contribution from fish (Dugan et al. 2007, Molden et al. 2010). However, where there is water there are fish, and fish can often be integrated into water management systems built primarily for agriculture, such as reservoirs or irrigation schemes. Wetlands, floodplains and small shallow-water bodies are the most productive aquatic systems per unit area, and combined they outsize the larger lakes, reservoirs and rivers (Lehner & Döll 2004, Welcomme et al. 2010), but their relative contribution to food security in

A History of Water

Africa is largely unknown. Freshwater fish are probably among the most resilient harvestable natural resources, provided their habitat, including the quantity, timing, and variability of river flow are maintained (Welcomme & Petr 2004, Dugan et al. 2007). Thus, maintenance of wetlands and their livelihood values should be taken into consideration in future appraisals of food production. Water productivity increases with seasonal fluctuations in water level (Kolding & van Zwieten 2012), which indicates that it is important that wetlands and lakes maintain their natural cycles, while man-made dams and reservoirs can be regulated to mimic natural fluctuations, and thus increase productivity.

The livelihood benefits of fisheries are vital but mostly ignored or underestimated, although there is a dawning recognition of the importance of fish in the human diet composition. Small species and juvenile fish have the highest level of biological production (Morgan et al. 1980), the highest level of essential micronutrients since they are consumed whole, and are energy friendly since they only need sun-drying for preservation. Catching fish in proportion to their productivity, so-called 'balanced harvest' (Garcia et al. 2012), which favours small fish, is also the most sustainable and least disturbing to ecosystem structure and resilience (Law et al. 2012, 2013). Catching and consuming small fish traditionally has been practised in Africa for centuries, and there is still no significant price differential between small and large fish at local markets. Catching small fish, however, requires fishing methods, such as small mesh sizes, which are largely prohibited over most of the African continent. Unfortunately, the gear and size regulations are based on theoretical assumptions, which have rarely been verified empirically (Kolding & van Zwieten 2011). On the contrary, there is increasing evidence that the highly selective fishing pattern that current management regulations promote, are having serious adverse effects on the fish communities (Garcia et al. 2012).

Paradoxically, the currently favoured management policy is the so-called 'ecosystem approach' to fisheries (EAF, FAO 2003), where a key feature is conservation of the ecosystem structure and functioning (UNEP/CBD/COP 1998). EAF is now internationally accepted, and commitments to adopt an EAF are already written into many policy documents (Jennings et al. 2014). However, any selective removal of any part of the fish community will inevitably change the structure. There is thus an underlying incompatibility between the observed results of regulatory imposed selectivity and the aim of EAF. On the other hand, unregulated small-scale fisheries in Africa, or fisheries that defy administrative regulations, seem to adopt a fishing pattern that is largely unselective at the ecosystem level, and which provides high sustainable yields (Kolding & van Zwieten 2014). As fishing effort increases and the individual catch rate decreases, fishers will naturally adapt by diversifying their fishing methods and target species, so more gears types and different species, and smaller mesh sizes, thus smaller individuals. Together these two mechanisms will approach a more balanced harvest that will increase both the ecosystem resilience and the amount of food that can be harvested.

CONCLUSION

African small-scale inland fisheries are characterized by a high adaptability to seasonal and long-term environmental and climatic changes. They are essential providers of protein and micronutrients to millions of people, and some of them appear to be the best empirical examples we have of an ecosystem approach to fisheries in terms of keeping structure and function. The latter is mainly because the African market has no particular size preferences, and because small nutritious and prolific fish are in high demand. The anthropogenic threats to African inland fisheries are not so much poor fisheries management in the conventional sense, but rather a profoundly wrong ecosystem management approach that needs revision. The most important anthropogenic threats, however, are physico-chemical changes of the aquatic environments in the form of water abstraction, flood regulations, land drainage and eutrophication. As long as the aquatic environments are kept intact and productive, there will always be food. Where there is water there are fish.

REFERENCES

Allan, J.D., Abell, R., Hogan, Z., Revenga, C., Taylor, B. Welcomme, R.L., Winemiller, K. 2005. 'Overfishing in inland water'. *BioScience* 55: 1041–51.

Anderson, A. 1989. 'The development and management of fisheries in small water bodies'. In Giasson, M., Gaudet, J.L. (eds), Summary of *Proceedings and selected papers. Symposium on the development and management of fisheries in small water bodies. Accra, Ghana, 7–8 December 1987*. FAO Fisheries Report, 425: 15–19.

Béné, C., MacFadyen, G., Allison, E.H. 2006. 'Increasing the Contribution of Small-Scale Fisheries to Poverty Alleviation and Food Security'. *Technical paper 481*, FAO Fisheries.

Benson, T. 2008. 'Improving nutrition as a development priority; addressing undernutrition in national policy processes in Sub-Saharan Africa'. *Research Report 156*, International Food Policy Research Institute.

Beveridge, M.C.M., Thilsted, S.H., Phillips, M.J., Metian, M., Troell, M., Hall, S.J. 2013. 'Meeting the food and nutrition needs of the poor: the role of fish and the opportunities and challenges emerging from the rise of aquaculture'. *Journal of Fish Biology* 83: 1067–84.

Beverton, R.J.H. 1959. 'Report on the State of the Lake Victoria Fisheries'. Fisheries Laboratory, Lowestoft: Mimeo.

Beverton, R.J.H., Holt, S.J. 1957. 'On the dynamics of exploited fish populations', Fishery Investigations, Series II, Vol. 19. London, G.B.: Ministry of Agriculture 553.

Bonhommeau, S., Dubroca, L., LePape, O., Barde,J., Kaplan, D.M., Chassot, E., Nieblas, A.-E. 2013. 'Eating up the world's foodweb and the human trophic level'. Proceedings of the National Academic of Sciences: 20617–20.

Brummett, R.E. 2000. 'Factors influencing fish prices in Southern Malawi'. Aquaculture 186: 243–51.

de Graaf, G., Bartley, D., Jorgensen, J., Marmulla, M. 2012. 'The scale of inland fisheries, can we do better? Alternative approaches for assessment'. Fisheries Management and Ecology 22(1): 64–70.

Delgado, C.L., Wada, N., Rosegrant, M.W., Meijer, S., Ahmed, M. 2003. 'Outlook for fish to 2020: meeting global demand'. IFPRI Food Policy Report. Penang: WorldFish Center; Washington, DC: IFPRI.

Duarte C.M., Holmer M., Olsen Y., Soto D., Marbà, N. et al. 2009. 'Will the Oceans Help Feed Humanity?' BioScience 59: 967–76.

Dudgeon, D., Arthington, A.H., Gessner, M.O., Kawabata, Z.-I., Knowler, D.J., Lévêque, C. et al. 2006. 'Freshwater biodiversity: importance, threats, status and conservation challenges'. Biological reviews of the Cambridge Philosophical Society Cambridge Philosophical Society 81: 163–82.

Dugan, P., Sugunan, V.V., Welcomme, R.L., Béné, C. et al. 2007. 'Inland fisheries and aquaculture'. In Molden, D., Water for Food, Water for Life: A Comprehensive Assessment of Water Management in Agriculture. London: Earthscan; Colombo: International Water Management Institute, pp. 459–83.

Essington, T.E., Beaudreau, A.H., Wiedenmann, J., 2006. 'Fishing through marine food webs'. Proceedings of the National Academy of Sciences 103: 3171–5.

FAO 2003. 'The ecosystem approach to fisheries'. FAO Technical Guidelines for Responsible Fisheries 4(Suppl. 2). Rome: FAO.

———— 2006. The State of World Aquaculture. FAO Fisheries technical paper 500. Rome: FAO.

———— 2012. The State of World Fisheries and Aquaculture. Rome: FAO.

———— 2014. Fisheries and Aquaculture topics. Small-scale and artisanal fisheries. Topics Fact Sheets. Text by Jan Johnson. Bibliographic citation [online]. Rome. Updated 27 May 2005. [Cited 14 November 2014]. http://www.fao.org/fishery/topic/14753/en

Friend, R., Arthur, R., Keskinen, M. 2009. 'Songs of the doomed: the continuing neglect of capture fisheries in hydropower development in the Mekong'. In Molle, F., Tira Foran, T., Käkönen, M. (eds), Contested waterscapes in the Mekong Region. London: Earthscan.

Garcia, S.E., Kolding, J., Rice, J., Rochet et al. 2012. 'Reconsidering the consequences of selective fisheries'. Science 335:1045–7.

Heck, S., Béné, C., Reyes-Gaskin, R. 2007. 'Investing in African fisheries: building links to the Millennium Development Goals'. Fish and Fisheries 8(3): 211–26.

HLPE, 2014. 'Sustainable fisheries and aquaculture for food security and nutrition'. A report by the High Level Panel of Experts on Food Security and Nutrition of the Committee on World Food Security. Rome.

Hoffman, A. 2010. 'Social Networks and Power Relations in the Dagaa Business in Mwanza, Tanzania'. MSc thesis Law and Governance Group, Wageningen: Wageningen University.

IOC 2012. 'Regional fish trade in eastern and southern Africa – products and markets: a fish trade guide'. Smart Fish Working Papers 13. Indian Ocean Commission, Ebène.

IPCC 2014. 'Intergovernmental Panel on Climate Change. Impacts, Adaptation and Vulnerability'. IPCC Working Group II Contribution to AR5.

Jennings, S., Smith, A.D.M, Fulton, E.A., Smith, D.C. 2014. 'The ecosystem approach to fisheries: management at the dynamic interface between biodiversity conservation and sustainable use'. *Annals of the New York Academy of Sciences* 1322: 48–60.

Jul-Larsen, E., Kolding, J., Nielsen, J.R., Overa, R., van Zwieten, P.A.M. 2003. 'Management, co-management or no management? Major dilemmas in southern African freshwater fisheries. Part 1: Synthesis Report'. *FAO Fisheries Technical Paper* 426/1.

Junk, W.J., Bayley, P.B., Sparks, R.E., 1989. 'The flood pulse concept in river-floodplain systems'. *Canadian Journal of Fisheries and Aquatic Sciences* 106: 110–127 (Special Publication).

Junker, J., Boesch, C., Mundry, R., Stephens, C., Lormie, M., Tweh, C., Kühl, H.S. 2015. 'Education and access to fish but not economic development predict chimpanzee and mammal occurrence in West Africa'. *Biological Conservation*. http://www.sciencedirect.com/science/article/pii/S0006320714004595

Kawarazuka, N., Béné, C. 2011.' The potential role of small fish species in improving micronutrient deficiencies in developing countries: building evidence'. *Public Health Nutrition* 14(11): 1927–38.

Kolding, J., van Zwieten, P.A.M., Mkumbo, O., Silsbe, G., Hecky, R. 2008. 'Are the Lake Victoria fisheries threatened by exploitation or eutrophication? Towards an ecosystem based approach to management'. In Bianchi, G., Skjoldal, H.R. (eds), *The Ecosystem Approach to Fisheries*. Rome: CAB International, pp. 309–54.

Kolding, J., van Zwieten, P.A.M. 2014. 'Sustainable fishing in inland waters'. *Journal of Limnology* 73: 128–144.

—— 2012. 'Relative lake level fluctuations and their influence on productivity and resilience in tropical lakes and reservoirs'. *Fisheries Research* 115–6: 99–109.

—— 2011. 'The tragedy of our legacy: how do global management discourses affect small-scale fisheries in the South?' *Forum for Development Studies* 38(3): 267–97.

—— 2006. *Improving Productivity in Tropical Lakes and Reservoirs*. Cairo: Challenge Programme on Water and Food – Aquatic Ecosystems and Fisheries WorldFish Center.

Kolding, J., Béné, C., Bavinck, M. 2014a. 'Small-scale fisheries – importance, vulnerability, and deficient knowledge'. In Garcia, S., Rice, J., Charles A. (eds), *Governance for Marine Fisheries and Biodiversity Conservation. Interaction and coevolution*. Chichester: Wiley-Blackwell, pp. 317–31.

Kolding, J., Jacobsen, N.S., Andersen, K.H., van Zwieten, P.A.M. 2015. Maximizing fisheries yields while maintaining community structure. *Can. J. Fish and Aquatic. Sci.* 73(4): 644–655.

Kolding, J., Law, R., Plank, M., van Zwieten, P.A.M. 2014b. 'The Optimal Fishing Pattern'. In Craig, J. (ed.) *Freshwater Fisheries Ecology*. Chichester: Wiley-Blackwell, pp. 524–540.

Kolding, J., Ticheler, H and Chanda, B. 2003a. 'The Bangweulu Swamps – a balanced small-scale multi-species fishery'. In Jul-Larsen, E., Kolding, J., Nielsen, J.R., Overa, R., van Zwieten, P.A.M. (eds), *Management, Co-Management or No Management? Major Dilemmas in Southern African Freshwater Fisheries. Part 2: Case Studies*. Rome: FAO, pp. 34–66.

Kolding, J., Musando, B., Songore, N. 2003b. 'Inshore fisheries and fish population changes in Lake Kariba'. In Jul-Larsen, E., Kolding, J., Nielsen, J.R., Overa, R., van Zwieten, P.A.M. (eds), *Management, Co-Management or No Management? Major Dilemmas in Southern African Freshwater Fisheries. Part 2: Case Studies*. Rome: FAO, pp. 67–99.

Law, R., Plank, M.J., Kolding, J. 2014. 'Balanced exploitation and coexistence of interacting, size-structured, fish species'. *Fish and Fisheries*. DOI: 10.1111/faf.12098.

——— 2013. 'Squaring the circle: Reconciling fishing and conservation of aquatic ecosystems'. *Fish and Fisheries* 16: 160–174. DOI: 10.1111/faf.12056.

——— 2012. 'On balanced exploitation of marine ecosystems: results from dynamic size spectra'. *ICES Journal Marine Sciences* 69: 602–14.

Lehner B., Döll P. 2004. 'Development and validation of a global database of lakes, reservoirs and wetlands'. *Journal of Hydrology* 296: 1–22. Available at: http://www.worldwildlife.org/science/data/item1877.html (accessed January 2015).

Longley et al. 2014. *The Role of Fish in the First 1,000 Days in Zambia*. IDS Special Collection. Brighton: Institute of Development Studies. Available at www.ids.ac.uk

Lynch, A.J., Cooke, S.C., Deines, A.M., Bower, S.D., Bunnell, D.B., Cowx, I.G., Nguyen, V.M., Nohner, J., Phouthavong, K., Riley, B., Rogers, M.W., Taylor, W.W., Woelmer, W., Youn, S-J. and Beard, T.G. 2016. The social, economic, and environmental importance of inland fish and fisheries. *Environ. Rev.* 24: 1–7 dx.doi.org/10.1139/er-2015-0064

Marshall, B., Maes, M. 1994. 'Small water bodies and their fisheries in southern Africa'. *CIFA Technical Paper* 29.

Mills, D.J., Westlund, L., de Graaf, G., Willmann, R., Kura, Y., Kelleher, K. 2011. 'Underreported and undervalued: Small-scale fisheries in the developing world'. In Andrew, N.L., Pomeroy, R (eds). *Small-scale Fisheries Management: Frameworks and Approaches for the Developing World*. Wallingford: CABI, pp. 1–15.

Molden, D. (ed.). 2007. *Water for Food, Water for Life: A Comprehensive Assessment of Water Management in Agriculture*. London: Earthscan; Colombo: IWMI.

Molden, D., Oweis, T., Steduto, P., Bindraban, P., Hanjra, M.A., Kijne, J. 2010. 'Improving agricultural water productivity: Between optimism and caution'. *Agricultural Water Management* 97: 528–35.

Morgan, N.C, Backiel, T., Bretschko, G., Duncan, A. et al. 1980. 'Secondary production'. In Le Cren, E.D., Lowe-McConnel, R.H. (eds), *The functioning of freshwater ecosystems. International Biological Programme 22*, Cambridge: Cambridge University Press, pp. 247–341.

Mosepele, K. 2014. 'Classical fisheries theory and inland (floodplain) fisheries management: is there need for a paradigm shift? lessons from the Okavango Delta, Botswana'. *Fisheries and Aquaculture Journal* 5(3).

Nicholson, S.E. 1996. 'A review of climate dynamics and climate variability in Eastern Africa'. In Johnson, T.C., Odada E. (eds), *The limnology, climatology and paleoclimatology of the East African lakes*. Amsterdam: Gordon and Breach, pp. 25–56.

Pauly, D., Christensen, V., Dalsgaard, J., Froese, R., Torres, F.J. 1998. 'Fishing down marine food webs', *Science* 279: 860–3.

Plank, M.J., Law, R. Kolding, J. 2015. 'Do unregulated, artisanal fisheries tend towards balanced harvesting?' In Garcia, S.M. et al. (eds), 'Balanced Harvest in the Real World. Scientific, Policy and Operational Issues in an Ecosystem Approach to Fisheries'. *Report of an international scientific workshop of the IUCN Fisheries Expert Group (IUCN/CEM/FEG) organized in close cooperation with the Food and Agriculture Organization of the United Nations (FAO)*, Rome.

Postel, S.L. 1998. 'Water for Food Production: Will There Be Enough in 2025?' *BioScience* 48: 629–37.

Richter, B.D., Postel, S.L., Revenga, C., Scudder, T., Lehner, B., Churchill, A. et al. 2010. 'Lost in development's shadow: The downstream human consequences of dams'. Water Alternative 3: 14–42.

Roos, N., Islam, M.M., Thilsted, S.H. 2003. 'Small indigenous fish species in Bangladesh: Contribution to vitamin A, calcium and iron intakes'. *Journal of Nutrition* 133(11): 4021–6.

Tamatamah, R.A., Hecky, R.E., Duthie, H.C. 2005. 'The atmospheric deposition of phosphorus in Lake Victoria (East Africa)'. *Biogeochemistry* 73: 325–433.

Thilsted, S.H. 2012. 'The Potential of Nutrient-rich Small Fish Species in Aquaculture to Improve Human Nutrition and Health', in Subasinghe, R.P., Arthur, J.R., Bartley, D.M., De Silva, S.S. et al. (eds), *Farming the Waters for People and Food. Proceedings of the Global Conference on Aquaculture*. Rome: FAO; Bangkok: Network of Aquaculture Centres in Asia-Pacific (NACA).

Tsikliras, A.C., Polymeros, K. 2014. Fish market prices drive overfishing of the 'big ones'. *PeerJ* 2:e638

Tweddle. D., Cowx, I.G., Peel, R.A., Weyl, O.L.F. 2015. Challenges in fisheries management in the Zambezi, one of the great rivers of Africa. *Fisheries Management and Ecology* 22:99–111.

UNEP/CBD/COP. 1998. *Convention of Biological Diversity. Report of the Workshop on the Ecosystem Approach. Lilongwe, Malawi*, 26–8 January 1998. Available from http://www.cbd.int/doc/meetings/cop/cop-04/information/cop-04-inf-09-en.pdf.

van Zalinge, N., Thuok, N., Tana, T.S. 1998. 'Where there is water, there is fish? Fisheries issues in the Lower Mekong Basin from a Cambodian perspective', *Project for Management of the Freshwater Capture Fisheries of Cambodia*. Phnom Penh: Mekong River Commission.

van Zwieten P.A.M., Béné, C., Kolding, J., Brummett, R., Valbo-Jørgensen, J. 2011. 'Review of tropical reservoirs and their fisheries in developing countries: The cases of Lake Nasser, Lake Volta and Indo-Gangetic Basin reservoirs'. *FAO Fisheries Technical paper 557*. Rome: FAO.

Welcomme, R.L., 1979. *The Fisheries Ecology of Floodplain Rivers*. London: Longman.

Welcomme, R.L. 1999. A review of a model for qualitative evaluation of exploitation levels in multi-species fisheries. *Fisheries Management and Ecology* 6(1):1–19.

Welcomme R.L. and Petr T. (eds.) 2004. Proceedings of the Second International Symposium on the Management of Large Rivers for Fisheries. Vols I and II, FAO Regional Office for Asia and the Pacific, Bangkok, Thailand. RAP Publication 2004/17. 356 and 309pp.

Welcomme, R.L., Béné, C., Brown, C.A., Arthington, A., Dugan, P., King, J.M., Sugunan, V. 2006. 'Predicting the water requirements of river fisheries –

ecological studies'. In Verhoeven, J.T.A., Beltman, B., Bobbink, R., Whigham, D.E.S. (eds), *Wetlands and Natural Resources Management 190*. Berlin/Heidelberg: Springer-Verlag, pp. 123–54.

Welcomme, R.L., Cowx, I.G., Coates, D., Béné, C. Funge-Smith, S., Halls, A., Lorenzen, K. 2010. 'Inland capture fisheries'. *Philosophical Transactions of the Royal Society* B(365): 2881–96.

Welcomme, R.L., Lymer, D. 2012. 'An audit of inland capture fishery statistics – Africa'. *FAO Fisheries and Aquaculture Circular 1051*. Rome: FAO.

Wilkie, D.S., Starkey, M., Abernethy, K., Effa, E.N., Telfer, P., Godoy, R. 2005. 'Role of prices and wealth in consumer demand for bushmeat in Gabon, central Africa'. *Conservation Biology* 19: 268–74.

19 Climate Change Adaptation Strategies and Food Security: A Case of the Chewa People in Central Malawi

Jessica Kampanje-Phiri and Dean Kampanje-Phiri

INTRODUCTION

Climate change models indicate that weather patterns in the Zambezi River Basin, Malawi included, are sensitive to climate change as small changes in temperature have significant effects in precipitation and runoff patterns. While a high degree of uncertainty remains concerning the actual implications of climate change and variability in the short, medium and long-term, most analysts agree that the Zambezi River Basin will become much drier in the future. Farmers may therefore need to adopt early maturing varieties, switch to crops that are drought resistant such as tubers, employ soil and water conservation techniques, enhance irrigation agriculture and much more in order to adapt to climate change.

This chapter explores the predominant agricultural practices, food preferences and post-harvest handling of food crops among Chewa people in Lilongwe District, Central Malawi. In particular, Chewa foodways will be used as a case in point to elaborate how food security may be impacted, particularly under changing climate and increased climate variability. This chapter argues that incorporation of social and cultural aspects in relation to food production, preferences and distribution in devising adaptation strategies that mitigate the impact of climate change in this area and beyond is important particularly as climate becomes more uncertain.

BACKGROUND

Malawi lies in the north-eastern part of the Zambezi River Basin. Rainfall in the Zambezi River Basin exhibits spatial variations with the northern and eastern parts of the basin receiving much higher rainfall and decreases progressively towards the southern and western parts of the basin (Tumbare 1999). Even though Malawi lies in the north-eastern part of the basin, precipitation in the country also exhibits significant variations with annual averages ranging from 500 mm in the low lying areas to as high as 1,800 mm in the high areas. Rainfall in Malawi is influenced by four main rain-bearing systems which include the movement of the Inter-Tropical Convergence Zone (ITCZ), monsoons, the Congo Air Boundary and tropical cyclones (Malawi Government 2010: 215). The lakeshore districts of Nkhotakota, Nkhatabay and Karonga have the highest rainfall totals on record due to the effect of south-easterly winds on the western shores of the lake. Runoff is high in the southern region districts of Zomba, Thyolo and Mulanje due to the dominating features of the Zomba Mountain and Shire Highlands (Malawi Government 2010: 188). Nevertheless, rainfall in Malawi and the Zambezi River Basin in general is largely influenced by the Inter-Tropical Convergence Zone, resulting in distinct rainfall seasons between November and March or December and April (Malawi Government).

The central region of Malawi lies on a central plateau that has an escarpment eastward. Lilongwe District in Central Malawi is situated in the Linthipe River Basin which is 8,641 km^2 in size with the Lilongwe and Diamphwe rivers as main tributaries. The annual average rainfall in the basin is around 964 mm and the average runoff of 151 mm with a mean flow of 41 m^3/s (Malawi Government 2010: 190, 212). The Lilongwe River, which is the source of water for the Lilongwe city, the seat of government in Malawi, has a mean annual flow of 8.70 m^3/s.

CLIMATE CHANGE OVERVIEW

The average rainfall received in Malawi is around 850 mm which is sufficient for rainfed agricultural production (Malawi Government's National Adaptation Programmes of Action (NAPA), 2006). Over the last decades, Malawi has, however, experienced climate hazards including droughts and floods. Some of these hazards have increased in frequency, intensity and magnitude, in particular drought and flood (Malawi Government 2013, Asfaw et al. 2013: 2, Nangoma 2007: 1). The onset and offset of the rains as well as rainfall distribution within each season have also increasingly become irregular owing to increased climate variability and climate change. Malawi is generally vulnerable to climate change because agriculture which provides livelihoods to more than 80 per cent of the population and contributes to as much 90 per cent of export earnings, relies heavily on rainfall as more than 90 per cent of agriculture productivity is rainfed (see also Nangoma 2007: 2). With increasingly erratic rainfall patterns, agricultural productivity is becoming unreliable. The situation is made worse by the low economic capacity of the nation to adapt to the effects of climate change, the high levels of rural poverty and poor infrastructure which hinders smooth access to markets and inputs as a way of providing

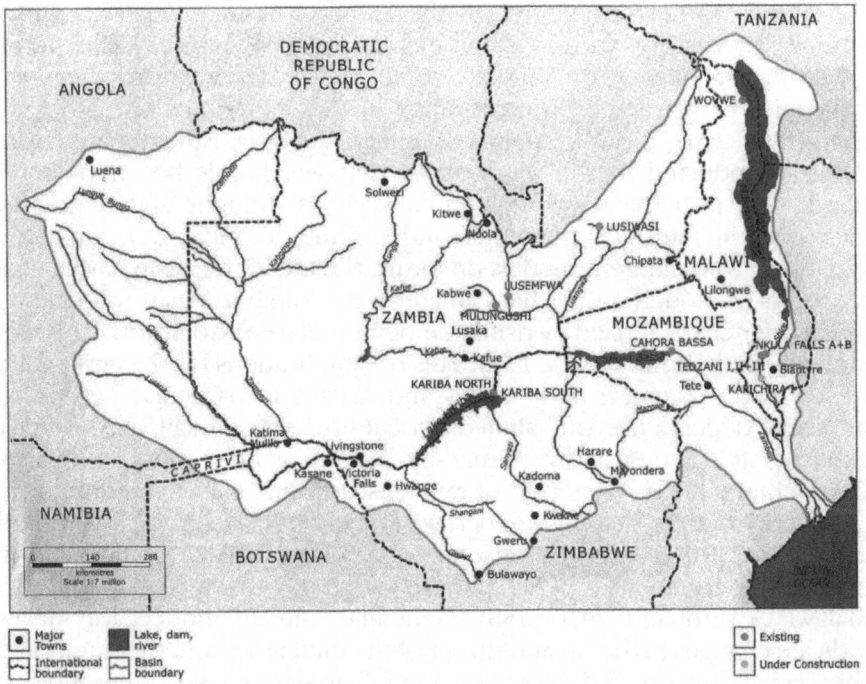

Figure 19.1. Map of the Zambezi River Basin (© Zambezi River Authority).

alternative livelihoods and enhanced agricultural productivity. Moreover, the high population density in Malawi, 160 inhabitants/km^2 (Arndt, Schlosser, Strzepek & Thurlow 2014: ii85), signifies low acreage per household family which may impact yields and rural livelihoods.

Owing to climate variability in the Zambezi basin, predicting future climate in the basin is challenging. Assessments on river flows in the basin by Tumbare and others show that the basin exhibits cyclic patterns with some of the decades being relatively wet while others relatively dry. Using differential mass curves for the Zambezi River main channel at the Victoria Falls and Kafue River at the Kafue Hook Bridge, Tumbare shows that the basin experienced a dry sequence between 1908/1909 and 1937/1938, a total of 30 years, an average sequence between 1938/1939 and 1947/1948, a total of ten years, a wet sequence between 1948/1949 to 1979/1980, a total of 32 years, and a dry sequence between 1980/1981 and 1995/1996, a total of 15 years (Tumbare 2000: 202). This importantly suggests that temperatures and precipitation which, among other things, influence runoff have been shifting back and forth in the basin to produce the cyclic patterns.

Climate change studies normally consider a 30-year period for analysis to track changes in weather patterns (see Oestigaard 2011: 2). However, considering the cyclic patterns in the Zambezi basin which are of long nature, predicting future climate will be challenging. The Inter-governmental Panel on Climate Change (IPCC) acknowledges the difficulties of predicting future climate in the short, medium and long-term (IPCC 2007, Bie, Mkwambisi & Gomani 2008: 1). Nevertheless, many studies conducted in the basin agree that weather is becoming more erratic and also that climate hazards such as drought and floods are increasing in intensity, frequency and magnitude. Moreover, while the variations from different models on future climate in the basin have a wide range, most models agree that climate in the Zambezi basin will become much drier characterized by rising temperatures (see also Arndt, Schlosser, Strzepek and Thurlow 2014: ii84). This has implications for the agricultural sector in the Zambezi basin states and Malawi in particular because of its heavy dependence on rain-fed agriculture. The rainfall seasonality is expected to change in the future by becoming much shorter (Malawi Government 2013). The onset of rainfall season in Malawi is shifting from October/November to November/December (see Malawi Government 2010: 188, 219). This suggests also that the normal peak of rainfall intensity in Malawi in the months of February to March or early-April may shift (See Malawi Government 2010: 188). Generally, climate models for Malawi indicate changes in the on-and-off-set of the rainfall season, the duration of the rainfall season and increased rainfall intensity (Bie, Mkwambisi & Gomani 2008: 1). Moreover, agronomic experiments indicate that maize yields will decline by up to 22 per cent due to a combination of heat and drought between now and 2050 (Lobell et al. cited in Cook, Ricker-Gilbert & Sesmero 2013: 1). This means that long maturing varieties will increasingly become less suitable without drastic changes to the water-resource system either through soil and water conservation technologies or increased irrigation agriculture. The major emphasis on maize production as a staple food crop, which historically has received political backing, increases the vulnerability of the rural farmers to climate change because of the agro-climatic conditions that suit the crop.

Studies on climate change indicate that a 2°C increase in temperature reduces agriculture productivity by around 5 to 10 per cent in the tropics, while a 3°C increase will increase the number of people at risk of hunger by 150–550 million people and a 4°C rise in temperature will decrease agricultural productivity by between 15 and 35 per cent in Africa (Malawi Government 2013: 7). Since average temperatures in Malawi are expected to rise by 1°C by 2030 and between 1.1 and 3.0°C by the 2060s, productivity will likely go down if no measures are taken to address the impacts of climate change (Malawi Government 2013: 4–5, see also Arndt et al. 2014: ii85, ii95). Several climate change models for Malawi indicate also that due to climate change, the country's GDP will likely decrease because of the effects of climate change on water, food, energy, environment, etc. (Arndt et al. 2014: ii99). On the one hand, a study on climate change adaptation in

Malawi by Cook, Ricker-Gilbert and Sesmero has found out that an increase of intra-seasonal variability in precipitation by 1 per cent reduced household income by 1.5 per cent (Cook, Ricker-Gilbert & Sesmero 2013). On the other hand, household profit can be reduced by as much as 19 per cent for every 1 per cent reduction in growing season precipitation assuming that other factors remain constant since there is a lot of uncertainty concerning climate change (Cook et al. 2013: 31).

Streams and rivers are the largest source of irrigation water for smallholder irrigation schemes, and because of that, rising temperatures can reduce further the amount of water that is available in the system, if precipitation in the basin remains largely unchanged from today's values. This should be less surprising since the Zambezi River Basin has high evaporation rates where only 10 per cent of precipitation is converted to runoff even though Malawi fairs better with an average of 19 per cent against the global mean of 30 per cent (Chenje 2000, Malawi Government 2010: 188). Increase in average temperatures can potentially turn many streams into ephemeral streams where streams will have no water flowing during dry periods when the need for water for irrigation is actually at its peak. In terms of agricultural productivity, the cereal harvested area for maize is around 1.06 MT/ha in the Zambezi basin against a potential irrigated yield of 7.5 MT/ha (World Bank 2008: 18). While most of the irrigation activities are from surface water sources, irrigation from these sources is limited and as such increasing production under rainfed agriculture must incorporate other adaptation technologies. The World Bank study on water for agricultural development in the Zambezi River basin suggests that scaling of irrigation in the basin may only benefit 12 to 20 per cent of the rural population, meaning that the remaining 80 per cent must be reached by other means (World Bank 2008: xv). Most studies on climate change adaptation strategies in Malawi have singled out three adaptation strategies that have been shown to have positive results and which may be promoted to create climate resilient communities. These include adoption of improved seed variety, soil and water conservation measures and increased use of fertilizers. Other strategies include crop-diversification, intercropping, horticulture, improved post-harvest technologies as well as utilizing remittances to cushion the effects of low productivity (see also Bie, Mkwambisi & Gomani 2008: v).

Increasing cereal production under rainfed cultivation will therefore entail incorporating soil and water conserving measures, adopting improved seed varieties, increased use of fertilizers, embracing conservation farming and strengthening extension services (See World Bank 2008: 23). The government of Malawi, through its key documents on climate change as well blue-prints for national social and economic development such as the Malawi Growth and Development Strategies (MDGS), list a number of projects and strategies for climate change mitigation and adaptation. The 2006 National Adaptation Programmes of Action (NAPA)

Figure 19.2. Waiting for rains to plant crops. Rainfed agriculture is predominant in Malawi (© Jessica Kampanje-Phiri and Dean Kampanje-Phiri, at Traditional Authority Kalolo, in Lilongwe Rural).

developed by the government of Malawi under the United Nations Framework Convention on Climate Change (UNFCCC), for instance cite the lack of willingness by the local communities as a potential limitation to the successful roll-out of climate change adaptation programmes (Malawi Government 2006). It is in this context that we need to explore how the Chewa people of Lilongwe, in Central Malawi, conceptualize food security and how their food preferences and post-harvest handling of food relate to the proposed adaptation strategies by the government and other lead institutions.

Food policy makers in Malawi regard it as crucial for farmers to be encouraged to diversify their crops countrywide. Moreover, they emphasize that it is important to promote appropriate cultural and food practices that do not jeopardize 'food security' (see Malawi Government 2008: 11–15). This is pertinent now as the need to adapt agricultural practices to climate change is emphasized. So what are the implications of how the Chewa conceptualize food and food security? In other words, how would the Chewa foodways impact on the efforts by government and other stakeholders to promote adaptation to climate change in order to attain food security? 'Foodways' throughout this chapter refers to the different ways and processes through which food not only sustains life but also articulates, mediates or embodies the meanings, values and interests of a particular group of people and/or institutions. This also encompasses the different ways in which food mediates various groups of people or social arenas, and hence includes all processes of food production, consumption/preferences and distribution.

NOTES ON THE CHEWA PEOPLE

There is a general lack of pre-colonial anthropological data on the matrilineal Chewa in Malawi. Historians suggest that the Chewa were the largest group of Bantu migrants who settled within the Central Region of Malawi and some parts of Zambia (see also Kachapila 2006, Phiri 1983). Overtime, the Chewa people have undergone various transformations, including the nature of domestic authority and control (Phiri 1983: 257). Phiri shows that changes among the Chewa since the nineteenth century have come about as a result of the slave trade, the intrusion of patrilineal groups such as the Ngoni from southern Africa and the Swahili from the East Coast, and from Christian missionaries' teachings and European techniques of education (Phiri 1983: 258). Furthermore, from the 1890s, the Chewa people were not only subjected to colonial rule, but were also drawn into a capitalist economy, both of which contributed to significantly change Chewa people's way of life (Phiri 1983, Kachapila 2006).

Like elsewhere in Malawi, maize is the main food crop grown by the Chewa people, and it is supplemented by a wide range of produce such as groundnuts, sweet potatoes, sorghum, millet and a variety of vegetables. Its production, as is the case elsewhere in Malawi is also largely rainfed. The main cash crop is tobacco, introduced through the plantation economy. According to Phiri (Phiri 1983: 271), the effect of this addition to the crop pattern was an erosion of the matrilineal system of tenure where the means of production ideally lies in the hands of adult women as matrilineal inheritors. As is also common in most parts of Malawi, the British colonialists who were mostly familiar with a patrilineal system, gave men exclusive access to tobacco and, thus, to the only cash crop (Phiri 1983: 271). With such changes in tenure, and the historical labour migration of men, in conjunction with the strain these changes placed on the matrilineal kinship system, women have been placed in a key position in terms of food production. With the cessation of international labour migration in the 1970s, and the introduction of market liberalization policies allowing smallholder farmers to grow tobacco as a cash crop without quota restrictions in the 1990s, other changes also occurred within the matrilineal structures (see Phiri 1983, Englund 1999). These changes are fundamental in understanding the trajectory of Chewa foodways, especially in the context of change (be it social, political or climatic).

FOODWAYS AMONG THE CHEWA

There have always been many *foodways* in Malawi beyond food's physiological capacity to sustain life. A more refined meaning of foodways as introduced earlier in this chapter encompasses the ways which relate to what we eat, how we eat, and who we eat with, as well as how we acquire, prepare and share our food. In other words, foodways includes all aspects of food that are imbued with values and meanings that define who we are (see also Adema 2007, Counihan 1999, Harris 1987). In this way, foodways are expressive of rich meanings, beliefs and practices that are pertinent to a specific group of people. Among the Chewa, not only is food central to their celebrations, funerals and rituals, but it is also the foundation of their cosmological understanding and what it means to be Chewa. Whether exchanged as gifts or as commodities for sale, foods are deeply embedded in Malawian culture and maize holds a special significance.[1]

Symbolic values are attached to different varieties of maize and to different stages in its production, consumption and distribution. For example, pure white maize grain and flour symbolize high status and wealth, while dark coloured maize is associated with low rank and 'poverty'. The strength of social relations and their tensions are also expressed through the idiom of food, which is manifested daily through the interpretation of hospitality. For important visitors, most Malawians share the custom of preparing a thick porridge (*nsima*) made with white maize flour served with chicken and vegetables as a symbolic gesture of welcoming the visitors. Those failing to serve their visitors with such hospitality risk being labelled stingy and at times may even risk being accused of practising witchcraft. These rich socio-cultural aspects of food are rarely considered by policy makers. However, they are issues that have great implications in influencing how effective implemented policies or interventions are since it is obvious that cropping patterns are not solely affected by the character of the prevailing water system.

CONCEPTUALIZING 'FOOD SECURITY' AND THE COMPLEX CHEWA FOODWAYS

Through information obtained in a yearlong ethnographic study among the Chewa people of T.A. Kalolo, Lilongwe District in Central Malawi, interpretations of what the local people understood by the concept of 'food security' differed among individuals and/or institutions. According to several Chewa informants, the concept of 'food security,' based on the information they got from extension services, meant ensuring that they had food supplies all year round. Nevertheless, due to the lack of a local name or translation of the term, many understood the term food security as the ability to have food, specifically maize, throughout the year, taking care of food supplies, not wasting food, diversifying food sources to make the maize last longer, access to dambos for irrigation agriculture so as to produce crops for sale and home consumption. (Field notes 29/11/07).

Contrary to the definitions of 'food security' given above, people in this area normally prepared large quantities of food (large amounts of *nsima*) specifically for their two main meals (lunch and dinner) but also, sometimes, for breakfast, before setting out for their daily chores. Moreover, diversification of main meals with other foods was largely non-existent. Other foods – such as rice, potatoes, roasted groundnuts, pumpkins or fruits – were eaten between main meals as snacks, or else given to their children. Those who could not have *nsima* due to lack of maize flour, perceived themselves as lacking food despite the fact that they had other foods like rice or sweet potatoes. It is also in the post-harvest period when the people of this area tend to conduct the majority of their cultural activities pertinent to the Chewa way of life. This is when Chewa people prepare a lot of food and beer whose main ingredient is maize. Despite their knowledge of food security as specified above, foregoing these significant annual festivities and practices is unexpected except in food crisis situations. Important cultural festivities and practices which contribute to food wastage as outlined by chief John Eliasi, of one of the villages under study, include *rituals on gule wamkulu*,[2] *chieftainship ceremonies, weddings, friendship exchanges, wedding agreements, initiations, memorials and funerals*. Large quantities of foods such as nsima, relish and beer are a central part of the ceremonies. Not so much thought is given to the amount of food wasted in these ceremonies because these festivities and practices are part the Chewa traditions and they are of great value (personal conversation with Chief John Elias, Fieldnotes 22/12/08).

As this suggests, to understand the Chewa conceptualization of 'food security' is to understand their socio-cultural context, and the values and meanings that contribute to the ways in which they make decisions on food production, consumption and distribution. Food practices and foodways have also been found to be an informative tool for understanding local conceptualizations of 'food security' (Mzamu 2012). As Weismantel (1988) and Mintz (1986) argue, food is a class maker and different types of foods can be used to distinguish the rich from the poor. This was also found to be relevant in the course of our research, but in subtle and sometimes contradictory ways. To the Chewas, *nsima*, made from maize flour is a central part of their dietary intake. Those who can afford *nsima* at all times are considered better off and other foods like fruits, tubers including cassava and sweet potatoes, rice, tea, cooking oil, or bread are consumed in between the main meals consisted of *nsima*. Those only eating the other types of food are considered poor and taken pity on. Dambo (wetland) cultivation also plays a major part in differentiating the food habits of the households as those who cultivate in dambos are able to grow different types of crops which increase their harvest. Furthermore, the ability of

Figure 19.3. Preparing meals for cultural ceremonies. Such large meals in post-harvest periods are blamed for its contribution to food insecurity (© Jessica Kampanje-Phiri).

different households to rent extra land for cultivation also plays a significant factor in household level yields. The quality of the flour, i.e. refined white maize flour which is considered a delicacy as it requires a lot of maize, rather than regular maize flour differentiates the households in terms of class meaning that those who can afford the refined white maize flour are better off than those who cannot.

This exploration of the variety of food practices among the Chewa emphasizes the complexity of food issues in this region. Although there is general understanding of the concept of food security as advanced by extension workers, food practices among the Chewa usually challenge this received wisdom. How the Chewa conceptualize food issues tends to be a complex interaction of socio-cultural values, meanings, individual preferences, land issues, natural forces, market force, and policy processes (most of which tend to be beyond their control). To understand more fully these aspects of Chewa people's everyday lives, and the policy processes that contribute to their predicament, it is also important to analyse such relations within a hunger crisis scenario. This is significant not only because the 'food security' policy in Malawi was developed during the lengthy hunger crisis of the 2001–6 growing seasons, but also because donor interventions during this period ignited different local responses towards humanitarian assistance in general and to the templates of 'poverty', 'the poor' and 'food security'.

'FOOD SECRURITY' AND CLIMATE CHANGE RELATED CRISES: THE CASE OF THE MALAWI 2001–6 HUNGER CRISIS

The locally embodied meanings of food tend to become particularly tense in the advent of hunger, droughts or any natural crises that affects local peoples' foodways (Mzamu 2012: 282). Climate variability (erratic weather patterns; persistent droughts, dry spells and floods among others) is among the most prominent factors that disrupt local peoples' foodways. When such climate variability-related crises arise, it is common that foreign food arrives as part of food aid. For instance, the Malawi hunger crisis (2001–6) that occurred due to persistent droughts, dry spells and floods in different parts of the country is a case in point.

Climate variability related food crises, and the various forms of intervention intended to address them, forms part of the history of the Chewa people of T.A. Kalolo and in Malawi generally. Whilst food is a straightforward matter in policy, humanitarian or development discourse, as argued, food among the Chewa constitutes thick, complex and paradoxical issues that are pertinent to their ways of life. As already stated, food is one of the strongest ethnic and class markers among the Chewa tribe. It provides endless metaphorical references not only to hostility among rival groups but also in terms of relations of domination that make use of food symbolism (Weismantel 1988: 143–67). More specifically, how food symbolism, relations or choices becomes a complex issue in times of hunger, is relevant in answering some of the pertinent questions to be addressed in this chapter. Some pertinent questions addressed within this chapter include: What happens to food decisions and established foodways at such times? How does imported food aid articulate local perceptions and practices? Most importantly, to what extent do the social cultural conditions related to food production, consumption and preferences affect adaptation strategies aimed at mitigating the impacts of climate change and variability among the Chewa of Malawi?

A call to develop a specific national 'food security' policy was made by donors, the Malawi government and NGOs in response to the hunger crisis that occurred during the 2001–6 growing seasons. This policy was to be concerned with 'poverty', food and agricultural issues in Malawi. Previous food policies were formulated and implemented either as part of other institutional legislation (in the colonial era) or as part of other 'poverty' and agricultural regulatory policies (in post-colonial administrations).[3] This being the case, the significance of this particular hunger crisis is not limited to having initiated the development of a specific 'food security' policy in Malawi.

The contradictions associated with policy templates, which in turn guide the direction of policy interventions, can be shown by the example of the distribution of yellow maize during the hunger crisis described above. Although white maize was generally distributed during the hunger crisis, yellow maize was the most common ration distributed during this hunger crisis. Due to the colour of maize given out as an intervention for the crisis and its foreign origin, Malawian hunger victims were suspicious of rations of yellow maize (a surplus produce of the USA farmers, pushed into the aid orbit by multi-national corporations such as Cargill). Ranging from the rations being labelled as feed for animals, rumours surrounding its foreign origin as comprising chemicals that cause infertility were not uncommon. For humanitarian aid workers, their main concern was to provide a local staple food to those suffering from hunger, such that the main distress was on how best they could procure the staple food cheaply during emergency situations. For the Chewa people however, the distribution of yellow maize became a complex social-cultural issue rather than just being about survival.

As the above suggests, the distribution of yellow maize during this hunger crisis provides us with a good example of the kinds of divergent interpretations that are induced by standardized policy interventions. While for humanitarian aid policy makers the concern was simply to provide food that would help to eradicate the hunger crisis, the Chewa people's response to it reveals that there were cultural and social meanings attached to the food aid, rather than it being a simple matter of survival. A multiplicity of aspects of food aid was overlooked, on the whole, by the humanitarian workers, in a situation that consequently worsened the experience of the hunger situation among the Chewa. For instance, for the aid workers, it was important that the staple food for most Malawians is maize; it did not matter to them whether they distributed yellow or white maize. In fact, the distribution of yellow maize during the hunger crisis was partly justified by aid workers due to its high vitamin A content and that it was also cheap to procure. Procured yellow maize from the subsidized USA farmers' surplus was cheaper than the white maize that could be purchased anywhere within southern Africa, where most farmers' crop production remains largely unsubsidized. Because yellow maize was considered to be advantageous by both aid workers and 'food security' policy makers, they considered it to be the best food aid available for aid purposes (see also Muzhingi et al. 2008, Tschirley & Santos 1995).

Contrary to the thinly conceived aspects of yellow maize outlined above, the Chewa people perceive maize in relation to all its social and cultural complexities. White maize in particular has deeply engrained meanings in Chewa people's lives. Not only does white maize contain social, symbolic, cosmological and economic values, it is also part of their culturally and historically shaped sensorium. Because yellow maize does not contain and manage these fundamental meanings and values, it is considered to be an anomalous and unclassifiable food among the Chewa (Mzamu 2012: 286).

For the Chewa, it is difficult to imagine or place yellow maize flour into any category in the way different types of flour made from white maize are categorized. For example, pure refined white maize flour (*ufa oyela*) is used symbolically and physically consumed in Chewa rituals and during their important annual festivities. It is also part of a classificatory system whereby those who mostly use white flour for their *nsima* throughout the year are considered rich, as it signals that they have larger maize stocks in comparison to others. This is evident in a number of significant aspects of Chewa way of life. For instance, idioms of hospitality are expressed by the Chewa people through serving their visitors *nsima* made from white flour. *Nsima* made from *white maize flour* is generally preferred among the Chewa because of its taste, texture and aroma in comparison with that made from other types of flour. Even though *nsima* made from non-refined white maize flour is mostly preferred by men rather than women, the symbolic value and higher status that white maize flour accords is generally acknowledged among the Chewa. *Nsima* made from white maize husks is generally perceived as food for those who do not have maize and who are, thus, considered to be poor. While this type of *nsima* can be tolerated in times of crisis such as hunger, without raising many moral questions, *nsima* made from yellow maize is generally placed outside the realms of Chewa food systems and values.

In line with Douglas (1966) observations on food abominations of Leviticus, yellow maize among the Chewa is considered to be 'matter out of place' (Douglas 1966: 44), which Douglas explains in the context of an ordered system and classification of matter. For instance, 'dirt', to the Western mind, 'is the by-product of a systematic ordering and classification of matter, in so far as ordering involves rejecting inappropriate elements' (Douglas 1966: 44). This idea of dirt proposed by Douglas allows a connection between symbolic fields with more obvious systems. For example, among the Chewa, pure white maize flour holds symbolic and cosmological significance in their rituals; in addition, the taste and colour of white maize constitute Chewa people's socio-culturally mediated sensorium. The sweet taste of local maize variety (locally referred to as the maize of the ancestors) is usually preferred among the Chewa to other white hybrid varieties, let alone yellow maize. As related by, and observed among, the Chewa, white maize flour made from the local maize variety is brighter, sweeter smelling and has a good taste, and its pure whiteness also contributes significantly to Chewa symbolism, and to their cosmological as well as socio-culturally mediated values and meanings. In contrast, the darker colour, and Chewa people's perception of the flour made from yellow maize as smelling and tasting awful, contributes to the incompatibility of yellow maize with Chewa people's way of life. In short, white maize's capacity to articulate, constitute and manage their symbolic, cosmological and socio-culturally mediated sensorium, values and meanings makes it a significant aspect of Chewa way of life. In this regard, yellow maize amongst the Chewa is, as Douglas (1966) terms it, 'matter out of place' because of its inability to contain and elaborate key symbolic, cosmological and sensorial values pertinent to the Chewa people's way of life.

The Chewa revulsion against yellow maize is not an intrinsic result of their conservatism in not wishing to accept an alien or unfamiliar food. Historically, Chewa people have adopted different types of food introduced by traders, explorers and colonial administrators, and white maize is just one of these imports. They have managed to adopt and incorporate white hybrid maize varieties into their rituals and diets, even though *the local maize variety* remains their preferred variety. Thus, Chewa people's ability to give local meanings to white maize, which were previously imbued in millet or sorghum flour, contributed greatly to the widespread adoption of white maize by the Chewa ancestry. However, the incompatibility of yellow maize with Chewa people's symbolic, cosmological, sensorial and social realm contributes significantly to its being conceptualized as 'matter out of place'. In fact, one of informants to this study precisely indicated that 'yellow maize flour smells like dirt' (Fieldnotes 29/11/2007).

As will be clear by now, for the Chewa yellow maize is an anomalous food that certainly arouses anxiety and negative feelings. However, it is not considered as *abject* since it was consumed during the hunger crisis without causing Chewa people to vomit. It is doubtful, however, that yellow maize could be converted into the 'delicious', as Korsmeyer proposes, because of other symbolic, cosmological and social cultural values that yellow maize fails to acquire. Korsmeyer also points out that 'no one can stand out of culture and proclaim a neutral list of disgusting foods' (Korsmeyer 2002:219): what might be disgusting in one culture might be pleasurable in another. The revulsion felt towards yellow maize by Chewa people should rather be understood in terms of the totality of Chewa people's way of life. As dirt or animal feed would be perceived as 'matter out of place' on western dinner tables, so yellow maize in Chewa people's kitchens and ritual arenas is considered in the same way.

Local complaints about yellow maize tend to be dismissed by both the humanitarian aid agencies and the government practitioners involved in 'food security' policy processes. In fact, the kinds of objections against yellow maize exemplified in the discussion above are usually regarded as an example of Chewa people's conservative nature and reluctance to change their over-reliance on white maize.[4] Food aid monitors at the distribution centres generally talked about yellow maize as being more nutritious than white maize, to try to convince beneficiaries to change their food habits. Such conservative local attitudes also tend to be conceived by policy practitioners as contributing to chronic food insecurity across the region. In this way, most Chewa socio-cultural practices and annual festivities that involve cooking a lot of food soon after harvest fall into the category of cultural practices that jeopardize 'food security' and are considered as stumbling blocks to development policies. These observations are in agreement with other scholars such as

Pottier who have also discussed how local people's understanding of hunger, the term possibly closest to the concept of 'food insecurity', may well contain viewpoints that do not interest the officials whose job is to improve 'food insecurity' (see Pottier 1999).

IMPLICATIONS FOR ADAPTATION TO CLIMATE CHANGE

The preference for maize among the Chewa as well the majority of rural people in Malawi has the potential to limit the adaptation of the agricultural system to current and future climate trends. Since maize is deeply embedded in the social and cultural values and meanings of the Chewa, adaptation strategies that promote maize substitution with tubers will not be acceptable among the Chewa. This is despite the fact that some of the soils and climate variability hardly suit the wide production of maize. Moreover, the definition of food security is problematic because it caters for 'food preferences' other than just 'dietary needs'. The 1996 World Food Summit adopted a definition of food security which states that:

> Food security, at the individual, household, national, regional and global levels is achieved when all people, at all times, have physical and economic access to sufficient, safe and nutritious food to meet their dietary needs and food preferences for an active and healthy life (FAO 2003: 28).[5]

While this definition was revised in the State of Food Insecurity in 2001, both components of *dietary needs* and *food preferences* were maintained (FAO 2003: 28). This means that without maize, which is the preferred food among the Chewa, they basically become food insecure despite having other food stuffs like tubers that can adequately meet their dietary requirements for an active and healthy life. This explains partly why more than 90 per cent of Malawi's rural population grow maize, mainly for subsistence and there is reluctance to shift to other crops even where maize has shown not to do well (see Bie, Mkwambisi & Gomani 2008: 12).

Pragmatic adaptation strategies to climate change therefore means that intercropping has to be encouraged alongside other measures such as soil and water conservation, increased use of fertilizer and others. This is not to suggest that maize substitution should not be encouraged as the climate changes. It is, however, vital as part of the policy processes to acknowledge that maize will be central in the foodways of the Chewa in the short to medium-term. As history suggests, the Chewa have adopted maize over time from what was predominantly millet and sorghum as staple foods. Over time maize has been socio-culturally mediated and given meaning. As Wolf (1982) reminds us, 'meanings are not imprinted into things by nature; they are developed and imposed by human beings' (Wolf 1982: 388). This suggests that as the climate changes and if it drastically affects the production of maize, the Chewa will eventually adapt. However, cultural change is a slow and enduring process. This denotes that promotion of intercropping of tubers which are drought resistant as well as promoting the increased production of sorghum and millet which are drought-resistant crops will be crucial in the short and medium-term. Efforts are already underway to promote millet and sorghum as an adaptation strategy to climate change and climate variability (see Malawi Government 2010: 145). The expectation is that increased production of tubers, sorghum and millet will slowly be internalized particularly if maize production continues to decline and in such a way that the community may reconstitute itself and embed new meanings to these substitute crops. Research may also be employed to develop maize varieties that at least maintain the important traits of the local maize variety such as durability in storage and taste while improving its performance to drought conditions. From the people's experience in the area, irrigation and access to irrigation land along the dambos plays a significant part in improving food availability and thus security at a household level. Nevertheless, irrigable areas are limited which importantly suggests that soil and water conservation in the upland farms will be important to secure water for improved agricultural productivity.

People in the rural areas have lived under changing climate and variability and employed several mechanisms to cope with such changes (Nangoma, 2008: 1). Nevertheless, such copying mechanisms are considered not to be effective particularly as climate continues to be more variable and rainfall patterns more erratic. Cook et al. suggest that people will respond to environmental changes through behavioural adjustments which may include adoption of adaptation strategies (Cook et al. 2013: 9). The authors further indicate that at least four adaptation strategies are prominent in adaptation studies and these include: switching crop varieties, crop diversification, crop and livestock diversification, and farm and non-farm diversification (Cook et al. 2013: 10). Important to note is also that climate change is going to be rather slow and thereby generates a slow response among the rural farmers.

A study by Asfaw et al. on climate change adaptation and food security in Malawi presents some interesting findings. Their findings show that the declining rainfall trend in Malawi is accompanied by adoption of soil and water conservation (SWC) activities, increased use of inorganic fertilizers, intercropping and planting of trees. On the other hand favourable precipitation encourages the adoption of organic fertilizers and improved seeds (Asfaw, McCarty, Lipper, Arslan and Cattaneo 2013: 1). Climate-smart agriculture at a micro-level can importantly be incorporated to increase the resilience of farmers to climate change and may include adopting drought resistant crops, modifying the agricultural activities like planting time, soil and water conservation techniques, irrigation, increased use of fertilizer, etc. (Asfaw, McCarty, Lipper, Arslan & Cattaneo 2013: 2). Perennial trees, particularly nitrogen fixing trees may be increasingly promoted to help with improving soil quality, providing land cover, preventing soil erosion but also in a way improving water filtration into the soil as well as water retention capabilities of the soil. Asfaw et al. state that promotion of perennial trees has been part of the sustainable agriculture efforts in Malawi (Asfaw et al. 2013: 4).

Adoption of soil and water conservation techniques as well as use of inorganic fertilizers is highest in the central region where Lilongwe is located while trails in terms of use of improved seed varieties (Asfaw et al. 2013: 4–5). Planting perennial trees in farms as a way of enhancing sustainable agriculture is also the lowest in the Central Region of Malawi; the region is second in intercropping. The Chewa food production systems may also partly explain why these particular adoption strategies have taken ground in the Central Region. One reason may be due to the fact that because of matrilineal system and the fact that food production is largely in the realm of women, it facilitates the adoption of organic fertilizers, SWC, and other long run strategies rather than those in rented plots which require or are facilitated by owning land (see Asfaw et al. 2013: 12). Based also on information obtained from informants from Kalolo, renting land is also an option for those who can afford it. This is also where inorganic fertilizers are used.

This suggests that farmers are already adjusting their farming practices – whether that is solely in response to climate variability and climate change may benefit from more studies. Nevertheless, farmers are more likely to adopt some technologies based on what technologies they have used in the past (Asfaw et al. 2013: 7). Intercropping for instance has been long promoted to address dwindling land size issues, increase yields and improve soils. It is nevertheless currently promoted as one of the key strategies of climate change adaptation, among others, which farmers may easily adopt since they have already been exposed to it for a long time. What studies reveal when it comes to the different adaptation strategies is that farmers will normally adopt them as a package other than as single methods. Use of improved seed for instance is accompanied with the use of inorganic fertilizer while use of organic fertilizer is accompanied by intercropping and soil and water conservation measures (Asfaw et al. 2013: 10). More importantly, Asfaw et al. state that the adoption of inorganic fertilizer is more likely with men than women while on the other hand use of organic fertilizers and intercropping maize with legumes is more likely among women. Furthermore land ownership promoted the adoption. Irrigation farming tends to increase maize productivity and according to the study maize yields are comparatively higher by 40–75 per cent than in farms without irrigation (Asfaw et al. 2013: 15).

Soil and water conservation measures have been promoted and largely adopted by people in the central region. This is despite the fact that statistics from the irrigation department in Malawi indicate that Lilongwe dominates in smallholder irrigation agriculture in terms of irrigation equipped area. Nevertheless, the size of irrigated area per beneficiary is several times smaller compared to irrigation schemes along the lakeshore areas and Shire valley (Department of Irrigation 2013). On average, each beneficiary has an irrigation-equipped area the size of 0.07 ha compared to 0.16 ha per beneficiary in the lower Shire Valley, for instance. Moreover, streams dominate as the primary source of water for irrigation. This may suggest that the equipped irrigation area is very limited to the areas closet to these perennial streams. Thus, improved maize seed varieties, soil and water conservation, mulching, intercropping and other forms of climate-smart agriculture are promoted as the surest way to adapt to climate change.

While several adaptation strategies may be promoted and adopted to increase food production, what will remain challenging is addressing post-food handling that is intrinsically linked to important Chewa culture and symbolism. The government may develop campaigns to address post-handling of food to avoid wastage and promote food security. Whether the Chewa people will carry out their ceremonies with food security issues in mind is yet to be seen but has an obvious impact on what food and how much remains at household level.

CONCLUSION

As the climate becomes increasingly uncertain, food security perceptions and efforts and strategies to attain the same are facing an increased challenge. This is mainly due to the fact that rainfall has increasingly become unreliable and onset and offset of rains becomes unpredictable. The received wisdom suggests that farmers have a better chance to adapt to such unpredictability by adopting drought resistant crops such as tubers, millet and sorghum as well as adopting other adaptation strategies that include early maturing varieties, crop diversification, climate-smart agriculture, etc.

What this chapter has argued is that cropping patterns are not solely influenced by the character of the prevailing water system. In Malawi this has resulted in the predominant cultivation of maize regardless of the fact that some of the cultivated lands are not suitable for maize production. Moreover, as the case study of the Chewa people suggests, the mere fact of having more food at household level does not translate the household into a food secure one. This is due to the fact that food among the Chewa holds such complex and deep cultural and symbolic meanings. This suggests that foods that do not hold such meanings may be considered out of place. Maize will therefore remain central in the short to medium term in the foodways of the Chewa. Assuming that climate change forces the people to switch to other crops other than maize, the symbolic and cultural roles of maize may cease and such meanings may be ascribed to other foods that may replace it. While continual production of maize increases the vulnerability of the people to erratic weather patterns because of the changing agro-ecological conditions which disfavour its production, policy makers should not ignore the embedded and symbolic importance of the crop. Policy makers must therefore promote other foods alongside maize in order to lessen communities' vulnerability to climate change and allow the slow and enduring processes of cultural change to take effect before maize can be significantly replaced if and when that step needs to be taken.

NOTES

1 'Malawian culture' refers to a historical ideology that was advocated by the first Malawian president after Independence. It still has an effect on Malawians to this day. Dr Kamuzu Banda's concept of 'Malawian culture' was based on the Chewa matrilineal tradition and this was enforced on those who attempted to resist his concept of oneness. See also see also D. Kaspin (1995) and P. G. Foster (1994) on further discussions on 'Malawian Culture'.
2 *Gule wankulu* rituals involve the performance of masked dancers. It is a ritual that singles out the Chewa as a distinct group of people.
3 See also Government of Malawi (2006:14, fn 9), where it is stipulated that food security policies carried out by the government throughout the 1990s were primarily driven by ADMARC's capacity to regulate domestic maize production, which avoided increases in maize prices to consumers, especially during the lean season of November to March.
4 The over-reliance on maize in Malawi, especially the preference for white flint maize varieties, which take longer to mature than most hybrid varieties, is constantly referenced as one of the causes of food insecurity by policy practitioners (see Menon 2007, Sahley et al. 2005). The preference for this particular maize variety, also known as the 'maize of the ancestors' among the Chewa, is regarded as one of the factors that sets back their adoption of new hybrid maize varieties (see also Smale 1993, Smale et al. 1995).
5 This was part of the Rome Declaration on World Food Security and World Food Summit Plan of Action.

REFERENCES

Adema, P. 2007. 'Foodways'. In Smith, A.F (ed.), *The Oxford Companion to American Food and Drink*. Oxford and New York: Oxford University Press, pp. 232–3.
Arndt, C., Schlosser, A., Strzepek, K., Thurlow, J. 2014. 'Climate change and economic growth prospects for Malawi: An uncertainty approach', *Journal of African Economies* 23, AERC Supplement 2: ii83–ii107.
Asfaw, S., McCarty, N., Lipper, L., Arslan, A., Cattaneo, A. 2013. 'Adaptation to climate change and food security: Micro-evidence from Malawi'. Invited paper

presented at the *4th International Conference of the African Association of Agricultural Economists, 22–5September*, Hammamet, Tunisia.

Bie, S.W., Mkwambisi, D., Gomani, M. 2008. *Climate change and rural livelihoods in Malawi: Review study report of Norwegian support to FAO and SCC in Malawi, with a note on some regional implications.* NORAGRIC report 41: Department of International Environment and Development Studies.

Chenje, M. (ed). 2000. *State of the Environment Zambezi Basin 2000*. Maseru/Lusaka/Harare: SADC/IUCN/ZRA/SARDC.

Cook, A.M., Ricker-Gilbert, J.E., Sesmero, J.P. 2013. 'How do African households adapt to climate change? Evidence from Malawi'. Selected paper prepared for presentation at the *Agricultural & Applied Economics Association's 2013 AAEA & CAES Joint Annual Meeting, 4–6 August*, Washington DC.

Counihan, C. 1999. *The Anthropology of Food and Body: Gender, Meaning and Power*. New York and London: Routledge.

Department of Irrigation. 2013. *Department of Irrigation Annual Report 2012–2013*. Lilongwe: Department of Irrigation.

Douglas, M. 1966. *Purity and Danger: An Analysis of Concepts of Pollution and Taboo*. London: Routledge & Kegan Paul.

Englund, H. 1999. 'The Self in Self-interest: Land, Labour and Temporalities in Malawi's Agrarian Change'. *Journal of the African Institute* 61(1): 138–59.

FAO. 2003. *Trade Reforms and Food Security: Conceptualizing the linkages*. Rome: FAO

Harris, M. 1987. 'Foodways: Historical Overview and Theoretical Prolegomenon'. In Harris M., Ross, E.B. (eds), *Food and Evolution: Towards a Theory of Human Food Habits*. Albuquerque: Temple University Press, pp. 57–90.

Kachapila, H. 2006. 'The Revival of Nyau and Changing Gender Relations in Early Colonial Central Malawi'. *Journal of Religion in Africa* 34(3/4): 319–45.

Malawi Government. 2006. *Malawi's National Adaptation Programmes of Action (NAPA), under the United Nations Framework Convention on Climate Change (UNFCC)*. Lilongwe: Environmental Affairs Department.

—— 2008. *Food Security Policy*. Lilongwe: Ministry of Agriculture, Irrigation and Food Security.

—— 2010. *Malawi state of the environment and outlook report: Environment for sustainable economic growth*. Lilongwe: Environmental Affairs Department.

—— 2013. *National Climate Change Investment Plan*. Lilongwe: Environmental Affairs Department.

Muzhingi, T., Langyintuo, A.S., Malaba, L.C., Banziger, M. 2008. 'Consumer Acceptability of Yellow Maize Products in Zimbabwe'. *Food Policy* 33: 352–61.

Nangoma, E. 2007. 'National adaptation strategy to climate change: A case study of Malawi'. *Human Development Report 2007/2008, Fighting climate change: Human solidarity in a divided world*, Human Development Office Occasional paper, 2007/52.

Oestigaard, T. 2011. 'Water and Climate Change in Africa – from causes to consequences', *Policy Notes* 2011(4), Nordic Africa Institute.

Phiri, K.M. 1983. 'Some Changes in the Matrilineal Family System Among the Chewa of Malawi since the Nineteenth Century'. *Journal of African History* 24(2): 257–74.

Pottier, J. 1999. *Anthropology of Food: The Social Dynamics of Food Security*. Cambridge: Polity Press.

Tschirley, D.L., Santos, A.P. 1995. 'Who Eats Yellow Maize? Preliminary Results of a Survey of Consumer Maize Preferences in Maputo, Mozambique'. *MSU International Development Working Paper 53*, Department of Agricultural Economics and Department of Economics, Michigan State University.

Tumbare, M.J. 1999. 'Equitable Sharing of the Water Resources of the Zambezi River Basin'. *Physics and Chemistry of the Earth (B)* 24(6): 571–8.

—— (ed.) 2000. 'Cyclic hydrological changes of the Zambezi River basin: Effects and mitigation measures'. *Management of River Basin And Dams: The Zambezi River basin*, Rotterdam: A.A Balkema Publishers, pp. 201–8.

Weismantel, M.J. 1988. *Food, Gender and Poverty in the Ecuadorian Andes*. Philadelphia: University of Pennsylvania Press.

Wolf, E. 1982. *Europe and the People without History*. Berkeley: University of California Press.

World Bank. 2008. *Zambezi River Basin: Sustainable Agriculture Development*, Washington, D.C.: International Bank of Reconstruction and Development.

20   Land and Water for Drugs, Cash for Food: Khat Production and Food Security in Ethiopia

Gessesse Dessie

INTRODUCTION

Consumed for the amphetamine-like properties of its fresh leaves, khat is an agricultural cash crop, perishable product and high-value commodity. Both production and consumption of khat have been the source of heated debate in developing countries as well as globally. In the nearest past khat emerged from an obscure, wild, backyard bush to become an intensively farmed crop of global importance. This according to several studies (Gebissa 2004, Hailu 2005, Anderson et al. 2007) is attributed to a number of factors including its adaptability to wide environmental variables; its comparative economic advantage for small land holdings; improvement of transport infrastructure and increasing diaspora community from East African and the Arabian Peninsula.

Although it is not mentioned as one of the 12 farming systems in sub-Saharan Africa (Dixon et al. 2001), khat is an important part of the agricultural landscapes of Ethiopia, Kenya, Madagascar and Uganda. Studies from these countries (Gebissa 2010, Njiru et al. 2013, Gezon 2012) have shown the food security connotations of khat. Two contrasting sides of the link between khat and food security have been underlined: first, khat is considered a drug hence it seems straightforward to favour food crops instead of growing khat; and secondly, khat is a non-food cash crop like that of cotton, tea, coffee and cocoa worth considering for its benefit to farmers' livelihood.

Comparative cash crop (food and non-food) and subsistent crop (mainly food crops) analysis in accordance with food security at smallholder farmers level exists (Govereh & Jayne 2003, Babu et al. 2009, Schneider & Gugerty 2010). According to Govereha and Jayne (2013) the cash crop vs food crop debate has two comparative arguments: (1) cash crops are drivers of agricultural growth and development through which food security can be attained; and (2) cash crops compromise smallholders' food security. The issue of khat in this debate has extra moral, political and legal dimensions (Dessie 2013a). The increasing interest of farmers growing khat, a plant with limited nutritional value and yet classified as a drug in a country where food security is high on the agenda, is worth scrutiny. According to FAO (2013) Ethiopia is still challenging food security issues to address undernourishment (minimum level of dietary energy consumption). There are several determinant factors regarding food security in Ethiopia, including significant amount of food availability gap (Adenew 2004, Berhane et al. 2013); apparent yield gap for major food crops including wheat, sorghum and maize (Schneider 2010), and frequent drought (Adimassu et al. 2014).

The case of khat can be examined from a commercial agriculture point of view. The relationship between commercialization of agriculture and food security is based on the influence of the increased household income of food consumption and nutritional adequacy (Babu et al. 2009). The assumption is that increased income can be used for increased food expenditures. However, the income–food consumption relationship is influenced by many factors such as who controls the income, the proportion of money spent on food and non-food items and whether the increased income results in a higher intake of calories or intake of more expensive calories (Babu et al. 2009).

Food security is better understood when examined holistically with a clear understanding of food systems, the food chain and food sovereignty components (Ingram 2009). Food security exists when all people, at all times, have physical, social and economic access to sufficient, safe and nutritious food which meets their dietary needs and food preferences for an active and healthy life (FAO 2003). Food security outcomes are described in terms of four components and their sub-components: food availability (production, distribution and exchange); food access (affordability, allocation and preference); food utilization (nutritional and social values, and food safety); and stability (extent and exposure to risks) (Ingram 2009, FAO 2013). Food security analysis must include two primary components: (1) physical availability of food in proximity of household, regardless of the process through which it was made available; and (2) the level and type of resources expended to attain household food security relative to the total resources available at the household level (Jonsson & Toole 1991). One of the ways to approach such food security analysis is to make spatial analysis of landscapes to determine the fraction of khat out of other land and water uses containing food and others.

Khat land use is critically linked with sufficiency and the timely availability of water. Increasing production of khat in parts of Yemen and Ethiopia is linked to acute water scarcity (Heffez 2003, Alemayehu 2006). Khat consumes significant amount of water throughout the whole value chain from production to consumption. The fact that it is consumed fresh and green (perishable otherwise) means the final product contains high moisture content. Cultivation of this crop is water intensive due to the fact that the frequency of harvest of the fresh leaves is determined by water supply. Moreover, if nutritional security is the goal of interest, estimates of access to food should be combined with estimates of access to clean water and good sanitation (Pinstrup-Andersen 2009).

From khat production points of view examining the food security issues need to take into account the global banning directive issued against khat and its age old controversies in Ethiopia. Both can have a detrimental impact on the future of khat production. Studies so far emphasize the affordability factor (Mulatu & Kassa 2001, Hailu 2005, Gebissa 2008, Gezon 2013). Holistic analysis requires the link between elements of food security (affordability, availability, quality and safety) with characteristics of khat production that include high financial return, competition for land with other lands uses and anorexic effect on the consumers. Until we understand the plight of the drug-producing farmers, we will never successfully address one of the most difficult problems of our time, the global drug trade (Steinberg et al. 2004).

The objective of this chapter is to analyse food security implications of the increasing production of khat on the producers of the stimulant in Ethiopia. More specifically the study addresses the link between khat and food security in three categories: (1) quality and safety aspects in line with nutrition and calories; (2) food affordability aspects from khat income analysis; and (3) food availability aspects from land cover frequency and agroecosystem points of view. The chapter contextualizes the khat food security link in the context of land and water within a production-consumption continuum.

STUDY APPROACH

The study employed secondary data sources and spatial analysis techniques. Sources of secondary information were statistical abstracts and literature review. Statistical abstracts were used to extract quantitative means of comparison between khat and food crops requirements. Careful reading and cross-referencing of literature was done to situate khat in the wider socio-ecologic landscapes. The literature review was based on original research papers from different spatial and temporal settings which have made rigorous analysis of primary data. With regards to spatial analyses, digital photographs were visually interpreted. A series of ground photographs of khat landscapes in Ethiopia which were taken during several research field work trips during the past five years were used to qualify and quantify agroecosystem properties. This was done after the photographs were displayed on a computer screen, divided into equal size of square grids and the frequency of crops in each square was counted. Finally the frequency of khat farm abundance among other landscape components was determined.

The visual interpretation of digital photographs was chosen because the method has been used to assess vegetation covers and for social survey (Sykes et al. 1983, Schwartz 1989, Zhen-qi 2007), and visual interpretation of digital photographs provide cheap and reasonably reliable alternatives to digital interpretation of remote sensing products. This is particularly important due to the limitations of using satellite images to discriminate the spectral signature of khat from other crops with which it is mixed.

At the same time this study also recognized the limitations of visual interpretation of still ground pictures. While the ground photo shows the real appearance of the crops there are two limitations. First, nearly horizontal (low horizon picture) focus has varying scale along the depth of field hence area estimation is complicated, and secondly, objects in the foreground can block the view of the ones in the background. These two limitations were addressed by emphasizing frequency counts instead of area classification, and background objects are determined with a best guess based on the frequency of the surrounding objects.

CONTEXTUAL FOUNDATION

The contextual foundation of this study rests on khat production in accordance with land allocation, land potential and level of intensification (see Figure 20.1). Availability, sufficiency and temporal distribution of water characterize land potential. Food security can be achieved through producing enough food for one's own needs or having enough money to buy food or combine the two to supplement one with the other (FAO 1996). Food production is directly linked with the way land is divided between subsistent and commercial purposes. Khat production competes with food production for land and water. The size of land holding is an important determinant of an individual farm level decision to grow khat (Hailu 2005). From a food security perspective, the decision rests on which crop to emphasize: food crops or khat? In Ethiopia where land size is generally very small and declining over time (Headey et al. 2014), growing khat cannot avoid limiting food production space (Getahun & Krikorian 1973, Dessie & Kinlund 2008).

In khat production and consumption continuum demand and supply have direct bearing on land and water use. The market demand for khat determines the supply which is the amount produced based on yield, land area and level of intensification which is directly linked with water. Unlike storable and convertible agricultural produces, the perishability of khat affects its market supply (Carrier 2005). Unless the right speed is maintained within the perishability time frame post-harvest loss of the leaves is likely to be very high. Hence, the supply side is negatively affected; more production is needed to compensate for the loss; and more land or a high level of intensification is needed to do that. Intensification is particularly important here, because of land scarcity, Ethiopian khat

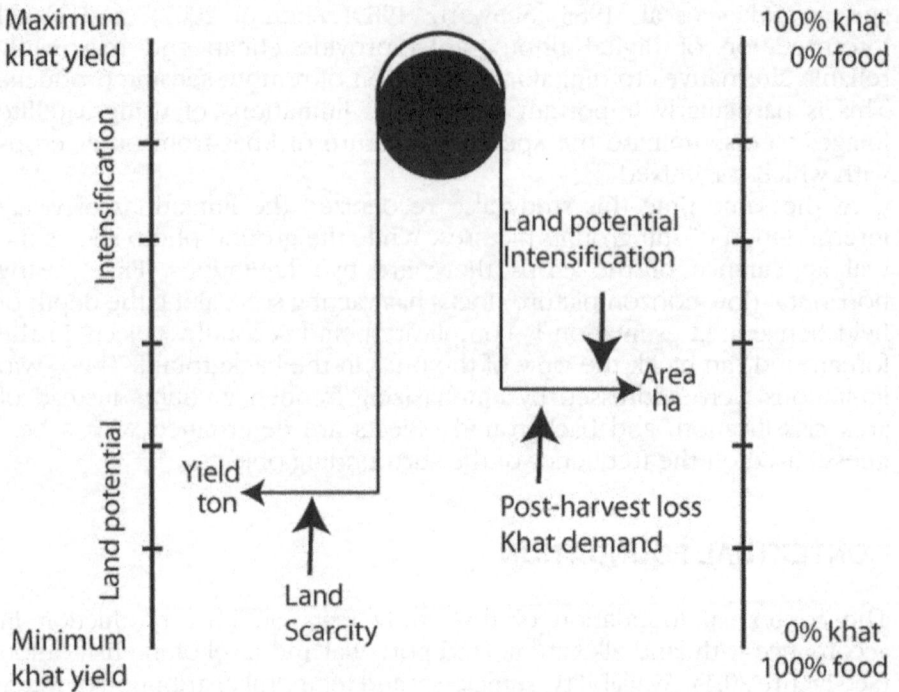

Figure 20.1. Opposing forces that determine competition for land between khat and food crops. It is assumed that khat and food production share the finite rural land. In theory when yields get too low, farmers tend to move to other land use types, mainly forests and natural cover areas. Increase in demand can be satisfied by extensification, adding new land to khat farming or improving the productivity of the existing khat land by intensification (© Gessesse Dessie).

farmers resort to intensifying productivity. Researchers have established a positive relationship between intensification, land scarcity and population density (Headey & Jayne 2014, Muyanga 2014, Headey et al. 2014). There is a positive relationship between farm size and productivity in both labour-scarce and land scarce smallholder farming (Dorward 1992). With regards to khat one measure of intensified production is crop frequency. Cropping frequency is a measure of land and water use intensity and allows an analysis of the dynamic relations of intensification and land expansion (Boserup 1965, Erb et al. 2014).

The shift towards a khat-based farm economy was unavoidable and in the absence of other feasible alternatives, khat production and marketing would remain important to the livelihood systems of farmers (Mulatu & Kassa 2001). Within given agroecosystems the expansion of khat can theoretically take over all the land occupied by food crops (Figure 20.1). The scale between no khat and all khat is determined by farmers' decisions making variables that balance the need for subsistent food production and monitory gain from khat farming. The mid point could be a point at which both khat and food coexist without a negative impact on each other. This can also be a point at which land areas fail to support a feasible yield of food while khat can still thrive.

As the production of khat is explained by the size of land area and yield per given land, the commercial part of it is the reflection of the increasing demand for the leaves. If competition for land and water cannot be avoided between khat and food crops, the extent of expansion can be reduced through dynamics and effective khat production systems. In fact, land allocation decisions vary among farmers whose food security statuses differ. Food secure households are engaged substantially in non-staple cash enterprises like livestock rearing, cash crops, and trading, implying that diversification, based on local resources and market opportunities, is an essential component of food security (Nigatu 2004). More diversified production may relieve some of the pressure by reducing the degree to which these farmers produce solely for subsistence. Smallholders whose agricultural production is highly subsistence oriented have less capital to purchase food in times of economic hardship or food scarcity (Dorsey 1999).

As the size of land diminishes, its high financial benefit makes khat more feasible to grow than food crops. At the same time scarcity of land justifies intensification of farming to harvest more yield per available area and water. While high land potential and intensification limits the area for khat production by creating more yield for a given area, the level of post harvest loss and increasing demand can result in more areas for production. Yield is the combined effect of land potential for good growing environment and the level of intensification. Availability of water during the dry season increases the frequency of harvest of khat leaves hence, yield per year. Therefore, water by increasing the potential for increasing the frequency of harvest is an important input in the intensification of khat farming. By the same analogy water deficit is a determinant to strain khat production and productivity. Hence, water is a critical input in the khat-food security linkage: (1) water that would have been used to grow food is consumed by khat; and (2) water that could have remained in local agricultural landscapes is taken out leading to decreasing food production potential. Khat trade embedded virtual water (Allan 2003) like any other commodities including agricultural products. The virtual water content of khat is the amount needed for the production of the valuable fresh green leaves measured at the producing localities.

QUALITY, SAFETY AND AVAILABILITY ASPECT OF FOOD SECURITY AND KHAT

Today khat is a permanent fixture in several agricultural landscapes distributed all over the country. Analysis of an agricultural sample survey of

A History of Water

Ethiopia (CSA 2010) shows that khat occupies about 20 per cent of the average land holding of 0.5 ha per family of 7–8 members in the major khat growing areas of the country. Although its land area coverage seems small compared to other land and water uses, khat constitutes a high proportion of farmers' income; contributes significantly to the livelihood of actors, and stands among the top five most important foreign currency earners for Ethiopia (Gebissa 2004, Andersson et al. 2007, Gezon 2013). At the farm level, income per hectare from khat surpasses all major agricultural crops by several margins (Dessie 2013b). Today over 10 per cent of households in Ethiopia produce khat compared to over 23 per cent households of coffee farmers (Gebre-Ab 2006). Khat farming and trade has been expanding rapidly during the past couple of decades. For example, the most recent six months' export performance report (2013–14) of Ethiopia shows a 10.5 per cent increase for khat while other crops including coffee have declined (Fortune 2014).

Food quality and safety is one of the components of food security. The quality and safety of food security is mainly related to how nutritious and healthy the available food is. Habitual consumption of khat has strong implications for this component of food security. Khat is not only deficient in major nutrition types; its habitual chewing is associated with an anorexic effect (Belew et al. 2000). Moreover, khat consumption has a proven effect on appetite suppression and is known to fend-off hunger (see, for example, Murray et al. 2008). This suggests khat chewers tend to take less food or lack a diversified food intake to fulfil the normal nutritious requirements. Khat is too poor in nutritional content to be considered as food, which according to the *Encyclopaedia Britannica* (2004) is material consisting essentially of protein, carbohydrate and fat used in the body to sustain growth, repair cells, and vital processes and to furnish energy. Nonetheless, khat contains essential components including amino acids, vitamins and minerals most notably very high vitamin C, calcium and iron content (Kalix & Braenden 1983).

As shown in Table 20.1, the protein content of khat is low compared to maize (*Zea mays*) and teff (*Eragrostis tef*), but high compared to enset

Table 20.1. Protein, carbohydrate and fibre content of food crops in relation to khat. This table is calculated from known nutritional values of crops grown in Ethiopia.

Plants	Protein (%)	Carbohydrate (%)	Fibre (%)	Unique Threat
Maize	9.7	71.5	7.4	High phytonutrients (antioxidant)
Enset	4.4	76.3	7.1	High Calcium content
Teff	11	80	13.5	High Iron content
Khat	5.6	NA	2.5	High vitamin C content
Spinach	2.4	11.8	5.9	Low in cholesterol content

Source: © Gessesse Dessie.

Table 20.2. Calories/ha of selected food crops. This table is calculated from known calorific values of each crop, estimated water requirement for producing the crops, currently attained yield per hectare of the crops in Ethiopia

Crop	Calories/kg	Water FP lit/kg	Yield kg/ha	Calories/ha Kilo calorie
Wheat	3,583	800	1,610	5,769
Maize	3,241	1,222	1,807	5,856
Sorghum	3,390	540	1,690	5,729
Teff	2,340	NA	1,150	2,691
Enset	1,900	NA	49,000	93,100
Potato	1,131	287	8,770	9,919
Banana	2,175	860	8,430	18,335
Lettuce	150	237	4,760	714
Cane sugar	3,990	1,782	79,290	316,367
Khat			8,000	

Source: © Gessesse Dessie.

(*Ensete ventricosum* (*Welw.*)). Though its carbohydrate content is not determined, it can be assumed low as in most leafy vegetables. In this regard teff, enset and maize have a high amount of carbohydrate. Although it has limited food value, it may not mean that it is a hard drug. A rational scale designed by Nutt et al. (2007) to evaluate the harm level of the 20 most consumed drugs around the world showed that khat is the least harmful of all.

If khat is not food, it cannot be helpful in enhancing food security directly by supplying nutrition. The food security aspect is therefore related to its indirect effect on the production, consumption and distribution of food. Food availability is the function of available food crops farmland. The increasing expansion of khat farming in agricultural landscapes competes with land resources including land and water; therefore negatively affects local food crop production. With regards to land the simple assumption is that if food crops are grown in areas taken up by khat farms, carbohydrates and proteins could have been produced (see Table 20.2). Khat cultivation in each hectare of land is equivalent to the loss of 5,856 kcal from maize and 2,691 kcal of teff (see Table 20.2). At 2,000 cal/day normal calorie requirement of an average person replacing a hectare of these two crops alone means a daily feed for over 2,900 and 1,300 people.

WATER CONSUMPTION OF KHAT AND ITS CASH VALUE

Most khat farms in Ethiopia are located in high potential irrigable areas consuming water that could have irrigated food crop production. A research from Yemen reported that the average annual evapotranspiration of khat is about 553.6 mm/year (Atroosh & Al-Moayad 2012). The freshly consumed leaves that contain about 90 per cent water negatively affect water supply to other land uses. This high moisture content indicate that each tonne of khat taken away from a given farm drains 900 kg of water. This is without calculating the amount of water consumed by the plant during the growing process until it reaches the harvestable stage. Water appears repeatedly in the whole khat value chain. At a production stage rain water and or irrigation water are used to sustain the growth and increase the frequency of harvesting the leaves. With irrigation water khat can be harvested up to 12 times a year compared to rainfed farming which yields a maximum harvest of twice a year. Crop yield increases with the same factor of water flow in the photosynthesis process (Gerbens-Leenes & Nonhebel 2004). As a perishable substance khat require moist handling and green wrapping and, therefore, packing and transporting consume water. Water is also necessary to facilitate the chewing and to dissolve the stimulant chemicals during the mastication process. Khat production contributes to the shrinkage and disappearance of water sources and wetlands in Ethiopia (Dessie & Kinlund 2008, Lemma 2011).

The water resources degradation in the growing landscape is related to the trade continuously mining the green fresh leaves. This export of virtual water brings a high cash value compared to food crops. Most food crops in Ethiopia have a low cash value compared to khat. The estimate of the area required for producing 2000cal/day/capita and its monitory value in relation to khat income shows a low value of the three most important cereal crops (see Figure 20.2).

For example teff, a unique crop to Ethiopia, scored the lowest cash value compared to khat while bananas score the highest. Compared to the two most important crops in Ethiopia, maize and teff, the monitory value of khat per hectare is 12 times higher than maize and ten times more than teff. With such a high cash advantage it is easy to assume that khat farmers can afford to buy food items as long as they are available on the market.

AGROECOSYSTEMS OF KHAT LANDSCAPES AND FOOD SECURITY

The proportion of Ethiopian rural landscapes occupied by khat varies. Table 20.3 supported by Figure 20.4, shows the frequency of occurrence of khat is 13.3 per cent, 25.5 per cent, and 24.4 per cent in south central, eastern, and north eastern Ethiopian khat landscapes respectively. The relative distribution is 0.21, 0.43 and 0.37 in south central, eastern, and north eastern Ethiopian khat landscapes respectively. In a landscape fully covered by khat the relative distribution would be 1. Significant expansion of khat has occurred since 1960s. In the most khat dominant landscape of eastern Ethiopia khat constitutes 13 per cent of the farmland in the 1960s, to constitute 30–50 per cent of total cash income per year per family, or

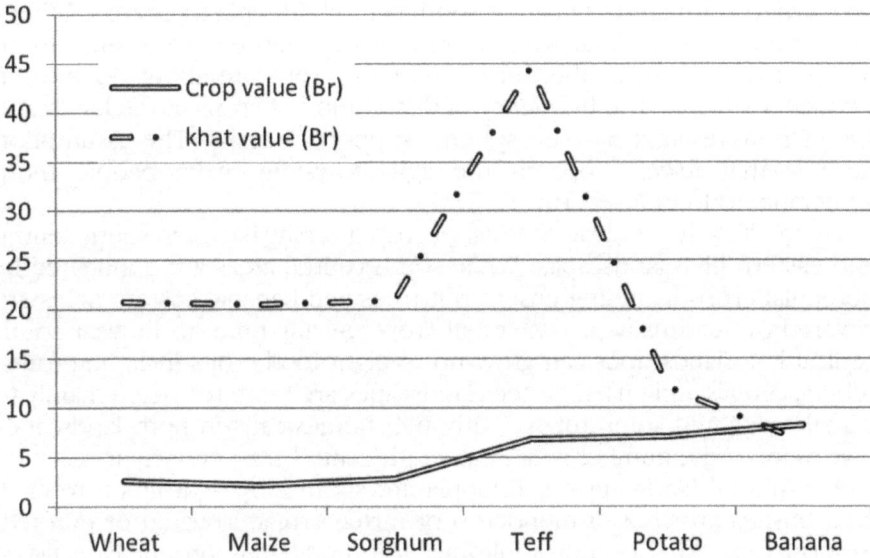

Figure 20.2. Value of food crops to produce 2000 cal daily per capita healthy human calories. This graph is constructed from the amount of area required to produce 2000 cal and converted to cash based on current crop prices. These areas are used to calculate the amount of money accrued if khat was grown instead of food crops (© Gessesse Dessie).

40–60 per cent of the total value of home-produced food used by the farm family (Getahun & Kridorian 1973).

Food crops are still important parts of khat landscapes and so are trees. Figure 20.3 shows that food crops surpass khat in south central and north eastern khat growing areas, but are lower in the eastern khat growing areas. In the east and north east open areas and settlements constitute small

Table 20.3. Relative frequency of khat in 3 different khat landscapes of Ethiopia. This table is calculated from visual count of crop frequency on digital photographs divided into square grids. The frequency estimation was done in two ways including percentage total: factor of each crop at a landscape level from the total count of all crops and per cent grid, and the number of grids that contain each crop. While per cent total shows the relative frequency of each crop in the landscape, per cent grid shows the concentration of crops in each grid (© Gessesse Dessie)

	South Central			Eastern			North Eastern Ethiopia		
Land use	Count	% Tot	% Grid	Count	% Tot	% Grid	Count	% Tot	% Grid
Trees	40	17.7	83.3	57	26.6	55.8	25	27.8	75.8
Homestead	23	10.2	47.9	22.5	10.4	19.1	10	11.1	30.3
Open land	23	10.2	47.9	25	11.7	26.3	2	2.2	6.1
Khat	30	13.3	62.5	54.5	25.5	58.9	22	24.4	66.7
Food crops	110	48.7	229.3	55	25.8	59.9	31	34.4	93.9

proportion compared to khat. In south central Ethiopia trees are 1.5 times more frequent than khat while food crops are more than 3 times more frequent than khat. In the north east food crops are about double the frequency of khat. The frequency of distribution of crops in the landscape shows how resilient agro-ecosystems support diversity. The assumption here is that fewer crops in the agroecosystem make people more vulnerable to food insecurity.

Figure 20.4 shows clear contrast of crop diversity between south central and eastern khat landscapes. While south central areas are dominated by perennial crops including enset, fruit trees and banana; eastern areas are covered by seasonally grown cereal crops. At any time of the year south central khat landscapes can grow up to eight food crops including fruits, tubers, cereals and enset. Eastern landscapes are more restricted mainly to cereal crops and some tubers. Although homesteads in both landscapes own animals the number is high in south central than eastern areas.

Agricultural landscapes in Ethiopia are spotted by dark green rows of khat bushes grown as a monocrop or mixed crops, irrigated or rain fed. In most cases khat occupies hillsides and moderately productive valleys. Existing khat landscapes (Figure 20.4) are interspaced food crop areas. Figure 20.4a shows intercropped khat within food crops while Figure 20.4b shows patches of khat fields within the landscapes. Both pictures show that khat and food could coexist. While cereals are dominant food crops in

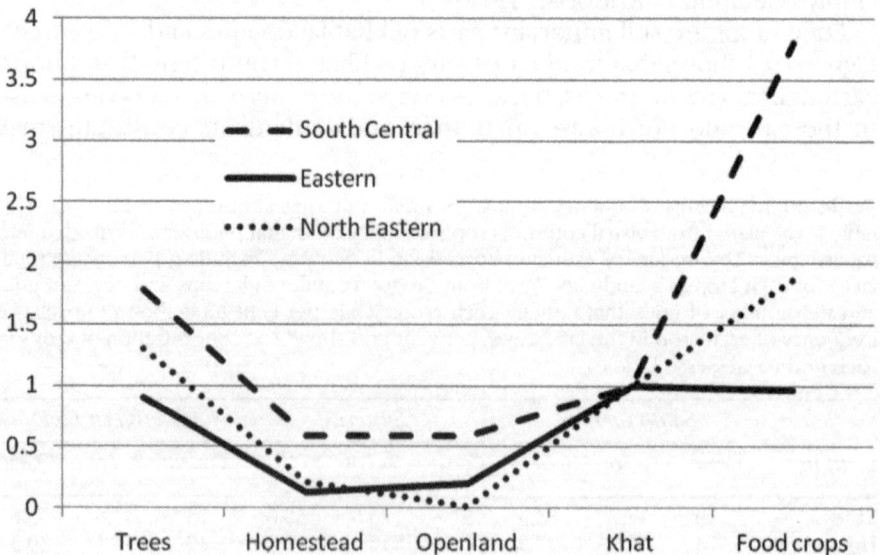

Figure 20.3. Relative abundance of land uses factored by khat frequency from major khat growing areas of Ethiopia. This graph is constructed from the relative abundance estimation based on visual interpretation of the digital photographs (© Gessesse Dessie).

Figure 20.4 (a and b). Two contrasting khat landscapes in Ethiopia 4a eastern Ethiopia and 4b south central Ethiopia (photo: Gessesse Dessie). The vegetation architecture of khat landscapes in Ethiopia has a dark green shed formed by orchards of khat farms. In southern Ethiopia where enset and fruit trees exist khat paint emerald green tone the yellowish green background formed by the upper canopy enset. Where khat is intercropped with maize and sorghum strips of whitish green and darker green alternatively exists where the latter is formed of khat strips unlike tapestry look of landscapes where crops are grown mixed in patches (© Gessesse Dessie).

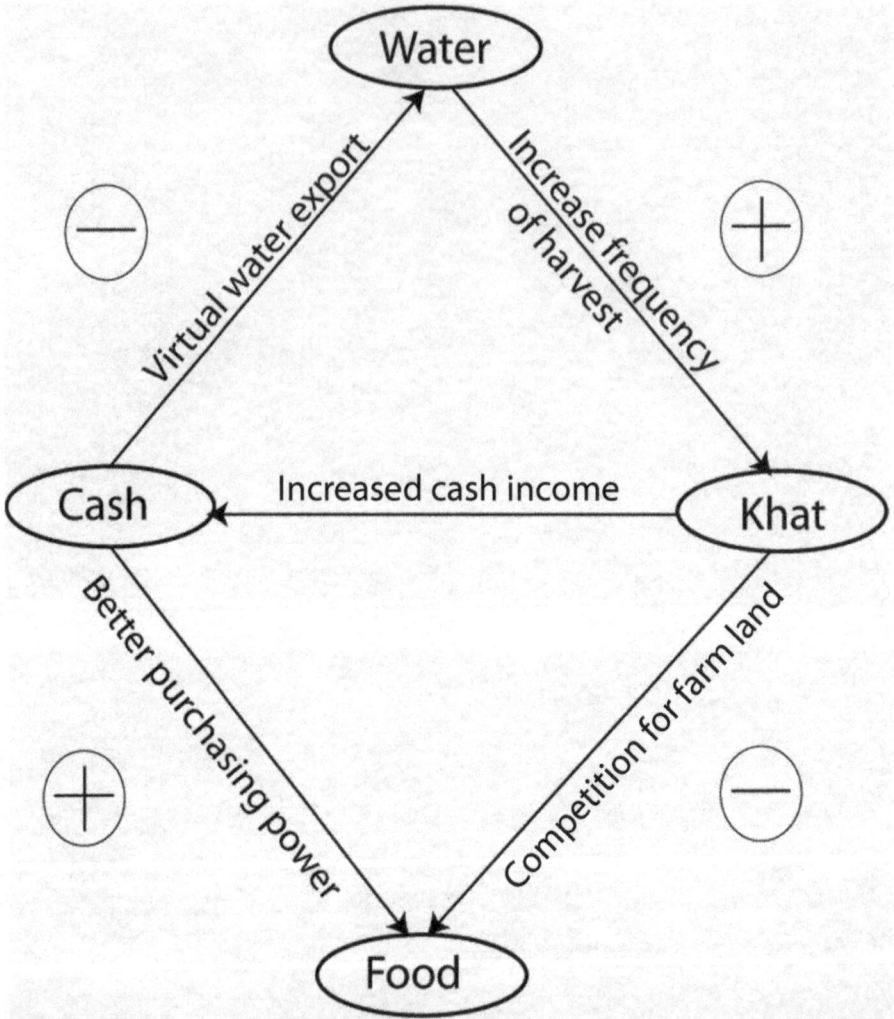

Figure 20.5. Links between water, khat production, cash income, and food security. Availability and annual distribution of water ensures high khat productivity through the increased frequency of leaves harvested. This led to improved cash income and better food purchasing power. Both productivity and affordability are positive food security elements. Khat production, competing for farmland that would otherwise produce food, can have a negative impact on food security. The cash incentive of khat production trade exacerbates local water degradation via virtual water export. This can have a negative impact on food security by decreasing the food production potential of local landscapes (© Gessesse Dessie).

Figure 20.4a enset stands out in 4b. In both landscapes, planted trees, in particular eucalyptus, are apparent.

Small proportions of open land (Table 20.3) restrict animal husbandry to intensive managment. In both landscapes farmers keep few animals, often with zero grazing management. Animals are needed for fertilizing fields and they are the main source of protein. In the case of south central Ethiopia where enset is important animals play a significant role in the farming of this crop. The enset system cannot operate without livestock and the enset production system also enables an intensification of livestock production since the leaves are often given to the cattle as food (Pankhurst 1993). Enset is a low protein food, but animals improve the basic diet of the people by supplying protein (Bezuneh 1993).

CONCLUSION

This study addressed the link between the increasing expansion of khat in Ethiopia and associated food security implications. It is with full cognizance of the global ban of khat as a classified drug that this study emphasized the food security aspect of it. The results confirmed that the expansion of khat production in the country has an impact on all elements of food security. However, this does not mean that the stimulant has a simple and direct impact on food security, since the causal relationship depends on several other factors. Putting aside the ethical dilemma of producing drugs for food security, systemic analyses of the competition for land provided a good context. From land management points of view, the growing of khat by smallholder farmers showed farmers' land related decision making behaviour at the background of attaining food security. The ways farmers prioritize the allocation of water and land to producing food or increasing cash income and their engagement in production intensification seem to have direct bearing on their food security issues (Figure 20.5).

A simple and direct method used to visually interpret the relative abundance of agricultural crops in this study has emphasized the importance of khat and the estimated values are in agreement with other khat landscape studies in Ethiopia (Feyisa & Aune 2003, Dessie & Kinlund 2008).

Food security is increasingly subjected to multiple pressures, on both the demand side and supply side of food availability, access and utilization (Misselhorn 2012). There is a common understanding that cash crop producing smallholder farmers are food secure (Govereha 2003, Schneider 2010). The same conclusion was reached about khat producing farmers in Ethiopia, Madagascar and Kenya (Gebissa 2008, Gezon 2012, Nijiru et al.). It is important to note that cash crop production and food security are not always positively related. There are studies indicating that cash crop production areas in isolated rural areas are exposed to multiple food insecurity risk factors (Casewell et al. 2012). Other studies like, Babu et al. (2009) are also cautious about the positive link, stating that it is not necessarily always the case.

The affordability factors seem to be at the centre of all the arguments that rationalizes the positive link between khat production and food security. Evaluation of Ethiopian food security programs has shown that food from own production as a primary household source of food has declined over time (Berhane et al. 2013). This means that food availability is limited and households rely on the market to balance the food gap, which is about 2.3 months per year. This is compounded by price fluctuations of khat. The price of khat is season dependent where ample moisture increase supply hence reduce price. Studies indicate that khat prices fluctuate up to 100 per cent across seasons (Dessie & Kinlund 2008).

Leaving the affordability factor aside, it is worthwhile to look into how khat production influences local food availability. As observed in khat growing areas in Ethiopia, farmers cultivate khat as well as food crops. This is not unique to khat growers; in general few farmers are willing only to cultivate non-food crops because food for sale or purchase could be scarce or unaffordable, seasonally and limited in years of poor harvests (Schneider 2010). In two of the most important khat growing areas in the country, namely in the eastern and north eastern regions, food insecurity is prevalent; areas like Hararghe (eastern Ethiopia) are prone to chronic food insecurity (Piquet 2003) and north Wello is a famine prone area.

In the case of khat, besides increasing farm income, it promotes crop-livestock integration and enhances sustainability of smallholder mixed farm systems (Mulatu & Kassa 2001). Khat farming facilitates an intensification system when more land is used in khat production, although it reduces animal feed and food crops while increasing manure demand. Livestock remain a critical livelihood system component with roles shifted from providers of food and draft power for tillage to resources for cash income, fertilizer, and transport services (Kassa et al. 2002).

Khat farming in a limited farm land environment cannot be done without replacing other land uses or other crops. The fact that khat compete for land with other crops has been mentioned since the 1960s in Ethiopia (Getahun & Kridorian 1973), which is a trend that still continues (Gebissa 2010). The area under khat cultivation has grown considerably, mainly in Ethiopia and Kenya, often at the expense of other cash or food crops (Seyoum et al. 1986, UN-ODCCP 1999, Andersson et al. 2007). In this process of prioritization, the agroecosystem is simplified and become inflexible. In each converted land use when cropland is replaced, the carbohydrate produced is lost; when grazing land is replaced the protein produced is lost; when wetland is replaced the water source dries out and when forest land is replaced firewood becomes scarce.

To accommodate khat in limited landscape spaces crop diversity is the likely compromise to ensure financial optimization. For example, a decrease in grazing land as a result of khat expansion can force farmers to limit the number of animals they own. This can have a food security implication by decreasing sources of protein and fat, and manure input in food crop productivity. In the southern Ethiopian enset growing areas where its farming system is strongly linked to animal rearing, a decline in grazing land may have severe implications. Enset is a resilient crop in the face of soil moisture deficit, buffers food slack periods and it is one of the main sources of carbohydrate in the diet. Moreover, enset leaves are important fodder for animals in the absence of sufficient grazing land.

Thus, to conclude, khat production has indirect impacts on all elements of food security: it competes for land with food crops and other land and water uses, it changes processes of land use dynamics and it improves producers' income and transforms agro-ecosystems. Despite the upsurge of demand for khat through time, its future has become complicated due to calls for banning. As food security has a temporal aspect, namely continuity in the supply of sufficient and nutritious food, the uncertainty of the future depends on the emerging demand for khat and trade systems and the evolution of the khat growing landscape and the process of land use transformation. Before concluding, if khat has a negative or positive implication on food security, it is important to comprehensively analyse the causal relationships between the elements of food security and the consequences of khat expansion in relation to water and land use.

REFERENCES

Adenew, B. 2004. 'The Food Security Role of Agriculture in Ethiopia'. *Journal of Agricultural and Development Economics* 1(1): 138–53.

Adimassu, Z., Kessler, A., Stroosnijder, A. 2014. 'Farmers' strategies to perceived trends of rainfall and crop productivity in the Central Rift Valley of Ethiopia'. *Environmental Development*.

Alemayehu, T., Furi, W., Legesse, D. 2006. 'Impact of water overexploitation on highland lakes of eastern Ethiopia'. *Journal of Environ Geology* 52: 147–54.

Allen, J.A. 2003. 'Virtual water-the water, food, and trade Nexus useful concept or misleading metaphor?' *Water International* 28(1): 4–11.

Anderson, D., Beckerleg, S., Hailu, D., Klein A. 2007. *The Khat Controversy Stimulating the Debate on Drugs*. Oxford: Bloomsbury.

Atroosh, K.B., Al-Moayad, M.A. 2012. 'Water Requirements of Qat (*Catha edulis*) Cultivation in the Central Highlands of Yemen'. *Journal of Scientific Research* 4(1): 77–82.

Babu, S., Sanyal, P., Babu, S. 2009. *Food Security, Poverty and Nutrition Policy Analysis Statistical Methods and Applications*. Burlington: Elsevier.

Belew, M., Kebede, D., Kassaye, M., Enquoselassie, F. 2000. 'The Magnitude of Khat Use and its Association with Health, Nutrition and Socio-Economic STATUS'. *Ethiopian Medical Journal* 38: 11–25.

Berhane, G., Hoddinott, J., Kumar, N., Seyoum, A., Michael, T., Diressie, T., Yohannes, Y., Sabates-Wheeler, R., Handino, M., Lind, J., Mulugeta, T., Sima, F.

2013. *Evaluation of Ethiopia's Food Security Program: Documenting Progress in the Implementation of the Productive Safety Nets Programme and the Household Asset Building Programme*, Washington, D.C.

Bezuneh, T. 1993. 'An overview on enset research and future technological needs for enhancing its production and utilization'. In Abate, T., Hlebsch, C., Brandt, S.A., Gebremariam, S (eds), *Enset-Based Sustainable Agriculture in Ethiopia*. Proceedings from the International Workshop on Enset held in Addis Ababa, Ethiopia, 13–20 December 1993.

Boserup, E. 1965. *The Conditions of Agricultural Growth: The Economics of Agrarian Change under Population Pressure*. London: Earthscan.

Carrier, N. 2005. 'The need for speed: contrasting timeframes in the social life of Kenyan Miraa'. *Africa* 75(4).

Caswell, M., Méndez, V.E., Bacon, C.M. 2012. Food security and smallholder coffee production: current issues and future directions. ARLG Policy Brief # 1. Agroecology and Rural Livelihoods Group (ARLG), Burlington: University of Vermont. Available online: http://www.uvm.edu/~agroecol/?Page=Publications.html

CSA (Central Statistical Agency). 2010. Agricultural Sample Survey 2008/2009. Report on Aarea and Production of Crops. *Statistical Bulletin* 446(1).

Dessie, G., Kinlund, P. 2008. 'Khat expansion and forest decline in Wondo Genet, Ethiopia', *Geografiska Annaler: Series B, Human Geography* 90(2): 187–203.

Dessie, G. 2013a. 'Is khat a social ill? Ethical arguments about a 'stimulant' among the learned Ethiopians'. *ASC Working Paper 108/2013*. Leiden: African Studies Centre.

——— 2013b. *Favouring a Demonised Plant. Khat and Ethiopian Smallholder Enterprises*. Uppsala: Nordic Africa Institute.

Dixon, J., Gulliver, A., Gibbon, D. 2001. *Farming Systems and Poverty: Improving Farmers' Livelihoods in a Changing World*. Washington, D.C.: World Bank; Roma: FAO.

Dorsey, B. 1999. 'Agricultural intensification, diversification, and commercial production among smallholder coffee growers in Central Kenya'. *Journal of Economic Geography* 75(2): 178–95.

Dorward, A. 1999. 'Farm size and productivity in Malawian smallholder agriculture', *The Journal of Development Studies* 35(5): 141–61.

Encyclopaedia Britannica. 2014. Food. Encyclopaedia Britannica Online. http://www.britannica.com/EBchecked/topic.

Erb, K.H., Haberl, H., Jepsen, M.R., Kuemmerle, T., Lindner, M., Mülle, D., Verburg, P.H., Reenberg, A. 2013. 'A conceptual framework for analysing and measuring land-use intensity'. *Current Opinion in Environmental Sustainability* 5: 464–70.

Ethiopia Strategy Support Program II (ESSP II) International Food Policy Research Institute, Institute of Development Studies, University of Sussex.

Fafchamps, M. 1992. 'Cash crop production, food price volatility, and rural market integration in the third world'. *American Journal of Agricultural Economics* 74(1): 90–9.

FAO. 1996. *Report of the World Food Summit*. Rome: FAO.

——— 2003. *Trade Reforms and Food Security*. Rome: FAO.

——— 2013. *The State of Food Insecurity in the World: The multiple dimensions of food security*. Rome: FAO.

Feyisa, H., Aune, J.B. 2003. 'Khat Expansion in the Ethiopian Highlands effects on the farming systems in Habro District'. *Mountain Research and Development* 23(2): 185–9.

Fortune 2014. Export paralysis. *Fortune* 14(721).

Gebissa, E. 2008. 'Scourge of Life or an Economic Lifeline? Public Discourses on Khat (Catha edulis) in Ethiopia'. *Substance Use & Misuse* 43(6): 784–802.

———— (ed.) 2010. *Taking the Place of Food: Khat in Ethiopia*. Asmara and Trenton: Red Sea Press.

Gebissa, E. 2004. *Leaf of Allah: Khat and Agricultural Transformation in Harerge, Ethiopia, 1875–1991*. Oxford: James Currey.

Gebre-Ab. 2006. 'Commercialization of smallholder agriculture in Ethiopia. Ethiopian Development Research Institute', *Notes and Papers Series 3*. Addis Ababa, Ethiopia.

Gerbens-Leenes, P.W., Nonhebel S. 2004. Critical water requirements for food, methodology and policy consequences for food security'. *Food Policy* 29: 547–64.

Getahun, A., Krikorian, A.D. 1973. 'Chat: Coffee's Rival from Harar, Ethiopia. I. Botany, Cultivation and Use'. *Economic Botany* 27(4): 353–77.

Gezon, L.L. 2012. 'Drug crops and food security: the effects of khat on lives and livelihoods in northern Madagascar'. *Culture, Agriculture, Food and Environment* 34(2): 124–35.

———— 2013. 'Leaf of paradise? The intricate effects of khat in Madagascar'. *Drugs and Alcohol Today* 13(3).

Gliessman, S.R. 2007. *Agroecology the Ecology of Sustainable Food Systems*. London: Taylor and Francis.

Govereha, J., Jayne, T.S. 2003. 'Cash cropping and food crop productivity: synergies or trade-offs?' *Agricultural Economics* 28: 39–50.

Hailu, D. 2005. 'Supporting a nation: khat farming and livelihood in Ethiopia'. *Drugs and Alcohol Today* 5(3).

Headey, D.D., Dereje, M., Taffesse, A.S. 2014. 'Land constraints and agricultural intensification in Ethiopia: A village-level analysis of high-potential areas'. *Food Policy*.

Headey, D.D., Jayne, T.S. 2014. 'Adaptation to land constraints: Is Africa different?' *Food Policy*.

Heffez, A. 2003. 'How Yemen Chewed Itself Dry, Farming Qat, Wasting Water'. *Foreign Affair*, http://www.foreignaffairs.com/articles/139596/adam-heffez/how-yemen-chewed-itself-dry

Ingram, J.S.I. 2009. 'Food system concepts'. In Rabbinge, R., Linneman, A. (eds), *ESF/COST Forward Look on European Food Systems in a Changing World*. Strasbourg: European Science Foundation.

Jonsson, U., Toole, D. 1991. *Household Food Security and Nutrition: A Conceptual Analysis*. New York: UNICEF.

Kalix P, Braenden O. 1985. 'Pharmacological aspects of the chewing of khat leaves'. *Pharmacological Review* 37: 149–64.

Kassa, H., Blake, R.W., Nicholson, C.F. 2002. 'The crop-livestock subsystem and livelihood dynamics in the Harar Highlands of Ethiopia'. In *Responding to the increasing global demand for animal products - Programme and Summaries. International conference organised by the British Society of Animal Science, American Society of Animal Science and Asociacion*

Mexicana de Produccion Animal, held at the Universidad Autonoma de Yucatan, Merida, Mexico, 12–15 November, 2002. Penicuik: British Society of Animal Production, pp. 74–5.

Lemma, B. 2011. 'The impact of climate change and population increase on Lakes Haramaya and Hora-Kilole, Ethiopia (1986–2006)'. In Brook, L., Abebe, G. (eds), *Impacts of climate change and population on tropical aquatic resources, proceedings of the Third International Conference of the Ethiopian Fisheries and Aquatic Sciences Association (EFASA)*. Addis Ababa: AAU Printing Press. pp. 19–42.

Misselhorn, A., Aggarwal, P., Ericksen, P., Gregory, P., Horn-Phathanothai, L., Ingram, J., Wiebe, K. 2012. 'A vision for attaining food security'. *Current Opinion in Environmental Sustainability* 4: 7–17.

Murray, C.D.R., Le Roux, C.W., Emmanuel, A.V., Halket, J.M., Przyborowska, A.M., Kamm, M.A., Murray-Lyon, I.M. 2008. 'The effect of Khat (Catha edulis) as an appetite suppressant is independent of ghrelin and PYY secretion'. *Appetite* 51: 747–50.

Mulatu, E., Kassa, H. 2001. 'Evolution of smallholder mixed farming systems in the Harar Highlands of Ethiopia: The shift towards trees and shrubs'. *Journal of Sustainable Agriculture* 18(4).

Muyanga, M., Jayne, T.S. 2014. 'Effects of rising rural population density on smallholder agriculture in Kenya'. *Food Policy*.

Negatu, W. 2004. *Reasons for Food Insecurity of Farm. Households in South Wollo, Ethiopia: Explanations at Grassroots*. Addis Ababa: Institute of Development Research (IDR), Addis Ababa University.

Njiru, N., Muluvi, A., Owuor, G., Langat, J. 2013. 'Effects of khat production on rural household's income In Gachoka Division Mbeere south district Kenya'. *Journal of Economics and Sustainable Development* 4(2).

Nutt, D., King, L.A., Saulsbury, W., Blakemore, C. 2007. Development of a rational scale to assess the harm of drugs of potential misuse. *Health Policy* 369: 1047–1053.

Pankhurst, A. 1993. 'Social consequences of Enset production'. In Abate, T., Hlebsch, C., Brandt, S.A., Gebremariam, S. (eds), *Enset-Based Sustainable Agriculture in Ethiopia*. Proceedings from the International Workshop on Enset held in Addis Ababa, Ethiopia 13–20 December 1993.

Pinstrup-Andersen, P. 2009. 'Food security: definition and measurement'. *Food Science* 1: 5–7.

Piguet, F. 2003. 'Hararghe Food Security hampered by long-term drought conditions and economic constraints'. *Assessment Mission: 2–13 March 2003*. Report from the UN-Emergencies Unit for Ethiopia.

Schneider, K. 2010. 'Yield Gap and Productivity Potential in Ethiopian Agriculture: Staple Grains & Pulses'. *EPAR Brief* 98. Evans School Policy Analysis and Research (EPAR), University of Washington.

Schneider, K., Gugerty, M.K. 2010. 'The impact of export-driven cash crops on smallholder households'. *EPAR Brief* 94, University of Washington.

Schwartz, U. 1989. 'Visual ethnography: using photography in qualitative research'. *Qualitative Sociology* 12(2).

Seyoum, E., Kidane, Y., Gebru, H., Sevenhuge, N. 1986. 'Preliminary study of income and nutritional status indicators in two Ethiopian communities'. *Food and Nutrition Bulletin* 8: 37–41.

Smith, M., Pointing, J., Maxwell, S. 1992. 'Household food security, concepts and definitions: an annotated bibliography' *Development Bibliography 8*. Brighton: Institute of Development Studies, University of Sussex.

Steinberg, M.K., Hobbs, J.J., Mathewson, K. 2004. *Dangerous Harvest Drug Plants and the Transformation of Indigenous Landscapes*. Oxford: Oxford University Press.

Sykes, J.M., Horrill, A.D., Mountford, M.D. 1983. 'Use of visual cover assessments as quantitative estimators of some British woodland taxa'. *Journal of Ecology* 71: 437–50.

UN-ODCCP 1999. *Drug Nexus in Africa. United Nations Office for Drug Control and Crime Prevention*. Vienna.

Zhen-qi, H., Fen-qin, H., Jian-zhong, Y., Xia, L., Shi-lu T., Lin-lin, W., Xiao-jing, L. 2007. 'Estimation of fractional vegetation cover based on digital camera survey data and a remote sensing model'. *Journal of China University of Mining & Technology* 17(1).

21   Emerging Water Frontiers in Large-Scale Land Acquisitions and Implications for Food Security in Africa

Atakilte Beyene and Emil Sandström

INTRODUCTION

Africa's agriculture is at a crossroad. On the one hand, there is the challenge of supporting its smallholder agriculture, which is the main source of livelihood for the continent's population. On the other hand, there are the challenges associated with the expansion of large-scale land acquisitions and the implications of these for Africa's food security. In recent years, arable land has become a highly demanded resource globally and investors often see Africa as an 'uncrowded space of opportunities'. The prospect of accessing abundant land resources is a focal point in many of the investors' agricultural business plans. Many scholars attribute the recent land rush to a number of interlinked global crises that emerged in 2007/8, for example, financial crisis, oil crisis and food crisis, but also to growing concerns related to climate change (Valdés & Foster 2012, Deininger & Byerlee 2011, De Schutter 2011, Friis & Reenberg 2010, Cotula et al. 2009).

Much of the literatures on the 'land grab' has focused on the global drivers of foreign agricultural investments and its impacts on land access. While acknowledging the significance of the interlinked global drivers for the present surge of land resources, there are also a number of conditions that have contributed to make Africa particularly attractive for large scale land acquisitions. Such conditions relate to, for example, Africa's unsettled tenure situation, preceding Structural Adjustment Programs (SAPs) and the recent liberalization of domestic agricultural policies. It has also become increasingly clear that access to water is both a target and a driver behind the current land rush (Mann & Smaller 2010, Mehta et al. 2012). Some of the large-scale land acquisitions also appear to be geopolitical, embedded in transboundary water politics, rather than being based purely on commercial principles (Sandström 2016, Warner et al. 2013, Franco et al. 2013).

By framing the present large scale land acquisitions in historical and contemporary contexts, this chapter seeks to contribute to the discussion on how the phenomena of large scale land acquisitions can be understood and highlight the implications for Africa's food security. More specifically this chapter aims to explore the general significance of the water dimensions of current land acquisitions and the underlying structural agricultural conditions and narratives that have contributed to make Africa the main target continent for large scale land acquisitions.

The chapter is based on a review of the current state of knowledge regarding land acquisitions in Africa, combined with empirical data from the Land Matrix database, one of the most comprehensive in existence, and from our own and research colleagues' field data mainly from eastern Africa. First, a brief overview of the scale of the current large scale of land acquisitions in Africa is presented. We review data on the top ten countries targeted for large scale land acquisitions and highlight its water dimensions. In the second part, we explore similarities and differences between past and present land acquisitions. The third part discusses the historical conditions of Africa's agriculture and argues that Structural Adjustment Programs introduced in the 1980s and 1990s in combination with unsettled tenure regimes have contributed to create certain conditions that have made Africa particularly vulnerable and attractive for large scale land acquisitions. In the fourth part, we review two central narratives that circumvent the current discussions on large scale land acquisitions in relation to Africa's food security – the *win-win* narrative and the *lose-win* narrative. We also discuss the emergence of a possible third narrative, the *lose-lose* narrative. The final part summarizes the main points made in the chapter.

LAND ACQUISITIONS AND THE WATER DIMENSIONS: OVERVIEW

There are widely different approaches and approximations of the scale at which land resources are being acquired. The World Bank report (Deininger & Byerelee 2011) estimates that about 56 million ha of land have been acquired across the globe by 2011. Oxfam estimates that investors acquired and were in the process of acquiring 227 million ha globally between 2001 and 2011 (Oxfam 2011). A more contemporary register on global land acquisitions is available on line, i.e. that of the Land Matrix project (accessed October 2014). The Land Matrix project documents 993 deals covering 37.5 million ha of land globally over the period 2007–14 out of which Africa account for about 20.2 million ha, more than half of the world's land deals in terms of area (Ibid). This size is equivalent to the country of Greece. Other studies for example Cotula et al. (2011) have come to similar conclusions, that Africa is the number one target continent in terms of large-scale land acquisitions.

The ten countries that have emerged as the major targets for land acquisitions in Africa include: South Sudan, Democratic Republic of Congo, Mozambique, Congo, Liberia, Sudan, Sierra Leone, Ethiopia, Madagascar and Zambia (Table 21.1). They account for approximately 70 per cent of all the land deals on the continent as reported by the Land Matrix project in October 2014. Most of the top ten counties targeted for large scale land deals have complex and unsettled land ownership systems and weak governance structures. Some deals are also extremely large, for example a 640,000 ha South Korean deal in Sudan and a 600,000 ha deal in South Sudan, where the US based Nile Trading and Development Inc. and the Mukaya Payam Cooperative have obtained a 49-year leasehold (Oakland Institute 2011).

The reporting on large-scale acquisitions is dominantly on land, which has distracted the attention given to the water dimension in the deals. However, it has become increasingly clear that water is both a target and a driver of the current large-scale land acquisitions (Mehta et al. 2012, Woodhouse & Ganho 2011, Mann & Smaller 2010). The water requirements of large-scale land acquisitions are considered to be enormous. A recent study estimates that about 450 billion m^3 per year of freshwater (both green and blue water) are potentially required in the current global land acquisitions (Rulli et al. 2013). In this crude estimate, Africa's share is about 65 per cent (Ibid). Considering that agriculture accounts for about 70–95 per cent of the water withdrawals on the continent (UN-Water 2007), the current large scale land acquisitions altogether suggests that a new water frontier is emerging in Africa.

The water frontier of land acquisitions in Africa occurs across political, socio-economic, and ecological contexts involving high levels of complexity across administrative, hydrological boundaries (Mehta et al. 2012). The complexities involve for example surface water and groundwater interactions, inter-annual water variability as well as connections between 'blue' and 'green' waters.[1]

Large-scale land and water acquisitions also encompass ecological complexity as water systems most often span over different ecological systems, for example, floodplains, inland rivers, lakes, semi-arid or desert areas, coastal lands, wetlands, and peri-urban areas. Land and water acquisitions are also intertwined in legal and administrative complexity, that often involve unclear administrative boundaries and jurisdictions relating to diverse property regimes including commons, customary, public and private tenure systems that often overlap each other. These complexities across landscapes and administrative boundaries contribute to make water into a contested resource (Ibid).

The emerging water frontier appears to concentrate in water abundant areas where water access is considered to be relatively easy. Table 21.1 indicates that most of the land and water deals take place in countries situated within Africa's major transboundary water resource systems, for example the Congo River basin, the Nile River Basin and along the Zambezi River system. Blue water systems thus seem to be the major target of the investors. Even the major part of the land investments in the arid and

Table 21.1. Displaying the top ten target countries for large scale land deals in Africa.

Target Country	Total Area (ha)	Main Investors	Main Crops
South Sudan	3,491,453	US, Saudi Arabia, United Arab Emirates, Egypt and Norway	Various cereals and teak
Democratic Republic of Congo (DRC)	2,717,358	US and DRC government	Trees and palm oil
Mozambique	2,167,882	Norway, US, UK and South Africa	Eucalyptus and pine
Congo	2,082,000	Spain, Malaysia and South Africa	Palm oil, various cereals and sugar cane
Liberia	1,340,777	Malaysia, Singapore, Italy and Netherlands	Palm oil, trees and rubber
Sudan	1,191,013	South Korea, Saudi Arabia, United Arab Emirates and Egypt	Wheat, maize, sunflower and sugar cane
Sierra Leone	1,184,002	UK, Portugal and US	Palm oil, sugar cane and rice
Ethiopia	959,892	India, Saudi Arabia and Ethiopia	Cotton, sugar cane, jatropha and various cereals
Madagascar	592,858	UK and India	Jathropha, sugar cane and various cereals
Zambia	360,944	South Africa and Germany	Jathropha, wheat and maize
Grand total	**14,214,379**		

Source: Adjusted and compiled from The Land Matrix (accessed October 2014). Only land deals above 500 hectares have been included in the dataset above.

semi-arid areas of Africa will require irrigation for their materialization. In a review of crops by Woodhouse and Ganho (2011), it appears that crops proposed for cultivation in major land deals in Africa require irrigation, because they focus on cultivation of crops with year-round water requirements (for example sugar cane, rice) or crops that need irrigation during the dry season (for example wheat, vegetable, fruit). A similar pattern emerges in the land deal dataset for the top ten African target countries that indicates that the proposed crops are mainly water-intensive (Table 21.1).

Beyond the consumption of large quantities of blue water in irrigation schemes, irrigation systems may in themselves change rural people's access and rights to water across catchment and basin systems. For example, downstream water users may face the risk of reduction of water flows during dry periods when most of the water is captured by upstream irrigation systems. Water irrigation systems often also go hand in hand with the commodification of common water resources with the introduction of, for example, water fees and redefinitions of water rights (Beyene 2015).

Figure 21.1. Large-scale irrigation schemes often involve huge investment in water infrastructures, both to hold and divert water. Photo shows a primary canal and a control station, Koga irrigation scheme, Ethiopia (© Atakilte Beyene).

Table 21.1 also reveals that many investors are water stressed countries from the Middle East or from large net food, or so called virtual water, importing countries, (UK, Singapore and South Korea) (Table 21.1). Countries from the Middle East (mainly Saudi Arabia, United Arab Emirates and Qatar) account for about 20 per cent (about 4.3 million ha) of the land acquisitions in Africa, mainly targeting Sudan, South Sudan and Ethiopia. Some of the reported investors also come from water stressed African countries like Egypt, South Africa, Libya and Djibouti who invest in other African countries. For instance, Egyptian companies and the Egyptian government have invested in Sudan, South Sudan and Ethiopia and companies from South Africa have acquired land in DRC and Mozambique (Sandström 2016, Dixon 2014, Warner et al. 2013). Water can thus be seen as an important factor in shaping regional economic and political relations.

PUTTING THE CURRENT LAND ACQUISITION IN A HISTORICAL PERSPECTIVE: WHAT'S NEW?

Since the 1884–5 Berlin Conference that marked the climax of European surge for territory in Africa, ownership and control of Africa's natural resources have had a troubled history. At the end of World War II, Africa saw another surge in land acquisition, where white settlers pursued agriculture on large tracts of land particularly in southern Africa. In the post-independence period (1965–90) state driven commercial schemes involving millions of hectares were launched across Africa, such as the peanut schemes in the Niger Basin, the sesame and cotton schemes along the Nile in Sudan, the wheat and ground nut schemes in Tanzania or the privatized ranching schemes in Kenya and Botswana along with forest and mining concession schemes throughout the Congo Basin (Wily 2013).

As in the past, several of the current land deals are justified with reference to narratives of *terra* and *aqua nullius* ('land and water of no one') (Makki & Geisler 2011). These narratives are echoed recently by the former Emergent Asset Manager, Susan Pain, of the Goldman Sachs Bank who stated in an investment conference that 'Africa is the final frontier, it is the one continent that remains relatively unexploited' (cited in McMahon 2013: 191).

While the narratives of *terra* and *aqua nullius* continue to play a key role in how actors perceive Africa as an uncrowded space of opportunities, there are also new emerging patterns. A distinct feature of land acquisitions in Africa during the twentieth century was the confinement to the production of tropical cash crops such as rubber, tea, coffee, sugar and bananas. Most land deals of today, however, also include staple crops such as rice, wheat and maize. A survey from 2010 shows that of 54 companies engaged in agricultural investments in Africa indicates that 83 per cent and 13 per cent of the farmland acquired or leased for long-term was dedicated to staple crops and livestock production respectively. The remaining 4 per cent was targeted to more traditional export crops, such as sugar cane (OECD 2010). Crops related to biofuel production, such as jathropha, palm oil, sunflower, castor beans and sugar cane, have been extensively promoted across Africa in relation to the climate change concerns and the oil peak in 2008 (see Table 21.1 and Matondi et al. 2011) but very few of them have materialized.

Compared to the past, current land acquisitions are also highly diverse in the geographical origins of the investors and the forms of capital involved (Table 21.1). In the past, large-scale agro-investments were confined to investors from countries of the former colonial rulers from Europe. Wealthy European individuals and families were the major investors, whereas today, new assemblages of actors, often organized in different public–private partnerships take on a catalytic role. African governments are also actively involved in promoting policies that are attractive to investors, including provision of incentive mechanisms, such as very low land lease rates, long period of lease terms, tax exemptions etc. In addition, global financial institutions and sovereign states are also providing support to companies in the form of equity finance, cheap loans, guarantees and/or insurances. Many companies are also supported by investment banks, hedge funds and public or private pension funds (Dixon 2014, Swedish FAO Committee 2014, Daniel 2012).

Today's large-scale land acquisitions also involve new forms of geopolitics as regards African land and water resources. The extensive size of some of the land acquisitions implies in practice that new assemblages of actors may become, deliberately or not, new shareholders and players in the hydro-politics of Africa (Sandström 2016). For example, Egypt's recent surge to buy large tracts of land upstream in the Nile Basin (Sudan, Ethiopia and Uganda) has been explained as attempts to prevent other actors from accessing the Nile (Ibid). Similarly, one of Qatar's recent investments along the Sudanese banks of the Nile River has been explained as being motivated as a means to increase Qatar's geopolitical sphere of interest in the Nile Basin, where Qatar could 'play its preferred role as benevolent mediator ... and enable the Gulf State to influence water and food politics in the Nile Basin' (Keulertz 2013). The reconfiguration of shareholders and the creation of new assemblages through large-scale agricultural deals thus add new dimensions to the hydro-political landscape of African trans-boundary water systems, which may provide new actors with increased political leverage in trans-boundary and regional hydro-politics (Sandström 2016).

Finally, current land and water acquisitions are to a large degree shaped by new forms of food politics and food markets. Betting on the price of staple crops, like wheat, maize and soya on international financial markets has become a major operation of several investment banks. As the report, 'the Great Hunger Lottery,' by the World Development Movement, indicates financial speculation on food items have almost doubled from US$65 billion in 2006 to US$126 billion in 2011 worldwide (Jones 2010). The major risk of speculation in food commodities, the report argues, is that future global food markets no longer involve those in the supply chain. Instead, financial speculation operates by 'turning commodity derivatives into just another asset class for investors, distorting and undermining the effective functioning of agricultural markets' (Worthy 2011: 6). This has had consequences among several affluent and food importing countries, particularly Gulf States, who have lost confidence in the global food trading system and instead resorted to acquire farmlands in developing countries (Trostle 2008).

UNDERSTANDING 'INTERNAL' CONDITIONS OF AFRICA'S AGRICULTURE

In order to understand why Africa has become the main target continent for large-scale land acquisitions and the implication of this for the continent's food security, it is important to also highlight some of the internal structural conditions in relation to Africa's agriculture. In this section we will discuss two issues; conditions that emerged from the implementation of Structural Adjustment Program (SAPs), and conditions related to Africa's unsettled land tenure system.

Liberal economic agendas and the weakening of African agriculture: The current land acquisitions were preceded by the era of the SAPs designed and promoted by the World Bank and the International Monetary Fund in the 1980s and 1990s (Gibbon et al. 1995). The SAPs were 'among the most important policy frameworks of the last century that have greatly influenced both strategies and programs for agriculture, food and nutrition security in Africa and therefore overall economic development' (Heidhues and Obare 2011: 57). The SAPs emphasized macroeconomic stabilization, privatization, state budget balance and liberalization of domestic and external trade (World Bank 1981 & Gibbon et al. 1993). The SAP 'forced' African governments to cut public sector expenditure and support for agriculture (Havnevik et al. 2007). While many countries in Africa were required to cut public expenditure and liberalize their agriculture by the programs, the OECD countries, on the other hand, continued to subsidize their own agricultural sector (Mshomba 2009, Mather 2008, Carmody 1998).

A major consequence of these programs was that Africa shifted from being a net food producer (as in the 1960s and 1970s) to net food importer (since the 1980s) (Rakotoarisoa et al. 2012) and the continent became increasingly food insecure and dependent on food aid (see also Valdés and Foster 2012, Raikes and Gibbon 2000). Thus, the agricultural sector was marginalized not only in terms of global trade, but also in domestic politics.

Figure 21.2. EcoEnergy (earlier SEKAB), a Swedish company, sugar seed cane farm/ nursery south of Bagamoyo. Workers are employed from neighbouring villages for planting, weeding, etc. (© Atakilte Beyene).

As the challenge of food insecurity and poverty continued throughout the 1990s, African governments made an attempt to address the conditions of the smallholder agriculture. In 2003 the Comprehensive Africa Agriculture Development Program (CAADP) was endorsed by the member states of the African Union who promised to allocate 10 per cent of their national budget to agriculture, particularly to the smallholder agricultural sector, by 2008 (CAADEP 2014). However, only eight (Burkina Faso, Ethiopia, Ghana, Guinea, Malawi, Mali, Niger and Senegal) out of the 54 countries have met the target of 10 per cent as of December 2014 (Ibid). Instead several African governments have emphasized their focus in pursuing policies that attract the private sector and foreign direct investment in large-scale commercial farming, assuming that such investments would better address the continent's development and food security needs.

In most African countries, land investments by foreign investors are thus generally welcomed by host governments and seen as important contributions to the modernization of these states. The current land investments are spurred by policies that aim to provide a conducive business environment for foreign investors by, for example, different kinds of tax reductions and infrastructural support (Friis & Reenberg 2010). Governments are also setting aside millions of hectares of land as a way to attract investments in agriculture for example, in Ethiopia over 3 million hectares and in Congo-Brazzaville over 10 million hectares. In Tanzania, the government has been trying to gain control over village land (presently covering about 70 per cent of the land in the country) in order to be able to transfer it to foreign investors. For this purpose, the government has established the Tanzanian Investment Center (TIC). The TIC is tasked to identify 'unused' village land and transfer it to the Land Bank in order to provide long-term leases to prospective investors. In this process, village land is re-categorized as general land, belonging to the central government. Once the transfer is done, rural residents often lose their user rights to the land (Beyene et al. 2013, see also Abdallah et al. 2014).

African land tenure and the large-scale acquisition of land: The complexity of African land and water tenure systems has remained persistent throughout modern African history. In most African countries, land tenure systems are customary and rural people rarely hold formal legal rights to the resources they use and depend on. About 90 per cent of the land is estimated to be under this sort of customary tenure (Wily 2011). Land and water resources are, however, also considered as state property. This means that there exist competing parallel tenure systems across Africa. This dual tenure structure has made many African governments key actors in the negotiation and settlement of large-scale land and water deals (Wolford et al. 2013). In Ethiopia, Tanzania, South Sudan and the Sudan, land identification, delineation, allocation and registration are done by the governments.

In general, there is also a lack of recognition of the customary tenure system, on which the majority of the rural people depend (for example, Boone 2014, Wily 2011, Nzioki 2006). This lack of recognition makes it possible for governments to expropriate land. This is one of the key reasons why large-tracts of land are often relatively easily transferred from smallholder and pastoralists to foreign investors in Africa. Smallholder farmers rarely enter into negotiations with large-scale investors as 'legitimate' representatives, or owners and users of land and water resources. When farmers are involved, their involvement has largely been symbolic. Instead, governments have acted on the farmers' behalf in settling deals with investors. This act of the government has often been justified on capacity and coordination grounds that farmers cannot carry such activities by themselves.

This complex nature of tenure arrangements in Africa has created a void and uncertainties in which current large scale land acquisitions take place. In Tanzania, for instance, land that was allocated to investors through the government, has neither guaranteed security for the investors nor have they addressed enough security for the farmers (Beyene et al.

2013, Abdallah et al. 2014). Generally, applications for land are managed through ad hoc procedures at various levels of government, contributing to a lack of transparency and accountability with regard to many of the deals. For example in South Sudan, the Land Act has not yet passed the parliament and there is currently no uniform procedure for managing large-scale land deals (Deng 2011).

To sum up, the SAPs that were introduced in the 1980s and 1990s in combination with unsettled and competing tenure systems in Africa have contributed to create certain conditions that have made Africa vulnerable to large scale land and water acquisitions. One could argue, that Africa's agriculture was 'opened up' for large-scale foreign direct investments in agriculture in a state of 'weakness'. This is partly substantiated by other studies showing how large-scale land acquisitions tend to concentrate in countries where land governance is weak (Deininger & Byerlee 2011, Oxfam 2011, Gerlach & Liu 2010).

WIN-WIN, WIN-LOSE OR PLAINLY LOSE-LOSE

The current large-scale land and water acquisitions have provided new fuel to the discussions of Africa's food security situation and highlight the existence of very divergent perspectives and interests. In this regard, it is possible to distinguish between three different narratives: win-win, win-lose and lose-lose. These narratives are partly mirrored in the terminology that different actors use when describing the phenomena of large scale land deals. While the supporters of the win-win narrative often use the term 'land investment', proponents within the win-lose narrative often use the term 'land grabs'. Recently, there are studies showing that many of the current large-scale land transfers and investments face failures where the scenario becomes a lose-lose for all actors involved (Abdallaha et al. 2014).

The win-win narrative, which is clearly the most dominant, often portray the current land transfers as an innate 'natural' trend of capital flows, by arguing that cheap land and fairly easy access to water have made Africa an attractive destination for large scale land investments. This narrative presupposes investments as a flow of resources from 'less-efficient' to 'more-efficient' land use systems and actors. The current land acquisitions are, thus, often framed and justified by arguing that there are vast areas of idle or vacant land suitable for cultivation and that large scale farms have the potential to deliver economics of scale that will be enough to produce benefits for all involved.

In the win-win narrative land is, thus, treated as a commodity and compensation schemes are embarked on as a means to 'transfer' the land from smallholders and pastoralists. It is the private sector and several African governments and development banks, who largely adhere to this narrative. Common key arguments within the win-win narrative are that large scale land investments will bring rural employment, technology transfer to the smallholder agricultural sector, increase tax revenues for governments, and profits for the investors. This win-win narrative basically justifies any private investments in agriculture (Da via 2011).

Also the FAO together with various aid agencies partly adhere to this narrative, but they also recognize the need to set some conditions on the investment. For example, the FAO have gone to great lengths to show that large-scale land acquisitions can lead to win-win situations, if actors just followed voluntary principles and guidelines on how these investments are carried out (for example, the principles for responsible agricultural investment (PRAI) that were endorsed in Rome on 15 October 2014).

The win-lose narrative highlights that smallholder farmers, pastoralists and rural people in general are likely to lose in the current land and water rush, while the investors are the winners. Proponents of the win-lose narrative argue that cheap land and abundant water resources, as well as advantages in production technology and access to high value markets are likely to benefit the large-scale commercial farmers at the expense of smallholder farmers.

The narrative also contains critique against large-scale commercial farming that tends to give preference for mechanization and argue that it will not bring sufficient rural employment. In addition, weak working and labour rights and low wage rates are other risks which rural people face when they are involved as wage labourers (Daniel & Mittal 2010). In the current land acquisitions, many investors promise to deliver benefits, such as social services, schools and clinics, to local people. Unfortunately, investors appear to circumvent such expectations of the investment-hosting rural communities. Proponents of the win-lose narrative also argue that the current land rush can be seen as a form of commodification of nature in which powerful agrochemical-pharma-food-transport corporations are taking control of agriculture (GRAIN 2010a, McMichael 2012, GRAIN 2008).

The lose-lose narrative highlights, based on empirical research, that several of the large scale investments in agriculture face high levels of failures that tend to generate negative outcomes for all actors involved. A study from Tanzania shows that among the 32 investors who requested land (over 2,000 ha) for biofuel production since 2011, only ten investors were able to receive land (amounting to 200,000 ha of the total 1 million ha requested) (Abdallah et al. 2014). The authors also show that the implementation of the investments on the already allocated land to be 'very slow and/or unsuccessful' (Ibid. 52). Furthermore, it is not uncommon that investors, who acquired land through government allocation, face challenges of protecting the allocated properties from the host communities. The case of SEKAB (now EcoEnergy), a Swedish-based biofuel company, and Safe Agro Productions, a Turkish food crop producing company, that tried to acquire

Figure 21.3. A smallholder farm household located close to the former Razaba Ranch in Bagamoyo/Tanzania face uncertain future as EcoEnergy (earlier SEKAB) took over the farm to establish sugar cane plantation (© Atakilte Beyene).

land in the Rufiji area in Tanzania (Beyene et al. 2013); and the case of Karaturi, an Indian-based food and flower company in Ethiopia (Rahmato 2011), are examples that have faced failures. According to a recent report (*The Reporter* 2015), Karaturi, which was granted 300,000 ha to develop commercial farming, is now 'on the verge of collapsing in Ethiopia'.

Unfortunately, failed land investments leave a trail of destruction in their wake. For instance, when investments fail and the investors withdraw; there are numerous legal complications that hinder proper reallocation of the land for other users. Therefore, even if investors withdraw, farmers may not be allowed to regain their land. Failed investments are thus likely to lead into cycles of conflicts, erosion of trust among the actors with negative outcomes for all involved.

CONCLUSION

The analysis presented in this chapter suggests that much of the investment models that are based on acquiring large tracts of land and water resources will most likely not address Africa's own food security needs. On the contrary, they may even cause more harm than benefits. Increasing evidence shows that several of the present large scale land and water acquisitions in Africa are likely to even spur resource struggles and contribute to the accumulation of resources at the expenses of rural livelihoods or they tend not to materialize at all. The prospect of addressing rural food security, sufficient employment, and integration of rural people's economic activities in large-scale farming is thus unlikely to occur.

While the global food and energy crisis in 2007/2008 triggered investors' interest to acquire land in Africa, the underlying explanation for the current land and water rush is also conditioned by weak land governance structures and failed agricultural policies from the past. The present land and water acquisitions combine elements of the nineteenth-century colonial adventure and are framed within a narrative of economic liberalization that made its major entry into Africa through the SAP in the 1980s and 1990s. The analysis presented in this chapter suggests that the SAPs in combination with unsettled tenure regimes and more recent domestic policies that aim to provide conducive business climate for foreign investors have contributed to create certain conditions that have made Africa particularly vulnerable and attractive for large scale land and water acquisitions.

This chapter also addressed the fact that water is defining many of the transnational land acquisitions and most acquisitions tend to target blue water resources. The pattern across the top ten countries we studied indicates that it is the water rich basins of the continent's transboundary rivers that are the main targets and that the investors mainly come from water stressed countries or from virtual water importing countries. This adds additional dimensions and complexities to the present land acquisitions on the continent, which contribute to make particularly the transboundary water rivers systems into contested geopolitical resources. Altogether this suggests that a new water frontier is emerging in Africa, which is likely to create new spheres of economic and political influence across parts of Africa.

Considering that Africa is expected to increase its population by 1 billion by 2050 and that about 70–80 per cent of the continent's population derive their livelihoods directly from agriculture that is operated by small-scale farmers,[2] the relevance of promoting large scale agricultural schemes mainly for export markets, needs to be reconsidered. Unless more attention is paid to how the recent land and water acquisitions affect people's every day practices and livelihoods, Africa will continue to be a volatile and food insecure region. A preliminary proposition is that conflicts and tensions over land and particularly water resources in Africa may well increase unless more attention is paid to the structural agricultural conditions and the underpinning narratives that have made Africa's agriculture weak in the first place.

ACKNOWLEDGEMENTS

We are grateful to Kjell Havnevik for his constructive comments and contributions that we received while writing this chapter. We would also like to acknowledge that this work is partly financed by the Swedish Research Council under a project entitled: 'Water politics in the Nile Basin – emerging land acquisitions and the hydropolitical landscape'.

NOTES

1 The terms green and blue water were coined by Malin Falkenmark in the 1990s. Green water refers to water from rain that enters the soil root zone and is transpired by the plants directly, without ever seeing a river, whereas blue water is fresh surface and groundwater, found in for example lakes, rivers and aquifers. A typical relationship cited with regard to green and blue water interaction is that an increase in green water flows (mainly due to increase in biomass of vegetation) affects the blue water flows in terms of reduced run-off (Lundqvist and Steen 1999, Falkenmark and Lannerstad 2004).
2 In sub-Saharan Africa about 80 per cent of the farmers operate with less than 2 hectares (Lowder et al. 2014).

REFERENCES

Abdallah, J., Engström, L., Havnevik, K., Salomonsson, L. 2014. 'Large-scale land acquisitions in Tanzania: a critical analysis of practices and dynamics'. In Kaag M., Zoomers A (eds), *The global land grab: beyond the hype*. London: Zed Books, pp. 36–53.

Beyene, A. (2015) Agricultural water institutions in East Africa, Current African Issues, no. 63, The Nordic Africa Institute, Sweden.

Beyene, A., Mung'ong'o, C., Atteridge, A., Larsen, R. 2013. *Biofuel Production and its Impacts on Local Livelihoods in Tanzania: A Mapping of Stakeholder Concerns and Some Implications for Governance*. Stockholm Environment Institute, Working Paper 2013-03.

Boone, C. 2014. *Property and Political Order in Africa: Land Rights and the Structure of Politics*. Cambridge: Cambridge University Press.

CAADP (Comprehensive Africa Agriculture Development Program-about) 2014. Available at http://www.nepad.org/foodsecurity/agriculture/about.

Carmody, P. 1998. 'Constructing alternatives to structural adjustment in Africa'. *Review of African Political Economy* 25(75): 25–46.

Cotula, L. 2011. *Land Deals in Africa: What is in the Contracts?* London: International Institute for Environment and Development (IIED).

Cotula, L., Vermeulen, S., Leonard, R., Keeley, J. 2009. *Land Grab or Development Opportunity? Agricultural Investment and International Land Deals in Africa*. Rome: IIED, FAO, IFAD.

Da Via, E. 2011. *The politics of 'win-win' narrative: land grabs as development opportunity?* International Conference on global Land Grabbing, IDS, 6–8 April 2011, University of Sussex. Available at http://www.iss.nl/fileadmin/ASSETS/iss/Documents/Conference_papers/LDPI/63_Elisa_Da_Via_2.pdf.

Daniel, S. 2012. 'Situating private equity capital in the land grab debate'. *Journal of Peasant Studies* 39(3–4): 703–29.

Daniel, S., Mittal, A. 2010. *Mis (Investmemt) in Agriculture: The Role of International finance Corporation in Global Land Grabs*. Oakland: Oakland Institute.

De Schutter, O. 2011. 'How not to think of land-grabbing: three critiques of large-scale investments in farmland'. *Journal of Peasant Studies* 38(2): 249–79.

Deininger, K., Byerlee, D. 2011. *Rising Global Interest in Farmland: Can it Yield Sustainable and Equitable Benefits?* Washington, D.C.: The World Bank.

Deng, D. 2011. *The New Frontier: A Baseline Survey of Large-Scale Land-Based Investment in Southern Sudan*, The Norwegian People's Aid, Report 1/11 (March).

Dixon, M. 2014. 'The land grab, finance capital and food regime restructuring: the case of Egypt'. *Review of African Political Economy* 41(140): 232–48.

Falkenmark, M., Lannerstad, M. 2004. 'Consumptive water use to feed humanity: curing a blind spot'. *Hydrology and Earth System Science Discussion* 1: 7–40.

FAO. 2012. *Trends and Impacts of foreign investment in Developing Country Agriculture: Evidence from Case Studies*. Rome: FAO.

——— *Principles for Responsible Investment in Agriculture and Food Systems.* Available at http://www.fao.org/cfs/cfs-home/resaginv/en/.

——— *Voluntary Guidelines on the Responsible Governance of Tenure*. Available at http://www.fao.org/nr/tenure/voluntary-guidelines/en/.

Franco, J., Mehta, L., Velderwisch, G.J. 2013. 'The global politics of water grabbing'. *Third World Quarterly* 34(9): 1651–75.

Friis, C., Reenberg, A. 2010. *Land Grab in Africa – Emerging Land System Drivers in a Teleconnected World*. GLP Report No. 1, Global Land Project, Dept of Geography and Geology, University of Copenhagen, Denmark.

Gerlach, A., Liu, P. 2010. *Resource-seeking Foreign Direct Investment in Africa: a review of country case studies*. FAO Commodity and Trade Policy Research Working Paper No. 31, Rome.

Gibbon, P. 1995. 'Towards a political economy of the world Bank 1970-90'. In Mkandawire, T., Olukushi, A. (eds), *Between Liberalization and Oppression: The Politics of Structural Adjustment in Africa*. Dakar: CODESRIA.

Gibbon, P., Havnevik, K., Hermele, K. 1993. *A Blighted Harvest: The World Bank and African Agriculture in the 1980s*. London and Trenton: James Currey and Africa World Press.

GRAIN 2008. 'Sized: the 2008 land grab for food and financial security'. Available at http://www.grain.org/article/entries/93-seized-the-2008-landgrab-for-food-and-financial-security.

GRAIN (2010a). *Land Grabbing in Latin America*. Retrieved on 25 November (http://www.grain.org/system/old/articles_files/atg-24-en.pdf).

Havnevik, K., Bryceson, D., Birgegård, E.L., Matondi, P. and Beyene, A. 2007. *African Agriculture and the World Bank – Development or Impoverishment? Critique of the World Development Report 2008*. The Nordic Africa Institute, Uppsala, Sweden.

Heidhues, F. and Obare, G. 2011. 'Lessons from the Structural Adjustment Programmes and their effects in Afica'. *Quarterly Journal of International agriculture* 50(1): 55–64.

Jones, J. 2010. *The Great Hunger Lottery: How Banking Speculation Causes Food Crises*. The World Development Movement, available at http://www.globaljustice.org.uk/sites/default/files/files/resources/hunger_lottery_report_6.10.pdf

Keulertz, M. 2013. 'Land and water grabs and the green economy', in Allan, T.; Keulertz, M.; Sojamo, S. and Warner, J. (eds) *Handbook of Land and Water Grabs in Africa: Foreign direct investments and food and water security*. Routledge: London and New York.

Lowder, S.K., Skoet, J., Singh, S. 2014. *What do we Really Know About the Number and Distribution of Farms and Family Farms Worldwide?* Background paper for The State of Food and Agriculture 2014. ESA Working Paper No. 14-02. Rome: FAO.

Lundqvist, J., Steen, E. 1999. 'Water. Contribution of Blue Water and Green Water to the Multifunctional Character of Agriculture and Land'. *Cultivating Our Futures*. Background Paper 6: available at http://www.fao.org/docrep/x2775e/X2775E08.htm#TopOfPage.

Makki, F., Geisler, C. 2011. 'Development by Dispossession: Land Grabbing as New Enclosures in Contemporary Ethiopia'. Paper presented at the international *Conference on Global Land Grabbing, 6–8 April 2011*. Available at http://www.iss.nl/fileadmin/ASSETS/iss/Documents/Conference_papers/LDPI/29_Fouad_Makki_and_Charles_Geisler.pdf

Mann, H., Smaller, C. 2010. 'Foreign land purchases for agriculture: what impact on sustainable development?' *Sustainable Development Innovation Briefs 8*. United Nations, Department of Economic and Social Affairs. New York.

Mather, C. 2008. 'Value chains and tropical products in a changing global trade regime'. *ICTSD Project on Tropical Products, Issue Paper 13*. Geneva: International Centre for Trade and Sustainable Development.

Matondi, P., Havnevik, K., Beyene, A. (eds) 2011. *Biofuels, land Grabbing and Food Security in Africa*. London: Zed Books.

McMahon, P. 2013. *Feeding frenzy: The New Politics of Food*. London: Profile Books.

McMichael, P. 2012. 'The land grab and corporate food regime restructuring'. *The Journal of Peasant Studies* 39(3–4): 681–701.

Mehta, L., Velderisch, G., Franco, J. 2012. Introduction to the special issue: 'Water Grabbing? Focus on the (Re)appropriation of Finite Water Resources'. *Water Alternatives* 5(2): 193–207.

Mshomba, R. 2009. *Africa and the World Trade Organization*. New York: Cambridge University Press.

Nzioki, A. 2006. *Land Policies in Sub-Saharan Africa*. Nairobi: The Centre for Land Economy and Rights of Women.

Oakland Institute 2011. *Understanding Land Investment Deals in Africa: Country Report South Sudan*. Oakland: Oakland Institute.

OECD 2010. 'Private financial sector investment in farmland and agricultural infrastructure'. *OECD Food, Agriculture and Fisheries Working Papers 33*. Paris: OECD.

OXFAM 2011. 'Land and power: the growing scandal surrounding the new wave of investment in land'. *Oxfam International Briefing Paper 151*. London: Oxfam.

Rahmato, D. 2011. 'Land to investors: large-scale land transfers in Ethiopia'. *Forum for Social Studies*, Addis Ababa.

Raikes, P., Gibbon, P. 2000. 'Globalisation' and African Export Crop Agriculture'. *The Journal of Peasant studies* 27(2): 50–93.

Rakotoarisoa, M., Iafrate, M., Paschali, M. 2012. *Why Africa Became a Net Food Importer? Explaining Africa Agricultural and Food Trade Deficit*. Rome: FAO.

Rulli, M., Saviori, A., D'Odorica, P. 2013. 'Global land and water grabbing'. *Proceedings of the National Academy of Sciences* 110(3): 892–7.

Sandström, E. 2016.'Dealing with water: Emerging Land Investments and the Hydropolitical Landscape of the Nile Basin'. In Sandström, E., Jägerskog, A., and Oestigaard, T. (eds), *Land and Hydropolitics in the Nile River Basin: Challenges and New Investments*. London and New York: Earthscan/ Routledge.

Swedish FAO Committee 2014. *Responsible Agricultural Investments in Developing Countries – How to Make Principles and Guidelines Effective*. Sweden: Ministry for Rural Affairs.

The land Matrix 2014. *The Online Public Database on Land Deals*. Available at http://www.landmatrix.org/en/.

The Reporter 2015. *Karaturi under the spotlight*. Available at http://www. thereporterethiopia.com/index.php/news-headlines/item/3045-karuturi-under-the-spotlight.

TIC (Tanzanian Investment centre), available at http://www.tic.co.tz/

Trostle, R. 2008. *Global Agricultural Supply and Demand: Factors Contributing to the Recent Increase in Food Commodity Prices*. Washington, D.C.: WRS-0801 Economic Research Service, USDA.

UN-Water 2007. *Coping with Water Scarcity: Challenge of the Twenty-First Century*. UN-World Water Day 2007. Available at http://www.fao.org/nr/water/docs/escarcity.pdf.

Valdés, A., Foster, W. 2012. 'Net Food-Importing Developing Countries: Who They Are, and Policy Options for Global Price Volatility', *ICTSD Programme on Agricultural Trade and Sustainable Development, Issue Paper 43*. Geneva: International Centre for Trade and Sustainable Development.

Warner, J., Sebastian, A., Empinotti, V. 2013. 'Claiming back the land: the geopolitics of Egyptian and South African land and water grabs'. In Allan, T., Keulertz, M., Sojamo, S., Warner, J. (eds), *Handbook of Land and Water Grabs in Africa: Foreign Direct Investments and Food and Water Security*. London and New York: Routledge.

Wily, L. 2011. *The Tragedy of Public Lands: The Fate of the Commons under Global Commercial Pressure*. Rome: The International Land Coalition.

Wily, L. 2013. 'The global land grab: the new enclosure'. Available at http://wealthofthecommons.org/essay/global-land-grab-new-enclosures.

Wolford, W., Borras, S., Hall, R., Scoones, I., White, B. 2013. 'Governing global land deals: the role of the state in the rush for land'. *Development and Change* 44(2): 189–210.

Woodhouse, P., Ganho, A. 2011. *Is water the hidden agenda of agricultural land acquisition in sub-Saharan Africa?* International Conference on Global Land Grabbing, Institute of Development Studies. University of Sussex, 6–8 April 2011.

World Bank 1981. *Accelerated Development in Sub-Sahara Africa: An Agenda for Action*. Washington, D.C.; The World Bank.

Worthy, M. 2011. *Broken Markets: How Financial Markets Regulation can help Prevent Another Global Food Crisis*. The World Development Movement, available at http://www.globaljustice.org.uk/sites/default/files/files/resources/broken-markets.pdf

Part V   The Hidden Waters of Africa

22   New Perspectives on Saharan Mega-Aquifers: History, Economic Value and Sustainability

Fridtjov Ruden

INTRODUCTION

The Nubian Sandstone Aquifer System (NSAS) of north-eastern Africa constitutes one of the largest onshore groundwater reservoirs in the world, with some 5 per cent of the planet's total freshwater reserves stored in sandstone aquifers under the Saharan sands. The NSAS covers a large tract of land, and could easily accommodate the Indian subcontinent within its boundaries. For a number of reasons the water potential of the NSAS has remained largely intact in spite of the earlier Libyan regime's well publicized Great Man Made River project. The NSAS is relatively well explored by the oil industry, but this sector tends to keep vital information close to its chest. Getting to grips with the NSAS is therefore not a straightforward task due to sheer dimensions of the aquifer system, inhospitable environment, transboundary complications, corruption and political instability.

In order to plan and exploit such a large groundwater system, a certain perspective is required. The aquifers extend over great areas, the depths are truly large, regional contamination from oil industry is a real threat, and use of the resource is presently unregulated. Some parts of the aquifer may be already over-exploited while others are both under-exploited and under-estimated. Estimates of NSAS water reserves vary wildly and indicate in many cases the perspectives of the beholder rather than the perspectives provided by the geology.

This chapter does not attempt to provide a complete treatise of the north-eastern Africa's hydrogeology, but will try to provide some fresh examples and perspectives of the groundwater resources of this region. In this context, three selected basins (Ghadames, Kufrah and Murzuq) have been used as examples only. These basins constitute possibly less than a third of the overall water potential found within the NSAS under the Sahara. Accurate figures do not exist, and volume estimates by previous workers vary wildly.

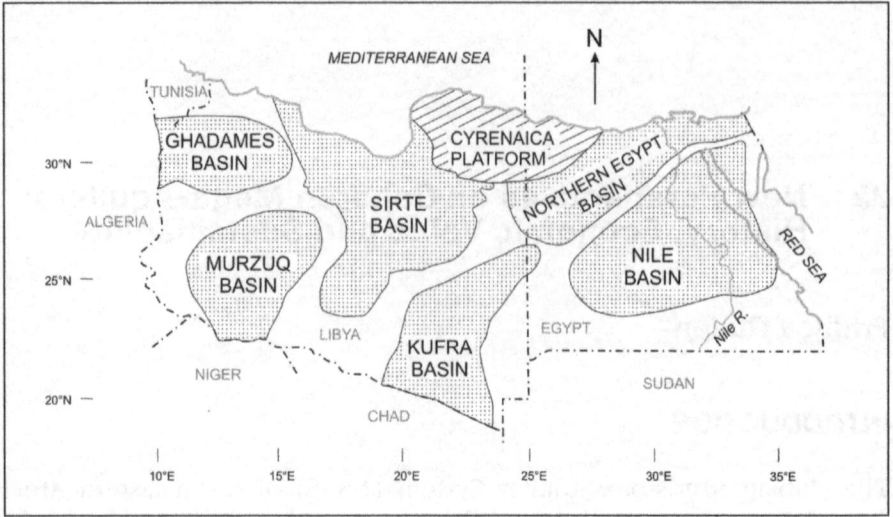

Figure 22.1. Principal Basins of the Nubian Sandstone Aquifer System (NSAS). In the northern part, the sandstones belonging to the Paleozoic-Mesozoic NSAS are overlain by successively younger (post-cretaceous) rocks of various categories, towards the North. The southern half of the NSAS is generally exposed, except for Pleistocene alluvial, lacustrine and eolian deposits. Only the NSAS is discussed in this chapter (© Krystyna Guzek k.guzek@10g.pl).

FORMATION OF THE NSAS BASINS

Schandelmeier (1988) outlines the genesis of the principal sandstone basins of north-eastern Africa. The present sedimentary basins were formed along pre-existing structures, as a tectonic rejuvenation along older structures. A rejuvenation of older pre-Cambrian structures, the basins are of Paleozoic-Mesozoic age, and attain thicknesses exceeding 4,000 m of clastic sediments. Park (1988) outlines a two-stage basin development which is common throughout Africa, in line with a pattern observed in basins of the same (Gondwana) age elsewhere in Africa. A typical example is the basal development of the Karroo-series of South and East Africa, where the older (glacial) deposits fill in depressions, and where successive younger strata become increasingly regional.

The nomenclature and classification of the principal deep Nubian basins under Libya, Chad and Egypt varies. In order to serve as illustrations for the water situation in north-eastern Africa, the following basins have been selected: Ghadames; Kufrah; and Murzuq.

Much of the (oil) reservoir rocks of the NSAS basins were formed in a glacial or periglacial environment during Carboniferous time, in a similar pattern as seen over most of Gondwana basins such as described by McKinley (1963) of the Karroo of Easdt Africa. The position of Libya on the

Gondwana continent during Carboniferous was in the Antarctic, and much of the present oil reservoirs are found in corresponding glacial deposits. A good example is the (oil) productive Murzuq Basin where oil is pumped from varved sediments and tillites; formations of a glacial environment. The water volume of the NSAS is in the order of a million times larger than the volume of oil found in the same formations. As an illustration, the same ratio of diesel and water, mixed in the same ratio ($1:10^6$) renders water useless for most purposes. Small quantities of oil can destroy very large volumes of water.

THE GHADAMES BASIN

The Ghadames basin illustrates clearly the potential for oil and water conflicts in the NSAS. In this context, water resources will invariably be relegated to a second rank commodity, except for injection in an oil/gas exploitation for pressure support. In the process, unknown aquifer volumes will be irreversibly destroyed by contamination during and after the oil exploitation phase. Unknown but significant volumes of contaminants are polluting freshwater aquifers in the wake of the oil industry, and due to lack of infrastructure and regulations, the process of well plugging and abandonment remains unregulated. A polluting and unplugged oil well with internal cross flow has the potential to contaminate and destroy regional aquifers, forever. For volume estimates, see Table 22.1.

THE MURZUQ BASIN

Pallas (1980) reports Murzuq oil concession well NC115 in sandstone with porosities around 25 per cent, and much of the reservoir showing otherwise excellent hydraulic properties. The oil play is normally in Cambrian–Ordovician glacial sandstone, with cap rock and source rock of the overlying Silurian shales. The oil and water conflict also dominates this reservoir, where significant quantities of water are required (and used) for injection purposes to enhance oil production. A deep exploitation of the central part of the basin (where no oil has been found) would have been feasible but would still meet resistance from the oil industry who wants to keep other interests and activities at bay. Libya has the world's fifth-largest oil reserves, by far the largest in Africa, and this invariably outweighs the importance of water. With social structures of the nation at present eroded or non-existent, environmental regulations and resource planning gives way to quick returns of investments, i.e. short-term oil revenues. In this context, the ancient water of the NSAS will be sacrificed. Oil production of the el Sharara reservoir on

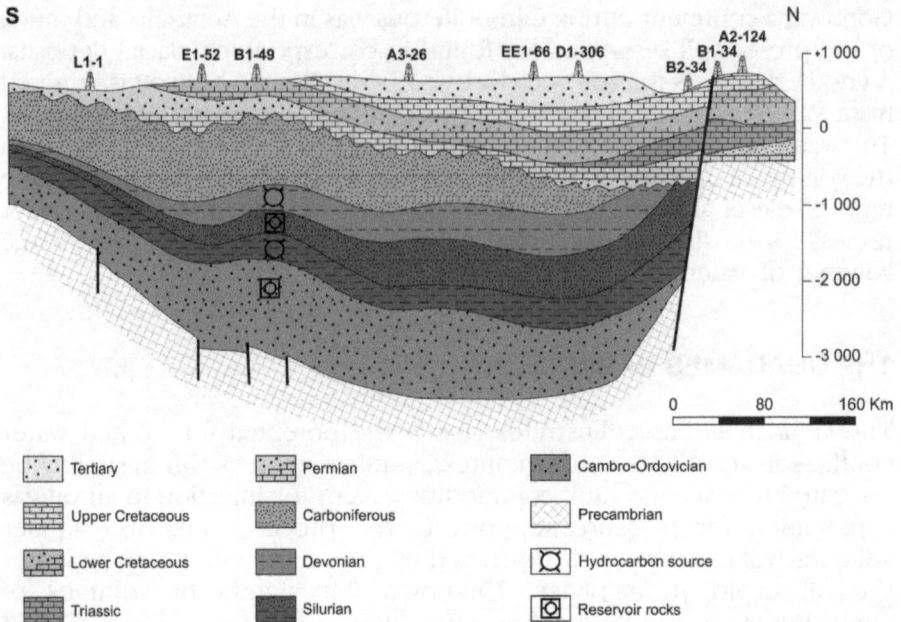

Figure 22.2. The Ghadames basin. Based upon Echik (1998) (© Krystyna Guzek k.guzek@10g.pl).

the western part of Murzuq has recently come online (2015), amidst continued political turbulence in the area. Water can only sustain relatively peaceful activities, in particular agriculture, whereas oil can finance anything, from guns to roses.

The Murzuq is a true continental basin where the aquifer has remained virtually intact for millions of years, not affected by geological events. The Murzuq basin is a main source for the Great Manmade River Authority supply system, where the GMRA has been tasked with exploitation and transport of water from the southern NSAS wellfields to the coastal area. The aquifers under GMRA have previously been equipped with monitoring devices; some 53 have been reported from the Murzuq basin alone. To what extent the system is still operational, after years of civil unrest remains unknown. Monitoring systems have not been installed for wellfields other than the GMRA. For volume estimates, see Table 22.1.

THE KUFRAH BASIN

Named after a famous oasis in central Libya, the Kufrah basin is the largest sedimentary basin in north-eastern Africa, and represents one of the world's largest aquifers. A true continental and transboundary aquifer, the Kufrah is

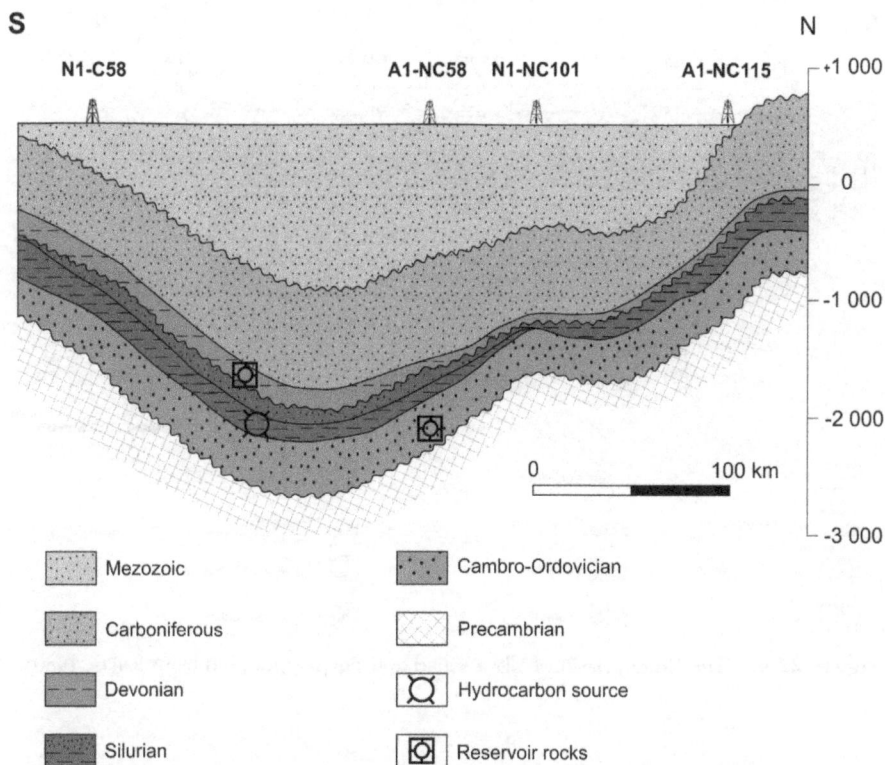

Figure 22.3. North–south cross-section of Muruzq Basin. Based upon Worsley, 2000 (© Krystyna Guzek k.guzek@10g.pl).

comprised of a sandstone aquifer with about 2/3 within Libya, a third within Chad, and including a smaller portion on the Egyptian side to the north-east. The Kufrah aquifer represents a large aquifer system, yet unspoilt by the oil industry. In spite of extensive prospecting, the oil industry has yet (2015) failed to come up with major plays in this particular basin. The basin has therefore escaped the potential confrontation between oil and water sectors as has been the case elsewhere in Libya. However, the petroleum sector still holds a firm grip on the basin in the form of oil con concessions, awaiting an improved search model and more secure environment. Until then, water will have to play second fiddle, or none, as usual.

The agricultural potential presented by the combination Kufrah oasis- and the underlying Kufrah aquifer is considerable. The development of agriculture and water abstraction of the vast underlying Kufrah aquifers could easily go hand in hand, as the aquifer must be exploited in a decentralized fashion. This implies that the traditional clustering of wells into well fields must be avoided due to the anticipated low diffusivity of the aquifer system (for volume estimates, see Table 22.1).

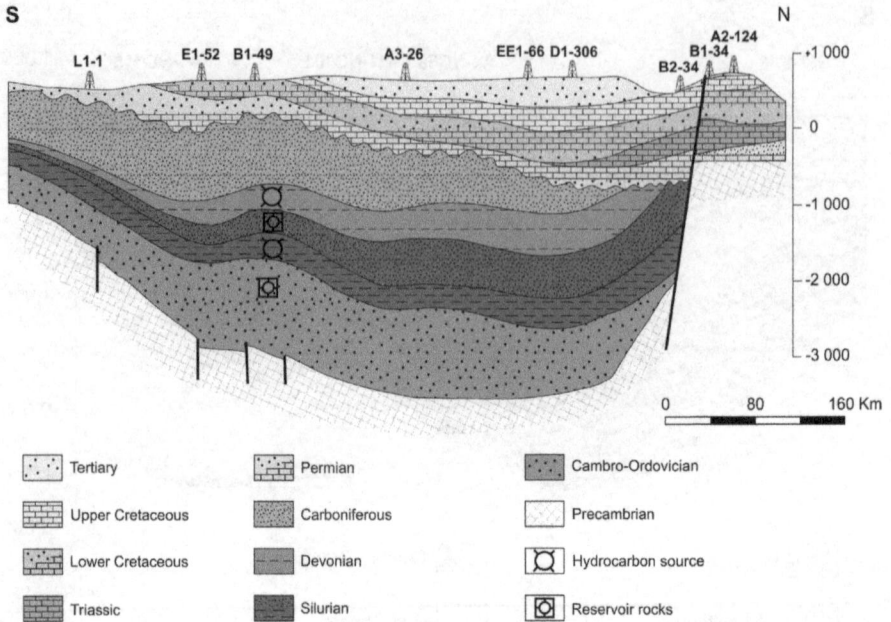

Figure 22.4. The Kufrah basin of Libya, Chad and Egypt (modified from Pallas, 1980).

IMPACT OF MESSINIAN SALINITY CRISIS ON NSAS

About 5.26 million years ago during the Pliocene, an event took place that had significant and lasting impacts on some (but not all of the north-eastern African aquifers). At this time, the Straits of Gibraltar closed, and the Mediterranean dried up in an event termed the Messinian Salinity Crisis (MSC). In the wake of the MSC, several thousand meters deep canyons were incised into the coastal formations, effectively draining coastal aquifers including the coastal Sirte and Ghadames basins. The same aquifers were subsequently re-filled with saline water during the rapid re-flooding of the Mediterranean, termed 'the Zanclean cataclysmic event' (Blanc 2002), enabling the costal aquifers to be replenished back to their pre-Pliocene level. Illustrated below is a schematic example along the Nile valley which was eroded to a depth of about 2.400 m below the modern delta during the Messinian Crisis, draining all adjacent aquifers, and removing most of the pre-Messinian Nile delta in the process. Following the Zanclean re-filling episode of the Mediterranean, all canyons were eventually backfilled, including the entire Nile valley. Saline water encroached the coastal aquifers from the Mediterranean, eventually encountering freshwater moving northwards from the hinterland. Hydrostatic head on the southern (fresh) part of the aquifer system

THE MESSINIAN SALINITY CRISIS

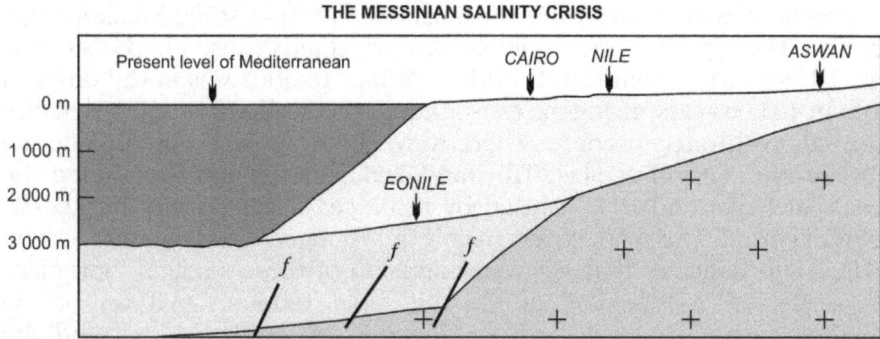

Figure 22.5. The Messinian salinity crisis caused a near complete desiccation of the Mediterranean, dewatering of parts of the NSAS, and the corresponding drop of the erosional basis caused deep erosional channels (modified from Said, 1962 and Krijgsman et al., 1999).

eventually forced the saline interface to a position close to the present Mediterranean shore line.

The Zanclean flooding of the Mediterranean took only eleven years to complete (Blanc 2002), considerably faster than the corresponding continental freshwater from the South could balance the saline intrusion of coastal aquifers. This imbalance shifted the saline/freshwater boundary in the aquifers southwards, until pluvial periods enabled a slow migration of the fresh/saline interface to the North. This explains remnants of saline water in parts of the northernmost (coastal) regions of the NSAS, including the occasional saline horizons found in the Sirte and Ghadames basins, as well as in aquifers on the Egyptian side. Land-locked basins such as Murzuq and Kufrah were not affected by these events.

IMPACTS FROM PLEISTOCENE FLUCTUATIONS OF MEDITERRANEAN SEA LEVELS

Later episodes, during the Pleistocene had similar albeit somewhat less dramatic effects on the coastal aquifers. These correspond to fluctuating sea levels as the result of glaciations, variations in climate, as testified by the important Marine Oxygen Isotope stages, and the north–south shifting of the Inter Tropical Convergence Zone (ITCZ) (Osmond & Dabous 2004). As the result of the latter, recharge of the NSAS dried up almost completely a number of times during Pleistocene. The shifting of the convergence zone and the corresponding cycles of wet episodes over Sahara seem to be linked to the 23,000-year precession cycles postulated by Milankovitch (1941), and to some extent modulated by the 100-year eccentricity cycle (Muller & MacDonald 1997). During the Pleistocene glaciations the level of the Mediterranean fluctuated around 130 m, but with far less consequences than during the Messinian salinity crisis, some 5 million years earlier. The typical rebound time, from low stands caused by Pleistocene glacial maxima to high stands during warmer periods was in the order of 100 m/10,000 years; giving the coastal aquifers considerable time to balance the salt/freshwater interface. There were about 40 such episodes during Pleistocene. Gossel et al. (2010) modelled seawater intrusion along the coast of Egypt and Libya caused by rising sea levels during the Abassia pluvial period. This work covers mainly the younger Nubian Aquifer System (NAS) and indicates that salt water invasion of these aquifers took place at a rate of 2–3 mm/day during the rapid rebound (120 m) of the Mediterranean sea level, between the end of the last glacial and the beginning of Holocene. Similar conditions may have been experienced within the NSAS.

THE AFRICAN HUMID PERIOD AND THE ITCZ

The African Humid Period (AHP) was the last of a long series of wet (pluvial) periods of the Sahara (DeMenocal et al. 2000). The AHP terminated abruptly some 5,900 years ago with the southward shifting of the Inter Tropical Convergence Zone (ITCZ) from Sahara into the South Sudan. The gigantic waterways that flowed through Sahara during the alluvial periods are now buried by relatively shallow eolian sands, but the old water ways still trickle feed some of the oases of Sahara, including the Kufrah (Coulthard, Barton et al. 2013). Most oases of the Egyptian desert (Kufrah, Farafra, Dhakla) are aligned along ancient waterways; the same waterways that have may have provided a human migration route from Africa to the rest of the world during earlier pluvial periods. The abrupt termination of humid climate caused by the shifting of the ITCZ took only a few years, less than a generation, and forced the Paleolithic inhabitants of the Sahara out of the Sahara, and to Egypt, among other destinations, during the Mousterian and Abbassia Pluvials. It can be argued that it was the shifting of ITCZ that forced humans out of the Sahara and into Egypt, as ancient climate refugees, and thus laid the foundation for the Egyptian agricultural culture.

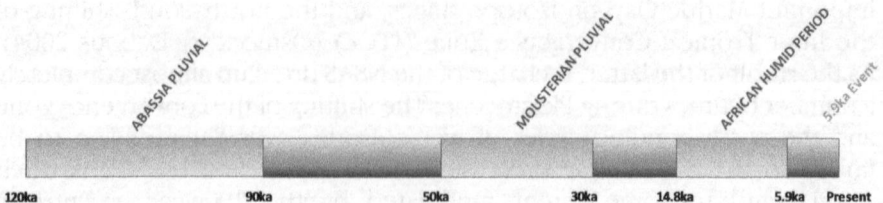

Figure 22.6. The last wet (pluvial) phases in Sahara (modified from Krijgsman et al. 1999).

THE CONCEPT OF FOSSIL WATER

Both concepts of 'sustainability' and of 'fossil' water can only be defined in a time- and space-related context. The term 'fossil' water must therefore be qualified both in terms of age and the vectors of water movement. Some water in the NSAS has been dated to be more than 1 million years (Sturchio et al. 2003) and is therefore automatically classified as 'fossil'. Any water extraction will be consequently classified as 'mining'. This may not necessarily hold true. Zekster and Everett (2004) provide the following interesting comment:

> Based on estimates of the hydraulic parameters (gradient 3×10^{-4}, hydraulic conductivity 10^{-5}m/sec, and effective porosity 10 per cent), groundwater flow velocity has been estimated at 1m/yr. Thus the groundwater needs about one million years to pass through the system from recharge areas in the southern boundaries to the Qattara discharge area.

Also in other contexts, the notion of 'fossil' water may not be relevant, for obvious but rather trivial reasons. Figure 22.8 shows a profile through the Libyan Ghadames basin (right) and Murzuq basin (left). Here are two scenarios: The Murzuq is a basin (to some extent) isolated by the Ghargaf (Al Quarqaf) ridge, forming a completely isolated continental basin. The Murzuq basin, like most other basins in north-eastern Africa has remained filled to the brim with groundwater, probably since its formation several hundred million years ago. Much of the water in Murzuq basin may even be remnants of syn-genetic connate water, i.e. the ancient water in which the sand particles were transported and deposited during the Paleozoic. A series of recharge events has come and gone since then, but few of these can significantly have rejuvenated the groundwater except in the upper reaches; an already full aquifer system is unable to accommodate (even more) water. In contrast, the Ghadames basin has a conduit to the Mediterranean, providing a certain potential for transfer between periods of recharge, in particular during the more dramatic dewatering and flooding events. In the case of Murzuq and Kufrah, a certain replenishment is therefore possible during wet periods, but essentially limited to the upper part of the aquifer. The maturity of the sand is high and water quality has remained surprisingly good over the eons.

As shown in Figure 22.8 the Murzuq basin (left) is isolated by the Gargaf (Al Quarqaf) ridge (centre), and has no natural outlets. The Ghadames basin (right) has hydraulic connections to the Mediterranean, allowing transfer of water to/from the aquifer. The transfer is controlled by a fluctuating sea level, causing varying discharge rates and in some cases even reversal and saltwater flow (encroachment) into the aquifer, depending on the relative elevations of aquifer/sea level. The level of the Mediterranean

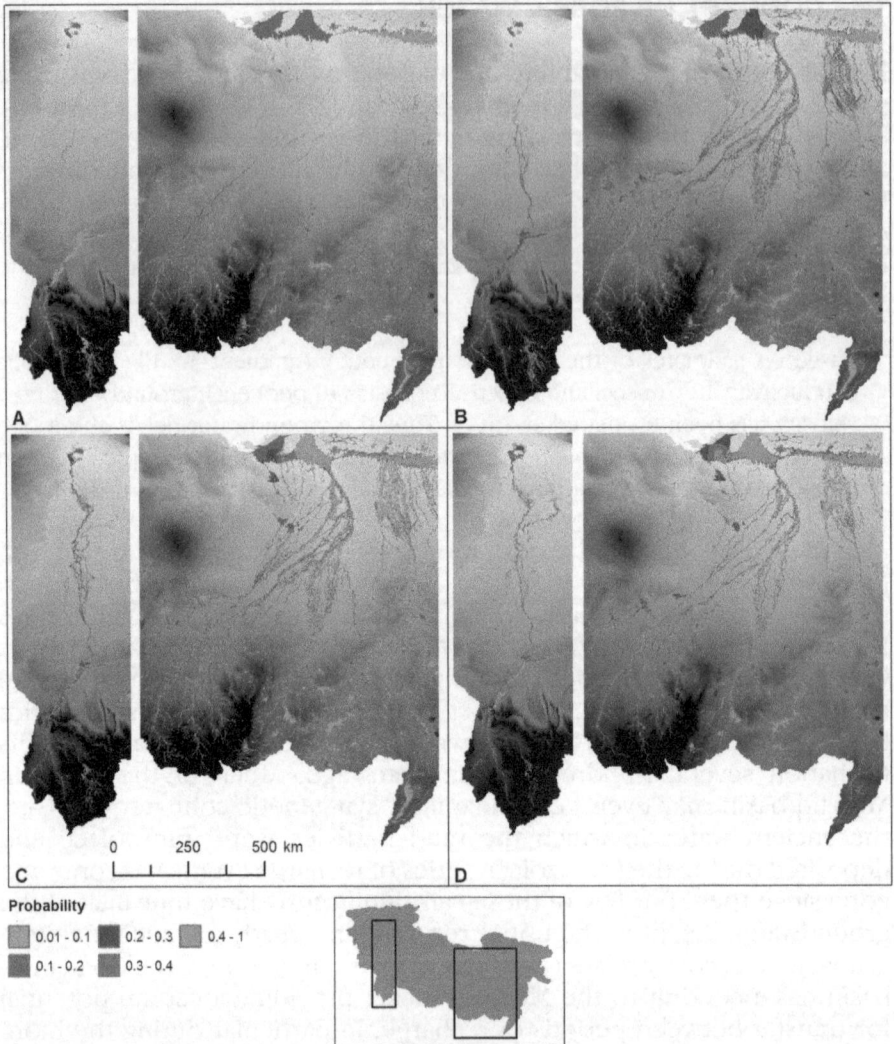

Figure 22.7. Large open waterways dominated the present desert landscape during the MIS 7, now buried under eolian sands of the Sahara. The same waterways provided ideal recharge condition for the NSAS. Similar paleo-hydrological developments are assumed during earlier pluvial phases (© Coulthard et al. 2013).

has experienced a series of fluctuations (between +5–130m) during Pleistocene, and much more dramatic and near complete desiccation during Pliocene (the Messinian salinity crisis) some 5.96 million years ago (Krijgsman et al. 1999). It can be assumed that most of the coastal aquifers were completely drained during the Messinian epoch, and that the age of the water present in the aquifers today should post-date this epoch.

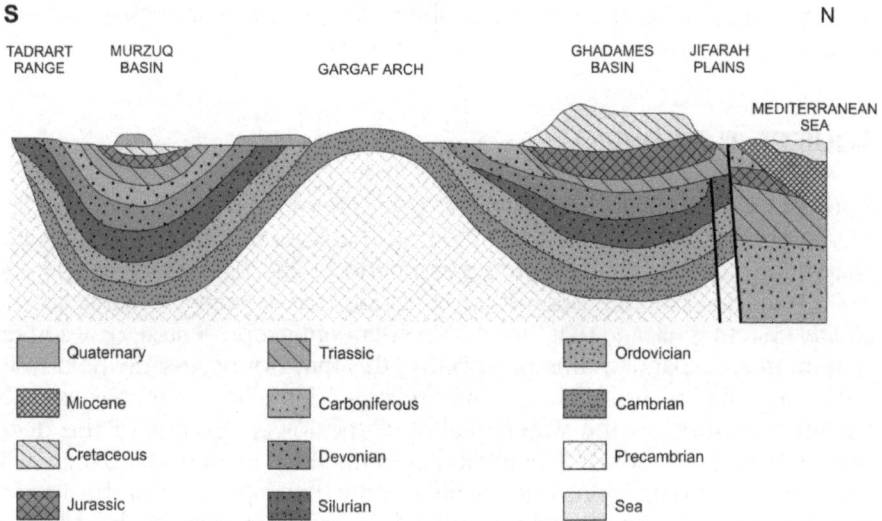

Figure 22.8. Schematic illustration showing the Murzuq basin (left) which is isolated by the Gargaf (Al Quarqaf) ridge (centre) (© Krystyna Guzek k.guzek@10g.pl).

In contrast, the isolated Ghadames and Kufrah basins can be assumed to have retained most if not all of its water intact during the Messinian crisis.

AGE AND GROUNDWATER QUALITY

The sandstones of the NSAS, making up around 80 per cent of the total rock volume, are remarkably mature, consisting of virtually pure silicates. Rock and water interaction over millions of years does not seem to make much impact on the water quality. Verbal (non-confirmed) reports from drill stem tests reported by oil crews drilling in the Khufra basin indicate a salinity in the lower parts (at 3,500 m depth) of around 600 mg/L, decreasing to some 300 mg/L towards the upper reaches. This implies that for some aquifers, in particular the Kufrah and Murzuq, groundwater quality is good and will not be a critical issue. In others, such as Ghadames, more consideration will have to be paid to the groundwater quality. In yet others, such as the Sirte basin, salt water intrusion (Gossel 2010) has been observed as a result of the Holocene sea level rise following the Last Glacial Maximum. Coupled with a high water table and high potential evapotranspiration rate (PET), the development of sabkhas render agriculture impossible in these areas. The possibility of reducing and even reversing the adverse effects of sabkha development around the oases could be envisaged from a principal perspective, by lowering the water table and introducing pulsed irrigation techniques to transport leached salts down through the soil profile. In order to claim more land for agriculture, this remains a possibility, but will require resources and infrastructure not feasible in Libya at present.

AQUIFER TURNOVER

Mathematical simulations of dynamic responses in the NSAS over the last 25,000 years (Gossel et al. 2008) indicate that (upper parts of) the aquifer system have been recharged during two periods: 25–20,000 years BP and 10–5,000 years BP. The authors further conclude that ongoing recharge to the aquifer system is nil, and that the NSAS consequently represents a fossil aquifer system. In an expanded time perspective, this may not necessarily hold true.

Varying climate and recharge over several 100 million years has actually left little imprint on the water quality of the lower reaches of the deep Nubian basins. Periods of drought as experienced in present-day Sahara will cause transient conditions but will affect only the upper part of the aquifer systems. As is the case today, much of the western part of the NSAS is actually discharging (artesian) water and sustaining life in the depressions (Quattara) and the oases (Siwa, Kufrah, Farafra, Dhakla, Kharga). At the same time the recharge under the present (the last 6,000 year) climatic regime is considered to be nil. In other words, the upper part of the NSAS is in a transient state, being depleted both by natural mechanisms and industrial-scale abstraction for irrigation. The lower parts remain unaffected. Present exploitation takes place in concentrated areas while the bulk of the NSAS remains unaffected. Voss and Soliman (2014) point out that transboundary effects from a future large-scale water abstraction of the NSAS will be minimal due to the low diffusivity through the aquifer system. At the same time, local drawdown around concentrated wellfields may be excessive, forcing a decentralized mode of exploitation. In particular, lowering of the water table may lead to loss of oases. On the one hand, water abstraction should be conducted in a decentralized fashion, dispersed over large areas, and conveyed in from afar to locations where conditions are suitable for agricultural activity. On the other hand, the natural high water table of most oases has led to extensive development of sabkha, which is barren land from an agricultural perspective, and a lowering of the water table may reduce this particular problem. Whether the effects of several thousand years of oasis water logging (including Siwa) realistically can be rehabilitated by lowering the water table should be looked into, and weighed against a host of other environmental effects.

REDEFINING SUSTAINABLE USE OF NSAS

The NSAS has been subject to a large number of scientific investigations, including mathematical modelling, groundwater exploration and exploitation, hydrocarbon activities and remote analyses involving paleo-hydrological reconstructions of Pleistocene and Holocene waterways. Yet surprisingly little is actually known regarding the NSAS groundwater volumes and quality. Most estimates are conservative, some merely reflect depth limitations of the available methodology, and some are possibly experiencing the sheer dimensions of the NSAS as an intellectual hurdle, difficult to pass. Few if any hydrogeologists have literally ventured to the bottom of these basins, in the fashion of the oil industry. This is illustrated by the following estimates of NSAS water volumes. Alker (2008a) provides insight into the varying historical volume estimates of the NSAS. These range from $15,000 km^3$ (Ambroggio 1966), $135,000 km^3$ (Gossel et al. 2004) and $457,000 km^3$ (CEDARE 2002) and $50,000 km^3$ (Sturchio 2003). Based upon Gleick (1993), the NSAS may contain up to 5 per cent of all groundwater on the planet.

As an initial and simplistic approximation for reservoir and water volume calculations, the principal basins of the NSAS have been considered as half-ellipsoids, with principal axes as indicated below. With porosity and net-to-gross ratio values as indicated in Table 22.1, and considering the aquifer systems as unconfined, the gross available water volume is calculated as a function of drawdown. The formula for aquifer water volume calculations is:

$$V = [porosity/100] * [net\text{-}to\text{-}gross/100]$$
$$* pi * (1/12) * A * B * z * (3 - (z/D)^2).$$

where:

- V: volume from ground surface to drawdown level 'z'
- A: largest horizontal diameter of ellipsoid at surface
- B: smallest horizontal diameter of ellipsoid at surface
- D: total depth of basin (vertical)
- z: distance to drawdown water level

The cost of abstracting water from an aquifer will increase in a linear fashion as the head (drawdown) increases. At one point, the head will pass a critical level and further pumping will be deemed uneconomical. The salient question is then: uneconomical on what criteria, since the presence of water only 500 m underground represents the very basis for human existence, and in particular for agricultural activity. The balance between the need for water and energy used for pumping will essentially be a political and social-economic issue, and will be based upon regulations. Cost estimates of water abstraction can therefore not be based solely on short-term market criteria. Under these circumstances, pumping water from depths up to 500 m is considered a realistic operating framework, in

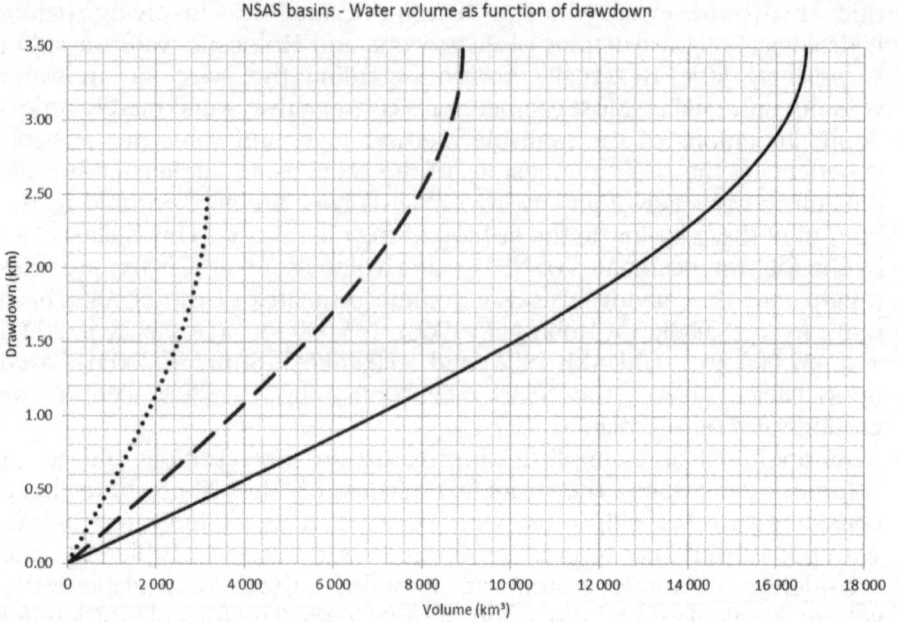

Figure 22.9. Water content as function of drawdown in three principal Libyan basins. Refer to Table 22.1. Left: Murzug, centre: Ghadames, right: Kufrah. 1 km^3 of water is considered sufficient to supply 5 million people for 1 year in a modern society (© Krystyna Guzek k.guzek@10g.pl).

particular in a region already awash with energy (oil). The suggested head is roughly corresponding to the operational heads already experienced in the Herodian Aquifer on the West Bank, Palestine. According to Alker (2008b) the annual consumption from the Libyan part of the NSAS was reported at 0.8 km^3, and about the same on the Egyptian side. NSAS. This figure tallies

Table 22.1. Dimensions and water content of three principal basins within the NSAS of Libya (for convenience, a hypothetical uniform (horizontal) drawdown (dd) has been assumed over the entire basin).

	Kufrah	Unit	Ghadames	Unit	Murzuq	Unit	H$_2$O	Unit
Depth	3.5	km	3.5	km	2.50	km		
Min. diameter	250	km	150	km	200	km		
Max. diameter	350	km	240	km	150	km		
Net-to-gross ratio	80	%	80	%	50	%		
Porosity	13	%	15	%	16	%		
H$_2$O at 250 m dd(m)	2,245	km^3	1,000	km^3	560	km^3	3,805	km^3
H$_2$O at 500 m dd(m)	**3,600**	**km^3**	**2,000**	**km^3**	**930**	**km^3**	**6,530**	**km^3**
H$_2$O at 750 m dd(m)	5,300	km^3	2,513	km^3	1,370	km^3	9,183	km^3
H$_2$O at 1,000 m dd(m)	7,000	km^3	3,240	km^3	1,760	km^3	12,000	km^3
Total H$_2$O	16,800	km^3	7,900	km^3	3,100	km^3	27,800	km^3

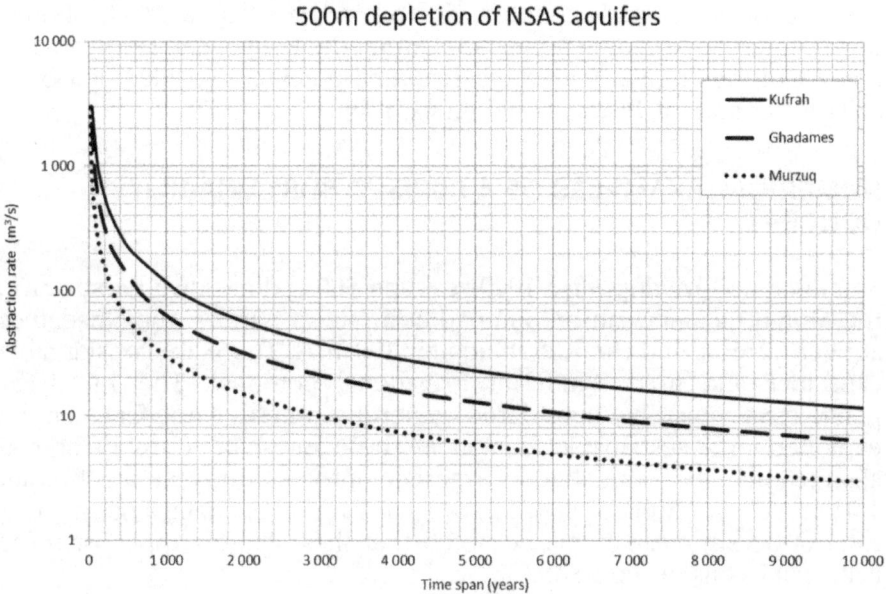

Figure 22.10. 500-m depletion scenarios (drawdown) of three Nubian aquifers in Libya. The hydraulic parameters of the aquifers are outlined in Table 22.1 (© Krystyna Guzek k.guzek@10g.pl).

with Vrba (2004) who reports the total abstraction of water from the NSAS by transboundary states (Egypt, Libya, Sudan, Chad) to be $1.6 \, km^3$ in 2004. Chad does not consume significant amounts of water from the NSAS. Obviously, there is much room for increase of water abstraction from the above basins, and in particular from the NSAS as a whole.

The Milankovitch rule will inevitably force another pluvial period upon the Sahara. If exploited by then, even the partially depleted landlocked aquifers will now have room for replenishment of water, in contrast to the previous recharge period during the African Humid Period almost 6,000 years ago.

Alternatively, even if one chooses to object to the generous time-frame adopted below (500, 1,000 years), sustainability can again be reformulated, this time based on technological optimism. One may assume that mankind, 500 or 1,000 years from now has progressed to a state where technological development has reached a stage where water can be obtained from sources other than the underlying NSAS aquifers.

In the case of the large NSAS aquifers, the definition of sustainability can therefore be reformulated in terms of volume and time. By shifting the time perspective, as indicated above, the aquifers be exploited with considerable water abstraction, over very long periods of time. In the practical world, demanding a 1,000-year perspective to qualify as being sustainable is extraordinary, to the point of being ridiculous. However, in the case of the Libyan aquifers, this can be accomplished, and with predictable consequences (in stark contrast, the oil industry has a much shorter horizon, rarely exceeding 50 years.). During the Roman era, Libya was the granary of the Empire. With the agricultural potential provided by Libya's soils, water and oil, history could repeat itself.

NUMERICAL EXAMPLES OF A REVISED SUSTAINABILITY CONCEPT

Summing up, by adopting a 1000-year and 500 m drawdown perspective, the Murzuq aquifer can be pumped at a rate of $30\,m^3/s$, the Ghadames aquifer $60\,m^3/s$, and the Kufrah aquifer $100\,m^3/s$. These are considerable capacities and hold significant irrigational potential. The 1,000-year perspective must be considered generous and the aquifers can be exploited in a sustainable fashion at these rates. In terms of annual abstractions, the same figures translate into $3.1\,km^3/year$ for Kufrah, $1.0\,km^3/year$ for Murzuq and $2.2\,km^3/year$ for the Ghadames basin, i.e. a total of $6.3\,km^3/year$. It will be noted that these figures apply for three principal basins of the NSAS.

VALUE OF WATER

The gross value of water in these basins can be based upon the current market value of water, ranging from $US\$1.00/m^3$ for domestic water, to US $\$0.25/m^3$ for irrigation water. The gross market value of water in the three basin aquifers (within the 500-m drawdown level) will consequently amount to a staggering $6.3\text{-}1.5\ 10^{12}$ USD, a figure which is several orders of magnitude higher than the corresponding market value of oil from the same reservoirs (2015).

Much as oil and water does not mix: the economic interests of water and oil may be conflicting. In this particular tug of war, water is the inevitable loser, for a number of reasons. Typical ingredients in this conflict may be sheer ignorance, greed, corruption, tangible conflict of interests, economic priorities, war-like situations, demand for quick returns, and lack of regulations, long-term vision and planning.

The current unit price of oil is about 600–1,800 times that of water, at current oil and water market prices (2014). A good oil well pumped at 10 L/s will produce a gross value (at the well head) of around US$6 /sec. The corresponding value for a water well pumped at 100 L/s is around US cents 3–6/sec; in practice around 1 per cent of the corresponding oil well. The life span of a properly constructed water well can easily outlast an oil well, two to three times or more, resulting in 3–4 per cent. Still, in both a short- and medium-term economic perspective, an oil well will provide the best return of investments. Whether this is the case in a greater socio-economic context is a completely different matter. Since oil and water often represents incompatible and mutually exclusive commodities, a choice between the two will have to be taken during early planning stages. Once the oil industry gets a foot in the door, the water sector will play second fiddle, or none at all.

THE CONTAMINATION LEGACY OF THE OIL INDUSTRY

Around every oil/gas well in north-eastern Africa there is a number of dedicated water supply wells, installed for the sole purpose of providing water for drilling operations. Water is used on site, for circulation systems during drilling, and sometimes in vast quantities for underground injection, for pressure support, in order to enhance oil/gas production. The consumption of water required to sustain the full life span of an oil well varies, but the volume of consumed water for the actual drilling operations is nominal. More serious problems may arise from circulation fluids made up of hyper-saline brines, toxic chemicals and diesel. However, the real problems start during the operational (exploitation) phase, and in particular during the indefinite post-production period following well abandonment.

During pressure support, in order to stimulate the return of hydrocarbons from the production well, large volumes of water are pumped into the formations. Not only does this process consume large amounts of water; the chemicals added to the water in order to stimulate production will permanently contaminate and destroy an aquifer, forever. In a North Sea context there are several thousand wells waiting to be plugged (2015). Here, deep underground aquifers (some of which contain freshwater) are not considered a resource other than as a potential for storage of CO_2.

The emerging shale gas industry has often been criticized for using a wide array of toxic ingredients in fluids used for fracking and stimulation of gas shale extraction. The same fracking techniques are also envisaged in the Nubian (Cambrian) shale reservoirs once the traditional and highly producing sandstone reservoirs have been depleted. The shale fracking procedure also requires great volumes of water during the drilling and exploitation phases, and with modern deviated drilling technology the contaminants are now spread in a 360° fashions, at any depth, and up to 5 km radius in any direction. Since fracking takes place in non-aquifer (impermeable) formations such as shale, there are no primary conflicts between fracking procedures and water interests. The problem occurs when highly toxic fracking fluid migrates along tectonic conduits and into under- or overlying aquifers. The fracking technology therefore represents a quantum leap in spreading contaminants in a regional context, and represents the next generation of threat to the vast aquifers of north-eastern Africa.

Possibly the most serious threat to aquifers is related to the oil industry's well abandonment practices, or the lack thereof. As with fracking; also with injection procedures for pressure support, abandonment practices can cause widespread aquifer contamination and destruction. Improperly sealed and abandoned wells continue to contaminate, indefinitely, and have the potential of inflicting regional and permanent damage to vast water resources, thereby causing destruction that may exceed the value of the extracted oil, by far. An abandoned oil well may leak oil and/or saline water into aquifers penetrated by the oil well due to improper plugging, neglect or corrosion.

As a numerical example, an improperly cemented or plugged well will convey water from high to low head, i.e. saline water from an aquifer may be transported up along the well and migrate into fresh water aquifers. A hypothetical and minor crossflow scenario of $10\,L/s$ within an improperly plugged well will inject more than $300,000\,m^3$ of saline water per year into the aquifer. Dispersion and dilution of such a scenario has the potential of rendering more than 30 million $m^3/year$ of freshwater unfit for consumption, and for irrigation. Damage will be on a permanent basis. On the surface, nothing reveals what goes on, deep in the underground. All installations have a finite life span following completion, ranging from zero in the case of a 'dry' or exploratory well, to 20 years or more for production wells. In particular, oil wells in Libya have a reputation for having short life expectancies due to a highly corrosive subsurface environment. Cross flow caused by corrosion of casings and cementing has been known in the area since the onset of oil exploration in the early 1950s.

As per definition, all exploited oil reservoirs are being mined, as these reservoirs require geological time spans to form and mature. The life expectancy of a producing oil reservoir is rarely more than 50 years. In contrast, an aquifer system can be operated indefinitely, if handled properly. It is therefore imperative to protect aquifer systems from being permanently damaged by short lived oil ventures. In this perspective, as has been demonstrated, an aquifer is significantly more valuable than even the most productive oil reservoir.

The responsibility of underground environmental damages often terminates with the oil licensee moving out of area, if not before. To the author's knowledge, no serious regulations to safeguard the under-ground aquatic environment have ever been imposed upon the oil industry. Even if imposed, such regulations would cause an initial outcry, similar to the historic tobacco ban or compulsory wearing of seat belts. But, regulations will require a functioning society; in the case of Libya this is sadly not the case at the time of writing (2015). Recent details on the ongoing oil exploitation in the Murzuq and Ghadames basins have not been available.

REFERENCES

Alker, M. 2008a. 'The Nubian Sandstone Aquifer System'. In Scheumann, W., Herrfahrdt-Pähle, E (eds), *Conceptualizing Cooperation on Africa's Transboundary Groundwater Resources*. German Development Institute. http://www.isn.ethz.ch/ (Accessed on 21 May 2010).

———— 2008b. *The Nubian Sandstone Aquifer System. A Case Study for the Research Project 'Transboundary Groundwater Management in Africa'*. German Development Institute (DIE), pp. 237–73.

Ambroggio, R.P. 1966. 'Water under Sahara'. *Scientific American* 214(5): 21–49.

Blanc, O.L. 2002. 'The opening of the Plio-Quaternary Gibraltar Strait: assessing the size of a cataclysm'. *Geodinamica Acta* 15(5–6): 303–17.

Coulthard, T.J., Ramirez, J.A., Barton, N., Rogerson, M., Brücher, T. 2013. 'Were Rivers flowing across the Sahara during the Last Interglacial? Implications for human migration through Africa'. *PLoS One* 8(10). http://journals.plos.org/plosone/article?id=10.1371/journal.pone.0074834.

deMenocal, P., Ortiz, J., Guilderson, T., Adkins, J., Sarnthein, M., Baker, L., Yarusinsky, M. 2000. 'Abrupt onset and termination of the African Humid Period'. *Quaternary Science Reviews* 19(1–5): 347–61.

Echik, K. 1998. Geology and hydrocarbon occurrences in the Ghadames Basin, Algeria, Tunisisa, Libya. Geological Society, London, Special Publications 1998, v. 132, pp. 109–129.

Gleick, P.H. (ed.). 1993. *Water in Crisis: A Guide to the World's Fresh Water Resources*. New York: Oxford University Press.

Gossel, W. et al. 2004. 'A very large scale GIS based groundwater flow model for the Nubian sandstone aquifer in eastern Sahara'. *Hydrogeology Journal* 12: 698–713.

———— 2008: 'A GIS-based flow model for groundwater resources management in the development areas in the eastern Sahara, Africa'. In Adelana, S., MacDonald, A. (eds), *Applied Groundwater Studies in Africa*. Leiden: CRC Press, pp. 43–64.

———— 2010.'Modelling of paleo-saltwater intrusion in the northern part of the Nubian Aquifer System, Northeast Africa'. *Hydrogeology Journal* 18(6): 1447–63.

Krijgsman, W., Hilgen, F.J., Raffi, I., Sierro, F.J., Wilson, D.S. 1999. 'Chronology, causes and progression of the Messinian salinity crisis'. *Nature* 400: 652–5.

McKinlay, A.C.M. 1963. 'The coalfields and the coal resources of Tanzania'. *Geological Survey of Tanzania* 28.

Milankovitch, M. 1941. *Canon of Insolation and the Ice Age Problem*. Belgrade: Zavod za.

Muller, R.A., MacDonald, G.J. 1997. 'Glacial cycles and astronomical forcing'. *Science* 227: 215–8.

Osmond, J.K., Dabous, A.A. 2004. 'Timing and intensity of groundwater movement during Egyptian Sahara pluvial periods by U-series analysis of secondary U in ores and carbonates'. *Quaternary Research* 61(1): 85–94.

Pallas, P. 1978. Water resources of the Socialist Peoples Libyan Arab Jamahiriya. In: Salem and Busrewil, 2nd Symposium Geology of Libya, Academic Press, pp. 539–594.

Park, R.G. 1988. *Geological Structures and Moving Plates*. London: Blackie & Sons.

Said, R. 1962: The River Nile. Geology, Hydrology and Utilization, Pergamon, p. 319.

Schandelmeier, H. 1988. 'Pre-Cretaceous Intraplate Basins of north-eastern Africa'. *Episodes* 11(4): 270–3.

Sturchio, N.C. et al. 2004: 'One million year old groundwater in the Sahara revealed by krypton-81 and chlorine-36'. *Geophysical Research Letters* 31(5).

Sturchio, N.C., et al. 2003. 'Million year old Nubian aquifer groundwater dated by atom counting of Kr[81]'. *Paper 166-7*, 2003 Annual Meeting, Geological Society of America.

Voss, C., Soliman, M. 2014. 'The transboundary non-renewable Nubian Aquifer System of Chad, Egypt, Libya and Sudan: classical groundwater questions and parsimonious hydrogeologic analysis and modeling'. *Hydrogeology Journal* 22(2).

Vrba, J. 2004. 'The world's groundwater resources'. *Report IP2004-1*, International Groundwater Resources Centre.

Worsley, D, 2000: Geological exploration in Murzuq Basin. National Oil Corporation, Elsevier, p. 519.

Zekster, I.S., Everett, L.G. (eds). 2004. 'Groundwater resources of the world and their use'. *IHP Series on groundwater No 6*, UNESCO, p. 221.

Contributors

EDITORS

Terje Tvedt is Professor of Geography at the University of Bergen and Professor of Political Science and of Global History at the University of Oslo. He has published extensively on water related topics and made three television documentary series on water, shown in 150 countries worldwide. His books include *The River Nile in the Age of the British*, *A Journey in the Future of Water* and *Water and Society*, all published by I.B.Tauris.

Terje Oestigaard is a senior researcher at the Nordic Africa Institute, Uppsala, and docent in archeology at the Department of Archaeology and Ancient History, Uppsala University. His recent books include: *Framing African Development: Challenging Concepts* (2016), edited with Havnevik, K., Virtanen, T. & Tobisson, E; *Dammed Divinities: The Water Powers at Bujagali Falls, Uganda* (2015); *Religion at Work in Globalised Traditions: Rainmaking, Witchcraft and Christianity in Tanzania* (2014); and *A History of Water, Series III, Vol. 1: Water and Urbanization* (2014), edited with Terje Tvedt.

CHAPTER 1

Elena A.A. Garcea is an Africanist archaeologist and teaches prehistoric archaeology at the University of Cassino and Southern Latium, Cassino, Italy. She has worked on various projects in different parts of Sudan for 30 years, since 1986, and at Gobero in Niger in 2005 and 2006. Her research interests include the shift of pottery-bearing foragers to pastoralism in North Africa, sedentism/mobility dynamics and the emergence of complex societies. She is author of 7 books and over 190 articles on African archaeology.

CHAPTER 2

Randi Haaland is Professor Emerita of Middle Eastern and African Archaeology at Bergen. She has been coordinator for projects on research and competence building in Sudan, Tanzania, Zimbabwe, Palestine and Nepal. Her main research interest is on transition to food production in the Sudanese Nile valley, and she has undertaken ethnographic fieldwork in Sudan, Tanzania, Ethiopia and Nepal on technological and symbolic aspects of food-ways.

CHAPTER 3

Louise Bertini is Affiliate Assistant Professor of Egyptology at the American University in Cairo. She is a zooarchaeologist who has worked in the field at numerous sites including Giza, Saqqara, Mendes, Amarna, Abydos, and Merimde Beni-Salame. She is the author of numerous articles on Ancient Egyptian animals, economy and the ancient environment including: 'Killing Man's Best Friend?' and 'The Size of Ancient Egyptian Pigs: A Biometrical Analysis Using Molar Width'.

CHAPTER 4

Katherine Blouin is Associate Professor in Roman History at the University of Toronto. She works on the socio-economic and environmental history of Roman Egypt, with a special focus on the Nile Delta. Her publications include *Le conflit judéo-alexandrin de 38-41: l'identité juive à l'épreuve* (2005) and *Triangular Landscapes = Environment, Society, and the State in the Nile Delta under Roman Rule* (2014).

CHAPTER 5

Johann Tempelhoff is Professor of History at North-West University's Vaal Campus in Vanderbijlpark, Gauteng Province, South Africa. He is a water historian and works in transdisciplinary research strategies on contemporary and historical issues related to the cultural dynamics of water in South Africa.

CHAPTER 6

Alan Mikhail is Professor of History at Yale University. He is the author of *The Animal in Ottoman Egypt* (2014) and *Nature and Empire in*

Ottoman Egypt: An Environmental History (2011) and also editor of *Water on Sand: Environmental Histories of the Middle East and North Africa* (2013).

CHAPTER 7

Matthew V. Bender is Associate Professor of History and Director of the International Studies Program at The College of New Jersey. He is a specialist in modern African history, environmental history and water history. He has published several articles on the history of water management in East Africa, and is now completing a book entitled *Water Brings No Harm: Knowledge, Power, and the Struggle for the Waters of Kilimanjaro*.

CHAPTER 8

Brock Cutler is Assistant Professor of History at Radford University. His work focuses on the appearance of modern sovereignty in North Africa, especially through environmental relations in the colonial period. His work has appeared in the *Journal of North African Studies*, *Colonialism and Colonial History* and the *Journal of the Economic and Social History of the Orient*, among others.

CHAPTER 9

Maurits W. Ertsen is an associate professor within the Water Resources Management group of Delft University of Technology, the Netherlands. He works on human agency in (ancient, colonial and modern) irrigation to understand human-environmental interactions. In his recent book *Planning Improvised Development on the Gezira Plain, Sudan, 1900–1980* (2016), he discusses the huge Gezira Scheme in terms of daily interactions between all different kinds of interested parties. Maurits is President of the International Water History Association (IWHA) and one of two main editors of *Water History*, the official journal of IWHA.

CHAPTER 10

Andrew Ogilvie works at the Institut de Recherche pour le Développement (IRD) in the G-EAU (Water management, actors and uses) joint research group in Montpellier, France. His research focuses on the evolution of semi-arid and West African hydro-social systems under climatic and human influences. Working in the Niger Basin since 2008, he notably collaborated on several CGIAR water and food research programmes. He holds a joint PhD in geography and hydrology from King's College London and Montpellier universities.

Jean Charles Clanet is a university professor and specialist of Saharan desert regions, where he studied for over 30 years the extensive livestock systems, their access to resources and local climatic changes. Working within the Chad, Volta and Niger river basins, he notably collaborated on the CGIAR BFP Volta before coordinating the BFP Niger, whilst being seconded from the University of Reims, France, to the G-EAU research group at the Institut de Recherche pour le Développement (IRD) in Montpellier, France.

Georges Serpantié is a researcher in agronomy at the Institut de Recherche pour le Développement (IRD) where he belongs to the IRD-UM3 GRED (Governance, Risk, Environment, Development) unit in Montpellier, France. He works in West Africa and in the Indian Ocean on rural development issues and the adaptation of agriculture to drought and environmental policies. His current research topics are watershed-payment for ecosystem services (PES) and adaptation to climate change in rice-growing lowlands.

Jacques Lemoalle started his research in limnological studies on Lake Chad before being involved in programmes dealing with a number of other lakes and coastal lagoons of tropical and northern Africa. He is now an emeritus scientist at the Institut de Recherche pour le Développement (IRD) where he has spent all of his career.

CHAPTER 11

Andrew Reid is a senior lecturer at the Institute of Archaeology, University College London, and has previously taught at the University of Dar Es Salaam, Tanzania, and the University of Botswana. He specialises in the emergence of states in eastern and southern Africa and in the social manipulation of cattle. He has co-edited *Ancient Egypt in Africa* (with David O'Connor) and *African Historical Archaeologies* (with Paul Lane).

CHAPTER 12

Marcel Rutten is a geographer based at the African Studies Centre, Leiden University. His research activities concentrate on natural resource management, notably of land and water, in (semi-)arid Africa. Studies address the effects of land tenural changes; famine and drought coping strategies; and conflicts and co-operation over natural resources (land and water). Results have been published in *Selling Wealth to Buy Poverty* (1992), *Inside Poverty and Development in Africa* (2008) and 'How natural is natural' (2014), among others.

CHAPTER 14

Raphael M. Tshimanga received his PhD in Hydrology from the Institute for Water Research, Rhodes University. He works at the Department of Natural Resources Management, Faculty of Agricultural Sciences, University of Kinshasa, Kinshasa, Democratic Republic of Congo, and he is a member of Congo Basin Network for Research and Capacity Development in Water Resources (CB-HYDRONET), Kinshasa.

CHAPTER 15

Tobias Haller (Institute of Social Anthropology) holds an extraordinary professorship in social anthropology and is Co-Director of the Institute of Social Anthropology at the University of Bern. He specialises in economic, ecological and political anthropology. His research focuses on common pool resource governance and institutional change. Theoretically, he combined and further developed a new institutionalism and political ecology perspective. He did fieldwork in Cameroon on agricultural change and in Zambia on the management of the commons in river floodplains. He has coordinated comparative research projects on the commons in Africa and has published on African floodplain areas (*Disputing the Floodplains*, 2010, with a foreword by Elinor Ostrom).

CHAPTER 16

Tor A. Benjaminsen is a human geographer and professor at the Department of Environment and Development Studies at the Norwegian University of Life Sciences. As a political ecologist he has carried out research and published widely on land and environmental management, land-use conflicts and land tenure in several African countries.

CHAPTER 17

Pierre Morand is a senior researcher in biostatistics and environmental sciences. His recent work focuses on the development and changes in

548

A History of Water

small scale fisheries, especially regarding vulnerability and adaptation of fishing communities to various dimensions of global change. He is currently Deputy Director of the research unit 'resilience' at the French Institute of Research for Development (IRD).

Famory Sinaba is a researcher in agro-economy at the Institute of Rural Economy (IER) of Mali. His research deals with the dynamics in production and consumption systems. He also considers the organization and performance of value chains as well as the evaluation of the impact of agricultural policies. He is currently studying the vulnerability of socio-ecosystems that have undergone climate change, market fluctuations and changes in public policies.

Awa-Niang Fall is a senior lecturer at Cheikh Anta Diop University, Dakar, Senegal. She contributes to the design, implementation and management of the Doctorate School on Water, Water Quality and Uses. Recently, she has worked on continental hydrology, Integrated Water Resources Management and remote sensing applied to water resources management. She also leads a research project on the resilience of coastal social-ecosystems, especially in the Senegal River delta.

CHAPTER 18

Jeppe Kolding is a professor at University of Bergen, specializing in small-scale fisheries. He has lived and worked for several years in Africa and has experience from shorter term engagements in more than 25 developing countries. His primary research interests are in fisheries ecology, harvest strategies and the management of small-scale tropical fisheries. He has published more than 100 papers, chapters and technical reports, many of which can be found here: https://uib.academia.edu/JeppeKolding.

Paul A.M. van Zwieten is an assistant professor at Wageningen University, the Netherlands, with a special interest in the information requirements for management of small-scale fisheries. He has worked and lived in southern Africa for several years and lead several multidisciplinary socio-ecological research projects on Lake Victoria, coastal mangrove areas in Vietnam and Indonesia and small-scale tuna fisheries in Indonesia and the Philippines. He has published over 100 papers, book-chapters and technical reports, many of which can be found here: https://wur.academia.edu/PvanZwieten.

Ketlhatlogile Mosepele is a senior research scholar in fisheries biology and management at the Okavango Research Institute of the University of Botswana, based in Maun, and currently a PhD student with the University of Bergen, Norway. Working in the Okavango Delta and with floodplain fisheries management has been the main focus of his research for the past 14 years. During this period he has published more than 40 publications (journal articles, book chapters and books) on various aspects of African inland fisheries biology and management.

CHAPTER 19

Jessica Kampanje-Phiri is a social anthropologist specializing in understanding the cultural dimension of food systems in Malawi and beyond. Her specific areas of academic and research expertise include: food and nutrition policy analysis; the natural, social, political, institutional, economic, cultural and technological aspects of food; poverty and livelihood security; power and gender relations; social-cultural inequalities; humanitarian interventions; and development assistance. She is currently a lecturer and a deputy head for the Department of Human Ecology at the Lilongwe University of Agriculture and Natural Resources (LUANAR).

Dean Kampanje-Phiri is a researcher in water resources, natural resource and climate change management. His research focus is on the water-food-energy-environment nexus and how to maximise the benefits of water exploitation for multiple uses. He also lectures at the Lilongwe University of Agriculture and Natural Resources on water, natural resource and climate change management on a part-time basis.

CHAPTER 20

Gessesse Dessie worked in Ethiopia at the Wondo Genet College of Forestry and Natural Resources as an assistant professor and as a research fellow at the United Nations University Institute for Natural Resources in Africa, in Accra, Ghana. He is currently United States Forest Service International Programs/Ethiopian Climate Change Advisor Ministry of Environment, Forestry and Climate Change, Government of Ethiopia; and United States Forest Service International Programs USAID Contractor. He completed a PhD in Geography at Stockholm University and holds an MSc from the International Institute for Aerospace Survey and Earth Science. He was born and raised in Koffele, Ethiopia.

CHAPTER 21

Atakilte Beyene is a researcher at the Nordic Africa Institute. He holds a PhD in Rural Development Studies. He has been working in universities and research institutes in Ethiopia and Sweden. His research focuses on agrarian and rural institutions, natural resources management, food security and gender studies. He has conducted extensive field studies in Ethiopia and Tanzania, and has both coordinated and worked on many interdisciplinary research projects and networks in the Nordic and African countries. His current research includes studies on large-scale agricultural investments, irrigation management and smallholder agriculture in Africa.

Emil Sandström is a senior researcher at the Nordic Africa Institute in Uppsala, Sweden, where he conducts research on Nile Basin related issues. He is also Senior Lecturer and former head of the Department of Urban and Rural Development at the Swedish University of Agricultural Sciences in Uppsala. His research and teaching interests are inter-disciplinary in character and embrace perspectives found in political ecology and rural sociology in Eastern Africa and Sweden. His recent book is entitled *Land and Hydropolitics in the Nile River Basin: Challenges and New Investments* (2016), edited together with Terje Oestigaard and Anders Jägerskog.

CHAPTER 22

Fridtjov Ruden is a hydrogeologist and geophysicist, who graduated from the University of Oslo in 1978, and has worked in more than 25 countries of South East Asia, Africa, Caribbean and Central America. He has concentrated his activities around aspects of hydrogeology, ranging from village level to mega-city water supply projects. Ruden has worked in several areas of conflict and disasters, including Ethiopia (1983), Rwanda (1994), West Bank (2000) and Mozambique (1988). The discovery of the large Kimbiji coastal aquifer near Dar es Salaam, Tanzania, in 2006 is attributed to Ruden and his team. Ruden runs a geo-science company in Norway, specializing in geophysical innovations.

Index

khat, 482, 486, 495, 496
large-scale land acquisition, 16, 502, 510, 511, 513, 515
poverty, 335
Smith, M., 31
Soliman, M., 534
Soper, R., 145, 146
sorghum, 43, 60–1, 63, 66, 70
  beer, 60, 63, 67
  flood irrigation, 260
  porridge, 21, 55, 57, 59, 60, 63, 65, 66, 67
  Southern Africa, 128, 134, 141, 147–8
  see also cereal
South Sudan, 503, 504, 505, 507, 511, 512, 530
Southern Africa, 11, 121–5
  Bantu-speakers, 121, 122, 123, 132, 140
  drought, 122
  drying up, 11, 121–2, 149–50
  ethnic variety, 121–2
  ITCZ 123–4
  Khoisan, the, 121, 122, 125, 140–1
  Namib Desert, 122, 124
  Neolithic period, 125, 126, 143
  rainfall, 123–5, 126
  San/Bushmen, the, 121, 122, 125, 139, 140–1
  Stone Age, 125, 143
  water availability, 11, 122–3
  see also Southern African Iron Age
Southern African Iron Age, 11, 122, 125–49, 150
  Bokoni, 142–3, 144, 146–9, 150
  climate change, 126, 132, 133, 136–7, 139, 150
  disease, 132
  domestic dwelling, 131
  drought, 122, 135, 137, 141
  Early Iron Age, 127, 128–33, 145
  Indian Ocean trade, 127, 134, 138
  irrigation, 143, 144–5
  Late Iron Age, 127, 136–41, 146
  Limpopo River valley, 126, 127, 128, 130–40 *passim*, 142
  Mapungubwe, the, 130–7, 138, 142, 150
  Middle Iron Age, 127, 133–6, 146

migration, 126–7, 132, 136–7, 139, 140
Mpumalanga, 134, 142–3, 147–9, 150
Nguni movements, 139–40
Nyanga, 144–6, 147, 150
pottery, 130, 145
rainfall, 125, 126, 131, 133
rainmaking ritual, 126, 131, 133, 134–6, 139
Sotho, the, 138, 139, 140, 148, 150
stone wall architecture and terracing, 143–4, 145–6, 147–9
temporal/spatial classification of the Iron Age, 127–8
Tswana, the, 133, 138, 139, 140–1, 150
urban development, 133–6, 138, 147
water availability, 11, 128, 130, 133, 140, 150
waterhole, 140–1
Zimbabwe/Great Zimbabwe, 11, 134, 136, 138–9, 142, 144–5, 150
Zulu, the, 137, 150
Southern African Iron Age, livelihood
  cattle, 128, 130, 134, 141, 142, 146, 147, 148, 149
  central cattle pattern, 136, 139–40, 144
  farming, 128, 130, 131, 132–3, 134, 138, 140, 141, 144, 146, 149
  farming of wetlands, 141–2
  floodplain agriculture, 128, 134, 142–3
  hunting/gathering, 131, 135, 137, 141
  mining/ironwork, 130, 131, 138, 139, 141, 145, 148
Spear, T., 402–3
Speke, J.H., 281
Stanley, H.M., 281
Stone Age, 2, 3, 125, 143, 358
Stump, D., 144, 146, 149
sub-Saharan Africa, 333
  hunger, 5, 333, 335
  poverty, 335, 337
  rainfed agriculture, 13, 332, 333, 335–6, 352
Sudan:
  large-scale land acquisition, 503, 504, 505, 507, 508, 511, 537